LOW-DIMENSIONAL
SEMICONDUCTORS

THE PHYSICS OF
LOW-DIMENSIONAL
SEMICONDUCTORS

AN INTRODUCTION

JOHN H. DAVIES

Glasgow University

CAMBRIDGE
UNIVERSITY PRESS

CAMBRIDGE UNIVERSITY PRESS
Cambridge, New York, Melbourne, Madrid, Cape Town, Singapore, São Paulo

Cambridge University Press
40 West 20th Street, New York, NY 10011-4211, USA

www.cambridge.org
Information on this title: www.cambridge.org/9780521481489

First published 1998
Reprinted 1999, 2000, 2004, 2005

Printed in the United States of America

A catalog record for this publication is available from the British Library.

Library of Congress Cataloging in Publication Data

Davies, J. H. (John H.)
The physics of low-dimensional semiconductors : an introduction / John H. Davies.
   p. cm.
Includes bibliographical references and index.
ISBN 0-521-48148-1 (hc). – ISBN 0-521-48491-X (pbk.)
1. Low-dimensional semiconductors.   I. Title.
QC611.8.L68039 1997
537.6'221 – dc21   97-88

ISBN-13   978-0-521-48148-9 hardback
ISBN-10   0-521-48148-1 hardback

ISBN-13   978-0-521-48491-6 paperback
ISBN-10   0-521-48491-X paperback

To Christine

CONTENTS

PREFACE

I joined the Department of Electronics and Electrical Engineering at Glasgow University some ten years ago. My research was performed in a group working on advanced semiconducting devices for both electronic and optical applications. It soon became apparent that advances in physics and technology had left a gap behind them in the education of postgraduate students. These students came from a wide range of backgrounds, both in physics and engineering; some had received extensive instruction in quantum mechanics and solid state physics, whereas others had only the smattering of semiconductor physics needed to explain the operation of classical transistors. Their projects were equally diverse, ranging from quantum dots and electro-optic modulators to Bloch oscillators and ultrafast field-effect transistors. Some excellent reviews were available, but most started at a level beyond many of the students. The same was true of the proceedings of several summer schools. I therefore initiated a lecture course with John Barker on nanoelectronics that instantly attracted an enthusiastic audience. The course was given for several years and evolved into this book.

It was difficult to keep the length of the lecture course manageable, and a book faces the same problem. The applications of heterostructures and low-dimensional semiconductors continue to grow steadily, in both physics and engineering. Should one display the myriad ways in which the properties of heterostructures can be harnessed, or concentrate on their physical foundations? There seemed to be a broad gap in the literature, between a textbook on quantum mechanics and solid state physics illustrated with semiconductors, and an analysis of the devices that can be made. I have aimed towards the textbook, a fortunate decision as there are now some excellent books describing the applications. The experience of teaching at a couple of summer schools also convinced me that a more introductory treatment would be useful, one that concentrated on the basic physics. This book addresses that need.

Acknowledgements

Several colleagues contributed to the course out of which this book developed. John Barker, Andrew Long, and Clivia Sotomayor-Torres shared the lecturing at various times and helped to shape the syllabus. Several students and postdoctoral research assistants encouraged me to continue the course and learn some topics that were new to me. I would particularly like to thank Andrew Jennings, Michael and Frances Laughton, Alistair Meney, and John Nixon. It is also a pleasure to thank Andrew Long and my wife for their helpful comments on the manuscript.

Many colleagues have kindly provided data that I have been allowed to replot in a convenient way to illustrate the text. I am very grateful for their help, particularly to those who generously supplied unpublished measurements and calculations, and to Mike Burt, who also gave advice on effective-mass theory.

It has taken a long time to complete this book. I don't imagine that I am the first author who has sadly underestimated the effort required to turn a pile of lecture notes into a coherent manuscript. Most of the work has been done in evenings, between reading bedtime stories to my daughters and feeling exhaustion setting in. As most parents with young children will appreciate, this interval is short and frequently non-existent. I am very grateful to my family for their forbearance and encouragement. I would also like to thank the publishers for their tolerance, as they might well have despaired of ever receiving a finished manuscript. The final proofreading was carried out at the Center for Quantized Electronic Structures (QUEST) in the University of California at Santa Barbara. It is a pleasure to acknowledge their hospitality as well as the financial support of QUEST and the Leverhulme Trust during this period.

I would like to finish with a quotation from the preface by F. Reif to his book, *Fundamentals of statistical and thermal physics*. It must reflect many authors' feelings as their books approach publication.

> It has been said that 'an author never finishes a book, he merely abandons it'. I have come to appreciate vividly the truth of this statement and dread to see the day when, looking at the manuscript in print, I am sure to realize that so many things could have been done better and explained more clearly. If I abandon the book nevertheless, it is in the modest hope that it may be useful to others despite its shortcomings.

John Davies
Milngavie, September 1996

INTRODUCTION

Low-dimensional systems have revolutionized semiconductor physics. They rely on the technology of heterostructures, where the composition of a semiconductor can be changed on the scale of a nanometre. For example, a sandwich of GaAs between two layers of $Al_xGa_{1-x}As$ acts like an elementary quantum well. The energy levels are widely separated if the well is narrow, and all electrons may be trapped in the lowest level. Motion parallel to the layers is not affected, however, so the electrons remain free in those directions. The result is a two-dimensional electron gas, and holes can be trapped in the same way.

Optical measurements provide direct evidence for the low-dimensional behaviour of electrons and holes in a quantum well. The density of states changes from a smooth parabola in three dimensions to a staircase in a two-dimensional system. This is seen clearly in optical absorption, and the step at the bottom of the density of states enhances the optical properties. This is put to practical use in quantum-well lasers, whose threshold current is lower than that of a three-dimensional device.

Further assistance from technology is needed to harness low-dimensional systems for transport. Electrons and holes must be introduced by doping, but the carriers leave charged impurities behind, which limit their mean free path. The solution to this problem is modulation doping, where carriers are removed in space from the impurities that have provided them. This has raised the mean free path of electrons in a two-dimensional electron gas to around 0.1 mm at low temperature. It is now possible to fabricate structures inside which electrons are coherent and must be treated as waves rather than particles. Observations of interference attest to the success of this approach. Again, there are practical applications such as field-effect transistors in direct-broadcast satellite receivers.

As these examples show, complicated technology underpins experiments on low-dimensional systems. In contrast, it turns out that most of the physics can be understood with relatively straightforward concepts. The aim of this book is to explain the physics that underlies the behaviour of most low-dimensional systems in semiconductors, considering both transport and optical properties. The methods described, such as perturbation theory, are standard but have immediate application – the quantum-confined Stark effect, for example, is both a straightforward illustration of perturbation theory and the basis of a practical electro-optic modulator. The most advanced technique used is Fermi's golden rule, which marks a traditional dividing line between 'elementary' and 'advanced' quantum mechanics.

The disadvantage of this approach is that it is impossible to describe more than a tiny fraction of the applications of the basic theory. Many topics of current research, such as the chaotic behaviour of electrons in microcavities or the optical properties of self-organized quantum dots, have to be omitted. Fortunately there are several surveys of the applications of low-dimensional semiconductors, and also more advanced theoretical descriptions. This book provides the foundations on which they are built.

Outline

Chapters 1 and 2 provide the foundations of quantum mechanics and solid state physics on which the rest of the book builds. The presentation is intended only as a refresher course, and there are some suggestions for textbooks if much of the material is unfamiliar. A survey of heterostructures is given in Chapter 3, and Chapter 4 covers the basic theory of low-dimensional systems. This entails the solution of simple quantum-well problems, and an appreciation of how trapping in such a well makes a three-dimensional electron behave as though it is only two-dimensional. Chapter 5 is devoted to tunnelling, with applications such as resonant-tunnelling diodes. Electric and magnetic fields provide important probes of many systems and are discussed in Chapter 6. Perhaps the most dramatic result is the quantum Hall effect in a magnetic field, which is found to be of value as a standard of resistance while continuing to tax our understanding. Chapter 7 contains a range of approximate methods used to treat systems in a steady state. These have wide application, notably to band structure. Another example is the WKB method, which can be used to find the energies of allowed states in a quantum well or to estimate the rate of tunnelling through a barrier. Fermi's golden rule is derived in Chapter 8 and used to calculate the scattering of electrons by impurities and phonons. Optical absorption is another major application. The final chapters are devoted to the two principal low-dimensional systems. The two-dimensional electron gas in Chapter 9 is used primarily for its transport properties, whereas the optical properties of a quantum well in Chapter 10 find employment in devices such as semiconductor lasers.

The book is pitched at the level of beginning postgraduate and advanced undergraduate students. All the basic techniques in this book were in my undergraduate physics courses, although few of the applications had been invented! It is assumed that the reader has had a glimpse of quantum mechanics and solid state physics, but only at the level covered in many courses quaintly named 'modern physics'. The first two chapters will be rather hard going if this is not the case. My experience has been that students with a degree in physics find the basic theory familiar, but the applications improve their understanding immensely. The background of students from electrical engineering varies widely, but the lecture course evolved to address this, and the level of the book should be suitable for them too. The theoretical level has deliberately been kept low and should not provide any impediment.

Exercises

Each chapter has around twenty exercises. Their difficulty varies considerably; some are trivial, whereas others require numerical solution. I did almost all the calculations for this book with an obsolete spreadsheet (Trapeze, dating from 1988), so the numerical aspects are not serious. However, a symbolic-manipulation program such as Maple, Mathematica, or the like would make these tasks easier. Occasionally some integrals or special functions appear and references for these are given in the bibliography.

Units

I have used SI units throughout this book, with the exception of the electron volt (eV), which is far too convenient to abandon. The main problems for users of CGS units are in equations concerning electric and magnetic fields. Removing a factor of $4\pi\epsilon_0$ from equations for electrostatics in SI units should give the corresponding result in CGS units. The magnetic field or flux density $\mathbf{B}$ is measured in tesla (T) in SI units, and formulas should be divided by the velocity of light c wherever $\mathbf{B}$ or the vector potential $\mathbf{A}$ appears to give their form in CGS units. Finally, I use nanometres rather than Ångström units for lengths; $1\,\mathrm{nm} = 10\,\text{Å}$.

Notation for vectors

It is often necessary to distinguish between two- and three-dimensional vectors in low-dimensional systems, particularly for position and wave vectors. I have tried to follow a consistent notation throughout this book, the penalty being that some familiar formulas look slightly odd.

Most low-dimensional structures are grown in layers and the z-axis is taken as the direction of growth, normal to the layers. Vectors in the xy-plane, parallel to the layers, are denoted with lower-case letters. Thus the position in the plane is $\mathbf{r} = (x, y)$. Upper-case letters are used for the corresponding three-dimensional vector, so for position $\mathbf{R} = (\mathbf{r}, z) = (x, y, z)$. Similarly, wave vectors are written as $\mathbf{K} = (\mathbf{k}, k_z) = (k_x, k_y, k_z)$. The only other quantity that needs to be distinguished in this way is $\mathbf{Q}$, which is used for the wave vector in scattering.

This notation requires some familiar results to be written with upper-case letters for consistency. The energy of free electrons in three dimensions is $\varepsilon_0(K) = \hbar^2 K^2 / 2m_0$, for example. I hope that the clarity offered by consistent usage offsets this small disadvantage.

References and further reading

It is not appropriate in a textbook to give detailed references to original papers. Instead there is a bibliography of more advanced books and articles from review journals. In one or two cases I have referred to original papers when I was unable to find an appropriate review. There are also several summer schools on low-dimensional semiconductors that publish their proceedings; naturally I have provided a reference to one that I edited.

There are several books that develop the material in this book further. Among them, Bastard (1988) gives a lucid account of the electronic structure of heterostructures, including a thorough description of the Kane model, with their electronic and optical properties. Weisbuch and Vinter's book (1991) is an enlargement of an earlier review in Willardson and Beer (1966–). They describe the applications of heterostructures as well as their physics, with a particularly good section on quantum-well lasers. Their list of references exceeds 600 entries, which gives some idea of the activity in this field. Finally, Kelly's book (1995) is notable for the breadth of its coverage. He describes the technology of fabrication and an enormous range of applications of heterostructures in physics and engineering. Just a glance at the topics covered in this survey leaves one with no doubt that this field will advance vigorously into the next century.

FOUNDATIONS 1

This book is about low-dimensional semiconductors, structures in which electrons behave as though they are free to move in only two or fewer free dimensions. Most of these structures are really *heterostructures*, meaning that they comprise more than one kind of material. Before we can investigate the properties of a heterostructure, we need to understand the behaviour of electrons in a uniform semiconductor. This in turn rests on the foundations of quantum mechanics, statistical mechanics, and the band theory of crystalline solids. The first two chapters of this book provide a review of these foundations. Unfortunately it is impossible to provide a full tutorial within the space available, so the reader should consult one of the books suggested at the end of the appropriate chapter if much of the material is unfamiliar.

This first chapter covers quantum mechanics and statistical physics. Some topics, such as the theory of angular momentum, are not included although they are vital to a thorough course on quantum mechanics. The historical background, treated at length in most textbooks on quantum mechanics, is also omitted. There is little attempt to justify quantum mechanics, although the rest of the book could be said to provide support because we are able to explain numerous experimental observations using the basic theory developed in this chapter.

1.1 Wave Mechanics and the Schrödinger Equation

Consider the motion of a single particle, such as an electron, moving in one dimension for simplicity. Elementary classical mechanics is based on the concept of a point particle, whose position x and momentum p (or velocity $v = p/m$) appear in the equations of motion. These quantities are given directly by Newton's laws, or can be calculated from the Lagrangian or Hamiltonian functions in more advanced formulations.

Wave mechanics is an elementary formulation of quantum theory that, as its name implies, is centred on a *wave function* $\Psi(x, t)$. Quantities such as position and momentum are not given directly, but must be deduced from Ψ. Instead of

Newton's laws we have a wave equation that governs the evolution of $\Psi(x, t)$. In one dimension this takes the form

$$-\frac{\hbar^2}{2m}\frac{\partial^2}{\partial x^2}\Psi(x, t) + V(x)\Psi(x, t) = i\hbar\frac{\partial}{\partial t}\Psi(x, t), \quad (1.1)$$

which is the *time-dependent Schrödinger equation*. We shall see a partial justification of its form a little later. This equation describes a particle moving in a region of varying potential energy $V(x)$; forces do not enter directly. The potential energy could arise from an electric field expressed as a scalar potential, but the inclusion of a magnetic field is more complicated and will be deferred to Chapter 6.

A useful simplification is obtained by looking for separable solutions where the dependence on x and t is decoupled, $\Psi(x, t) = \psi(x)T(t)$. I shall use capital Ψ for the time-dependent wave function and the lower-case letter ψ for the time-independent function. Substituting the product into the time-dependent Schrödinger equation (1.1) and dividing by ψT gives

$$\frac{1}{T}i\hbar\frac{dT(t)}{dt} = \frac{1}{\psi}\left[-\frac{\hbar^2}{2m}\frac{d^2\psi(x)}{dx^2} + V(x)\psi(x)\right]. \quad (1.2)$$

Now the left-hand side is a function of t only, whereas the right-hand side is a function of x only, so the separation has succeeded. This makes sense only if both sides are equal to a constant, E, say. Then the left-hand side gives

$$T(t) \propto \exp\left(-\frac{iEt}{\hbar}\right) \equiv \exp(-i\omega t), \quad (1.3)$$

where $E = \hbar\omega$. This is simple harmonic variation in time. We have no choice about the complex exponential function: it cannot be replaced by a real sine or cosine, nor is $\exp(+i\omega t)$ acceptable. The form of the time-dependent Schrödinger equation requires $\exp(-i\omega t)$ and this convention is followed throughout quantum mechanics. It contrasts with other areas of physics, where $\exp(+i\omega t)$ may be used for the dependence of oscillations on time, or engineering, where $\exp(+j\omega t)$ is usual. Unfortunately this choice of sign has a far-reaching influence and crops up in unexpected places, such as the sign of the imaginary part of the complex dielectric function $\tilde{\epsilon}_r(\omega)$ (Section 8.5.1).

The spatial part of the separated Schrödinger equation becomes

$$-\frac{\hbar^2}{2m}\frac{d^2\psi}{dx^2} + V(x)\psi(x) = E\psi(x). \quad (1.4)$$

This is the *time-independent Schrödinger equation*. The equation takes the same form in three dimensions with $\partial^2/\partial x^2$ replaced by $\nabla^2 = \partial^2/\partial x^2 + \partial^2/\partial y^2 + \partial^2/\partial z^2$. Thus the solutions of the time-dependent Schrödinger equation take the form

$$\Psi(x, t) = \psi(x)\exp\left(-\frac{iEt}{\hbar}\right). \quad (1.5)$$

Later we shall justify the identification of the separation constant E as the energy of the particle. Thus the solutions of the time-independent Schrödinger equation describe states of the particle with a definite energy, known as *stationary states*. Again, this term will be justified later, but first we shall look at some simple but important solutions of the Schrödinger equation.

1.2 Free Particles

The simplest example is a particle (an electron, say) in free space, so $V(x) = 0$ everywhere. The time-independent Schrödinger equation is

$$-\frac{\hbar^2}{2m}\frac{d^2\psi}{dx^2} = E\psi(x). \tag{1.6}$$

This is the standard (Helmholtz) wave equation, and is simple enough that we can guess the possible solutions. One choice is to use complex exponential waves, $\psi(x) = \exp(+ikx)$ or $\exp(-ikx)$. Alternatively we could use real trigonometric functions and write $\psi(x) = \sin kx$ or $\cos kx$. It turns out that the choice of real or complex functions has important consequences. Substitution shows that any of these functions is a solution with

$$E = \frac{\hbar^2 k^2}{2m} \equiv \varepsilon_0(k). \tag{1.7}$$

Classically the kinetic energy can be written as $E = p^2/2m$ so we deduce that the momentum $p = \hbar k$. Combining this with the relation between energy and frequency yields the two central relations of old quantum theory:

$$E = \hbar\omega = h\nu \qquad \text{(Einstein)}, \tag{1.8}$$

$$p = \hbar k = \frac{h}{\lambda} \qquad \text{(de Broglie)}. \tag{1.9}$$

Dividing the energy by $\hbar$ gives the dispersion relation between frequency and wave number to be $\omega = (\hbar/2m)k^2$. This is nonlinear, which means that the velocity of particle waves is a function of their frequency and must be defined carefully. The two standard definitions are as follows:

$$\text{(phase velocity)} \qquad v_{\text{ph}} = \frac{\omega}{k} = \frac{\hbar k}{2m} = \frac{p}{2m}, \tag{1.10}$$

$$\text{(group velocity)} \qquad v_{\text{g}} = \frac{d\omega}{dk} = \frac{\hbar k}{m} = \frac{p}{m} = v_{\text{cl}}, \tag{1.11}$$

where v_{cl} is the classical velocity. The wave packet sketched in Figure 1.1 shows the significance of these two velocities. The wavelets inside the packet move along at the

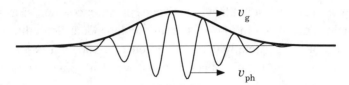

FIGURE 1.1. A wave packet, showing the envelope that moves at v_{g} while the wavelets inside move at v_{ph}.

phase velocity v_{ph} while the envelope moves at the group velocity v_{g}. If this packet represents a particle such as an electron, we are usually interested in the behaviour of the wave function as a whole rather than its internal motion. The group velocity is then the appropriate one and it is a relief that this agrees with the classical result.

Even if we use a wave packet to represent an electron, it is still spread out over space rather than localized at a point as in classical mechanics. This is inevitable in a picture based on waves, and means that we cannot give the location of a particle precisely. We'll look at this further in Section 1.5.3.

Finally, we found that $E = \hbar^2 k^2 / 2m$, which means that $E \geq 0$ if k is a real number. The wave number becomes imaginary, $k \to i\kappa$, if $E < 0$. Thus $\psi(x) \propto \exp(\kappa x)$, $\exp(-\kappa x)$, $\sinh \kappa x$, or $\cosh \kappa x$ with $E = -\hbar^2 \kappa^2 / 2m$. These wave functions are all real, and all diverge as $x \to \pm\infty$ in at least one direction. A divergent wave function is not physically acceptable, so these solutions can be used only in a restricted region of space.

1.3 Bound Particles: Quantum Well

We considered electrons that were free to propagate over all space in the previous section; now we shall see what happens when they are restricted to a finite region of space. This is called a *quantum well* or a *particle in a box*. The simplest example is the infinitely deep 'square' well and is illustrated in Figure 1.2. The electron has zero potential energy in the region $0 < x < a$, and infinitely high potential barriers prevent it from straying beyond this region. Note that a is the *full* width of the well here, but sometimes it is used for the *half*-width.

The Schrödinger equation for motion inside the well is identical to that for free space, equation (1.6), because there is no potential energy. The solutions are therefore the same, and can be written as $\exp(+ikx)$, $\exp(-ikx)$, $\sin kx$, or $\cos kx$, with $E = \hbar^2 k^2 / 2m$. Nothing yet requires E to be positive; k would be imaginary and the waves would be replaced by real exponentials and hyperbolic functions if E were negative. The imposition of boundary conditions removes this freedom.

The barriers present an infinitely high potential energy outside the well. The Schrödinger equation must remain well behaved, and the term $V(x)\psi(x)$ remains finite only if $\psi(x) = 0$ in the barriers. Discontinuities in the wave function are also unacceptable, so $\psi(x)$ must vanish at the edge of the barriers. This gives boundary

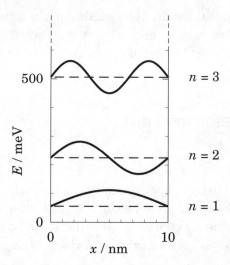

FIGURE 1.2. Infinitely deep square well in GaAs of width 10 nm along x, showing the first three energy levels and wave functions.

conditions $\psi(x{=}0) = 0$ and $\psi(x{=}a) = 0$. The first of these can be satisfied by choosing $\psi(x) = \sin kx$ or $\sinh \kappa x$ for any value of k or κ. The second condition cannot be satisfied by $\sinh \kappa x$ so we must choose $\sin kx$. This must vanish at $x = a$, so $\sin ka = 0 = \sin n\pi$. Thus $k = n\pi/a$, and the wave functions and energies are

$$\phi_n(x) = A_n \sin \frac{n\pi x}{a}, \qquad \varepsilon_n = \varepsilon_0(k_n) = \frac{\hbar^2 k_n^2}{2m} = \frac{\hbar^2 \pi^2 n^2}{2ma^2}. \qquad (1.12)$$

The integer $n = 1, 2, 3, \ldots$ is a *quantum number* that labels the states. Quantization, the restriction on the allowed energies, has come from the boundary conditions that constrain the motion of the particle. Particles that can move freely over all space have a continuous range of energy levels; those that are confined to a region of space have discrete levels. Often a particle may be bound in some range of energies and free in others, with a mixture of discrete and continuous energy levels.

The lowest state has energy $\varepsilon_1 > 0$. This contrasts with classical mechanics, where the state of lowest energy has the particle sitting still, anywhere on the bottom of the well, with no kinetic energy and $E = 0$. Such behaviour would violate the uncertainty principle in quantum mechanics, described in Section 1.5, so even the lowest state has positive *zero-point energy*.

The wave functions in the square well have an important symmetry property. Those with n odd are even functions of x about the centre of the well, whereas those with n even are odd functions of x (it would be neater if the numbering of states started with 0 rather than 1). This symmetry holds for any potential well where $V(x)$ is an even function of x. Symmetry properties like this are important in providing *selection rules* for many processes, such as optical absorption. This is a very simple example; group theory is needed to describe more complicated symmetries, such as those of a crystal.

Now that we have calculated the energy levels in a quantum well, an obvious question is how do we measure them experimentally. Optical methods provide the most direct techniques, so we shall next take a look at optical absorption in a quantum well.

1.3.1 OPTICAL ABSORPTION IN A QUANTUM WELL

The quantum well looks like an artificial model, which is at home in a textbook but has little application in the real world. Although an infinitely deep well cannot be made, it is simple nowadays to grow structures that are close to ideal finite wells. The energy levels in a finite well will be calculated in Section 4.2, but in practice the infinitely deep well is often used as an approximation because its results are so simple.

A heterostructure consisting of a thin sandwich of GaAs between thick layers of AlGaAs provides a simple quantum well, shown in Figure 1.3(a). ('AlGaAs' really means an alloy such as $Al_{0.3}Ga_{0.7}As$, but the abbreviation is universal.) To justify this, we need to anticipate some concepts that will be explained more fully later on.

First, look at the behaviour of electrons. Free electrons have energy $\varepsilon_0(k) = \hbar^2 k^2/2m_0$. Electrons in a semiconductor live in the *conduction band*, which changes their energy in two ways. First, energy must be measured from the bottom of the band

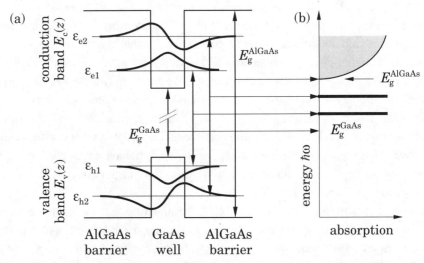

FIGURE 1.3. Optical absorption in a quantum well formed by a layer of GaAs surrounded by AlGaAs. (a) Potential well in conduction and valence band, showing two bound states in each; the energy gap of GaAs is really much larger than this diagram implies. (b) Transitions between states in the quantum well produce absorption lines between the band gaps of the GaAs well and AlGaAs barrier.

at E_c rather than from zero. Second, electrons behave as though their mass is $m_0 m_e$, where the *effective mass* $m_e = 0.067$ in GaAs. Thus $\varepsilon_c(k) = E_c + \hbar^2 k^2 / 2 m_0 m_e$. The sandwich acts like a quantum well because E_c is higher in AlGaAs than in GaAs, and the difference ΔE_c provides the barrier that confines the electrons. Typically $\Delta E_c \approx 0.2$–0.3 eV, which is not large. However, we shall approximate it as infinite to find the energy levels in a well of width a. Adapting equation (1.12) shows that the energy of the bound states, labelled with n_e, is

$$\varepsilon_{e n_e} \approx E_c^{\text{GaAs}} + \frac{\hbar^2 \pi^2 n_e^2}{2 m_0 m_e a^2}. \tag{1.13}$$

We could measure these energy levels by shining light on the sample and de-termining which frequencies were absorbed. A photon is absorbed by exciting an electron from a lower level to a higher one, and the energy of the photon matches the difference in electronic energy levels. We might therefore hope to see absorption at a frequency given by $\hbar \omega = \varepsilon_{e2} - \varepsilon_{e1}$, for example. Unfortunately this is a difficult experiment and a different technique is usually used.

Semiconductors have energy levels in other bands. The most important of these is the *valence band* which lies below the conduction band. The top of this is at E_v and the band curves *downwards* as a function of k, giving $\varepsilon_v(k) = E_v - \hbar^2 k^2 / 2 m_0 m_h$, which contains another effective mass m_h ($m_h = 0.5$ in GaAs). The conduction and valence bands are separated by an energy called the *band gap* given by $E_g = E_c - E_v$. Again there is a quantum well because E_v is at a different level in the GaAs well and AlGaAs barriers. The energies of the bound states are

$$\varepsilon_{h n_h} \approx E_v^{\text{GaAs}} - \frac{\hbar^2 \pi^2 n_h^2}{2 m_0 m_h a^2}. \tag{1.14}$$

Everything is 'upside down' in the valence band, as shown in Figure 1.3(a).

The valence band is completely full, and the conduction band completely empty, in a pure semiconductor at zero temperature. Optical absorption must therefore lift an electron from the valence band into the conduction band. In a bulk sample of GaAs this can occur provided that $\hbar \omega > E_g^{\text{GaAs}}$, the band gap of GaAs. Similarly we need $\hbar \omega > E_g^{\text{AlGaAs}}$ in AlGaAs. This process leaves behind an empty state or *hole* in the valence band, so a subscript h is used to identify parameters of the valence band.

Now look at the quantum well. Although the well is of GaAs, absorption cannot start at $\hbar \omega = E_g^{\text{GaAs}}$ because the states in the well are quantized. The lowest en-ergy at which absorption can occur is given by the difference in energy $\varepsilon_{e1} - \varepsilon_{h1}$ between the lowest state in the well in the conduction band and the lowest state in the well in the valence band. Absorption can occur at higher energies by us-ing other states, and we shall see later that the strongest transitions occur between corresponding states in the two bands, so set $n_e = n_h = n$. Therefore strong absorption occurs at the frequencies given by

$$\hbar\omega_n = \varepsilon_{en} - \varepsilon_{hn} = \left(E_c^{\text{GaAs}} + \frac{\hbar^2\pi^2 n^2}{2m_0 m_e a^2} \right) - \left(E_v^{\text{GaAs}} - \frac{\hbar^2\pi^2 n^2}{2m_0 m_h a^2} \right)$$

$$= E_g^{\text{GaAs}} + \frac{\hbar^2\pi^2 n^2}{2m_0 a^2} \left(\frac{1}{m_e} + \frac{1}{m_h} \right). \tag{1.15}$$

The energies look like those in a quantum well where the effective mass is m_{eh}, given by $1/m_{\text{eh}} = 1/m_e + 1/m_h$. This is called the optical effective mass; almost every process has its own effective mass!

If the wells really were infinitely deep, there would be an infinite series of lines with frequencies given by equation (1.15). The barriers in the semiconductor are finite, and absorption occurs in the AlGaAs barriers for all frequencies where $\hbar\omega > E_g^{\text{AlGaAs}}$. The resulting spectrum is shown in Figure 1.3(b), assuming that there are two bound states in both the conduction and valence bands. No absorption is possible for $\hbar\omega < E_g^{\text{GaAs}}$, and there is a continuous band of absorption for $\hbar\omega > E_g^{\text{AlGaAs}}$. Between these two frequencies lie two discrete lines produced by transitions between states in the quantum well, with energies given by equation (1.15). The width of the well can be inferred from the energy of these lines if the effective masses are known. This is a routine check to see that layers have been grown correctly.

In practice a slightly different experiment is usually performed, called *photoluminescence* (PL). Light with $\hbar\omega > E_g^{\text{AlGaAs}}$ is shone on the sample, which excites many electrons from the valence to the conduction band everywhere. Some of these electrons become trapped in the quantum well, and the same thing happens to the holes in the valence band. It is then possible for an electron to fall from the conduction band into a hole in the valence band and release the difference in energy as light. This luminescence is the reverse process to absorption and can occur at the same energies. Only the lowest levels are usually seen, so the PL spectrum should show a line at $\hbar\omega_1$. An example of a photoluminescence spectrum is shown in Figure 1.4. The sample has four wells of different widths, each of which contributes a peak to the PL. A detailed analysis of this spectrum is left as an exercise.

Unfortunately the true picture is slightly more complicated. One problem is that the valence band is not as simple as we have assumed. A better model is to assume that there are two varieties of holes, heavy and light. Thus two sets of spectral lines should be seen, although the heavy holes are much more prominent. A further complication is that electrons and holes bind together to form *excitons*, analogous to hydrogen atoms, and this modifies the energies slightly. This problem will be addressed in Section 10.7.

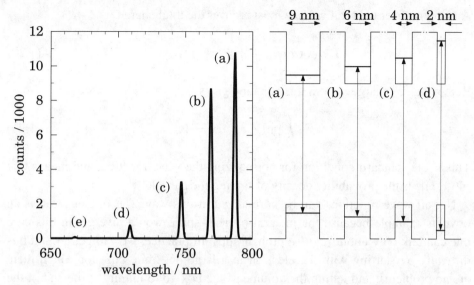

FIGURE 1.4. Photoluminescence as a function of wavelength for a sample with four quantum wells of different widths, whose conduction and valence bands are shown on the right. The barriers between the wells are much thicker than drawn. [Data kindly supplied by Prof. E. L. Hu, University of California at Santa Barbara.]

1.4 Charge and Current Densities

The Schrödinger equation yields a wave function $\Psi(x, t)$, and we should now consider how to deduce quantities of interest from it. First we would like to know the location of the particle. This follows from the squared modulus of the wave function:

$$|\Psi(x, t)|^2 \propto \text{probability density of finding the particle at } x. \tag{1.16}$$

This does not strictly imply that the particle itself is spread out: it should be interpreted as a statement about our knowledge of the particle. However, this distinction will not be important for the topics covered in this book, and we can put the relation in a more physically transparent form by using the charge density. If the particle has charge q, this becomes

$$q|\Psi(x, t)|^2 \propto \text{charge density of particle,} \tag{1.17}$$

or

$$q|\Psi(x, t)|^2 dx \propto \text{charge in a region } dx \text{ around } x. \tag{1.18}$$

If the electron is bound within some volume, we know that the total charge enclosed in that volume must be q. The proportionality can then be turned into an equality:

$$q|\Psi(x, t)|^2 = \rho(x) = \text{charge density of particle.} \tag{1.19}$$

An integral over the total volume must recover the total charge:

$$\int \rho(x)\,dx = \int q|\Psi(x,t)|^2 dx = q. \tag{1.20}$$

Removing the charge q from this equation gives

$$\int |\Psi(x,t)|^2 dx = 1. \tag{1.21}$$

This is the standard condition for normalizing the wave function, and means that $|\Psi(x,t)|^2$ is the probability density of finding the particle.

Not all wave functions can be normalized in this way. The free electron is an obvious example because the integral over all space would diverge. In this case one can talk only about *relative* probabilities. In practice, we can get around this difficulty by starting with the electron in a large but finite box, for which there are no problems, and letting the volume of the box go to infinity at the end of the calculation. We shall do this in Section 1.7 to calculate the density of states.

Normalization gives physical dimensions to the wave function. Take the infinitely deep well as an example. The wave functions $\phi_n(x) = A_n \sin(n\pi x/a)$, and the condition for normalization is

$$1 = \int_0^a |A_n|^2 \sin^2 \frac{n\pi x}{a} dx = \frac{a|A_n|^2}{2}. \tag{1.22}$$

Thus the normalized wave function, if A_n is taken to be real, is

$$\phi_n(x) = \sqrt{\frac{2}{a}} \sin \frac{n\pi x}{a}. \tag{1.23}$$

Normalization has given the wave function dimensions of (length)$^{-1/2}$ in one dimension. This is often useful as a check.

A plane wave such as $\phi_k(x) = Ae^{ikx}$ in an infinite volume can be normalized in a slightly different way. Here the density $|\phi_k(x)|^2 = |A|^2$, which can be set to a given density of particles.

Now that we have a charge density, there should be a current density J (or just current in one dimension) associated with it. These must obey the continuity equation

$$\frac{\partial J}{\partial x} + \frac{\partial \rho}{\partial t} = 0 \tag{1.24}$$

to ensure the conservation of charge and particles. In three dimensions $\partial J/\partial x$ becomes div $\mathbf{J}$.

To construct a current density, start with the time-dependent Schrödinger equation

$$-\frac{\hbar^2}{2m}\frac{\partial^2}{\partial x^2}\Psi(x,t) + V(x,t)\Psi(x,t) = i\hbar\frac{\partial}{\partial t}\Psi(x,t). \tag{1.25}$$

Multiply both sides on the left by the complex conjugate of the wave function, Ψ^*:

$$-\frac{\hbar^2}{2m}\Psi^*\frac{\partial^2}{\partial x^2}\Psi + \Psi^* V\Psi = i\hbar\Psi^*\frac{\partial}{\partial t}\Psi. \tag{1.26}$$

Obtain a second equation by going back to the Schrödinger equation, taking its complex conjugate, and multiplying from the left by Ψ. This gives

$$-\frac{\hbar^2}{2m}\Psi\frac{\partial^2}{\partial x^2}\Psi^* + \Psi V^*\Psi^* = -i\hbar\Psi\frac{\partial}{\partial t}\Psi^*. \tag{1.27}$$

Now subtract (1.27) from (1.26). The terms with the potential cancel provided that $V(x, t)$ is real. The two terms on the right-hand side add and are clearly the derivative of a product, so the difference becomes

$$-\frac{\hbar^2}{2m}\left(\Psi^*\frac{\partial^2}{\partial x^2}\Psi - \Psi\frac{\partial^2}{\partial x^2}\Psi^*\right) = i\hbar\frac{\partial}{\partial t}|\Psi|^2. \tag{1.28}$$

To simplify the left-hand side, use the rule for the derivative of a product:

$$\frac{\partial}{\partial x}\left(\Psi^*\frac{\partial}{\partial x}\Psi\right) = \left(\frac{\partial\Psi^*}{\partial x}\right)\left(\frac{\partial\Psi}{\partial x}\right) + \Psi^*\frac{\partial^2}{\partial x^2}\Psi. \tag{1.29}$$

When this is applied to (1.28), the products of single derivatives cancel and it reduces to

$$-\frac{\hbar^2}{2m}\frac{\partial}{\partial x}\left(\Psi^*\frac{\partial}{\partial x}\Psi - \Psi\frac{\partial}{\partial x}\Psi^*\right) = i\hbar\frac{\partial}{\partial t}|\Psi|^2. \tag{1.30}$$

Finally, moving the factor of $i\hbar$ to the left and multiplying throughout by q to turn the probability densities into charge densities gives

$$-\frac{\partial}{\partial x}\left[\frac{\hbar q}{2im}\left(\Psi^*\frac{\partial}{\partial x}\Psi - \Psi\frac{\partial}{\partial x}\Psi^*\right)\right] = \frac{\partial}{\partial t}\left(q\,|\Psi(x, t)|^2\right) = \frac{\partial\rho}{\partial t}. \tag{1.31}$$

Comparing this with the continuity equation (1.24) shows that the current density is given by

$$J(x, t) = \frac{\hbar q}{2im}\left(\Psi^*\frac{\partial}{\partial x}\Psi - \Psi\frac{\partial}{\partial x}\Psi^*\right). \tag{1.32}$$

In three dimensions the derivative $\partial\Psi/\partial x$ becomes the gradient $\nabla\Psi$.

The dependence on time vanishes from both ρ and J for a stationary state because $\exp(-i\omega t)$ cancels between Ψ and Ψ^*. This partly explains the origin of the term 'stationary', although the states may still carry a current (constant in time) so it is slightly misleading. However, a stationary state where $\psi(x)$ is purely real carries no current. This applies to a particle in a box and to bound states in general. A superposition of bound states is needed to generate a current. This feature emphasizes that the wave function in quantum mechanics is in general a 'genuine' complex quantity. This contrasts with the complex notation widely used for oscillations in systems ranging from electric circuits to balls and springs. Here the response is real and the complex form is used only for convenience.

As an example, consider $\Psi(x, t) = A \exp[i(kx - \omega t)]$, which describes a plane wave moving in the $+x$-direction. Its charge density $\rho = q|A|^2$, uniformly over all space, and the current $J = q(\hbar k/m)|A|^2$. Now $\hbar k/m = p/m = v$ so $J = \rho v$, which is the expected result (like '$J = nev$').

The Schrödinger equation is linear, so further wave functions can be constructed by superposing basic solutions. For example,

$$\Psi(x, t) = [A_+ \exp(ikx) + A_- \exp(-ikx)] \exp(-i\omega t) \qquad (1.33)$$

describes a superposition of waves travelling in opposite directions. The quantum-mechanical expression for the current gives the expected result

$$J = \frac{\hbar q k}{m}(|A_+|^2 - |A_-|^2). \qquad (1.34)$$

There is an interesting result for two counter-propagating *decaying* waves,

$$\Psi(x, t) = [B_+ \exp(\kappa x) + B_- \exp(-\kappa x)] \exp(-i\omega t). \qquad (1.35)$$

Neither component would carry a current by itself because it is real, but the superposition gives

$$J = \frac{\hbar q \kappa}{im}(B_+ B_-^* - B_+^* B_-) = \frac{2\hbar q \kappa}{m} \mathrm{Im}(B_+ B_-^*). \qquad (1.36)$$

The wave must contain components decaying in *both* directions, with a phase difference between them, for a current to flow. This effect is shown in Figure 1.5 for a wave hitting a barrier. We shall see in Chapter 5 that an oscillating wave turns into a decaying one inside a high barrier. If the barrier is infinitely long, it contains a single decaying wave and there is no net current. A finite barrier, on the other hand, transmits a (small) current and must contain two counter-propagating decaying waves. The returning wave ($\exp \kappa x$) from the far end of the barrier carries the information that the barrier is finite and that a current flows.

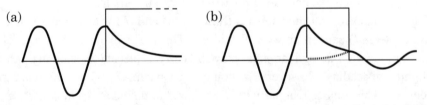

FIGURE 1.5. Current carried by counter-propagating decaying waves. (a) An infinitely thick barrier contains a single decaying exponential that carries no current. (b) A finite barrier contains both growing and decaying exponentials and passes current. (The wave function is complex, so the figure is only a rough guide.)

1.5 Operators and Measurement

It is now time to return to the theory of quantum mechanics in a little more depth, and to see how physical quantities can be deduced from the wave function.

1.5.1 OPERATORS

It is a postulate of quantum mechanics that observable quantities can be represented by operators that act on the wave function (although it is a further postulate that the wave function itself is *not* observable). Operators will be denoted with a hat or circumflex. The position, momentum, and total energy can be represented by the following operators on $\Psi(x, t)$:

$$x \rightarrow \hat{x} = x, \tag{1.37}$$

$$p \rightarrow \hat{p} = -i\hbar \frac{\partial}{\partial x}, \tag{1.38}$$

$$E \rightarrow \hat{E} = i\hbar \frac{\partial}{\partial t}. \tag{1.39}$$

An important feature is that the momentum $\hat{p}$ appears as a *spatial* derivative.

More complicated operators can be constructed from these components. For example, the Hamiltonian function $H = p^2/2m + V(x)$ gives the total energy of a classical particle in the type of system that we have studied, where energy is conserved. This becomes a Hamiltonian operator $\hat{H}$ in quantum mechanics and is given by

$$\hat{H} = H(\hat{x}, \hat{p}) = -\frac{\hbar^2}{2m}\frac{\partial^2}{\partial x^2} + V(x). \tag{1.40}$$

Equating the effect of this operator with that of the energy operator gives $\hat{H}\Psi = \hat{E}\Psi$, or

$$\left[-\frac{\hbar^2}{2m}\frac{\partial^2}{\partial x^2} + V(x) \right] \Psi(x, t) = i\hbar \frac{\partial}{\partial t}\Psi(x, t). \tag{1.41}$$

We are back to the time-dependent Schrödinger equation (1.1).

The time-independent Schrödinger equation can now be written concisely as $\hat{H}\psi(x) = E\psi(x)$, where E is a number, not an operator. This resembles a matrix eigenvalue equation: there is an operator acting on the wave function on one side, and a constant multiplying it on the other. The ideas of eigenvectors and eigenvalues work in much the same way for differential operators as for matrices, and similar terminology is used. Here ψ is called an eigenfunction or eigenstate, and E is the corresponding eigenvalue. This will be developed further in Section 1.6.

The current density can be rewritten in terms of the momentum operator, giving

$$J(x, t) = \frac{q}{2} \left[\Psi^* \left(\frac{\hat{p}}{m} \Psi \right) + \left(\frac{\hat{p}}{m} \Psi \right)^* \Psi \right]. \qquad (1.42)$$

This shows that the current is related to the velocity p/m. A more elaborate expression is needed in a magnetic field, which complicates the relation between velocity and momentum. This will be considered in Chapter 6.

Now look at the effect of the momentum operator on a wave function. A plane wave $\psi(x) = A \exp(ikx)$ gives

$$\hat{p} \psi = \left(-i\hbar \frac{d}{dx} \right) (Ae^{ikx}) = \hbar k A e^{ikx} = (\hbar k)\psi. \qquad (1.43)$$

Again this has reduced to an eigenvalue equation. We interpret this as meaning that the momentum has a definite value $p = \hbar k$, a result which we inferred earlier by analogy with classical mechanics.

A further postulate of quantum mechanics states that the only possible values of a physical observable are the eigenvalues of its corresponding operator. If the wave function is an eigenfunction of this operator, as in the case of the plane wave and momentum, the observable has a definite value. In general this is not the case. Consider the effect of the momentum operator on a particle in a box:

$$\hat{p} \phi_n(x) = -i\hbar \frac{d}{dx} A_n \sin \frac{n\pi x}{a} = \frac{-i\hbar n\pi A_n}{a} \cos \frac{n\pi x}{a}. \qquad (1.44)$$

These wave functions are not eigenfunctions of $\hat{p}$, and therefore do not have a definite value of momentum. Measurements of momentum would yield a range of values which we could characterize in terms of an average value (zero here) and a spread. Taking another derivative shows that $\phi_n(x)$ *is* an eigenfunction of $\hat{p}^2$. It therefore has a definite value of kinetic energy, whose operator $\hat{T} = \hat{p}^2/2m$.

Similar issues arise when we measure the position of a particle, which we shall consider next.

1.5.2 EXPECTATION VALUES

Suppose that we are given $\Psi(x, t)$ for some particle. Two simple quantities that we might wish to know are the average position of the particle and how well it is localized about that position. Note that we cannot say that the particle is *at* a particular point, unlike in classical mechanics, because we are using a picture based on waves.

We know that the probability density for finding a particle is $P(x, t) \propto |\Psi(x, t)|^2$, and can use this in the standard formula for finding a mean value. This gives

$$\langle x(t) \rangle = \int x \, P(x, t) \, dx, \qquad (1.45)$$

where angle brackets $\langle\rangle$ are used to denote expectation values. For normalized wave functions this becomes

$$\langle x(t) \rangle = \int x |\Psi(x,t)|^2 dx = \int \Psi^*(x,t)\, x\, \Psi(x,t)\, dx. \qquad (1.46)$$

To answer the question of how well the particle is localized, a common measure is the standard deviation Δx defined by

$$(\Delta x)^2 = \langle x^2 \rangle - \langle x \rangle^2, \qquad (1.47)$$

where $\langle x^2 \rangle$ is the expectation value of x^2, given in the same way by

$$\langle x^2 \rangle = \int x^2\, P(x)\, dx = \int \Psi^*(x,t)\, x^2\, \Psi(x,t)\, dx. \qquad (1.48)$$

Take the lowest state of a particle in a box as an example. Then

$$\langle x \rangle = \frac{2}{a} \int_0^a x \sin^2 \frac{\pi x}{a} dx = \frac{a}{2}, \qquad (1.49)$$

which is obvious from symmetry, and

$$\langle x^2 \rangle = \frac{2}{a} \int_0^a x^2 \sin^2 \frac{\pi x}{a} dx = a^2 \left(\frac{1}{3} - \frac{1}{2\pi^2} \right). \qquad (1.50)$$

Thus

$$\Delta x = a \sqrt{\frac{1}{12} - \frac{1}{2\pi^2}} \approx 0.18\, a. \qquad (1.51)$$

The particle is most likely to be found in the middle of the well, but with considerable spread around this (which increases for higher states).

The same questions can be asked about the momentum of the particle and can be answered in the same way using the momentum operator. The general expression for the expectation value $\langle q \rangle$ of some physically observable quantity q is

$$\langle q \rangle = \int \Psi^*(x,t)\, \hat{q}\, \Psi(x,t)\, dx, \qquad (1.52)$$

where $\hat{q}$ is the corresponding operator. For example, the average value of the momentum is given by

$$\langle p \rangle = \int \Psi^*(x,t)\, \hat{p}\, \Psi(x,t)\, dx = \int \Psi^*(x,t) \left[-i\hbar \frac{\partial \Psi(x,t)}{\partial x} \right] dx. \qquad (1.53)$$

This can be extended to quantities such as $\langle p^2 \rangle$ and Δp as was done for x.

The results of these expressions are physical quantities, and must therefore be mathematically real numbers. This requires that physical quantities be represented by *Hermitian* operators. Such operators have real eigenvalues, which guarantees that measurements on the wave function will yield real values. Their properties are reviewed briefly in Section 1.6. Non-Hermitian operators are important in other applications, notably as creation and annihilation operators in field theory, but will not be used in this book.

Expectation values of stationary states are constant in time, because their dependence on time cancels between Ψ and Ψ^*. For example, $\langle x \rangle$ is constant for any stationary state, so the particle appears to be 'stationary'. A superposition of states is required for the particle to 'move' in the sense that $\langle x \rangle$ varies with time. Going back to the one-dimensional well again, we can construct a moving wave function from the first two states,

$$\Psi(x, t = 0) = A_1 \phi_1(x) + A_2 \phi_2(x). \tag{1.54}$$

As this wave function evolves in time, the average position becomes

$$\langle x(t) \rangle = \frac{a}{2} - \frac{32 a A_1 A_2}{9\pi^2} \cos\left[\frac{(\varepsilon_2 - \varepsilon_1)t}{\hbar}\right]. \tag{1.55}$$

The particle oscillates back and forth in the well at angular frequency $(\varepsilon_2 - \varepsilon_1)/\hbar$ given by the difference in energy of the two levels. The analysis is left as an exercise.

1.5.3 MOTION OF A WAVE PACKET

Elementary classical mechanics rests on the concept of point particles, whose position and momentum can be specified precisely, but this is not tenable in wave mechanics. The natural analogue is a wave packet like that in Figure 1.1, a wave that is restricted to a finite region by an envelope. This also provides another illuminating example of expectation values. Start with a plain carrier wave $\exp(ip_0x/\hbar)$, and modulate it with a Gaussian envelope at $t = 0$:

$$\Psi(x, t = 0) = \frac{1}{(2\pi d^2)^{1/4}} \exp\left(\frac{ip_0x}{\hbar}\right) \exp\left[-\frac{(x - x_0)^2}{4d^2}\right]. \tag{1.56}$$

The probability density of this is a normalized Gaussian function with mean x_0 and standard deviation d:

$$|\Psi(x, t = 0)|^2 = \frac{1}{(2\pi d^2)^{1/2}} \exp\left[-\frac{(x - x_0)^2}{2d^2}\right]. \tag{1.57}$$

It is clear from this that $\langle x \rangle = x_0$ and $\Delta x = d$ at $t = 0$. We can make the wave packet as localized as we desire by choosing an appropriate value of d.

The carrier has definite momentum p_0 but we have had to mix many waves together to get the wave packet, so there is now a range of momenta in the wave function. There are two ways of extracting $\langle p \rangle$ and Δp. One is to use the definitions of the expectation values like equation (1.53) given earlier. The other way is to write the wave function as a function of momentum rather than position. Since we know that a plane wave $\exp(ipx/\hbar)$ has definite momentum p, the distribution of momenta within Ψ is given by resolving it into plane waves – just a Fourier transform. Thus the wave function in momentum space $\Phi(p, t)$ is related to that in real space by

$$\Psi(x, t) = \int \Phi(p, t) \exp\left(+\frac{ipx}{\hbar}\right) \frac{dp}{\sqrt{2\pi\hbar}}, \tag{1.58}$$

$$\Phi(p, t) = \int \Psi(x, t) \exp\left(-\frac{ipx}{\hbar}\right) \frac{dx}{\sqrt{2\pi\hbar}}. \tag{1.59}$$

The factors of $\sqrt{2\pi\hbar}$ ensure that Φ has the same normalization as Ψ.

Taking the Fourier transform of equation (1.56) for the Gaussian wave packet gives

$$\Phi(p, t = 0) = \frac{1}{[2\pi(\hbar/2d)^2]^{1/4}} \exp\left[\frac{-i(p - p_0)x_0}{\hbar}\right] \exp\left[-\frac{(p - p_0)^2}{4(\hbar/2d)^2}\right], \tag{1.60}$$

whose probability density is

$$|\Phi(p, t = 0)|^2 = \frac{1}{[2\pi(\hbar/2d)^2]^{1/2}} \exp\left[-\frac{(p - p_0)^2}{2(\hbar/2d)^2}\right]. \tag{1.61}$$

This is another normalized Gaussian with mean $\langle p \rangle = p_0$ from the carrier and standard deviation $\Delta p = \hbar/2d$.

An important result comes from the product of the standard deviations in space and momentum:

$$\Delta x \, \Delta p = d \frac{\hbar}{2d} = \tfrac{1}{2}\hbar. \tag{1.62}$$

Thus the better we localize the particle to fix its position in real space, the more waves we need and the wider the spread in momentum becomes. This is the famous *Heisenberg uncertainty principle*: we cannot measure both the position and momentum of a particle to arbitrary precision. It contrasts with the classical picture where both x and p could be known precisely. Gaussian wave packets happen to give the minimum uncertainty, and in general the result is

$$\Delta x \, \Delta p \geq \tfrac{1}{2}\hbar. \tag{1.63}$$

As one example, a plane wave $\exp(ip_0x/\hbar)$ definitely has momentum p_0, so $\Delta p = 0$, but it is spread evenly over all space giving $\Delta x = \infty$.

The uncertainty principle also forces the lowest state in a quantum well to have nonzero kinetic energy, unlike classical mechanics where it would be still. The momentum p would be known exactly (zero) if the particle were at rest, giving

$\Delta p = 0$, while Δx is finite because we know that the particle is somewhere in the well. Thus $\Delta x \, \Delta p = 0$, which is not allowed. The only way around this is for the particle to have a *zero-point energy* in the lowest state so that both Δx and Δp are non-zero.

This can be used to estimate the zero-point energy. Consider, for example, an infinitely deep potential well of width a. We know that the particle is in the well, so roughly $\Delta x \approx a/4$. This means that $\Delta p \approx \hbar/2\Delta x = 2\hbar/a$. Taking the kinetic energy as $(\Delta p)^2/2m$ gives an estimate of $(\hbar^2/2m)(2/a)^2$ for the energy of the ground state. The exact result has π instead of 2. This explains the dependence of the energy levels on the width a: making the well narrower reduces the spread of the particle in real space and therefore increases its range of momenta and hence the energy. This principle can be extended to estimate the zero-point energy in any well by including the mean potential energy.

Returning to the Gaussian wave packet, we found that it has minimum uncertainty (in the sense of the product $\Delta x \, \Delta p$) at $t = 0$, but this changes as it evolves in time. We know that a plane wave $\exp(ipx/\hbar)$ evolves in time like $\exp(-i\omega t)$ with $\hbar\omega = p^2/2m$. This applies to each Fourier component of the wave packet, so the wave function in Fourier space for $t > 0$ is

$$\Phi(p, t) = \frac{1}{[2\pi(\hbar/2d)^2]^{1/4}} \exp\left[\frac{-i(p - p_0)x_0}{\hbar}\right] \exp\left[-\frac{(p - p_0)^2}{4(\hbar/2d)^2}\right] \exp\left(\frac{-ip^2t}{2\hbar m}\right).$$

(1.64)

We must transform this back to real space to find $\Psi(x, t)$. A little rearrangement gives

$$\Psi(x, t) = \frac{1}{[2\pi(\hbar/2d)^2]^{1/4}} \exp\left[\frac{ip_0(x - p_0t/2m)}{\hbar}\right]$$

$$\times \int_{-\infty}^{\infty} \exp\left[\frac{i(p - p_0)(x - x_0 - p_0t/m)}{\hbar}\right]$$

$$\times \exp\left[-\frac{(p - p_0)^2}{4(\hbar/2d)^2}\left(1 + \frac{i\hbar t}{2md^2}\right)\right] \frac{dp}{\sqrt{2\pi\hbar}}.$$

(1.65)

The prefactor gives a carrier wave with momentum p_0, moving at the phase velocity $v_{\text{ph}} = p_0/2m$. Inside the integral, the first exponential shows that the wave packet is now centred on $x_0 + p_0t/m$ and therefore moves at the group velocity $v_g = p_0/m$. The second exponential, which controls the width of the wave packet, is also modified. Evaluation of the integral shows that

$$\Delta x(t) = \sqrt{d^2 + \left(\frac{\hbar t}{2md}\right)^2}.$$

(1.66)

The pulse spreads out in space as it propagates. The momentum remains unchanged if there are no forces acting on the particle, so the product $\Delta x \, \Delta p$ grows and our information about the particle deteriorates in time. This is the typical effect of dispersion as seen, for example, in communications.

Dispersion arises because a wave packet necessarily contains a range of momenta, each of which propagates at a different velocity causing the wave packet to spread. Eventually this overwhelms the initial width. The range of velocities is $(\Delta p)/m$ so at large times we expect $\Delta x \approx (\Delta p)t/m = \hbar t/2md$, in agreement with equation (1.66). A short pulse contains a wider range of momenta than a longer pulse and will eventually become longer.

1.5.4 FURTHER PROPERTIES OF OPERATORS

The uncertainty relation can be traced back to properties of the operators involved. We are trying to measure both the position and momentum of the particle described by the wave packet. A problem arises because of the order of these operations. Suppose we first measure the momentum, then the position. The operators acting on the wave function are

$$\hat{x}\hat{p}\Psi = x\left(-i\hbar\frac{\partial}{\partial x}\right)\Psi = -i\hbar x \frac{\partial \Psi}{\partial x}. \tag{1.67}$$

The opposite order gives

$$\hat{p}\hat{x}\Psi = \left(-i\hbar\frac{\partial}{\partial x}\right)x\Psi = -i\hbar\left(x\frac{\partial \Psi}{\partial x} + \Psi\right). \tag{1.68}$$

The last line follows from the derivative of the product. Clearly the results are different and the order of the operations is significant. Subtracting the two gives

$$\hat{x}\hat{p}\Psi - \hat{p}\hat{x}\Psi = i\hbar\Psi. \tag{1.69}$$

Since this equation holds for any Ψ, we can write it for the operators alone as

$$[\hat{x}, \hat{p}] \equiv \hat{x}\hat{p} - \hat{p}\hat{x} = i\hbar. \tag{1.70}$$

The notation $[\hat{x}, \hat{p}]$ is called a *commutator*. Two operators are said to commute if $[\hat{A}, \hat{B}] = 0$ since the order of their operation is unimportant. It is possible to measure two physical quantities simultaneously to arbitrary accuracy only if their operators commute. Clearly this does not apply to x and p, and their accuracy is limited by the uncertainty principle.

Similar relations apply to other coordinates and their corresponding momenta such as $[\hat{y}, \hat{p}_y] = i\hbar$. On the other hand $[\hat{y}, \hat{p}_x] = 0$, so these quantities can be measured simultaneously. Some further examples are given in the problems.

A quantity whose operator commutes with the Hamiltonian is called a constant of the motion because its value does not change with time. For example, $[\hat{p}, \hat{H}] = 0$ for a free particle so its momentum remains constant. These constants usually arise from some symmetry of the system, translational invariance in this case.

We have seen that the order of operators such as $\hat{x}$ and $\hat{p}$ is important and that they cannot be reordered like numbers. The same is true of matrices, and we shall see later that operators can be represented by matrices instead of the differential operators used here. Further, the choice of operators depends on the way in which the wave function is represented. We derived the wave function of a wave packet in momentum space before and could use the corresponding operators

$$\hat{p} = p, \qquad \hat{x} = i\hbar \frac{\partial}{\partial p}. \tag{1.71}$$

These obey the same commutation relation $[\hat{x}, \hat{p}] = i\hbar$ as the earlier forms in x and are therefore an equally valid choice.

1.6 Mathematical Properties of Eigenstates

This is a brief section on formal properties of eigenstates, which will be needed later in the construction of perturbation theory. Further details can be found in a book on mathematical methods for physics such as Mathews and Walker (1970).

We have already seen that the wave functions in the infinitely deep square well can be normalized. Assume that we are dealing with a finite system, so we can ignore the problems posed by plane waves and the like. Let the eigenstates (wave functions) of the Hamiltonian be $\phi_n(x)$ with corresponding eigenvalues (energies) ε_n, and normalize each state such that

$$\int |\phi_n(x)|^2 \, dx = 1. \tag{1.72}$$

The range of integration covers the region within which the particle can move, $0 < x < a$ for the particle in a box.

Eigenstates with different eigenvalues are *orthogonal*, that is,

$$\int \phi_m^*(x) \, \phi_n(x) \, dx = 0 \quad \text{if} \quad \varepsilon_m \neq \varepsilon_n. \tag{1.73}$$

This is a generalization of the concept of perpendicular vectors, and the integral is analogous to a scalar product. In the case of the quantum well this means that

$$\frac{2}{a} \int_0^a \sin \frac{m\pi x}{a} \sin \frac{n\pi x}{a} = 0 \quad \text{if} \quad m \neq n. \tag{1.74}$$

This relation is familiar from the theory of Fourier series.

Different states with a common eigenvalue are said to be *degenerate* (one of the many meanings of this term). In this case it is possible to choose the eigenstates so that they are orthogonal for $m \neq n$ even though $\varepsilon_m = \varepsilon_n$. Assuming that this has been done, equations (1.72) and (1.73) can be combined to define an *orthonormal* set:

$$\int \phi_m^*(x)\, \phi_n(x)\, dx = \delta_{m,n}. \tag{1.75}$$

The Kronecker delta is defined by $\delta_{m,n} = 1$ if $m = n$ and $\delta_{m,n} = 0$ otherwise. We shall assume that the ϕ_m are orthonormal, like the states in the quantum well (equation 1.23).

It can also be shown that the eigenstates form a *complete* set. This means that any wave function $\psi(x)$ that obeys the same boundary conditions can be expanded as a weighted sum over the ϕ_n,

$$\psi(x) = \sum_{n=1}^{\infty} a_n \phi_n(x). \tag{1.76}$$

The coefficient a_m is found by multiplying both sides by $\phi_m^*(x)$ and integrating:

$$\int \phi_m^*(x)\psi(x)dx = \int \phi_m^*(x) \sum_{n=1}^{\infty} a_n\phi_n(x)dx = \sum_{n=1}^{\infty} a_n \int \phi_m^*(x)\phi_n(x)dx. \tag{1.77}$$

The orthonormality condition (1.75) causes all terms in the sum on the right to vanish except for $m = n$, and we are left with

$$a_m = \int \phi_m^*(x)\psi(x)dx. \tag{1.78}$$

Again this technique is familiar from Fourier series.

A final relation is found by checking that the expansion is consistent. We should recover the initial function if we substitute the coefficients from equation (1.78) into the expansion (1.76). This leads to the requirement

$$\sum_{n=1}^{\infty} \phi_n(x)\phi_n^*(x') = \delta(x - x'), \tag{1.79}$$

which is known as the *closure* (or *completeness*) *relation*.

The expansion of a wave function is useful for finding the evolution in time of an arbitrary starting state $\Psi(x, t = 0)$. The method is to resolve $\Psi(x, t = 0)$ into eigenstates of the time-independent Schrödinger equation $\phi_n(x)$:

$$\Psi(x, t = 0) = \sum_{n=1}^{\infty} a_n \phi_n(x). \tag{1.80}$$

We know that each eigenstate evolves in time like $\phi_n(x)\exp(-i\varepsilon_n t/\hbar)$, and therefore the given state evolves as

$$\Psi(x,t) = \sum_{n=1}^{\infty} a_n\phi_n(x)\exp\left(-\frac{i\varepsilon_n t}{\hbar}\right).$$ (1.81)

This is how we followed the evolution of the wave packet in Section 1.5.3.

1.7 Counting States

A complete description of a system requires the energies and wave functions of all its states. Clearly this is an impossible task for anything but the simplest systems, and most of the information would in any case be unwanted. For many applications the density of states $N(E)$ is adequate. The definition is that $N(E)\,\delta E$ is the number of states of the system whose energies lie in the range E to $E+\delta E$. Clearly this tells us nothing about the wave functions at all, just the distribution of energies. We shall first calculate the density of states of a one-dimensional system before looking at more general results.

1.7.1 ONE DIMENSION

An immediate problem, as we saw in the previous section, is that the wave functions $\exp(ikx)$ cannot be normalized in the usual way if the particles travel through all space. The simplest way around this problem is to put the particles in a finite box of length L, and set $L \to \infty$ at the end of the calculation. Having put the particles in a box, we need to choose boundary conditions. Two are commonly used.

(i) *Fixed* or *box* boundary conditions, in which the wave function vanishes at the boundary:

$$\psi(0) = \psi(L) = 0.$$ (1.82)

(ii) *Periodic* or *Born–von Karman* boundary conditions, in which we imagine repeating the system periodically with the same wave function in each system. The wave function at $x = L$ must match smoothly to that at $x = 0$, which requires

$$\psi(0) = \psi(L), \qquad \frac{\partial\psi}{\partial x}\bigg|_{x=0} = \frac{\partial\psi}{\partial x}\bigg|_{x=L}.$$ (1.83)

Fixed boundary conditions are obviously the same as the particle in a box studied earlier. The energy levels are given by $\varepsilon_0(k) = \hbar^2 k^2/2m$ and the allowed values of k are

$$k_m = \frac{\pi m}{L}, \qquad m = 1, 2, 3, \ldots.$$ (1.84)

The wave functions are standing waves with this choice. They carry no current, and the allowed values of k are all positive.

Periodic boundary conditions require a different choice of k. We can use travelling exponential waves rather than sine waves, and they must obey $\exp(ikL) = \exp(ik0) = 1 = \exp(2\pi ni)$. This also satisfies the condition on the gradient, and the normalized states are $\phi_n(x) = L^{-1/2}\exp(ik_n x)$. The allowed values of k are

$$k_n = \frac{2\pi n}{L}, \qquad n = 0, \pm 1, \pm 2, \dots. \qquad (1.85)$$

These are twice as far apart as with fixed boundary conditions, but both signs of k are permitted and there are two degenerate states at each energy level (except for $k = 0$), with opposite signs of k and velocity.

This raises the following crucial question: does the density of states, which we are trying to calculate, depend on which boundary conditions we choose to apply to our artificial box? Fortunately it can be shown that the result is insensitive to boundary conditions as $L \to \infty$. It is usually more appropriate to treat free electrons as travelling rather than standing waves, so periodic boundary conditions are generally used.

To turn these allowed values of k and ε into a density of states, plot the allowed values of k along a line as in Figure 1.6. This is a simple one-dimensional version of 'k-space'. The values are regularly spaced, separated by $2\pi/L$. They become closer together as L increases and tend to blur into a continuum. In this case the number of allowed k-values in the range δk is just δk divided by the spacing of the points.

These points account for the different states that arise from motion in the box, but we must also consider the internal motion of the particle. Classically, free motion can be separated into translation and rotation about the centre of mass, and it is found that electrons carry an angular momentum that is known as *spin*. The treatment of angular momentum within quantum mechanics shows that the spin can take two states, which are conventionally labelled as *up* and *down*. Each spatial wave function can be associated with either spin, so the total number of states available to the electron should be doubled to take account of spin.

Spin enters only as a factor of 2 in the density of states for most of the topics considered in this book, but there are two areas where care must be taken. The first is the behaviour of electrons in a magnetic field **B**, because there is a magnetic moment associated with the spin which contributes an energy $-\boldsymbol{\mu} \cdot \mathbf{B}$. This will be discussed in Section 6.4.3. Second, the separation of velocity and spin holds only

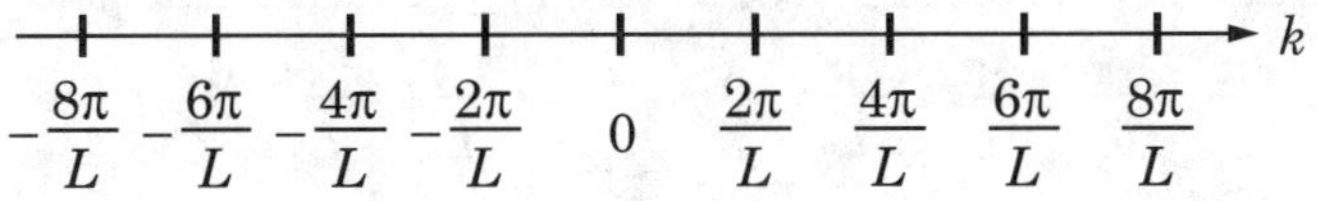

FIGURE 1.6. Allowed values of k for periodic boundary conditions in a system of length L plotted along a line, as a simple form of k-space.

within non-relativistic quantum mechanics. The conditions under which special relativity is important might seem far removed from ordinary semiconductors, but electrons move at a significant fraction of the speed of light when they pass close to a nucleus. This leads to an effect called spin–orbit coupling, which has a profound effect on the top of the valence band and therefore on the behaviour of holes. This will be described in Section 2.6.3.

Returning to the problem at hand, we can define a density of states in k-space such that $N_{1D}(k)\,\delta k$ is the number of allowed states in the range k to $k + \delta k$. It is given by

$$N_{1D}(k)\delta k = 2\frac{L}{2\pi}\delta k. \qquad (1.86)$$

The factor of 2 accounts for the spin, $L/2\pi$ is the density of points, and the range δk cancels to leave $N_{1D}(k) = L/\pi$. This is proportional to the volume (length) of the system, which makes sense: we would expect to double the number of states if we doubled the size of the system. Usually one takes this factor out to leave a density of states per unit length, which is $n_{1D}(k) = N_{1D}(k)/L = 1/\pi$.

The next task is to turn this into a density of states in energy. Figure 1.7 shows how the allowed values of k, which are evenly spaced, map to allowed values in energy through the dispersion relation $E = \varepsilon(k)$. These energies lie in a continuous band for $E \geq 0$ in a large system. The figure shows a parabola but the theory works for a more general dispersion relation. The resulting values of energy get further apart as k rises, so the density of states falls with increasing energy. A range δk in wave number corresponds to a range in energy of $\delta E = (dE/dk)\delta k$. The number of states in this range can be written in terms of $n_{1D}(k)$ or in terms of the density of states in energy per unit volume $n_{1D}(E)$. The two expressions must give the same number of states, so

$$n_{1D}(E)\delta E = n_{1D}(E)\frac{dE}{dk}\delta k = 2n_{1D}(k)\delta k. \qquad (1.87)$$

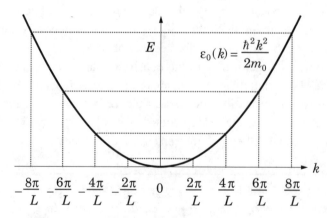

FIGURE 1.7. Dispersion relation $\varepsilon_0(k)$ for free electrons, showing how the allowed values of k map onto ε.

The factor of 2 in front of $n_{1D}(k)$ arises because of the two directions of motion; there is one range δk for $k > 0$ and another for $k < 0$. Note that the same symbol n_{1D} is used to represent the density of states in k and E; this is bad mathematics but typical usage in physics, where one rapidly tends to run out of variants of n and E to denote commonly used quantities. It shouldn't lead to confusion provided that the argument k or E is always included. Thus $n_{1D}(E) = (2/\pi)/(d\varepsilon/dk)$. This can be simplified using the group velocity $v = d\omega/dk = (1/\hbar)(d\varepsilon/dk)$, giving

$$n_{1D}(E) = \frac{2}{\pi \hbar v(E)}. \tag{1.88}$$

We shall see in Section 5.7.1 that the current depends on the product of the velocity and the density of states. Equation (1.88) shows that this is a constant in one dimension, which in turn leads to a quantized conductance.

Substituting the velocity for the special case of free electrons gives

$$n_{1D}(E) = \frac{1}{\pi \hbar} \sqrt{\frac{2m}{E}}. \tag{1.89}$$

The density of states diverges as $E^{-1/2}$ as $E \to 0$, a characteristic feature of one dimension.

1.7.2 THREE DIMENSIONS

In three dimensions, put the electrons into a box of volume $\Omega = L_x \times L_y \times L_z$. The wave functions are travelling waves in each direction with periodic boundary conditions, just as in the one-dimensional case, and their product gives

$$\phi_{lnm}(\mathbf{R}) = \frac{1}{\sqrt{L_x L_y L_z}} \exp[i(k_x x + k_y y + k_z z)] = \frac{1}{\sqrt{\Omega}} \exp(i\mathbf{K} \cdot \mathbf{R}). \tag{1.90}$$

Remember our convention that upper-case letters are used for three-dimensional positions and wave vectors. The product of three waves has been written as a three-dimensional plane wave using the scalar product. Similarly, the allowed values of K in each of the three directions can be combined into three-dimensional wave vectors

$$\mathbf{K} = \left(\frac{2\pi l}{L_x}, \frac{2\pi m}{L_y}, \frac{2\pi n}{L_z} \right), \qquad l, m, n = 0, \pm 1, \pm 2, \ldots. \tag{1.91}$$

These can be plotted as points in a three-dimensional $\mathbf{K}$-space with (k_x, k_y, k_z) as axes, where they form an evenly spaced rectangular mesh. Each unit cell encloses volume $(2\pi/L_x)(2\pi/L_y)(2\pi/L_z) = (2\pi)^3/\Omega$. Thus the density of allowed states is $N_{3D}(\mathbf{K}) = 2\Omega/(2\pi)^3$, where the 2 accounts for spin. Dividing by the volume gives the density of states in $\mathbf{K}$-space per unit volume of the system in real space, $n_{3D}(\mathbf{K}) = 2/(2\pi)^3$. This is again a constant, and generalizes in an obvious way to d dimensions as $n_d(\mathbf{K}) = 2/(2\pi)^d$.

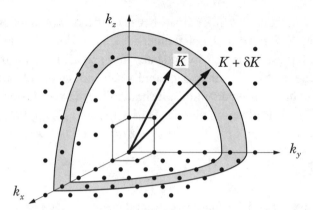

FIGURE 1.8. Construction in **K**-space to calculate the density of states for free electrons in three dimensions. The shells have radii K and $K + \delta K$, corresponding to energies ε and $\varepsilon + \delta\varepsilon$.

Now we need to derive the density of states as a function of energy. Consider free electrons only, because the calculation is more complicated for a general function $\varepsilon(\mathbf{K})$. Figure 1.8 shows two spheres about the origin in **K**-space, one with radius K and the other with radius $K + \delta K$. The volume of the shell between these two spheres is $4\pi K^2 \delta K$. The number of states in the shell is found from the product of this volume and the density of states $n_{3D}(\mathbf{K})$, giving $(K^2/\pi^2)\delta K$. The separation δK corresponds to a difference in energy of

$$\delta E = \frac{dE}{dK}\delta K = \frac{\hbar^2 K}{m}\delta K. \tag{1.92}$$

The number of states in the shell is given in terms of the density of states in energy by $n(E)\delta E$. Equating the two expressions yields $n_{3D}(E)\delta E = n_{3D}(E)(\hbar^2 K/m)\delta K = (K^2/\pi^2)\delta K$, whence

$$n_{3D}(E) = \frac{mK}{\pi^2\hbar^2} = \frac{m}{\pi^2\hbar^3}\sqrt{2mE}. \tag{1.93}$$

The square root is characteristic of three dimensions. Its singularity at the bottom of the band is much weaker than the one-dimensional result of $E^{-1/2}$. In general the density of states shows a stronger feature at the bottom of the band in fewer dimensions. Optical properties such as absorption are strongly influenced by the density of states, and low-dimensional systems are preferred for optoelectronic devices because their density of states is larger at the bottom of the band. The density of states for free electrons in one, two, and three dimensions is plotted in Figure 1.9. In all cases a low mass is associated with a low density of states.

The density of states for a three-dimensional crystal is more complicated because the surfaces of constant energy in **K**-space are not spheres. Further singularities of $n(E)$ appear inside bands, and provide fruitful material for optical spectroscopy. A simpler case arises if the energy depends on only the magnitude of **K** but not

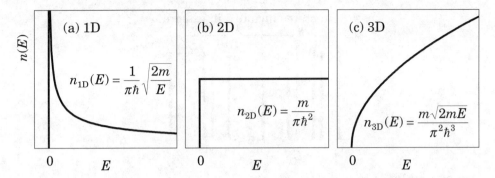

FIGURE 1.9. Densities of states for free electrons in one, two, and three dimensions.

its direction. In this case the surfaces of constant energy remain spherical and the derivation of $n(E)$ proceeds as before except for the form of $\varepsilon(K)$. For example, the conduction band of GaAs is often modelled by the expression

$$\varepsilon(K)[1 + \alpha\varepsilon(K)] = \frac{\hbar^2 K^2}{2m_0 m_e}. \tag{1.94}$$

This takes account of the fact that the band is not parabolic for high energies, with $\alpha \approx 0.6\,\mathrm{eV}^{-1}$.

1.7.3 A GENERAL DEFINITION OF THE DENSITY OF STATES

A general definition of the density of states is often useful. Let the states of a system have energies ε_n. Then the density of states in energy can be written as

$$N(E) = \sum_n \delta(E - \varepsilon_n), \tag{1.95}$$

where $\delta(E)$ is the Dirac δ-function. This is the total density of states, not that per unit volume. We shall now justify this definition and see how it is related to our previous calculations.

First, it is clear that equation (1.95) makes sense only if we integrate over it, because of the δ-functions. Consider

$$\int_{E_1}^{E_2} N(E)\,dE = \int_{E_1}^{E_2} \sum_n \delta(E - \varepsilon_n)\,dE = \sum_n \int_{E_1}^{E_2} \delta(E - \varepsilon_n)\,dE. \tag{1.96}$$

This is illustrated in Figure 1.10. If the energy of a state n lies within the range of integration from E_1 to E_2, the integral over $\delta(E - \varepsilon_n)$ gives unity by definition. If the energy ε_n lies outside the range of integration, on the other hand, there is no contribution because the weight of the δ-function is concentrated entirely at $E = \varepsilon_n$. Thus the integral gives 1 for all states in the range $E_1 \le \varepsilon_n \le E_2$ and zero for those

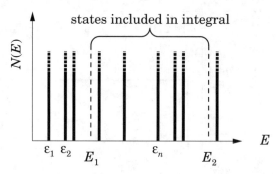

FIGURE 1.10. The 'δ-function' definition of the density of states, integrating to count the states between E_1 and E_2.

outside. Performing the sum next, we see that it adds up to the total number of states between E_1 and E_2. This is exactly what we would expect from an integral over $N(E)$, and shows that equation (1.95) is a valid definition of the density of states.

To confirm that it works, consider free electrons in one dimension again. In this case we can label the states by their wave number k and the definition (1.95) becomes

$$N(E) = 2 \sum_{k=-\infty}^{\infty} \delta[E - \varepsilon_0(k)]. \qquad (1.97)$$

The factor of 2 is for spin. Next, turn the sum into an integral, assuming a large system. We have already seen how to do this: the density of states in k-space is $L/2\pi$, so the sum becomes

$$N(E) = \frac{L}{\pi} \int_{-\infty}^{\infty} \delta[E - \varepsilon_0(k)] \, dk. \qquad (1.98)$$

There is still a function inside the δ-function, which is a nuisance, so change the variable of integration from k to $z = \varepsilon_0(k) = \hbar^2 k^2/2m$. This requires

$$dk = \frac{dk}{dz} dz = \frac{1}{\hbar} \sqrt{\frac{m}{2z}} \, dz, \qquad (1.99)$$

and the integral becomes

$$N(E) = \frac{2L}{\pi \hbar} \int_0^{\infty} \sqrt{\frac{m}{2z}} \, \delta(E - z) \, dz. \qquad (1.100)$$

The 2 in the prefactor comes from the two signs of k for each value of the energy z. Now the integral is trivial, as the only contribution is at $z = E$ when $E > 0$, and we finally get

$$N(E) = \frac{2L}{\pi \hbar} \sqrt{\frac{m}{2E}} = \frac{L}{\pi \hbar} \sqrt{\frac{2m}{E}} \qquad (1.101)$$

as before.

1.7.4 LOCAL DENSITY OF STATES

An important feature of the 'δ-function definition' of the density of states is that it can be extended to any system. The examples that we have studied so far are translationally invariant, which implies that the density of states is the same at each point. This is obviously a special case. As a simple example of a system that lacks translational symmetry, consider free electrons restricted to the region $x > 0$ by an impenetrable wall at $x = 0$. The wave functions are now $\sin kx$, so they are zero at $x = 0$. If the wave functions vanish, it seems logical that the density of states should do the same.

A *local* density of states can be defined to treat such situations, where the contribution of each state is weighted by the density of its wave function at the point in question. Thus equation (1.95) becomes

$$n(E, x) = \sum_n |\phi_n(x)|^2 \delta(E - \varepsilon_n). \tag{1.102}$$

This gives the previous result when integrated over the whole system:

$$\int n(E, x)\, dx = \sum_n \delta(E - \varepsilon_n) \int |\phi_n(x)|^2\, dx = \sum_n \delta(E - \varepsilon_n) = N(E). \tag{1.103}$$

The factor of $|\phi_n(x)|^2$ means that each state contributes to the local density of states only in the regions where its density is high. Clearly this is useful in inhomogeneous systems, and $n(E, x)$ contains more information than $n(E)$ alone.

For the one-dimensional system with a wall, the sine waves give

$$n_{1D}(E, x) = \frac{2}{\pi\hbar}\sqrt{\frac{2m}{E}}\sin^2 kx, \tag{1.104}$$

where $k = \sqrt{2mE}/\hbar$. An average over x restores the usual expression (1.89) for $n_{1D}(E)$ but oscillations persist at all distances from the wall. The corresponding result for a three-dimensional system restricted to $x > 0$ is $n_{3D}(E, x) = (1 - \text{sinc}\, 2kx)n_{3D}(E)$, where $\text{sinc}\,\theta = (\sin\theta)/\theta$. This is plotted in Figure 1.11 for electrons in GaAs. At $x = 20$ nm the local density of states is close to the square root that holds in an infinite system, but it shows strong oscillations for small x and vanishes for all energies at $x = 0$.

In fact we can generalize the local density of states still further to

$$n(E, x, x') = \sum_n \phi_n(x)\, \phi_n^*(x')\, \delta(E - \varepsilon_n). \tag{1.105}$$

This is a function of the variable in each wave function separately, and is called the *spectral function*. It is a natural point of contact with Green's functions in more advanced theory. The spectral function also provides a compact representation of

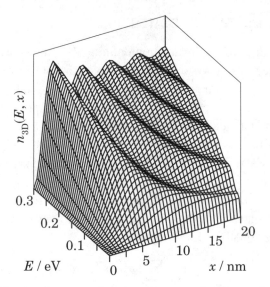

FIGURE 1.11. Local density of states as a function of energy and of distance from an impenetrable wall at $x = 0$ in GaAs.

some results that we shall derive later, such as optical absorption, but we shall not carry it further.

1.8 Filling States: The Occupation Function

The density of states tells us about the energy levels of a system, and the next job is to fill these levels with electrons or other particles.

In equilibrium, the average number of particles that occupy a state depends only on its energy and is given by an occupation function, which depends on the nature of the particles concerned. Electrons, protons, and other particles that carry a half-integer spin are called *fermions*. They obey the Pauli exclusion principle, which states that no more than one fermion can occupy a given state. We shall concentrate on the occupation function for fermions first, since it is the most important in semiconductors, and return to the others in Section 1.8.5.

1.8.1 FERMI–DIRAC OCCUPATION FUNCTION

The Pauli exclusion principle for fermions restricts the occupation number of a state to be either zero or one. The average occupation is governed by the Fermi–Dirac distribution function $f(E, E_F, T)$, or just $f(E)$ for short. It is given by

$$f(E, E_F, T) = \left[\exp\left(\frac{E - E_F(T)}{k_B T} \right) + 1 \right]^{-1}. \tag{1.106}$$

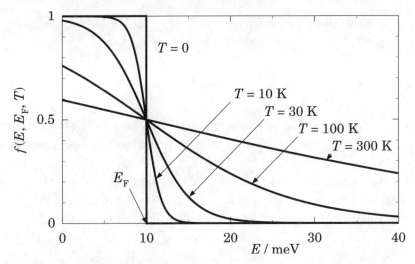

FIGURE 1.12. Fermi–Dirac distribution function at five temperatures with a constant Fermi level $E_F = 10\,\text{meV}$.

Here k_B is Boltzmann's constant and the shorthand $\beta = 1/k_B T$ is often used. The energy $E_F(T)$ is usually called the *Fermi level* in semiconductors, and it is important to remember that it varies with the temperature T. The Fermi–Dirac function is plotted for several temperatures in Figure 1.12, holding E_F constant.

The first important feature of the Fermi–Dirac distribution is that it takes values between zero and one, as we expect from the exclusion principle. It crosses 0.5 when $E = E_F$. Because the occupation number of a state may be either zero or one, $f(E)$ may also be interpreted as the *probability* of the state being occupied. It is a decreasing function of energy, which also makes sense: a state is more likely to be occupied if its energy is lower. The transition from one to zero becomes sharper as the temperature is lowered, and it becomes a Heaviside unit step function in the limit of zero temperature,

$$f(E, E_F, T=0) = \Theta(E_F^0 - E), \qquad (1.107)$$

where $\Theta(x) = 0$ if $x < 0$ and 1 if $x > 0$. Thus all states below E_F^0 are completely filled, and those above are empty. The superscript on E_F^0 is a reminder that this is the value of the Fermi level at zero temperature. Another notation is the Fermi temperature $T_F = E_F^0/k_B$.

In fact the limit E_F^0 of $E_F(T)$ at zero temperature is the strict definition of the Fermi level. The quantity that I have called $E_F(T)$ is really the chemical potential. Unfortunately, standard usage in semiconductor physics is to call both the Fermi level.

The transition from $f = 1$ to 0 broadens as the temperature rises, with a width of roughly $8k_B T$. For comparison, $k_B T \approx 25\,\text{meV}$ at room temperature (300 K). Although the thermal energy $k_B T$ sets the scale of the transition, it is worth remembering that the width is several times this.

For energies far above E_F, which means $(E - E_F) \gg k_B T$, the exponential factor is large and the $+1$ may be neglected to leave

$$f(E, E_F, T) \sim \exp\left(-\frac{E - E_F}{k_B T}\right).$$ (1.108)

This is the classical *Boltzmann distribution* and holds far away from the saturation of f at 1. Classical semiconductors are in this limit and will be discussed further in Section 1.8.3.

In the case of fermions, for which the occupation of a state can be only zero or one, we can describe the system in terms of *holes* rather than electrons. Here a 'hole' is defined simply as the absence of an electron. The distribution of holes $\bar{f}(E)$ is given by

$$\bar{f}(E, E_F, T) = 1 - f(E, E_F, T) = \left[\exp\left(\frac{E_F(T) - E}{k_B T}\right) + 1\right]^{-1}.$$ (1.109)

Note the reversal of energies. In this case the exponential becomes large for negative energies, $(E - E_F) \ll -k_B T$, in which case

$$\bar{f}(E, E_F, T) \sim \exp\left(-\frac{E_F - E}{k_B T}\right).$$ (1.110)

The holes now follow a Boltzmann distribution, as in the valence band of a classical semiconductor (Section 1.8.3).

1.8.2   OCCUPATION OF STATES

We now have expressions for the density of available states $N(E)$ and for the average number of fermions that occupy each state $f(E, E_F, T)$. The product of these gives the density of *occupied* states in the system. We can find such quantities as the total number of electrons using this:

$$N = \int_{-\infty}^{\infty} N(E) f(E, E_F, T) \, dE.$$ (1.111)

This becomes simple at zero temperature, where the Fermi function reduces to a step at E_F^0:

$$N(T=0) = \int_{-\infty}^{\infty} N(E) \, \Theta(E_F^0 - E) \, dE = \int_{-\infty}^{E_F^0} N(E) \, dE.$$ (1.112)

At higher temperatures the full integration (1.111) must be performed, and in most cases this cannot be done analytically.

Fortunately the two-dimensional electron gas is an exception, because its density of states is constant (Figure 1.9) at $m/\pi\hbar^2$ for $E > 0$. Thus the density of electrons per unit area is given by

$$n_{2D} = \int n(E)\, f(E, E_F, T)\, dE = \int_0^\infty \frac{m}{\pi\hbar^2} \left[\exp\left(\frac{E - E_F}{k_B T}\right) + 1\right]^{-1} dE.$$

(1.113)

We shall first look at the consequences of this, assuming E_F to be constant. This leads to an undesirable outcome, from which we deduce the behaviour of $E_F(T)$.

The integrand is proportional to the Fermi function plotted in Figure 1.12, where it was assumed that E_F remained constant. The density reduces to the simple result (1.112) in the limit $T \to 0$, to give $n_{2D}(T=0) = (m/\pi\hbar^2)E_F^0$. For temperatures above zero the integral can be simplified by substituting $z = \exp[-(E - E_F)/k_B T]$, which gives

$$n_{2D} = \frac{m k_B T}{\pi\hbar^2} \int_0^{\exp(E_F/k_B T)} \frac{dz}{1+z} = \frac{m k_B T}{\pi\hbar^2} \ln(1 + e^{E_F/k_B T}).$$

(1.114)

This shows that the density rises as a function of temperature if E_F remains constant. The reason is clear from Figure 1.12, because the tail of the Fermi distribution extends farther and farther out to higher energies and captures more electrons. This is not balanced by the decrease of f at low energies because the integral is cut off at the bottom of the band, $E = 0$.

Usually we do not expect the density of electrons to vary with temperature. Thus $E_F(T)$ must decrease instead to keep n_{2D} constant. We are now regarding the Fermi level as a quantity that we tune to keep the desired number of electrons in the system, and this is a good definition. The relation $E_F^0 = n_{2D}/(m/\pi\hbar^2)$ can be used to rewrite equation (1.114) in terms of E_F^0 rather than n_{2D} to give

$$E_F^0 = k_B T \ln(1 + e^{E_F/k_B T}).$$

(1.115)

This can be turned around to give an expression for $E_F(T)$:

$$E_F(T) = k_B T \ln(e^{E_F^0/k_B T} - 1) = k_B T \ln(e^{T_F/T} - 1).$$

(1.116)

This shows the expected fall of E_F as the temperature rises, becoming negative for $T > T_F/\ln 2$. The corresponding occupation functions are plotted in Figure 1.13. Several important points are shown by these plots.

For low temperatures, $T \ll T_F$, the distribution is near to the step function that holds when $T = 0$, and E_F is near its limit E_F^0. Most states are either completely filled or empty, except those in a region around E_F whose width is a few times $k_B T$. This distribution is called *degenerate* (another meaning of this overused word). It is characteristic of a real metal such as Al or Cu at room temperature (indeed, any

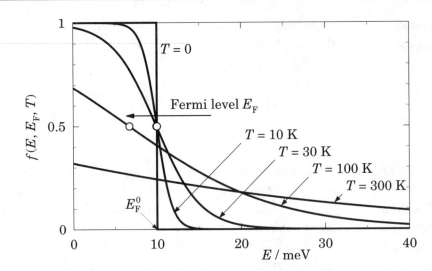

FIGURE 1.13. Fermi distribution function for a two-dimensional electron gas in GaAs at constant density $n_{2D} \approx 3 \times 10^{15}\,\mathrm{m}^{-2}$. The Fermi level E_F moves downwards from $E_F^0 = 10\,\mathrm{meV}$ as the temperature rises, as shown by the marker on each curve.

temperature at which it remains solid!). As far as low-energy phenomena such as ohmic transport are concerned, only the partly filled states near E_F are important. The fully occupied states far below E_F are unable to take part because any response would require them to change their state, but all states nearby in energy are also filled. Many quantities such as the conductivity therefore contain a factor of $-\partial f/\partial E$, which is peaked around the Fermi level. A little manipulation gives the explicit form

$$-\frac{\partial f}{\partial E} = \frac{1}{4k_BT}\,\mathrm{sech}^2\left(\frac{E - E_F}{2k_BT}\right). \tag{1.117}$$

In the limit of low temperatures, $f(E)$ becomes a step function whose derivative is a δ-function: $-\partial f/\partial E \to \delta(E - E_F^0)$. Here everything happens *at* the Fermi level. This behaviour is typical of metals but not of classical semiconductors.

Raising the temperature through T_F causes E_F to become negative. In this case the exponential in the denominator of the Fermi distribution is always large, even at its minimum value, which occurs at the bottom of the band at $E = 0$. We are always far into the tail of the distribution and can drop the $+1$ to get equation (1.108). This is the non-degenerate limit where the Boltzmann distribution holds over the whole band and $E_F \ll 0$. Classical lightly doped semiconductors are usually in this limit, and all electrons can contribute to processes such as conduction. The integral for the density of electrons in the two-dimensional gas takes the simplified form

$$n_{2D} \approx \frac{m}{\pi\hbar^2}\int_0^\infty \exp\left(-\frac{E - E_F}{k_BT}\right)dE = \frac{mk_BT}{\pi\hbar^2}\exp\left(\frac{E_F}{k_BT}\right) \tag{1.118}$$

(remember that $E_F \ll 0$). This can be written more compactly by defining $N_c^{(2D)} = m k_B T / \pi \hbar^2$, which is called the *effective density of states in the conduction band*, for reasons that will shortly be explained.

Unfortunately many low-dimensional electronic systems are in neither the degenerate nor the non-degenerate limits, separated roughly by the Fermi temperature T_F. This is typically around 50 K for a two-dimensional electron gas, so it is degenerate in liquid helium but non-degenerate at room temperature.

1.8.3 CLASSICAL LIGHTLY DOPED SEMICONDUCTORS

The formula for the density of electrons that we have just derived is close to those used in classical semiconductor physics. We will therefore take a brief diversion into classical semiconductor statistics, partly to emphasize that many of these results are *not* applicable to low-dimensional systems. Most of this discussion is restricted to homogeneous three-dimensional systems. The density of electrons n is given by an expression identical to equation (1.118) but with an effective density of states given by

$$N_c^{(3D)} = 2 \left(\frac{m_0 m_e k_B T}{2 \pi \hbar^2} \right)^{3/2}. \tag{1.119}$$

In classical semiconductors we have to include both the conduction and valence bands, with a density of states $n(E)$ as shown in Figure 1.14. The bottom of the conduction band is at E_c rather than zero, which modifies the expression for the density of electrons to

$$n = N_c^{(3D)} \exp \left(-\frac{E_c - E_F}{k_B T} \right). \tag{1.120}$$

The top of the valence band is at E_v, separated from the conduction band by the band gap of $E_g = E_c - E_v$. In an undoped semiconductor the valence band is completely full of electrons, so it is appropriate to consider the density of holes p

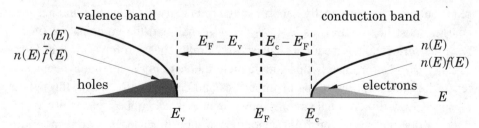

FIGURE 1.14. Conduction and valence bands in a lightly doped semiconductor, showing the Fermi level in the band gap and carriers in both bands.

rather than electrons and use the corresponding distribution (1.110). Similar arguments show that

$$p = N_{\mathrm{v}}^{(3\mathrm{D})} \exp\left(-\frac{E_{\mathrm{F}} - E_{\mathrm{v}}}{k_{\mathrm{B}} T}\right), \qquad (1.121)$$

where $N_{\mathrm{v}}^{(3\mathrm{D})}$ is the effective density of states in the valence band.

The values of n and p are usually controlled by adding small concentrations of so-
called impurities to the semiconductor. Donors release electrons into the conduction
band, while acceptors produce holes in the valence band. The other source of carriers
is thermal excitation of electrons from the valence band to the conduction band,
which is strongly influenced by temperature. Usually the carriers are dominated by
doping, giving *extrinsic* material, and $n \approx N_{\mathrm{D}}$ in n-type material with N_{D} donors per
unit volume. Similarly p-type material with a density N_{A} of acceptors has $p \approx N_{\mathrm{A}}$.
Equation (1.120) or (1.121) can then be turned around to extract the Fermi level.
The density of electrons and holes is equal in the opposite case of an undoped
semiconductor, that is, $n = p = n_{\mathrm{i}}$.

Taking the product of equations (1.120) and (1.121) shows that in general

$$np = N_{\mathrm{c}}^{(3\mathrm{D})} N_{\mathrm{v}}^{(3\mathrm{D})} \exp\left(-\frac{E_{\mathrm{c}} - E_{\mathrm{v}}}{k_{\mathrm{B}} T}\right) = N_{\mathrm{c}}^{(3\mathrm{D})} N_{\mathrm{v}}^{(3\mathrm{D})} \exp\left(-\frac{E_{\mathrm{g}}}{k_{\mathrm{B}} T}\right), \qquad (1.122)$$

where the band gap $E_{\mathrm{g}} = E_{\mathrm{c}} - E_{\mathrm{v}}$. The vital feature is that E_{F} cancels from this
product, which is only a function of temperature. Thus np does not depend on
doping, nor on the position of the Fermi level. This relation also holds in the special
case of undoped material where $n = p = n_{\mathrm{i}}$, so we can write the product in the
compact form $np = n_{\mathrm{i}}^2$. This relation is of such significance in classical devices that
it is called the *semiconductor equation*. For example, the majority carriers in an n-
doped semiconductor are electrons with concentration $n \approx N_{\mathrm{D}}$. The semiconductor
equation then shows that the density of minority holes is $p = n_{\mathrm{i}}^2/n \approx n_{\mathrm{i}}^2/N_{\mathrm{D}}$. An
important restriction is that its derivation holds only in the non-degenerate limit,
which rarely applies to low-dimensional systems (nor, indeed, to highly doped bulk
semiconductors).

To determine n and p precisely we need a further equation. A macroscopic sample should remain electrically neutral, so the total density of positive and negative
charges must be equal. Thus $p + N_{\mathrm{D}} = n + N_{\mathrm{A}}$, and the combination of this
with the semiconductor equation gives a quadratic equation for n or p in terms
of $N_{\mathrm{D}} - N_{\mathrm{A}}$. Unfortunately this simple picture holds only in a homogeneous system, where the concentrations of impurities and carriers are constant throughout
all space. Low-dimensional systems, however, are usually doped selectively to produce carriers in one region from donors elsewhere in space. It is a far more difficult task to determine the concentration of carriers in this case, as we shall see
in Chapter 9.

All these results for the occupation of states are restricted to systems in equilibrium. None of them holds precisely when a system is disturbed from equilibrium. For example, $np > n_i^2$ in the depletion region of a p–n diode under forward bias. Transport theory is now needed to describe the system. This is a vastly more difficult problem, which we shall barely touch. However, there is a notation that we shall use later. Suppose that we know n or p. Equation (1.120) or (1.121) can then be turned around to extract a 'Fermi level' out of equilibrium. For example,

$$E_F^{(n)} = E_c + k_B T \ln \frac{n}{N_c^{(3D)}}. \qquad (1.123)$$

This is no longer a true Fermi level because the system is not in equilibrium, so it is called a *quasi-Fermi level* or *imref* ('Fermi' spelt backwards – really!). The imref is different for electrons and holes, another statement of $np \neq n_i^2$. For example, $E_F^{(n)} > E_F^{(p)}$ in a p–n diode under forward bias. The use of imrefs can be misleading because it tends to give the impression that the carriers are distributed as in equilibrium, apart from a change in the Fermi level. This is often far from the case but the notation is too attractive to abandon.

1.8.4 THE ELECTRON GAS

After summarizing the nondegenerate limit of classical semiconductor statistics, we shall look briefly at the opposite case of the degenerate free-electron gas and its Fermi surface.

Assume that the system is at zero temperature, so $f(E)$ is a step function at $E = E_F^0 = k_B T_F$. The cutoff is given as an energy, but we know that $E = \hbar^2 k^2 / 2m$ for free electrons. Therefore there is a *Fermi wave number* corresponding to E_F^0 given by $k_F = (2m E_F^0)^{1/2} / \hbar$. In three-dimensional **K**-space this defines a spherical *Fermi surface*: all states inside this surface are filled at $T = 0$, while all those outside are empty. The surface becomes a circle in two dimensions, and a pair of points in one dimension. The energy E_F^0 measured from the bottom of the band is called the *Fermi energy* and is a measure of the density of electrons.

The Fermi wave number can be derived more directly using the uniform distribution of states in k-space. In two dimensions, the density of states in **k** is $2/(2\pi)^2$, the area of the Fermi circle is πk_F^2, so the density of electrons is

$$n_{2D} = \frac{2}{(2\pi)^2} \times \pi k_F^2 = \frac{k_F^2}{2\pi}. \qquad (1.124)$$

Therefore $k_F = (2\pi n_{2D})^{1/2}$. This is the easiest way of finding k_F and E_F^0. Other 'Fermi' quantities are also used, such as the Fermi velocity $v_F = \hbar k_F / m$ for free electrons (the group velocity at k_F in general). Given a density and effective mass for an electron gas, one can use these formulas to find T_F and determine whether the gas is degenerate or not at a given temperature.

The Fermi surface takes a more complicated shape if $\varepsilon(\mathbf{K})$ lacks spherical symmetry, with spectacular shapes in some materials. It can be defined as a surface of constant energy whose volume holds the correct density of electrons. Even for the common semiconductors, we shall see in the next chapter that the spherical approximation holds only in special cases such as the conduction band of GaAs.

1.8.5 OTHER DISTRIBUTION FUNCTIONS

There are other distribution laws for particles that are not fermions. Classical particles obey the Boltzmann distribution

$$f_{cl}(E, \mu, T) = \exp\left(-\frac{E - \mu}{k_B T}\right), \tag{1.125}$$

where μ is the chemical potential (which should really appear instead of E_F in the Fermi–Dirac distribution). We have already seen this as the non-degenerate limit of the Fermi distribution, widely used for classical semiconductors.

The remaining class of particles consists of bosons, which obey Bose–Einstein statistics

$$f_{BE}(E, \mu, T) = \left[\exp\left(\frac{E - \mu}{k_B T}\right) - 1\right]^{-1}. \tag{1.126}$$

This differs from the Fermi–Dirac distribution only in the sign of the '1'. Bosons are particles that carry zero or integral angular momentum in units of $\hbar$, unlike fermions, which have half-integral values. An example of a Bose particle is a ^{4}He atom, but many bosons are not 'particles' in the traditional sense of the word. Phonons and photons are the most important bosons for us.

We saw earlier that the chemical potential was tuned to keep the number of particles constant as the temperature changes. Although the same argument applies to ^{4}He, it does not apply to phonons or photons, which can be created or destroyed at will, so these have no chemical potential. For example, the number of phonons in a mode of frequency ω_q is given by

$$N_q = \left[\exp\left(\frac{\hbar\omega_q}{k_B T}\right) - 1\right]^{-1}. \tag{1.127}$$

This will be important later when we look at scattering rates in Section 8.4, which depend on the number of phonons in each mode. It has the simple limit $N_q \sim k_B T/\hbar\omega_q$ at high temperature, the thermal energy divided by the energy of each phonon.

The form of the Bose distribution is quite different from the Fermi function at low energy. It is defined only for $E > \mu$ and goes to infinity as $E \to \mu$ (or $E \to 0$ for phonons and photons). This means that conserved bosons such as ^{4}He undergo *Bose condensation* to end up in the ground state as the temperature falls to zero.

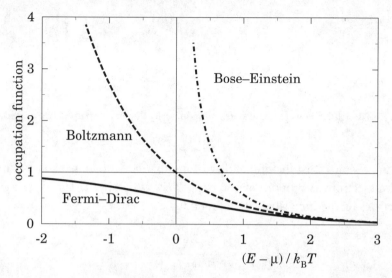

FIGURE 1.15. Fermi–Dirac, Boltzmann, and Bose–Einstein distribution functions plotted on a common scale against $(E - \mu)/k_B T$.

The behaviour of fermions and bosons is opposite in this limit: bosons try to clump together and squeeze as many as possible into the lowest state, whereas fermions are forced to keep apart and can have only one particle per state. The Fermi, Boltzmann, and Bose functions are plotted together for comparison in Figure 1.15.

There are some further restrictions on the validity of these occupation functions. First, they apply only to large systems, or to small ones that are in contact with a reservoir that can supply particles and energy. They do not apply to the electrons inside a quantum dot, for example, which is effectively an isolated artificial atom. A second restriction is that the particles must not interact with one another, because we have assumed that the probability of one electron occupying a state is unaffected by the electrons that occupy the other states. This is always an approximation, as electrons are charged and repel one another, and there are several cases where it fails.

One example is the occupation of an impurity state in a semiconductor that can trap up to two electrons with opposite spins. There are therefore four possible states, illustrated in Figure 1.16:

  (i) Empty, energy 0;
  (ii) one electron, spin up, energy ε;
 (iii) one electron, spin down, same energy ε;
 (iv) two electrons, one spin up and one spin down, energy $2\varepsilon + U$.

The extra energy U in state (iv) is called the *Hubbard U* and represents the Coulomb repulsion between the two electrons. It means that the second electron is bound less tightly than the first; the binding energy of a second electron is typically only about 5% of that of the first for a simple donor in a semiconductor.

(i)   (ii)   (iii)   (iv)

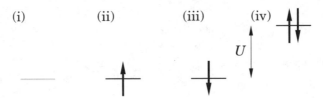

FIGURE 1.16. Four possible states of an impurity that can trap up to two electrons.

It can be shown that the probability of finding the system in a state with energy E and n electrons is proportional to $\exp[-(E - n\mu)/k_BT]$. The average number of electrons on the impurity is then found to be

$$\langle n \rangle = \frac{1 + e^{-(\varepsilon-\mu+U)/k_BT}}{1 + \frac{1}{2}e^{(\varepsilon-\mu)/k_BT} + \frac{1}{2}e^{-(\varepsilon-\mu+U)/k_BT}}. \tag{1.128}$$

To check this, ignore the interaction U between the electrons. Then $\langle n \rangle = 2f(\varepsilon, \mu, T)$, with f being the Fermi function for non-interacting electrons. Often the opposite limit holds, with a repulsion U so large that the probability of two electrons occupying the impurity is negligible. In this case the average occupation simplifies to

$$\langle n \rangle = \frac{1}{1 + \frac{1}{2}e^{(\varepsilon-\mu)/k_BT}} = \frac{1}{e^{(\varepsilon-\mu-k_BT\ln 2)/k_BT} + 1}. \tag{1.129}$$

Although the impurity is likely to be occupied by only one electron at a time, the probability is not given by the usual Fermi function because of the two possible values of the spin. The second form of equation (1.129) shows that the occupation can be made to resemble a Fermi function if a term $k_BT\ln 2$ is added to the chemical potential, and this is often used in calculations involving the occupation of impurities.

Further Reading

The methods described in this chapter are all standard, and further details can be found in any book on quantum mechanics. There is a vast range of approaches from the engineering to the mathematical. Two textbooks that have long been well regarded are by Merzbacher (1970) and Gasiorowicz (1974). A more recent suggestion is Bransden and Joachain (1989).

Reif (1965) is an excellent book on statistical physics, going far beyond this chapter.

1.1   What sort of wavelengths are involved (using $\lambda = 2\pi/k$) for electrons in typical solids? Some examples are Al ($E = 11.7\,\text{eV}$; use the free-electron mass m_0), highly doped n-GaAs ($E = 50\,\text{meV}$ and $m = 0.067\,m_0$), and a two-dimensional electron gas in GaAs ($E = 10\,\text{meV}$). These lengths are important because they set the size of structures that must be fabricated if we wish to manipulate the waves. The wavelength of electrons in an electron microscope ($E = 100\,\text{keV}$) sets the ultimate limit of resolution, although aberrations of the lenses reduce this drastically in practice. Finally, as a (very) classical object, consider the Clapham omnibus.

1.2   Use the model of an infinitely deep potential well to estimate the first few energy levels for an electron in GaAs in wells of widths $10\,\text{nm}$ and $4\,\text{nm}$. Remember the mass m in the equations should be replaced by $m_0 m_e$, where $m_e = 0.067$ is the effective mass for electrons at the bottom of the conduction band of GaAs. How accurate do you expect the results to be, given that real wells in GaAs/AlGaAs are about $0.3\,\text{eV}$ deep?

Repeat the calculation for holes whose well is about $0.2\,\text{eV}$ deep. These come in two varieties, heavy and light, with effective masses of $m_{hh} = 0.5$ and $m_{lh} = 0.082$.

1.3   Figure 1.4 shows some experimental measurements of the energy levels in quantum wells of varying thickness. The sample has several quantum wells of GaAs, of varying thickness a, grown in a thick layer of $\text{Al}_{0.35}\text{Ga}_{0.65}\text{As}$. The traces show the intensity of the PL, described in Section 1.3.1. The energy of photons emitted is the difference of the energies of the electron and hole in one of the wells. The narrower wells have bound states of higher energy and therefore emit light of higher frequency (shorter wavelength), giving rise to the set of four lines shown on the PL trace. Given the thicknesses of the four wells, how well does the particle-in-a-box model predict the energy of the PL? Assume that the electrons and holes have effective mass $m_e = 0.067$ and $m_{hh} = 0.5$ (heavy holes). What might be responsible for peak (e)?

1.4   The current and charge densities derived in Section 1.4 are guaranteed to conserve charge provided that $V(x, t)$ is real. Show that particles will be created or absorbed if V is complex. This trick is sometimes used for a source or sink of particles in modelling.

1.5   Construct a wave function to describe electrons moving in the $+x$-direction at $10^5\,\text{m s}^{-1}$ and carrying a current density of $-1\,\text{A mm}^{-2}$.

1.6   Derive the current density for counter-propagating travelling and decaying waves, equations (1.34) and (1.36).

1.7   Show that the root-mean-square width Δx of an electron in state n of an infinitely deep square well of width a is given by

$$(\Delta x)^2 = \frac{a^2}{12}\left(1 - \frac{6}{n^2\pi^2}\right). \qquad (E1.1)$$

Classically the particle is equally likely to be found anywhere in the well (unless it is at rest). Show that $\Delta x = a/\sqrt{12}$ in this case, and note that the quantum-mechanical result approaches this value for high energies (high n).

1.8   Consider the propagation of wave packets of initial width $d = 25\,\mathrm{nm}$ and $d = 50\,\mathrm{nm}$ in GaAs ($m = 0.067m_0$). Sketch $\Delta x(t)$ for the two packets. How long does it take before they double in length?

1.9   Using $[\hat{x}, \hat{p}] = i\hbar$, show that $[\hat{x}, \hat{p}^2] = 2i\hbar\hat{p}$. (Hint: add and subtract $\hat{p}\hat{x}\hat{p}$.) For the common form of Hamiltonian $\hat{H} = \hat{p}^2/2m + V(\hat{x})$, show that $[\hat{x}, \hat{H}] = (i\hbar/m)\hat{p}$. This result will be useful when treating optical response in Chapter 8.

1.10   Orbital angular momentum is defined in classical mechanics by $\mathbf{L} = \mathbf{r} \times \mathbf{p}$. Similarly, in quantum mechanics, the operator $\hat{L}_x = \hat{y}\hat{p}_z - \hat{z}\hat{p}_y$ and so on. Show that different components of $\hat{\mathbf{L}}$ do not commute with each other, with $[\hat{L}_x, \hat{L}_y] = i\hbar\hat{L}_z$.

   The total angular momentum can be defined by $\hat{L}^2 = \hat{L}_x^2 + \hat{L}_y^2 + \hat{L}_y^2$. Show that $[\hat{L}_x, \hat{L}^2] = 0$, so it is possible to know both one component and the total angular momentum.

1.11   The lowest three states in a parabolic potential (Section 4.3) can be written in the form $\phi_n(x) \propto H_{n-1}(x)\exp(-\frac{1}{2}x^2)$, where $H_n(x)$ is a Hermite polynomial. The first few of these are $H_0(x) = 1$, $H_1(x) = 2x$, and $H_2(x) = 4x^2 - 2$. Verify that the states are orthogonal (symmetry helps).

1.12   The wave function $\phi_1(x) \propto x\exp(-\frac{1}{2}bx)$ is often used as an approximation to the lowest eigenstate in a triangular potential for $x > 0$ (Section 7.5.2). Occasionally an approximation is needed for the second state. Write this in the form $x(c - bx)\exp(-\frac{1}{2}bx)$ and choose c to make it orthogonal to the lowest state.

1.13   Derive the expectation value of x for the wave function (1.54), containing a mixture of two states in a quantum well. First, show that the coefficients A_1 and A_2 must obey $|A_1|^2 + |A_2|^2 = 1$ for the state to be normalized (remember that the functions $\phi_n(x)$ are orthonormal). Take the coefficients to be real (complex values just change the phase of the oscillation). For $t > 0$, the wave function is given by

$$\Psi(x, t) = A_1\phi_1(x)\,e^{-i\varepsilon_1 t/\hbar} + A_2\phi_2(x)\,e^{-i\varepsilon_2 t/\hbar}. \qquad (E1.2)$$

Calculate the expectation value of $x(t)$ using

$$\langle x(t) \rangle = \int_0^a \Psi^*(x,t)\, x\, \Psi(x,t)\, dx. \qquad (E1.3)$$

This gives four terms when the wave function (E1.2) is inserted. Fortunately those in which both ϕs are the same give $a/2$ by symmetry. This leaves only the cross-terms, which can be evaluated with the aid of the integral

$$\frac{2}{a}\int_0^a x \sin\frac{\pi x}{a}\sin\frac{2\pi x}{a}dx = -\frac{16a}{9\pi^2}. \qquad (E1.4)$$

The final result is quoted in the main text (equation 1.55).

1.14 Consider the dispersion relation (1.94), which is used to model non-parabolicity in the conduction band of GaAs. Estimate the form of the density of states at small and at large energies. What happens to the velocity in these two limits?

1.15 Show that the density of states for free electrons in two dimensions (the simplest case) is

$$n_{2D}(E) = \frac{m}{\pi\hbar^2}\Theta(E). \qquad (E1.5)$$

(Again, the result is more complicated for a realistic dispersion relation, with further singularities inside the band.)

1.16 Derive the effective density of states for a three-dimensional system (equation 1.119). You will need the integral $\int_0^\infty x^{1/2}e^{-x}dx = \Gamma(3/2) = \frac{1}{2}\sqrt{\pi}$.

1.17 Combine the semiconductor equation with that for neutrality to show that

$$n = \frac{1}{2}\left[(N_D - N_A) + \sqrt{(N_D - N_A)^2 + 4n_i^2}\right] \qquad (E1.6)$$

in a classical semiconductor. What is the corresponding result for p?

1.18 Show that the Fermi wave number $K_F = (3\pi^2 n_{3D})^{1/3}$ in three dimensions and derive the corresponding result for one dimension.

1.19 For the following examples, calculate the Fermi temperature and determine whether the system is degenerate at room temperature, in liquid nitrogen (77 K), and in liquid helium (4.2 K).

(a) Al has 18.1×10^{28} m^{-3} electrons, about the maximum for a common metal, and an effective mass near unity. Cs is a less typical metal with a much lower density, 0.91×10^{28} m^{-3}.

(b) Highly doped n-GaAs has 5×10^{24} m^{-3} electrons with $m_e = 0.067$. One could also dope it p-type with a higher effective mass $m_h = 0.5$ for heavy holes. Lightly doped material might have something like 10^{21} m^{-3} carriers.

(c) A two-dimensional electron gas in GaAs has $(1 - 10) \times 10^{15} \, \mathrm{m}^{-2}$ electrons (don't forget the dimensions!).

1.20 Show that the average energy of electrons in a three-dimensional electron gas at zero temperature is $\bar{E}_{3D} = \frac{3}{5} E_F^0$, with $\bar{E}_{2D} = \frac{1}{2} E_F^0$ in two dimensions. Show also that the corresponding results are $\frac{3}{2} k_B T$ and $k_B T$ at high temperatures where the electrons are non-degenerate.

1.21 Plot the occupation of a donor described by equation (1.129) as a function of μ, for a range of values of the Hubbard U. Although it seems obvious that U should be positive in this example, there are other systems where it behaves as though it is negative. What effect does this have on $\langle n \rangle$?

ELECTRONS AND PHONONS IN CRYSTALS

2

Few low-dimensional systems are periodic (superlattices provide an obvious exception), but they all consist of relatively large scale structures superposed on the structure of a host. This may be a true crystal such as GaAs or a random alloy such as (Al,Ga)As; we shall ignore the complications introduced by the alloy and treat it as a crystal 'on average'. We must understand the electronic behaviour of the host before treating that of the superposed structure.

This chapter deals first with one-dimensional crystals, followed by three-dimensional materials. The final section is devoted to phonons, lattice waves rather than electron waves, which also have a band structure imposed by the periodic nature of the crystal. Photons are the third kind of wave that we shall encounter, and structures that display band structure for light have recently been demonstrated. Their behaviour can be described with a similar theory but we shall not pursue this.

2.1   Band Structure in One Dimension

The potential energy in a real crystal is clearly far more complicated than the systems that we have studied in the previous chapter. In Section 5.6 we shall solve the simple example of a square-wave potential in detail, but the most important results follow from the qualitative feature that the potential is *periodic*. In one dimension this means that $V(x + a) = V(x)$, where a is the lattice constant, the size of each unit cell of the crystal. A fictitious example of a periodic potential is sketched in Figure 2.1. It can be expanded in a Fourier series, like any periodic function, to give

$$V(x) = \sum_{n=-\infty}^{\infty} V_n \exp\left(\frac{2\pi i n x}{a}\right) \equiv \sum_{n=-\infty}^{\infty} V_n \exp(i G_n x). \tag{2.1}$$

We shall soon see the significance of the reciprocal lattice vectors $G_n = (2\pi/a)n$ (the name is more appropriate in higher dimensions).

What does the periodicity of the potential imply for the wave functions? We know that the wave functions of an infinite system with a uniform potential are plane waves, $\phi_k(x) = \exp(ikx)$, setting the normalization factor to unity. Their

FIGURE 2.1. An example of a periodic potential showing the lattice constant a.

density $|\phi_k(x)|^2 = 1$, which is constant everywhere. This is expected because the potential is the same everywhere, so there is no reason for an electron to prefer being at any one point rather than any other. In short, the system is translationally invariant. This condition is relaxed in the presence of the periodic potential, but it seems reasonable to expect the density to vary in the same way within each unit cell. In other words, we expect the density $|\psi(x)|^2$ to be a periodic function like the potential, that is, $|\psi(x + a)|^2 = |\psi(x)|^2$. This can be achieved if we multiply the plane waves of free space by a periodic function $u_k(x)$. The result is called a *Bloch function*:

$$\psi_k(x) = u_k(x)\exp(ikx), \qquad u_k(x + a) = u_k(x). \qquad (2.2)$$

The density $|u_k(x)|^2$ is periodic as required, although $u_k(x)$ is different for each value of k.

Equation (2.2) is a statement of *Bloch's theorem* for the wave functions in a crystal. An equivalent form is

$$\psi_k(x + a) = \exp(ika)\psi_k(x). \qquad (2.3)$$

The label k is now called the *Bloch wave number*, and the second form of Bloch's theorem shows that ka gives the change in phase of the wave function between unit cells. Note that k is not an 'ordinary' wave number in the sense that $\hbar k$ is the mechanical momentum of the particle, because the presence of $u_k(x)$ means that the wave function contains many momenta. Instead $\hbar k$ is called *crystal momentum*. It behaves in many ways like an ordinary momentum, as we shall see in Section 2.2.

The definition of k has now become somewhat ambiguous. Suppose that k lies in the range $\pi/a < k < 3\pi/a$. It can be rewritten as $k = (2\pi/a) + k'$ with $|k'| < \pi/a$. Then

$$\psi_k(x) = u_k(x)\exp(ikx) = u_k(x)\exp\left(\frac{2\pi i x}{a}\right)\exp(ik'x). \qquad (2.4)$$

Now, $\exp(2\pi i x/a)$ is periodic in x with the same period a as $u_k(x)$, so we can group these together to make a new periodic function:

$$\psi_k(x) = \left[u_k(x)\exp\left(\frac{2\pi i x}{a}\right)\right]\exp(ik'x) = u'_k(x)\exp(ik'x). \qquad (2.5)$$

We have reduced the Bloch wave number to the range $-\pi/a < k' < \pi/a$, which is called the *first Brillouin zone*. This process is related to aliasing in communication theory. The same can clearly be done for any value of k by subtracting an appropriate multiple n of $2\pi/a$. In other words, we add an appropriate reciprocal lattice vector G_n to k. We can then use the pair nk' to label the state instead of k. The wave function becomes

$$\psi_{nk'}(x) = u_{nk'}(x) \exp(ik'x), \qquad (2.6)$$

where k' is restricted to the range $(-\pi/a, \pi/a)$. The first scheme, where k could take any value, is called the *extended zone scheme* for labelling the wave function, and the second, where we use $|k'| < \pi/a$ and n, is the *reduced zone scheme*. A third *repeated zone scheme* exploits the periodicity in k to repeat all functions in every $2\pi/a$. Their use is illustrated for the energy bands in Figure 2.2.

2.1.1 FORMATION OF BAND GAPS

We have seen qualitatively what happens to wave functions in the presence of a periodic potential and now pass on to the energies. Like the wave functions, they are usually labelled with the wave number in the first Brillouin zone ($|k| < \pi/a$, dropping the prime) and the band index n, as $\varepsilon_n(k)$. This is plotted in Figure 2.2 in the extended, reduced, and repeated zone schemes for a weak periodic potential. Concentrate first on the extended zone scheme. For most wave numbers the energy $\varepsilon_n(k) \approx \varepsilon_0(k)$ for free electrons. The exceptions are near the zone boundaries, where $k = \frac{1}{2}G_n = n\pi/a$. Here $\varepsilon_n(k)$ turns away from the free-electron parabola to become flat where it meets the zone boundary. This happens both above and below $\varepsilon_0(\frac{1}{2}G_n)$, leaving a gap between where the crystal has no propagating states. These band gaps divide $\varepsilon_n(k)$ into energy bands. They are a characteristic feature of a periodic system, and their origin can be viewed in several ways.

Band structure is usually plotted in the reduced zone scheme. This is constructed by translating sections of the extended zone through reciprocal lattice vectors G_n. For the free-electron parabola, the central part with $|k| < \pi/a$ stays where it is. The part between $-3\pi/a$ and $-\pi/a$ is translated through $G_1 = 2\pi/a$ to bring it into the first Brillouin zone, while that between π/a and $3\pi/a$ is shifted by G_{-1}. The sections at higher energy are treated similarly to give the criss-crossing structure shown by the thin line in Figure 2.2(b). This zone folding, arising from the periodicity of the crystal, already makes the energy look complicated! The energy in the crystal $\varepsilon_n(k)$ is treated in the same way. The segments with opposite signs of k join to give smooth curves filling the zone, with gaps at the boundary or centre.

Finally, the energy bands in the reduced zone can be repeated to give the repeated zone scheme in Figure 2.2(c). This is less frequently used but clarifies some aspects of transport involving the zone boundary.

We now return to the origin of band gaps. A wave scatters from the atoms in all directions as it propagates through the crystal lattice. In most cases the contributions from different atoms tend to cancel one another, but in certain directions they interfere constructively and a strong scattered beam results. This is Bragg reflection due to the periodic nature of the lattice, illustrated in Figure 2.3. One incident beam

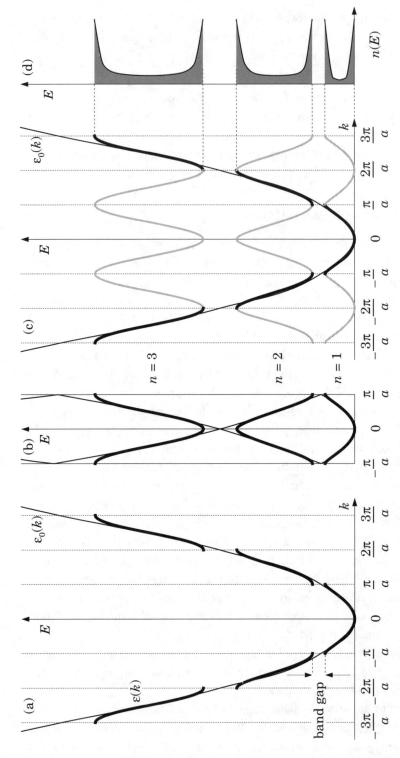

Figure 2.2. Band structure of a one-dimensional crystal in the (a) extended, (b) reduced, and (c) repeated zone schemes, and (d) the density of states as a function of energy. The thick lines show $\varepsilon(k)$ in a weak periodic potential, with bands labelled by n, while the thin parabola is $\varepsilon_0(k)$ for free electrons. The grey lines are periodic repeats.

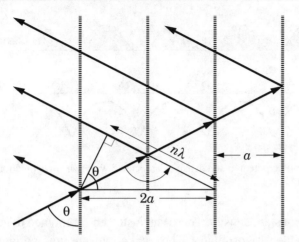

FIGURE 2.3. Condition for Bragg reflection of a wave of wavelength λ incident at angle θ onto the planes of a crystal, separated by a.

is shown, giving a scattered beam from each plane of the lattice. The difference in path length between adjacent beams must be an integral number of wavelengths if they are to add constructively. The triangle shows that this difference is $2a \sin \theta$ so the condition for Bragg reflection is

$$n\lambda = 2a \sin \theta, \qquad (2.7)$$

where $\lambda = 2\pi/k$. In a one-dimensional crystal the electrons travel normal to the 'planes', so $\theta = \pi/2$ (backscattering) and $\sin \theta = 1$. Thus the Bragg condition becomes $k = n\pi/a = \frac{1}{2}G_n$. When this condition is fulfilled, backscattering becomes so strong that the electron is unable to propagate through the crystal and has to become a standing wave.

Scattering from the lattice occurs for all wave numbers, although it is particularly strong at the Bragg condition. Consider a plane wave of wave number k and energy $\varepsilon_0(k)$ travelling through a lattice. Scattering admixes components with wave number $k + G_n$ for all n. This is where the reciprocal lattice vectors enter. In general the added components have different energy, so mixing is weak. It becomes strong only when the waves are of equal energy, which in turn requires the magnitudes of the wave numbers to be equal and fixes k:

$$\varepsilon_0(k + G_n) = \varepsilon_0(k), \qquad |k + G_n| = |k|, \qquad k = -\tfrac{1}{2}G_n. \qquad (2.8)$$

Again the Bragg condition for strong reflection, and the opening of band gaps, occurs whenever $k = \frac{1}{2}G_n$, emphasizing the importance of the reciprocal lattice vectors. This picture of mixing plane waves is the basis of nearly free electron theory, to be discussed in Section 7.8.

Another equivalent way of looking at the formation of band gaps is to note that there is always a pair of degenerate states in the free-electron case, $\exp(\pm ikx)$, both

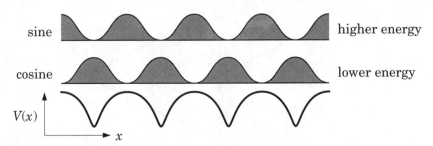

FIGURE 2.4. Periodic potential and densities associated with two wave functions at the first zone boundary $k = \pm \pi/a$.

of which have constant density. These can instead be combined to give $\sin(kx)$ and $\cos(kx)$, whose density oscillates. The oscillations are unimportant in free space, where the potential is constant, but become significant when superposed on the periodic potential, as illustrated in Figure 2.4. In this case the 'cosine' density is weighted to a more favourable part of the periodic potential, while the 'sine' density is weighted towards the hills rather than the valleys. Thus the cosine wave now has a lower energy and the degeneracy between the two states is broken to produce the band gap. This argument works only when the density and potential have the same period and interact cooperatively, which again needs $k = \frac{1}{2}G_n$.

2.2   Motion of Electrons in Bands

Free electrons have group velocity $v(k) = p/m_0 = \hbar k/m_0$, in agreement with classical physics. Remarkable new results are found for a crystal. To be specific, consider a simple cosine approximation to the shape of a band (dropping the subscript n that labels the band),

$$\varepsilon(k) = \tfrac{1}{2}W(1 - \cos ka) = W \sin^2 \frac{ka}{2}. \qquad (2.9)$$

This has the same general shape as the lowest band $n = 1$ or any other odd-numbered band for positive W; the sign of W should be changed for even n. The energy starts from zero and has full width $|W|$ as shown in Figure 2.5(a). This example has $W = 5\,\text{eV}$ and $a = 0.5\,\text{nm}$, rough values for a typical semiconductor. Expansion of the cosine for small k gives $\varepsilon(k) \approx \tfrac{1}{4}a^2 W k^2$. This is a parabola, and comparing with the usual form $\hbar^2 k^2/2m_0 m^*$ shows that the cosine band has an effective mass $m^* = 2\hbar^2/m_0 a^2 W$. Thus a broad band gives a small effective mass. A similar expansion holds near the top of the band around $k = \pm \pi/a$.

The group velocity is given by the usual derivative,

$$v(k) = \frac{1}{\hbar}\frac{d\varepsilon(k)}{dk} = \frac{aW}{2\hbar} \sin ka. \qquad (2.10)$$

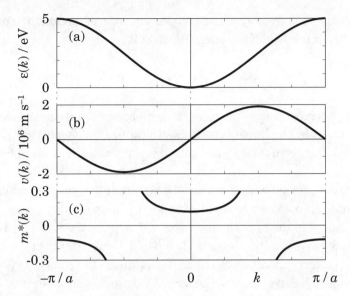

FIGURE 2.5. (a) Energy, (b) velocity, and (c) effective mass as a function of wave number for a cosine band of width $W = 5\,\mathrm{eV}$ in a crystal of lattice constant $a = 0.5\,\mathrm{nm}$.

This reduces to $\hbar k/m_0 m^*$ for small k where the parabolic approximation holds, but its general behaviour is more interesting. The velocity drops to zero at both extremities of the band, as shown in Figure 2.5(b). Recall Figure 2.4, which showed that the corresponding wave functions are standing waves that carry no current.

We can also define an effective mass throughout the band, not just around $k = 0$, through a further derivative. For free electrons $dv/dk = \hbar/m_0$, which can be turned around to define the effective mass as

$$m^*(k) = \frac{\hbar}{m_0} \bigg/ \frac{dv}{dk} = \frac{\hbar^2}{m_0} \bigg/ \frac{d^2\varepsilon(k)}{dk^2} \qquad (2.11)$$

$$= \frac{2\hbar^2}{m_0 a^2 W} \sec ka. \qquad (2.12)$$

The energy, velocity, and effective mass are plotted in Figure 2.5. This effective mass is most useful near the edges of the band, where it is roughly constant.

The top of the band is particularly interesting. The effective mass here is *negative*, which follows from the behaviour of $v(k)$: an increase of k leads to a decrease in v. This is quite unlike free electrons and looks as though it could lead to some odd behaviour, but we first need to know the equation of motion for these electrons. Newton's law tells us that 'force equals rate of change of momentum', but how is the momentum defined? Is it $mv(k)$, for example, or $\hbar k$? These are equal for free electrons, but behave quite differently in a band where k is the Bloch wave number. The answer, which will be justified for electric fields in Chapter 6, is that crystal momentum $\hbar k$ is the correct quantity. Thus the equation of motion for a particle of charge q in an electric field $\mathbf{F}$ and magnetic field $\mathbf{B}$ is

$$\hbar \frac{d\mathbf{K}}{dt} = q[\mathbf{F} + \mathbf{v}(\mathbf{K}) \times \mathbf{B}]. \tag{2.13}$$

This is written for the three-dimensional case because a magnetic field doesn't make much sense in one dimension. The electric field is denoted by $\mathbf{F}$ to avoid confusion with energy. In three dimensions the group velocity is given by $\mathbf{v}(\mathbf{K}) = \mathrm{grad}_\mathbf{K} \varepsilon(\mathbf{K})/\hbar$.

The velocity is zero at both the top and the bottom of the band. A particle moves in one direction in response to an electric field if it starts at the bottom of the band, but it moves in the *opposite* direction if it starts at the top. This is what is meant by the negative effective mass at the top of the band. However, another view is that this is perfectly normal behaviour (for positive mass) for a particle of the *opposite charge*. Thus the negatively charged electrons with a negative effective mass at the top of the band can instead be described as positively charged *holes* of positive mass. This remarkable result is of course second nature for anybody used to semiconductors.

Strange things also happen if we consider a single electron starting from $k = 0$ in an empty band, accelerated by a constant electric field $-F$ (the minus sign is to cancel that for the electronic charge). The result is shown in Figure 2.6. The equation of motion tells us that $k(t) = (eF/\hbar)t$, which is uniformly increasing. Figure 2.5(b) shows that its velocity increases first, but then reaches a maximum as the wave number passes through $k = \pi/2a$ and then *decreases*, falling to zero when $k = \pi/a$. This is equivalent to $-\pi/a$, using the periodicity of k-space or the repeated zone scheme. Thereafter the velocity becomes negative before falling back

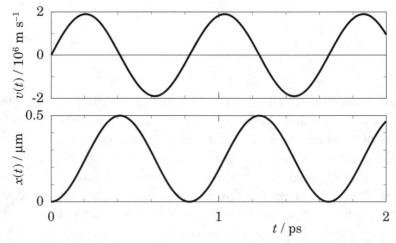

FIGURE 2.6. Bloch oscillations as a function of time induced by an electric field $F = 10^7 \, \mathrm{V \, m^{-1}}$ in a cosine band of width 5 eV.

to zero again when k returns to zero and repeating periodically. In equations,

$$v(t) = \frac{aW}{2\hbar} \sin\left(\frac{eFat}{\hbar}\right), \qquad x(t) = \frac{W}{2eF}\left[1 - \cos\left(\frac{eFat}{\hbar}\right)\right]. \qquad (2.14)$$

This result shows that the electron oscillates in both k- and real space, rather than accelerating uniformly as it would in the absence of a periodic potential. It is clear that the electron cannot behave classically because that would require its kinetic energy to increase without limit, and the band in a crystal has a finite width. This phenomenon of periodic motion rather than uniform acceleration is known as *Bloch oscillation*. There have been many attempts to observe and harness it as a possible source of microwave radiation. The frequency $\omega = eFa/\hbar$ does not depend on the bandwidth, only on the energy lost by the electric field across each unit cell.

An important result that follows from Bloch oscillations is the behaviour of electrons in a full band. These electrons cannot scatter anywhere in their band, because all states are filled, so they are forced to Bloch oscillate and the band as a whole carries no current.

Next we must explain why electrons in a partly filled band carry a current in an electric field rather than performing Bloch oscillations. The answer is that electrons in a real crystal are scattered (by impurities, phonons, or other electrons, for example) and the oscillation is interrupted long before a cycle is complete. The electrons start again but are repeatedly scattered to mask the oscillation. Near the bottom of the band, where $v \approx \hbar k/m$, the equation of motion in an electric field can be written

$$\frac{dv}{dt} = \frac{eF}{m} - \frac{v}{\tau}. \qquad (2.15)$$

The term $-v/\tau$ is a simple relaxation-time approximation to the scattering, giving the *Drude model*. Now the velocity approaches a steady value of $v = eF\tau/m = \mu F$ at large times, where the mobility $\mu = e\tau/m$. This can in turn be written as the familiar result $\sigma = ne^2\tau/m$ for the conductivity, where n is the number density of carriers. Here τ should properly be the *momentum* relaxation time, which we shall calculate in Section 8.2.

As for band gaps, they are often called 'forbidden gaps', but how 'forbidden' are they? It turns out that a strong electric or magnetic field can induce electrons to tunnel through the band gap to the next band. This is called Zener or magnetic breakdown and is illustrated in Figure 2.7 for an electric field. The edges of the

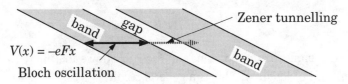

FIGURE 2.7. Bands of a one-dimensional system in a uniform electric field plotted in real space, showing Bloch oscillation and Zener tunnelling.

bands are tilted because of the potential energy $-eFx$ from the field. An electron of constant energy is constrained to move between the band edges as long as it stays in its band. This allows it to move along a length W/eF, the same as the excursion of Bloch oscillations found in equation (2.14). However, when it gets to the end of its permitted range, it is possible for the electron to tunnel through the gap into the next band rather than 'reflect' from the edge.

For a crude estimate of the rate of tunnelling, take the barrier to be of height E_g and thickness $d = E_g/eF$. The decaying wave number in the gap is then $\kappa \approx (2m_0 m^* E_g/\hbar^2)^{1/2}$. The usual formula (equation 5.25, to be derived later) then gives a probability of tunnelling in each cycle of Bloch oscillation of

$$T \approx \exp(-2\kappa d) = \exp\left[-\frac{(8m_0 m^* E_g^3)^{1/2}}{eF\hbar}\right]. \tag{2.16}$$

Better approximations can be developed with the WKB method (Section 7.4). The rate goes to zero in weak fields but may become significant if the field is very large or the band gap small. A well-known application of this is the Zener diode, which relies on tunnelling through the bandgap to cause reverse breakdown in highly doped p–n diodes (although many so-called Zener diodes involve avalanche breakdown).

2.3   Density of States

The band structure also modifies the density of states. Its qualitative form can be deduced from our previous results. We have seen that $\varepsilon_n(k)$ can be approximated as a parabola at both extremities of the band, like free electrons but with a different effective mass. Free electrons in one dimension have an $E^{-1/2}$ density of states at the bottom of the band, so we expect this at the bottom of the periodic band and also at the top. The relation (1.88) between $n(E)$ and velocity gives

$$n(E) = \frac{2}{\pi\hbar v(E)} = \frac{2}{\pi a}\frac{1}{[(\frac{1}{2}W)^2 - (\frac{1}{2}W - E)^2]^{1/2}} \tag{2.17}$$

for the cosine band. This shows the expected inverse–square-root divergences at top and bottom. It applies to each band in the crystal, and the result is plotted in Figure 2.2(d). The integral of the density of states in the cosine band (per unit length) is

$$\int_0^W n(E)\, dE = \frac{2}{a}. \tag{2.18}$$

The factor of 2 comes from the spin, and the $1/a$ shows that there is one spatial state per unit cell of length a. This is a general result.

This result for the number of states in a band allows us to predict whether a one-dimensional material should be a metal or an insulator. An odd number of electrons per atom results in a half-filled top band. There are free electrons at the Fermi level, giving a metal. On the other hand, an even number of electrons exactly fills some bands leaving the rest empty. The Fermi level lies in a band gap between the highest occupied state and the lowest empty state, and the material will be an insulator or semiconductor.

This picture can be applied to many one-dimensional materials, notably polymers. A famous example is polyacetylene $(CH)_n$. This has an odd number of electrons per unit cell (CH) and therefore ought to be a metal. In chemical terms, the carbon contributes one electron to a bond with hydrogen, leaving three electrons to be shared among the two bonds with neighbouring CH units. Thus this model predicts that there are $1\frac{1}{2}$ bonds between neighbours ($-\overset{...}{C}H\overset{...}{-}CH-$).

In fact this configuration is not stable. The chain distorts such that single and double bonds alternate along the length of the carbon backbone, giving a pattern of $-CH=CH-$. The unit cell has doubled in length, halving the Brillouin zone, so there is now an even number of electrons per unit cell and the result is a semiconductor. The deformation, known as the Peierls distortion, lowers the energy of the system and is a general property of one-dimensional metals. Lattice distortion is less widespread in three dimensions.

Polyacetylene can be doped, with iodine for example, giving high conductivity. Many highly conducting one-dimensional materials are now known, based on polymers or highly anisotropic crystals of organic salts. They can be doped and used to construct devices, notably light-emitting diodes.

Now we shall extend these principles to band structure in higher dimensions.

2.4 Band Structure in Two and Three Dimensions

The principles of band structure in two and three dimensions are the same as in one dimension, but inevitably the result is more complicated. The reciprocal lattice is now two- or three-dimensional, and the dispersion relation must be plotted as a function of a vector **k**. Again **k** can be reduced to the first Brillouin zone by adding reciprocal lattice vectors.

Start with a two-dimensional example, a rectangular lattice with spacing a along x and b along y. The periodic potential $V(x, y)$ can be expanded as a Fourier series

$$V(x, y) = \sum_{m,n=-\infty}^{\infty} V_{m,n} \exp\left(\frac{2\pi i m x}{a} + \frac{2\pi i n y}{b}\right)$$

$$= \sum_{m,n} V_{m,n} \exp(i\mathbf{G}_{m,n} \cdot \mathbf{r}), \qquad (2.19)$$

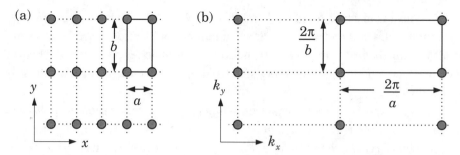

FIGURE 2.8. Two-dimensional rectangular lattices in (a) real and (b) reciprocal space.

where $G_{m,n} = (2\pi m/a, 2\pi n/b)$ are the reciprocal lattice vectors. These may be plotted out to form a reciprocal lattice in reciprocal or **k**-space in the same way as the points of the real or direct lattice, as shown in Figure 2.8. Both lattices are rectangular but have opposite aspect ratios (if one is tall and thin, the other is short and fat).

Again the periodic potential introduces band gaps and breaks **k**-space into Brillouin zones. The simplest argument for the formation of gaps is that based on the mixing of wave functions. A wave function with wave vector **k** gains components **k** − **G** from each Fourier coefficient −**G** of the potential. In general these have different energies and the mixing is small. Strong mixing, leading to band gaps, occurs when the energies are equal, which requires $|\mathbf{k}| = |\mathbf{k} - \mathbf{G}|$. The left-hand side is the distance of **k** from the origin, while the right-hand side is the distance of **k** from **G**, so this equation defines a plane bisecting the line from the origin to **G**. This is illustrated in Figure 2.9.

The periodic potential opens a band gap on each of these planes, which therefore divide **k**-space into Brillouin zones. Each point in the reciprocal lattice generates a plane and the result for the rectangular lattice is shown in Figure 2.10. The first zone is a simple rectangle as expected. Beyond this, **k**-space is divided into fragments of increasing complexity. In the reduced zone scheme, these fragments must be translated into the first zone by adding an appropriate reciprocal lattice vector. The result for the first four zones is shown in the figure. If the periodic potential is weak, an estimate of the band structure can be found by wrapping back the free-electron parabola into the first zone.

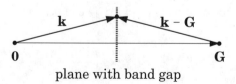

FIGURE 2.9. Generation of a band gap by a reciprocal lattice vector **G**. The gap appears in the plane where $|\mathbf{k}| = |\mathbf{k} - \mathbf{G}|$.

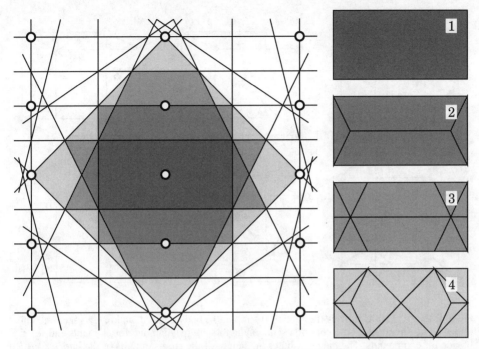

FIGURE 2.10. First four Brillouin zones of the rectangular lattice, displayed as the shaded regions in the reciprocal lattice. These have been translated through reciprocal lattice vectors to build successive rectangles, as shown on the right.

2.5 Crystal Structure of the Common Semiconductors

It is clear that band structure is complicated in three dimensions, and any tool that simplifies it is valuable. Fortunately the common semiconductors form highly symmetric crystals, and their symmetry is an essential ingredient in advanced theory. For example, it is often possible to show which optical processes are allowed or forbidden from the symmetry alone, without detailed knowledge of the band structure or wave functions. Group theory provides the mathematical framework for analyzing symmetry, but is beyond the scope of this book.

The common semiconductors have cubic crystals. Three cubic lattices are shown in Figure 2.11(a)–(c). The simple cubic lattice is primitive, meaning that it has only one atom per unit cell of volume a^3. A further atom can be added in the middle of each cube, giving a body-centred cubic (BCC) lattice with two atoms, or at the centre of each face to give a face-centred cubic (FCC) lattice with four atoms in each cubic unit cell.

Sometimes it is more convenient to use a unit cell with only one atom. The Wigner–Seitz cell contains all points that are closer to the atom at the origin than

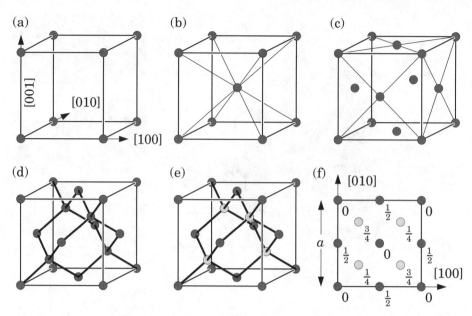

FIGURE 2.11. Cubic crystal structures: (a) simple, (b) body-centred, and (c) face-centred cubic lattices. The structures of the common semiconductors are (d) diamond for elements and (e) zinc blende for compounds. The plan (f) shows the height of the atoms in units of the lattice constant for the zinc-blende structure.

to any other atom. This is identical to the way in which the first Brillouin zone was defined in Figure 2.10. The result for the body-centred lattice is shown in Figure 2.12(a), where the cell surrounds the atom in the middle of the cell. The shape is a truncated octahedron, although you can also view it as a cube with its corners planed off. It is bounded by the planes halfway between the origin and the near points of the lattice; the eight hexagonal faces arise from the planes halfway to the atoms at the corners, while the six smaller square faces are halfway to the atoms in the middle of the next cells. Figure 2.12(b) shows how these unlikely unit cells fit together to fill space.

The common semiconductors are based on the FCC structure but with twice the number of atoms. The structure of elemental semiconductors such as Si is shown in Figure 2.11(d) and is known as the *diamond* lattice. There is an additional atom displaced by $(\frac{1}{4}, \frac{1}{4}, \frac{1}{4})a$ with respect to each atom in the FCC lattice. Another way of viewing this is to note that the four atoms in the FCC lattice (at $(0, 0, 0)$, $(\frac{1}{2}a, 0, 0)$, $(0, \frac{1}{2}a, 0)$, and $(0, 0, \frac{1}{2}a)$) lie at the vertices of a tetrahedron. Another atom is then added at the centre of this tetrahedron, where it can form a bond with each of the four atoms at the vertices. Thus a network of tetrahedral bonding can fill space. This corresponds to the chemical view that each atom forms four sp^3 bonds to its neighbours. The cubic unit cell of diamond contains eight atoms.

Compounds such as GaAs have the *zinc-blende* structure shown in Figure 2.11(e). The arrangement of the atoms is the same as in diamond but the two species

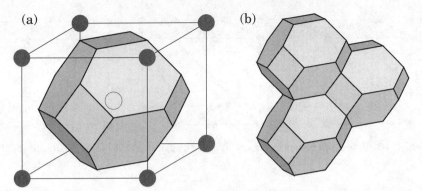

FIGURE 2.12.   (a) Wigner–Seitz cell of a body-centred cubic crystal, showing (b) the way in which they pack to fill space.

alternate: Ga occupies the original sites of the FCC lattice, while As occupies the tetrahedral sites, to give a unit cell that contains four atoms of each species. The length of each side is the lattice constant $a \approx 0.565$ nm in GaAs at 300 K, and the total number density of atoms (atoms per unit volume) is $8/a^3$ (data for the common semiconductors are summarized in Appendix 2). The length of each bond is $a\sqrt{3}/4 \approx 0.24$ nm.

There is a standard notation for describing planes and directions in crystals. Fortunately it is simple for cubic materials. Directions are specified by writing the Cartesian components of a vector without commas inside square brackets. For example, the z-axis is [001]. Negative entries are denoted by an overbar, so [00$\bar{1}$] is the $-z$-direction.

Planes are denoted in a similar way by *Miller indices*, writing a normal vector to the plane in parentheses, such as (001) for the xy-plane. The notation {001} means the plane (001) and all those related by symmetry, including (00$\bar{1}$), (100), and so on.

In fact the notation is more subtle because it is intended to describe sets of equally spaced planes. If one such plane passes through the origin, the set of planes is described by the reciprocals of the intercepts of the next plane on the axes, in units of the lattice constant. Consider, for example, the left-hand plane in Figure 2.13(c), taking the origin at the bottom left corner. This intercepts the axes at $(a, 0, 0)$, $(0, a, 0)$, and $(0, 0, a)$ and is therefore denoted by (111). If the planes were half as far apart the intercepts would be reduced to $(\frac{1}{2}a, 0, 0)$ and so on, giving Miller indices of (222). This specification may seem laborious but is essential for non-cubic crystals.

We can now use the notation to explore important planes in the crystal structure. Planes such as (001), seen in Figure 2.13(a), are normal to the principal axes. Each contains one species of atom, which alternates in successive planes of zinc blende, separated by $\frac{1}{4}a$. Growth usually proceeds by adding (001)-planes, and each pair comprises a monolayer, so there are two monolayers per unit cell.

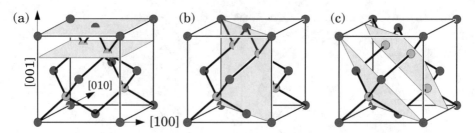

FIGURE 2.13. Important planes in the zinc-blende structure: (a) (001), (b) (110), and (c) (111).

GaAs usually cleaves along a (110)-plane, shown in Figure 2.13(b). These planes are parallel to the [001]-axis and cut the [100]- and [010]-axes at 45°. They contain both species of atom. Note that the figure shows an internal plane; an exposed surface is reconstructed after cleavage, adopting a different structure.

A third important plane is (111), which is equally inclined to all three principal axes. Each plane contains only one species of atom as for (100), and a plane of each kind is shown in Figure 2.13(c). It is perhaps clearer from Figures 2.11(d) and 2.11(e) that the [111]-direction is polar in a compound semiconductor, because the bonds all point in the same direction from Ga to As. This distinction is not present in an element, which therefore has a higher symmetry. Consider the point $(\frac{1}{8}, \frac{1}{8}, \frac{1}{8})a$, which lies halfway along the bond from the atom at the origin to that at $(\frac{1}{4}, \frac{1}{4}, \frac{1}{4})a$. Every atom in Si can be reflected through this point to leave the structure unchanged. This point is therefore described as a centre of symmetry. In a compound, however, the reflection interchanges the two species of atom and this operation is therefore not a symmetry of the crystal.

Symmetries such as rotation and reflection are summarized mathematically in the point group of the crystal. Elemental semiconductors show the highest cubic symmetry, described as $m3m$ in international notion or O_h in Schönflies notation. A mental picture is that this describes a perfect cube. The lack of a centre of symmetry in the compounds reduces their point group to $\bar{4}3m$ or T_d. This can be thought of as a tetrahedron oriented as in Figure 2.14.

The lower symmetry of the compounds has some important consequences. For example, they are piezoelectric, which means that certain strains generate an electric

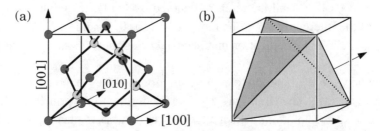

FIGURE 2.14. A tetrahedron oriented to show its relation to the crystal structure of the compound semiconductors.

field. Compress the structure in Figure 2.13(c) along the [111]-direction, which we have just found to be polar. This will change the relative spacing of the planes of Ga and As. The two species of atom are not precisely neutral but carry opposite electric charges. Thus a relative displacement of the planes sets up a polarization **P**, which may in turn lead to an electric field. Not all strains have this effect; compression along [001], for example, causes no polarization. The reduced symmetry of the compounds also permits them to show some optical nonlinearities that are forbidden in elements. The polar nature of directions such as [111] means that exposed (111)- and ($\bar{1}\bar{1}\bar{1}$)-surfaces have different properties. Figure 2.13(c) shows that the spacing between successive planes of atoms alternates. Growth proceeds by adding pairs of the more closely spaced planes. Thus, if GaAs is grown on the (111)-plane, it is found to terminate in a plane of Ga atoms, which is called a (111)A-surface. On the other hand, growth on the ($\bar{1}\bar{1}\bar{1}$)-plane terminates in a plane of As atoms known as (111)B. The distinction is important in processing, because growth or chemical etches often operate on the two surfaces with different rates.

2.6 Band Structure of the Common Semiconductors

We can now put the results in the preceding sections together, and study the band structures of three-dimensional semiconductors. First we need to know the reciprocal lattice and Brillouin zone. Remember our convention that upper-case vectors such as **K** are used to denote three-dimensional quantities.

2.6.1 BRILLOUIN ZONE

A simple cubic lattice with the atoms separated by a has a simple cubic reciprocal lattice of separation $2\pi/a$. The addition of extra atoms to form a face-centred cubic lattice in real space removes three-quarters of the points in reciprocal space, to leave a body-centred cubic lattice whose unit cell is $4\pi/a$ across. The first Brillouin zone is given by the Wigner–Seitz cell of this lattice, which we have already seen in Figure 2.12. This truncated octahedron is repeated in Figure 2.15(a), which also shows the standard notation for the points of high symmetry. Generally, Greek letters are used to denote points inside the zone, with Roman letters for the surface. The most important points are as follows:

- Γ is the origin of **K**-space.
- Δ is a direction such as [100] and meets the zone boundary at X, in the middle of a square face, with coordinates such as $(2\pi/a)(1, 0, 0)$.
- Λ is a direction such as [111], normal to the close-packed planes of the face-centred structure. It cuts the middle of the hexagonal faces of the zone at L, with coordinates such as $(2\pi/a)(\frac{1}{2}, \frac{1}{2}, \frac{1}{2})$.

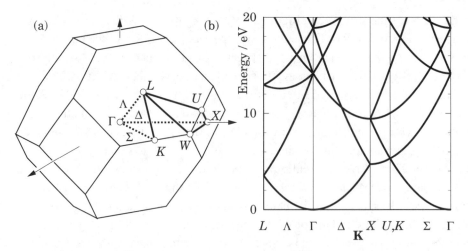

FIGURE 2.15. (a) Brillouin zone for a face-centred cubic crystal, showing the notation for special points and directions. Solid lines are on the surface with broken lines inside the zone. (b) Band structure in the free-electron model, showing the effect of folding back the parabola into the reduced zone.

- Σ is a direction such as [110] and meets the boundary at K, in the middle of an edge shared by two hexagons, with coordinates such as $(2\pi/a)(\frac{3}{4}, \frac{3}{4}, \frac{3}{4})$.

- U lies in the middle of edges shared by a hexagon and a square. The line S joins U to X.

- W lies at the vertices, each shared by two hexagons and a square.

These labels are needed to interpret the standard plots of band structure.

It is more difficult to present band structure in three dimensions than in one. Virtual reality may provide the ideal method but a less ambitious approach has to be taken for the pages of a book. The usual way of displaying band structure is to plot $\varepsilon_n(\mathbf{K})$ along selected directions in the reduced Brillouin zone, leaving the rest to the reader's imagination. This is shown in Figure 2.15(b) for free electrons in the GaAs lattice. The left-hand segment shows the bands between L and Γ, the Λ-direction. The plot then sets off in the Δ-direction to the zone boundary at X. It next turns along the surface of the zone towards U. This point is equivalent to K, as will be described shortly, from where the plot returns back along the Σ-direction to Γ.

The equivalence of U and K is not obvious because the [111]-direction has only threefold rotational symmetry, not sixfold. In fact the points are related by a translation, not a rotation. Consider $U = (2\pi/a)(1, \frac{1}{4}, \frac{1}{4})$. Subtracting the reciprocal lattice vector $(2\pi/a)(1, 1, 1)$ gives $(2\pi/a)(0, -\frac{3}{4}, -\frac{3}{4})$, which is a K-point.

The figure shows the effect on the free-electron dispersion of wrapping the parabola back through reciprocal lattice vectors into the first Brillouin zone. Some of the curves look like a parabola bouncing back and forth as in one dimension (Figure 2.2). In general the additional freedom in three dimensions makes the picture much more complicated and many of the bands are degenerate. It turns out, however, that the bands of the common semiconductors are not far removed from those for free electrons, particularly if the symmetries are considered more carefully.

2.6.2 GENERAL FEATURES

Band structures for Si, Ge, GaAs, and AlAs are shown in Figure 2.16. These are calculated using a relatively simple method and do not include the spin–orbit coupling which is important for the top of the valence band, to be discussed later. The general feature is a set of valence bands, which is full in a pure semiconductor at $T = 0$, separated by a band gap from the empty conduction bands above. The zero of energy is conventionally taken as the top of the valence band.

A simple chemical picture gives a broad explanation for the form of the valence bands. Si has four valence electrons, two in a filled s orbital and two in the three p

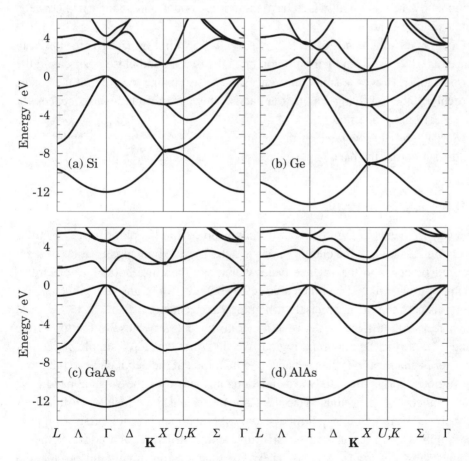

FIGURE 2.16. Band structure of four common semiconductors: silicon, germanium, gallium arsenide, and aluminium arsenide. The calculations do not include the spin–orbit coupling. [Results kindly supplied by Prof. G. P. Srivastava, University of Exeter.]

orbitals, which can accommodate a total of six. These orbitals can hybridize, giving four sp^3 orbitals that stick out like legs to the corners of a tetrahedron and join to form the diamond lattice. This elementary picture can be related to the bands of Si by regarding the valence band as being made from 'bonding' combinations of sp^3 orbitals and the conduction band from 'antibonding' combinations. An important feature that we shall use subsequently is that the top of the valence band has p-like symmetry. The bottom of the conduction band is s-like in GaAs but the picture of orbitals does not work nearly as less well for the conduction band because the wave functions spread further over the crystal and retain less of their atomic character.

The valence bands of the compounds are slightly more complicated than those of the elements because their lower symmetry permits a small transfer of charge from Ga to As, which gives an ionic character to the bond. This weakens the hybridization, and the valence band splits at around -8 eV into a single band from the s orbitals below a triple band from the three p orbitals. This trend is carried further in II–VI compounds such as ZnSe, which have more ionic bonds and a larger gap between flatter valence bands.

Qualitatively, the bands of all four materials shown in Figure 2.16 are similar in shape, as expected from their common crystal structure and close positions in the periodic table. However, small differences in energy change the order of the bands, particularly the low-lying conduction bands. This has profound consequences for their electronic properties.

Most interest is focussed on the small regions of the conduction and valence bands that border the band gap, so we shall now concentrate on these.

2.6.3 VALENCE BAND

The valence bands in Figure 2.16 are recognizable from the plot for free electrons in Figure 2.15(b) and the energy scales match. The top branch is double, so there are four branches to the valence band, which can therefore hold eight electrons. Each primitive unit cell (not the cubic cell!) contains two atoms with eight valence electrons between them, precisely filling the valence band.

There is little difference between materials except very close to the top of the band – but this is the important region. We shall first develop a simple picture for the overall shape of the valence band, and then look at the detail close to the top.

We noted earlier that the wave functions at the top of the valence band have the symmetry of p orbitals. This does not mean that the Bloch wave functions actually look like atomic p orbitals; only the symmetry is important. Suppose for simplicity that the crystal has a simple cubic structure, and consider the p_z orbitals shown in Figure 2.17(a). The wave functions are highly anisotropic and overlap strongly in the z-direction. This makes it easy for electrons to travel from atom to atom along z, so the effective mass in this direction is low. Overlap

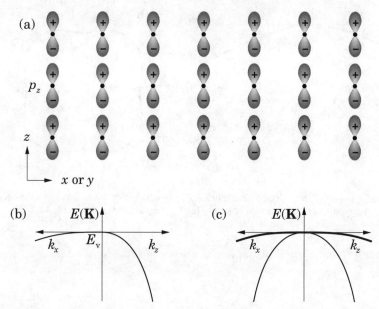

FIGURE 2.17. Valence bands constructed from p orbitals. (a) Lattice of p_z orbitals. (b) Band structure of the p_z orbitals only; the band is 'light' along k_z to the right and 'heavy' along k_x (or k_y) to the left. (c) Total bands from all three p orbitals, showing a doubly degenerate 'heavy' band and a single 'light' band.

is much weaker in the xy-plane, so electrons travel less freely and the effective mass is higher. Thus the band structure for the p_z orbital alone is strongly anisotropic, as shown in Figure 2.17(b). The other two orbitals behave in a similar way about their polar axis, and adding all three gives the bands in Figure 2.17(c) where cubic symmetry is restored. There is a single 'light' band, whose energy (from the point of view of the holes) increases rapidly with K, and a doubly degenerate 'heavy' band. This picture will be made quantitative using the tight-binding model in Section 7.7. It explains the overall shape of the valence bands, although a little subterfuge has occurred and will be remedied in Section 10.2 on the Kane model.

Unfortunately this simple picture fails for energies very close to the top of the band. A sketch of this region is shown in Figure 2.18(a). The bands are labelled with their symmetry using group-theoretical notation. Two bands, called Γ_8, meet at the top with a third split-off band Γ_7 just below. They are separated by the spin–orbit splitting Δ. This is a relativistic effect that will be discussed further in Section 10.2.2. The splitting increases as the fourth power of the atomic number of the element, and is about 0.29 eV for Ge but only 0.044 eV for Si; it is larger than the band gap in extreme cases. The split-off band can be neglected if the splitting is large, although this is not applicable to light materials such as Si.

The two bands that meet have different effective masses for small $\mathbf{K}$ and are known as *light* and *heavy* holes. They converge for large $\mathbf{K}$ to give the 'heavy' band

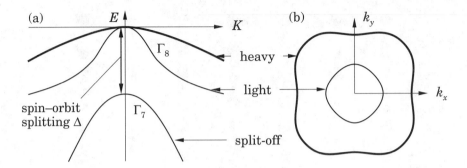

FIGURE 2.18. (a) Top of the valence band showing light and heavy holes, and the split-off band. (b) Constant-energy surfaces or 'warped spheres' for light and heavy holes in GaAs.

derived already. Near Γ their dispersion is *very roughly* described by

$$\varepsilon(\mathbf{K}) = E_v - \frac{\hbar^2 K^2}{2m_0 m_h}, \tag{2.20}$$

where $m_h = m_{hh}$ for heavy holes or m_{lh} for light holes. It must be emphasized that this approximation, although widely used for its simplicity and convenience, is poor. Superior approximations show that the bands are both non-parabolic and anisotropic, depending on the direction of $\mathbf{K}$ as well as its magnitude. This is shown by the constant-energy surfaces in Figure 2.18(b), which are loci of $\mathbf{K}$ that give a particular value of energy. The smaller effective mass of light holes puts their surface inside that for heavy holes. Neither surface is the sphere predicted by the simple approximation (2.20). The surfaces are known as *warped spheres* (equation 10.35) and will be described more fully in Section 10.2.2.

The valence bands are clearly complicated, and become worse if the cubic symmetry is broken. Straining the material, or growing a quantum well, lifts the degeneracy between the light and heavy holes at Γ, and also destroys the approximate isotropy. These effects will be considered a little more in Section 10.3. In general we will adopt the simplest model (2.20), despite its shortcomings, since the more faithful descriptions are too cumbersome.

2.6.4   CONDUCTION BAND

The valence bands are qualitatively similar in the common semiconductors but this is not true for the conduction bands, where small shifts relative to one another change the nature of the lowest minimum. Three situations are commonly found, with the lowest point in the conduction band at Γ, towards X (along Δ), or at L. We shall look at these in turn.

2.6.4.1 Γ minimum – GaAs

The bottom of the conduction band is at Γ in GaAs, labelled Γ_6. This is the same point in **K**-space as the top of the valence band, so GaAs has a *direct* gap, and light can excite electrons across the minimum band gap (Section 2.7). The dispersion relation is parabolic at low energies,

$$\varepsilon(\mathbf{K}) = E_c + \frac{\hbar^2 K^2}{2m_0 m_e}, \tag{2.21}$$

with an effective mass $m_e \approx 0.067$ in GaAs. Its slope increases less rapidly at higher energies and other approximations are used (equation 1.94); eventually it loses isotropy too. The low effective mass leads to a high mobility through $\mu = e\tau/m$ and explains the use of GaAs in high-frequency transistors.

2.6.4.2 X minima – Si and AlAs

The lowest points in the conduction band of Si lie along the Δ-directions such as [100], about 85% of the way to the zone boundary (X). These points are far removed in the Brillouin zone from the top of the valence band at Γ, so the gap is *indirect* (Section 2.7).

There are six Δ-directions and therefore six equivalent minima in the conduction band, each of which gives rise to a *valley*. A parabolic approximation can again be used for each valley, but it is not isotropic and is centred on the minimum k_0 rather than Γ. For the [100]-valley we can write

$$\varepsilon(\mathbf{K}) = E_c + \frac{\hbar^2}{2m_0}\left[\frac{(k_x - k_0)^2}{m_L} + \frac{k_y^2}{m_T} + \frac{k_z^2}{m_T}\right]. \tag{2.22}$$

Symmetry demands that the behaviour along the transverse directions y and z be identical so they share the same effective mass m_T. The longitudinal x-direction need not be the same and shows a different effective mass m_L. For Si these masses are $m_L \approx 0.98$ and $m_T \approx 0.19$, so the valley is highly anisotropic, as is usually illustrated with a constant-energy surface. Figure 2.19(a) shows the positions of the six valleys in the Brillouin zone with an enlarged view of one valley in Figure 2.19(b). The valley is cigar-shaped, elongated along the longitudinal direction, which has a high mass.

The density of states remains parabolic but a *density-of-states effective mass* given by $(m_L m_T m_T)^{1/3} \approx 0.33$ must be used. The usual formula must also be multiplied by the degeneracy $g_v = 6$ to account for the equivalent valleys; the degeneracy $g_s = 2$ due to spin remains too.

The material would be highly anisotropic if electrons could be confined to a single valley. For example, the electrons would have a high mobility $\mu = e\tau/m$ in directions where the effective mass was small, but a lower value where the effective

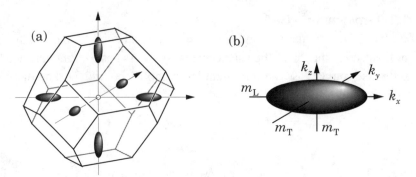

FIGURE 2.19. (a) Brillouin zone of Si with constant-energy surfaces for the six equivalent X-valleys. (b) Enlarged view of one of the valleys, showing the longitudinal and doubly degenerate transverse masses m_L and m_T.

mass was high. It is possible to apply strain to break the cubic symmetry and reveal some of this behaviour, but in normal material the whole set of valleys contributes. This gives isotropic conductivity, which conforms with the cubic symmetry of the crystal.

AlAs also has the lowest minima in its conduction band at X. This is different from GaAs, so the nature of the minimum changes as a function of x in the alloy $Al_x Ga_{1-x} As$. This leads to some interesting effects in heterostructures of GaAs and AlAs.

2.6.4.3 L minima – Ge

In Ge the minima of the conduction band are at the L-points, in the middle of the hexagonal faces of the Brillouin zone along directions such as [111]. The points on opposite faces are equivalent, giving four equivalent valleys or eight half-valleys. Again, the gap is indirect. Each valley can be depicted with a constant-energy surface as in Figure 2.19(b), with the long axis oriented along its [111]-direction. The masses are more anisotropic than in Si, with $m_L \approx 1.64$ and $m_T \approx 0.082$ for motion in the transverse plane.

2.6.4.4 Higher minima

Although the conduction band of GaAs has its lowest minimum at Γ, there are 'satellite' valleys at L and X that are not much higher. Those at L are only about 0.3 eV above Γ_6 with a set at X about 0.2 eV farther up; another minimum at X is only 0.4 eV higher still. These energies are small enough that energetic electrons can enter the L-valleys from Γ. The density of states in L is much higher, because of both the effective mass and the degeneracy, and the mobility is lower. This **K**-space transfer results in a region of negative differential mobility in the velocity–field characteristics of GaAs, harnessed in the Gunn effect.

A final comment before leaving band structure: Bloch's theorem applies rigorously only to an infinite crystal. No real crystal is infinite, of course, and this

approximation is particularly questionable near surfaces. There may be extra electronic states associated with surfaces that do not obey Bloch's theorem and bear
no relation to the band structure of the rest of the crystal; in particular, they may
lie in the middle of the band gap. Such surface states can often be associated with
reconstructions that take place on surfaces, so their structure no longer resembles
the bulk crystal simply cut off. Surface states may be due to defects in other cases,
and sometimes their origin remains unclear. They may be of great practical importance, notably the high density of surface states near the middle of the gap on
(100)-surfaces of GaAs. The behaviour of devices such as field-effect transistors is
strongly influenced by the surface states, as we shall see in Section 9.1.

2.7 Optical Measurement of Band Gaps

Band gaps can be measured by simple optical techniques such as absorption. The
interaction of light with the crystal can be viewed as an elastic scattering process
involving a photon and an electron, in which both energy and (crystal) momentum
must be conserved. Let the electron have initial energy and momentum $(E_i, \hbar \mathbf{K}_i)$
and final state $(E_f, \hbar \mathbf{K}_f)$. A photon of wave vector $\mathbf{Q}$ has energy $\hbar \omega = \hbar c Q$ and
momentum $\hbar \mathbf{Q}$, where c is the velocity of light inside the material. Conservation of
energy and momentum requires

$$E_f = E_i + \hbar c Q, \qquad \mathbf{K}_f = \mathbf{K}_i + \mathbf{Q}. \tag{2.23}$$

Typically the difference in energy $\hbar \omega \approx 1 \, \text{eV}$ for the band gap of a semiconductor,
giving $Q \approx 10^7 \, \text{m}^{-1}$. The size of the Brillouin zone is roughly π/a with $a \approx 0.5 \, \text{nm}$,
so $\pi/a \approx 10^{10} \, \text{m}^{-1}$. Thus Q is tiny on the scale of the Brillouin zone, so one
normally sets $\mathbf{K}_f = \mathbf{K}_i$, neglecting the change in momentum during an optical
transition. Such transitions are called *vertical* because they appear as vertical lines
on the band structure $\varepsilon(\mathbf{K})$.

In a semiconductor with a conduction band such as GaAs, the transition from
the top of the valence band to the bottom of the conduction band is vertical at Γ
(Figure 2.20(a)). It can therefore be induced by light, and absorption occurs as soon
as the energy of the photon exceeds the band gap: $\hbar \omega > E_g$. This is the significance
of a direct gap. Some measurements are shown in Figure 8.4.

The process also works in reverse. In devices such as a forward-biassed diode,
excess electrons near the bottom of the conduction band, and holes near the top of
the valence band, are injected into a semiconductor. The excess electrons tend to
recombine by 'falling' into the holes to restore equilibrium. Energy is released in
this process and can be either lost to the lattice as phonons, often with the aid of an
impurity, or emitted as a photon. The latter process, radiative recombination, dominates in semiconductors with direct gaps. They are consequently efficient sources

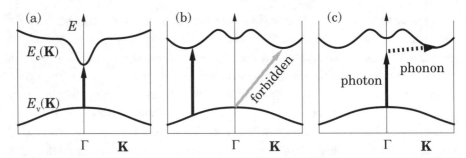

FIGURE 2.20. Optical absorption across the band gap in different types of semiconductor. (a) Absorption across a direct band gap at Γ. (b) Absorption across an indirect band gap is forbidden but vertical transitions occur for all **K**. (c) Transition across an indirect band gap with absorption of both a phonon and a photon.

of light. Most optoelectronic devices, and lasers in particular, are therefore based on materials such as GaAs with a direct gap.

In a material with an indirect gap such as AlAs, Si, or Ge, the extrema of the valence and conduction bands do not occur at the same **K** (Figure 2.20(b)). A transition between these points requires a large change in momentum as well as energy and cannot occur with a photon alone. Vertical transitions can occur at all points in the Brillouin zone, but the lowest energy of these optical transitions is greater than the minimum band gap.

Optical transitions across indirect gaps can occur if another process provides the change in momentum. Phonons can do this and either carry away or contribute a small amount of energy to the transition, as shown in Figure 2.20(c). Impurities may also be involved. These processes produce a tail in the optical absorption which starts around the energy of the indirect gap. Radiative recombination can occur by the same routes, but is much less efficient than that in materials with a direct gap. Thus Si, Ge, and their alloys are not generally efficient emitters of light. However, some light-emitting diodes (LEDs) use GaP, which has an indirect gap, and employ radiative recombination mediated by impurities to generate light.

2.8   Phonons

Phonons are the quanta of vibrations of the ions of the lattice away from their positions of minimum energy. The periodicity of the lattice means that phonons have band structure in much the same way as electrons. In some ways the band structure of phonons is more complicated than that for electrons, because phonons are polarized. Consider vibrations of very long wavelength, much greater than the lattice constant. These can be treated as classical sound waves, and there are three of them that propagate in a given direction. If we choose a simple direction, such as [100] in a cubic material, one mode is longitudinal and the other two are transverse.

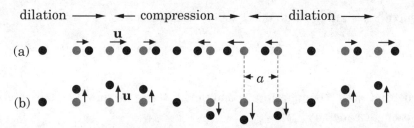

FIGURE 2.21. (a) Longitudinal and (b) transverse modes of the same wavelength $8a$ in a one-dimensional monatomic crystal. The light grey discs show the positions of the atoms at equilibrium, and the dark discs are their positions when displaced through **u**.

These are illustrated for a one-dimensional crystal in Figure 2.21. The longitudinal mode is like a sound wave in air: the ions vibrate back and forth in the same direction **q** as the wave travels, so the displacement **u** is parallel to **q**. This gives alternating zones of compression and dilation, in quadrature with the displacement. The ions vibrate in the plane normal to the direction of travel in the transverse modes, with **u** perpendicular to **q**, like an electromagnetic wave in free space.

In other respects phonons are simpler than electrons. For example, the number of bands is given by the number of atoms per unit cell, multiplied by three for the polarizations, unlike the infinite number of bands for electrons. We shall solve a simple one-dimensional model that displays most of the main features of phonons, before looking briefly at phonons in a semiconductor.

2.8.1 PHONONS IN ONE DIMENSION

The simplest model is a one-dimensional row of atoms of mass m along x, separated by a distance a at rest, shown in Figure 2.22. They are connected by springs of elastic constant K. Vibration is restricted to the x-direction so there are only longitudinal modes. Let the displacement of atom j be u_j. Newton's law gives

$$m\frac{d^2 u_j}{dt^2} = K[(u_{j-1} - u_j) - (u_j - u_{j+1})],\qquad(2.24)$$

where the terms in parentheses are the decreases in length of the bonds between atoms j and $j \pm 1$. We expect the solutions to be waves and can try a solution $u_j = U_0 \cos(qx - \omega_q t) = U_0 \,\mathrm{Re}\{\exp[i(qx - \omega_q t)]\}$, where $x = ja$ for atom j. The complex notation simplifies manipulation but is not essential as in the Schrödinger

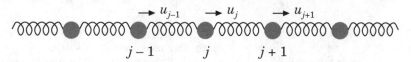

FIGURE 2.22. A one-dimensional row of springs and masses, labelled by j, allowed to vibrate longitudinally through displacements u_j.

equation. Substituting it into the equation of motion gives

$$-m\omega_q^2 U_0 e^{i(qja-\omega_q t)} = U_0 K[(e^{iq(j-1)a} - e^{iqja}) - (e^{iqja} - e^{iq(j+1)a})]e^{-i\omega_q t}.$$
(2.25)

The dependence on both position j and time t cancels to show that this solution is consistent, leaving

$$m\omega_q^2 = -K[(e^{-iqa} - 1) - (1 - e^{iqa})] = 4K \sin^2 \frac{qa}{2}.$$
(2.26)

We could have reached this more quickly by using Bloch's theorem in the form $u_{j+1} = e^{iqa}u_j$. The dispersion relation is

$$\omega_q = 2\sqrt{\frac{K}{m}} \left|\sin \frac{qa}{2}\right|.$$
(2.27)

The wave number should be restricted to the first Brillouin zone $|q| < \pi/a$, as with electrons. The dispersion is plotted in Figure 2.23(a). The relation is linear for small q, in contrast to the parabolic relation for electrons, and can be written $\omega_q = v_s q$, where the velocity of sound $v_s = a\sqrt{K/m}$. Adjacent atoms have similar displacements in this region of small q and long wavelength, as in Figure 2.21(a). The slope of ω_q decreases as q increases and becomes flat at the zone boundary $q = \pm\pi/a$, where the group velocity of the waves drops to zero as for electrons.

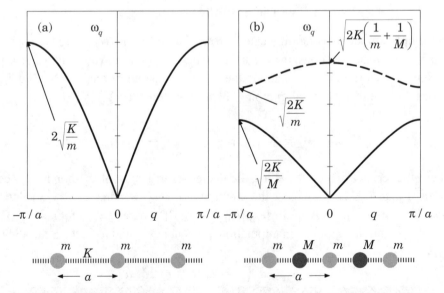

FIGURE 2.23. Dispersion relation ω_q for a one-dimensional crystal with longitudinal vibrations. (a) Monatomic crystal with atoms of mass m. (b) Diatomic chain of masses m and $M = 2m$, showing the acoustic branch (full line) and the optic branch (broken line). The scales are the same for both plots. The atomic configurations are shown underneath; both chains have the same lattice constant a and all springs have the same elastic constant K.

Also, atoms in adjacent cells vibrate in antiphase at this point. Alternate springs are fully compressed and stretched, giving the maximum frequency of vibration.

2.8.2 AMPLITUDE OF VIBRATION

We need a value for the amplitude of vibration U_0 to complete the analysis. Unfortunately we have hit a fundamental problem. The equations of motion are linear in the displacement, so the amplitude is arbitrary within classical mechanics. In quantum mechanics, however, the vibrations should be quantized into phonons. This can be performed by a standard procedure where the motion of the atoms is resolved into normal modes. Each mode resembles a harmonic oscillator, which we shall solve in Section 4.3. Unfortunately this procedure is beyond the scope of this book, and we shall take a shortcut.

The amplitude U_0 can be found for a single phonon by calculating the total energy of the vibration and equating it to the quantized value $\hbar\omega_q$. For the monatomic crystal, the velocity of atom j is given by

$$\frac{du_j}{dt} = U_0\omega_q \sin(qaj - \omega_q t), \tag{2.28}$$

so the kinetic energy is $\frac{1}{2}mU_0^2\omega_q^2 \sin^2(qaj - \omega_q t)$, whose average value is $\frac{1}{4}mU_0^2\omega_q^2$. This equals the average potential energy for harmonic motion, so the total average energy is $\frac{1}{2}mU_0^2\omega_q^2$ per atom. Multiplying by the number of atoms N_{atoms} yields the total energy, and equating this to the quantum $\hbar\omega_q$ fixes U_0. We can eliminate N_{atoms} in favour of the 'volume' Ω (really a length), the mass density ρ, and the mass of each atom m, as $N_{\text{atoms}} = \Omega\rho/m$. Thus

$$U_0 = \sqrt{\frac{2\hbar}{N_{\text{atoms}}m\omega_q}} = \sqrt{\frac{2\hbar}{\Omega\rho\omega_q}}. \tag{2.29}$$

For small wave vectors, where $\omega_q = v_s q$, this becomes $U_0 = \sqrt{2\hbar/\Omega\rho v_s q}$. The motion of an atom is finally

$$u_j(t) = \sqrt{\frac{2\hbar}{\Omega\rho\omega_q}} \cos(qaj - \omega_q t). \tag{2.30}$$

Note the appearance of the volume Ω of the crystal. A larger sample appears to have 'weaker' phonons but this is cancelled by the increase in the total density of states. We shall see that Ω vanishes from physical results such as scattering rates.

2.8.3 DIATOMIC CHAIN

Many crystals, including the common semiconductors, have more than one atom per unit cell, which introduces another feature. Consider the chain shown in Figure 2.23(b). This has alternating heavy and light masses, M and m, along its length,

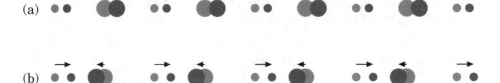

FIGURE 2.24. Motion of the atoms in (a) acoustic and (b) optic modes of very long wavelength for a one-dimensional chain of alternating light and heavy atoms. The light grey discs mark the resting positions of the atoms.

giving two atoms per unit cell. The lattice and spring constants retain their previous values a and K. The dispersion relation is

$$\omega_q^2 = K\left(\frac{1}{m} + \frac{1}{M}\right) \pm K\sqrt{\left(\frac{1}{m} + \frac{1}{M}\right)^2 - \frac{4}{mM}\sin^2\frac{qa}{2}}. \qquad (2.31)$$

There are now two bands or branches to the dispersion relation. The lower or acoustic branch looks much like the monatomic case, except that the increased mass has lowered the frequencies. However, the upper or optic branch is quite different. It has a maximum at $q = 0$ and slopes down gradually to the zone boundary, where there is a gap between the acoustic and optic branches. The amplitudes of vibration of the two atoms in each unit cell of the diatomic crystal can be calculated as for the monatomic chain.

The natures of the vibrations near $q = 0$ in the two branches are contrasted in Figure 2.24. The displacements in the acoustic branch are like sound waves, as the name implies. The two atoms in each unit cell move in the same direction by almost the same distance, and over a small region it appears as if the whole crystal has simply been compressed or stretched. We shall use this picture when calculating the interaction between such phonons and electrons in Section 8.4.1. The distortion of the bonds between the atoms becomes smaller for a given amplitude of vibration as $q \to 0$, sending $\omega_q \to 0$.

The heavy and light atoms move in opposite directions in the optic case, as in Figure 2.24(b). This is very similar to the top of the band for the monatomic chain. Again there is maximum distortion of every spring, so the frequency takes its maximum value.

A further feature appears in compounds because of the transfer of charge between the two species. Relative displacement of the two atoms, as in the optic mode, sets up an electric dipole **p**. This in turn sets up a polarization field **P** and an electric field which can interact with electromagnetic waves, hence the name 'optic'. Charged particles such as electrons also feel the electric field, and we shall see in Section 8.4.2 that polar longitudinal optic (LO) phonons scatter electrons rapidly. This polar nature is not seen in elements because all the atoms are identical and there is no transfer of charge.

Only one type of atom moves in each of the modes at the zone boundary, the other remaining stationary. For example, only the lighter atoms vibrate in the upper mode, with adjacent cells in antiphase. Thus only m appears in ω_q, and the expression is the same as that for the highest frequency in the monatomic chain, except that the effective value of K is halved because there are now two springs separating each light atom. This behaviour is analogous to that for electrons at zone boundaries, shown in Figure 2.4.

2.8.4 PHONONS IN THREE DIMENSIONS

The band structure for phonons can be plotted in exactly the same way as that for electrons. A small part of the dispersion is sketched in Figure 2.25 for Si and GaAs. The semiconductors have two atoms in each primitive unit cell, so the overall structure is close to that for the diatomic chain in Figure 2.23(b). The frequencies are higher for Si largely because of its lighter mass, which is 28, compared with an average of 72 for GaAs. The main difference from the one-dimensional case arises from the three possible polarizations. The two transverse modes are degenerate along [100] but in general there are three acoustic and three optic branches. Both acoustic modes become linear for small q, the longitudinal mode having a higher velocity.

The transverse optic (TO) and LO phonons in polar semiconductors have slightly different energies at Γ. This splitting is due to the electric field set up by LO phonons and is given by the Lyddane–Sachs–Teller relation (8.60). The energy of the LO phonon is 36 meV in GaAs; this high energy means that LO phonons can be

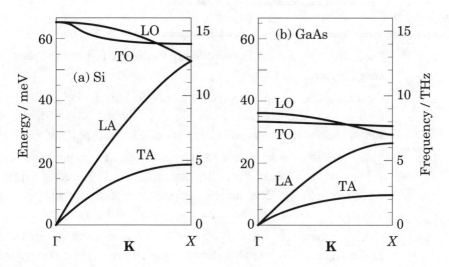

FIGURE 2.25. Sketch of the dispersion relation for phonons in (a) Si and (b) GaAs from Γ to X. The transverse branches are doubly degenerate along this direction.

neglected at low temperatures (below about 77 K), but they are an important mechanism for transferring energy between electrons and phonons at higher temperatures. The lighter mass of AlAs raises this energy to about 50 meV, but the energy of LO phonons in $Al_xGa_{1-x}As$ does not simply change smoothly from one to the other as a function of x.

This concludes our survey of the crystalline semiconductors. In the next chapter we shall see how these ingredients are put together to make heterostructures and investigate the properties that result.

Further Reading

Band structure is covered in all textbooks on solid state physics, including the classics by Kittel (1995) and by Ashcroft and Mermin (1976). They tend, however, to concentrate primarily on metals rather than semiconductors. A newer comprehensive introduction is by Myers (1990), and Bube (1992) is more biassed towards semiconductors.

Seeger (1991) gives a broad review of classical semiconductor physics including the effect of non-parabolic valence bands. Wolfe, Holonyak, and Stillman (1989) present a clear account of the physical properties of semiconductors. A thorough description of band structure and group theory, heavily biassed towards semiconductors, is given by Yu and Cardona (1996). They also cover phonons and electronic and optical properties with a chapter on heterostructures.

A wealth of data on semiconductors is tabulated by Madelung (1996).

EXERCISES

2.1   Show that the two statements of Bloch's theorem, equations (2.2) and (2.3), are equivalent.

2.2   Figure 2.5 was drawn for a typical band in a semiconductor. Repeat this for a superlattice, which might have $a = 10$ nm and $W = 50$ meV. What values of effective mass and maximum velocity are expected?

2.3   Many semiconductors have a saturation velocity for electrons of around 10^5 m s^{-1}. How much of the Brillouin zone do they need to explore to reach this velocity? The saturated velocity is reached at a field of around 10^6 V m^{-1}, so roughly what is the relaxation time?

2.4   Calculate the frequency of Bloch oscillation in as large an electric field as the material will tolerate before breaking down, and the corresponding excursion in space. Use the parameters of Figure 2.5. Compare this with a typical lifetime between scattering events (estimated from the mobility

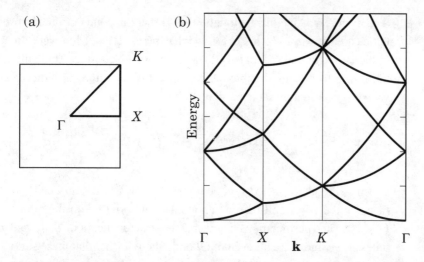

FIGURE 2.26. (a) Brillouin zone and (b) energy bands for free electrons in a square lattice.

$\mu = e\tau/m$, for example). Now repeat this for a superlattice. Which system is the better candidate for the observation of Bloch oscillations?

2.5 Estimate the probability of Zener tunnelling for the systems considered as Bloch oscillators in the previous problem. Assume that the gap is about 1 eV wide for the crystal and has the same width as the bands for the superlattice. Is there a window of 'large' electric fields where it should be possible to see Bloch oscillation before Zener tunnelling dominates?

2.6 Repeat for a square lattice the sketch of the first four Brillouin zones for a rectangular lattice (Figure 2.10).

2.7 The two-dimensional square crystal contains two electrons per unit cell. If they can be described as free electrons, they occupy a Fermi circle in **k**-space whose area is equal to that of the first Brillouin zone. Superpose this circle on the reciprocal lattice and show that parts of the first two zones are occupied. Sketch the Fermi line on these zones after they have been reduced to the square around the origin.

  Repeat this for the case of four electrons per unit cell, and show that the first four zones contain electrons. This sort of construction is used to explain the properties of some metals that can be described with the nearly free electron model.

2.8 Calculate the first few energy bands for free electrons in a square lattice. These are plotted along the directions of high symmetry in Figure 2.26.

2.9 The reciprocal lattice of a simple cubic crystal of lattice constant a is also simple cubic, with lattice constant $2\pi/a$. The coordinates of the reciprocal lattice are therefore $\mathbf{G}_{lmn} = (2\pi/a)(l, m, n)$. This result follows from the Fourier transform of a periodic potential, as in the one-dimensional case (equation 2.1). An equivalent definition is that the points of the reciprocal lattice are the wave vectors $\mathbf{G}$ for which $\exp(i\mathbf{G} \cdot \mathbf{R}) = 1$ for every point $\mathbf{R}$ of the lattice in real space. If further points are added to the cubic unit cell to make a face-centred cubic lattice, some of the reciprocal lattice points vanish. The structure factor is defined by

$$S(\mathbf{G}) = \sum_{j}^{\text{cell}} \exp(i\mathbf{G} \cdot \mathbf{R}_j), \qquad (\text{E2.1})$$

which is the phase factor summed over all atoms in the unit cell. For the face-centred cubic cell there is an atom at the origin with three others at $(\frac{1}{2}, \frac{1}{2}, 0)a$ and permutations. Show that the structure factor of $\mathbf{G}_{lmn}$ vanishes unless l, m, and n are all even or all odd. Show further that these surviving points define a body-centred reciprocal lattice with constant $4\pi/a$, as stated in Section 2.6.1.

2.10   Confirm that the first Brillouin zone of a face-centred cubic crystal contains one state per spin per atom of the lattice.

2.11   Show that the angle between any two of the bonds between adjacent atoms in the diamond lattice is $\arccos(-\frac{1}{3}) \approx 109°$.

2.12   Show that the hexagonal faces of the Brillouin zone of a face-centred cubic crystal are regular hexagons; this is not obvious, as pointed out in Section 2.6.1, because there is only a threefold rotational symmetry about the [111]-axis. This also shows that all edges of the zone have the same length.

2.13   Suppose that an electric field is applied at 45° to the x-direction of the [100]-valley in the conduction band of Si. Assume that only this valley contributes to the current. Resolve the force into components along the longitudinal and transverse directions, and calculate the direction of acceleration of the electron.

2.14   Suppose that a single impurity can be modelled as a lighter mass m' in the one-dimensional chain of balls and springs of Figure 2.22. Show that it gives rise to a localized mode of vibration with a frequency above the top of the band.

2.15   Calculate the dispersion relation of the diatomic chain, equation (2.31). Write down the equations that govern the motion of both atoms in cell j, u_j, and v_j, say, remembering that their displacements are different. These depend on the displacements in the adjacent cells, which can be replaced using Bloch's theorem. Assume that all coordinates oscillate in time as $e^{-i\omega t}$. The result is a 2×2 set of simultaneous linear equations for u_j and v_j, which can be solved in the usual way to yield the frequency and amplitudes.

2.16   Extend the model of a monatomic crystal to include springs between next-nearest neighbours. Let the spring constants between nearest and next-nearest neighbours be K_1 and K_2. Show that

$$m\omega_q^2 = 2K_1(1 - \cos qa) + 2K_2(1 - \cos 2qa). \qquad \text{(E2.2)}$$

This result can be generalized to generate a Fourier series for ω_q^2.

2.17   The diatomic chain whose vibrations were considered in Section 2.8.3 had identical springs but different masses. A model with alternating spring constants but the same mass might be more appropriate for elements like Si. Calculate the dispersion relation and show that it exhibits the same general features as in Figure 2.23(b).

2.18   The diatomic chain becomes monatomic in the limit $M \to m$. How is this reflected in the dispersion relation?

2.19   Young's modulus for GaAs is $8.5 \times 10^{10}\,\mathrm{N\,m^{-2}}$. Use this to estimate the spring constant and the dispersion curve for phonons. How well does it agree with Figure 2.25?

3 | HETEROSTRUCTURES

This chapter provides a review of the general properties of *heterostructures*, semi-conductors composed of more than one material. Variations in composition are used to control the motion of electrons and holes through *band engineering*. Knowledge of the alignment of bands at a heterojunction, where two materials meet, is essential but has proved difficult to determine even for the best-studied junction, GaAs–Al$_x$Ga$_{1-x}$As. Although effort was initially concentrated on materials of nearly identical lattice constant, current applications require properties that can be met only by mismatched materials, giving *strained layers*.

A huge variety of devices has been fabricated from heterostructures, for both electronic and optical applications, and we shall survey these before studying them in greater detail in later chapters. Finally, we shall look briefly at the effective-mass approximation. This is a standard simplification that allows us to treat electrons as though they are free, except for an effective mass, rather than using complicated band structure. It means, for example, that an electron in a sandwich of GaAs between Al$_{0.3}$Ga$_{0.7}$As can be treated as the elementary problem of a potential well.

We shall neglect the random nature of alloys such as Al$_x$Ga$_{1-x}$As, where there is assumed to be no ordering of the Ga and Al ions over the cation sites of the lattice. In principle Bloch's theorem does not apply to such materials because they lack translational invariance from cell to cell. Fortunately it turns out to be adequate to treat the alloy as a crystal whose properties can be interpolated between those of GaAs and AlAs. This approach has been given the dignified name of *virtual-crystal approximation*. The deviation from perfect periodicity leads to effects such as alloy scattering, a reduction of the mean free path of electrons and holes. These are usually small and can be treated within perturbation theory, in the same way as scattering from other kinds of impurity (Section 8.2).

3.1 General Properties of Heterostructures

A wide range of III–V materials has been investigated for their semiconducting properties, although only a few are commonly used in heterostructures. To increase

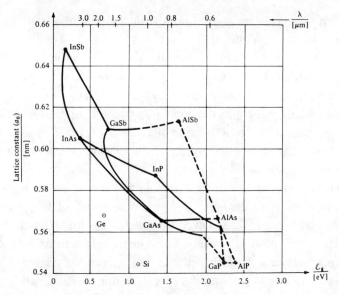

FIGURE 3.1. Plot of the lattice constant of various semiconductors against their minimum band gap E_g, expressed in eV and as a wavelength. Full lines show a direct band gap, with dashed lines for an indirect gap. [From Gowar (1993).]

the range of properties, alloys between the various compounds are also widely used, notably the alloy $Al_x Ga_{1-x} As$, which is usually abbreviated simply to AlGaAs – not to be read as a literal chemical formula! Two of the important properties of different materials, the minimum band gap and the lattice constant, are plotted in Figure 3.1. The band gap expressed as a wavelength is significant because many of the applications of the III–V materials are in optoelectronics, where particular wavelengths are desirable to make the best use of media such as optic fibres.

The active regions of heterostructures are typically at or close to interfaces. This is true not only for III–V compounds; electrons in a metal–oxide–semiconductor field-effect transistor (MOSFET), perhaps the most economically important semiconducting device, move along an interface between Si and SiO_2. In this case the Si is crystalline but the SiO_2 is amorphous (non-crystalline). It is impossible to join the two materials in a seamless way, and electrons are scattered by the resulting imperfections as they travel along the interface. This *surface-roughness scattering*, as well as scattering from charged defects in the oxide, keeps the mobility of electrons in a MOSFET at low temperature below about $4\,m^2\,V^{-1}\,s^{-1}$. III–V heterostructures must have nearly perfect interfaces if they are to perform better. This makes stringent demands on growth, which have been satisfied by the well-developed but expensive processes of molecular-beam epitaxy (MBE) and metal-organic chemical vapour deposition (MOCVD), to be described in the next section.

In principle it must be possible to join the two materials perfectly in an ideal heterostructure. This requires first that they have the same crystal structure (or at least symmetry), and this is satisfied for the common III–V compounds. A second requirement, if there is to be no strain in the final structure, is that the two lattice constants must be nearly identical.

The lattice constant of an alloy is usually given by linear interpolation between its constituents. This is known as *Vegard's law* and predicts, for example, that the lattice constant of $Al_xGa_{1-x}As$ is given by $xa_{AlAs} + (1 - x)a_{GaAs}$. A glance at Figure 3.1 immediately explains the popularity of these alloys: the lattice constant changes by less than 0.15% as a function of x. Thus it is possible to grow layers of GaAs and AlAs or any of the intermediate alloys $Al_xGa_{1-x}As$ on top of one another without 'significant' stress.

Unfortunately there are few other materials that can be grown on a GaAs substrate without strain. An alternative is to use a substrate of InP. The two alloys $Al_{0.48}In_{0.52}As$ and $Ga_{0.47}In_{0.53}As$ have the same lattice constant as InP, with direct gaps, and trap carriers more effectively. These materials are therefore used for high-speed electronic devices. A much wider choice becomes available if the demand of lattice matching is relaxed, and the special features of strained layers will be considered in Section 3.6.

3.2 Growth of Heterostructures

As we saw in the previous section, heterostructures must have interfaces of particularly high quality if they are to perform well: the atomic structures of the two materials must match one another, and the interface must not be contaminated with impurities or suffer from other defects. In addition the layers may be thin so we must be able to change the composition of successive layers very rapidly, preferably from one monolayer to another. Older methods of growing semiconductors such as liquid-phase epitaxy meet some of these criteria and have been used for coarser structures such as double-heterostructure lasers, but more specialized processes are generally needed. The most widespread methods are molecular-beam epitaxy and metal-organic chemical vapour deposition.

3.2.1 MOLECULAR-BEAM EPITAXY

Molecular-beam epitaxy or MBE is in principle a rather simple technique. The apparatus is sketched (with vast simplification!) in Figure 3.2. The substrate, on which the heterostructure is to be grown, sits on a heated holder in an evaporator that is evacuated to an ultrahigh vacuum (UHV), typically of 5×10^{-11} mbar or better (atmospheric pressure is about 1000 mbar). The elements that compose the heterostructure – Ga, As, and Al here – are vaporized in furnaces with orifices directed towards the substrate, but shielded from it by shutters. At this low pressure, the mean free path of molecules between collisions is much larger than the width of the chamber. This is the Knudsen or molecular-flow regime of a gas, and the furnaces are called Knudsen

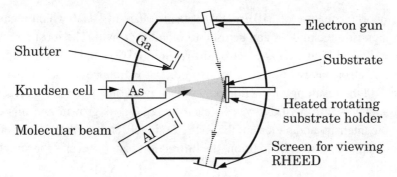

FIGURE 3.2. Highly simplified schematic diagram of an MBE machine showing three K-cells, the substrate on its rotating heated holder, and RHEED for analyzing the surface.

or K-cells. Molecules that emerge from the K-cells do not diffuse as in a gas at high pressure but form a molecular beam, travelling in straight lines without collisions until they impinge on the substrate or elsewhere. Growth commences when the shutters are opened, and the flux of each element can be controlled through the temperature of each furnace.

Dopants are added by using additional cells. The usual donor is Si. This is in group IV of the periodic table, so it is not obvious whether it should act as a donor or acceptor in a group III–V compound. In practice it usually acts as a donor, but this can be changed by growing on a different surface from the usual (100). It also tends to become amphoteric, giving both donors and acceptors, at very high concentrations (around 10^{25} m^{-3}). The usual acceptor is Be.

Although simple in principle, a few considerations show that MBE is not simple in practice. The wafers must be grown with extreme purity if their properties are not to be spoilt by contamination. This in turn requires pure starting materials, which must not be polluted by the K-cells. The background pressure in the evaporator must be kept low to reduce contamination, as well as to ensure that the Knudsen regime holds. The flux from the K-cells must be uniform across the substrate, or there will be variations in composition across the wafer (minimized by rotating the sample holder during growth); the temperatures of the furnaces must also be closely controlled to keep the flux constant. The temperature of the substrate is important: defects will not have time to be removed by annealing at low temperatures, while unwanted diffusion will occur and blur the interfaces if the temperature is too high. The morphology of the surface depends in a complicated way on temperature, and AlGaAs and GaAs grow better under different conditions.

MBE is a slow process, growing material at about 1 monolayer per second or 1 μm per hour. Some of these problems are analyzed in the exercises. Advantages of MBE include highly abrupt junctions between different materials, good control over the thickness of layers, and reasonable reproducibility. Obvious disadvantages include the cost and the fact that the process does not scale well for production.

An important feature of MBE is that it takes place in UHV, which means that many diagnostic techniques can be used to monitor growth. The most common of these is reflected high-energy electron diffraction (RHEED), shown in Figure 3.2. A beam of electrons is directed at nearly grazing incidence to the surface of the sample, and the resulting diffraction pattern is viewed on a fluorescent screen. The surface changes in a periodic way as each monolayer is grown, and this can be seen in the intensity and pattern of the RHEED signal. Thus growth can be counted precisely in monolayers. In addition, the diffraction pattern reveals the structure of the surface.

MBE may produce the best material for electronic applications. One measure of this is the peak mobility of electrons in nominally pure GaAs, which indicates a net concentration of ionized impurities $|N_D - N_A| < 5 \times 10^{19} \, \mathrm{m}^{-3}$. Another test is the mobility of electrons in a two-dimensional electron gas, which has exceeded $1000 \, \mathrm{m}^2 \, \mathrm{V}^{-1} \, \mathrm{s}^{-1}$, corresponding to a mean free path of nearly $0.1 \, \mathrm{mm}$. Some important problems remain, however; for example, it is relatively easy to grow a good interface of AlGaAs on top of GaAs, but the 'inverted' interface of GaAs on AlGaAs is problematic. Other materials, even the widely used $In_x Ga_{1-x} As$, are much less well controlled.

3.2.2 METAL-ORGANIC CHEMICAL VAPOUR DEPOSITION

Metal-organic chemical vapour deposition, or MOCVD (also known as metal-organic vapour-phase epitaxy or MOVPE), is another method capable of producing heterostructures of high quality. Figure 3.3 shows a highly simplified diagram of the apparatus, which usually operates near atmospheric pressure. The substrate sits on a heated block in a chamber through which different gases are passed in a carrier of hydrogen; the composition of the gases can be varied rapidly to control the composition of the material grown. The basic reaction is between a metal alkyl and a hydride of the group-V material, such as

$$(CH_3)_3 Ga + AsH_3 \overset{650°C}{\rightarrow} GaAs\downarrow + 3 \, CH_4. \tag{3.1}$$

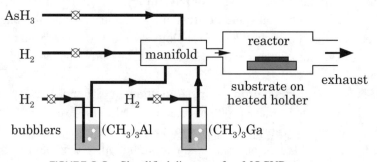

FIGURE 3.3. Simplified diagram of an MOCVD reactor.

Donors such as Si, S, or Se can be added as hydrides, and dimethyl Zn or Cd acts as an acceptor. Many other chemistries are in use, not least to avoid the poisonous group-V hydrides.

The apparatus is much simpler than that required for MBE, although a major practical problem is the handling of the highly toxic gases (which raises the overall cost to a similar level). Difficulties arise from the aerodynamics inside the reactor, where a stagnant boundary layer tends to form over the substrate. Depending on the temperature, growth may be limited by the rate at which reactants diffuse through this boundary layer, or by the rate of reaction on the surface of the sample. The volume of the system between the mixing of the gases and the substrate must be minimized to change the composition of the gas, and the resulting semiconductor, rapidly and give sharp interfaces. Further, the mixture of gases must be changed without affecting the flow through the reaction chamber.

MOCVD has a reputation for producing better optoelectronic devices than MBE. It is also faster and has been successfully scaled up for commercial production, growing on a large number of wafers simultaneously. Regrowth, where a substrate is patterned in some way and subsequently returned for further growth, is particularly successful with MOCVD. Spectacular structures have also been grown using substrates with exposed facets of different planes, where the dependence of growth rate on orientation as well as temperature has been exploited to produce 'self-organized' devices. Difficulties associated with MOCVD include contamination by carbon and the serious practical issue of safety.

A wide range of methods have blossomed from basic MBE and MOCVD. One example is chemical-beam epitaxy (CBE) or metal-organic molecular-beam epitaxy (MOMBE), where the solid sources of conventional MBE are replaced by similar precursors to those used in MOCVD. Another is atomic-layer epitaxy (ALE), where the reactants are supplied separately, forcing growth to occur layer by layer. All are aimed at improving control of the thickness and composition of layers to improve the performance of the resulting devices. The same methods are also used for the production of Si–Ge heterostructures.

3.3 Band Engineering

The whole point of growing heterostructures, which we have seen to be a complicated process, is the opportunity that they offer to manipulate the behaviour of electrons and holes through *band engineering*. Figure 3.1 showed that the different materials have different band gaps; now we need to look in more detail at the conduction and valence bands themselves, E_c and E_v. Consider a heterojunction between two materials A and B, with $E_g^A < E_g^B$.

The simplest theory yields *Anderson's rule*. This is based on the electron affinity χ of the materials, the energy required to take an electron from the bottom of the

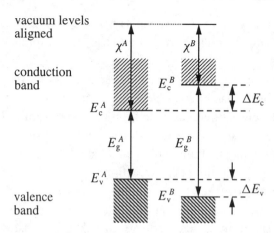

FIGURE 3.4. Anderson's rule for the alignment of the bands at a heterojunction between materials *A* and *B*, based on aligning the vacuum levels.

conduction band E_c to the vacuum level where it can escape from the crystal. The electron affinity is nearly independent of the position of the Fermi level, unlike the work function, which is measured from the Fermi level and therefore depends strongly on doping. Anderson's rule states that the vacuum levels of the two materials of a heterojunction should be lined up, as in Figure 3.4. This shows immediately that $\Delta E_c \equiv E_c^B - E_c^A = \chi^A - \chi^B$. For example, GaAs has $\chi = 4.07\,\text{eV}$ and $\text{Al}_{0.3}\text{Ga}_{0.7}\text{As}$ has $\chi = 3.74\,\text{eV}$, predicting $\Delta E_c = 0.33\,\text{eV}$. The band gap changes by $\Delta E_g = 0.37\,\text{eV}$, so $\Delta E_v = 0.04\,\text{eV}$. The fraction of the band gap that has gone into the conduction band $Q = \Delta E_c / \Delta E_g \approx 0.85$ according to this model.

Although it is easy to measure the difference in band gap, it is much more difficult to resolve the discontinuity in the individual bands. The value $Q = 0.85$ just given was long used for $\text{Al}_x\text{Ga}_{1-x}\text{As}$, but the smaller value $Q = 0.62$ is now firmly established for $x < 0.45$ where the gap remains direct. Many experiments, such as optical absorption in rectangular quantum wells, prove to be only weakly sensitive to Q. Other shapes of quantum wells, such as parabolic potentials (Figure 4.5), give better resolution.

In the case of GaAs–AlGaAs, the narrower band gap is enclosed within the wider band gap, as shown in Figure 3.4. Thus a sandwich with GaAs as the filling between two layers of AlGaAs traps both electrons and holes. This is called *type I* or *straddling* alignment. There are two other possibilities, illustrated in Figure 3.5 for two systems of three closely lattice-matched materials. The band gaps and discontinuities are shown; the discontinuity is positive if it has the sign expected for a type I interface. InP and $\text{In}_{0.52}\text{Al}_{0.48}\text{As}$ show *type II* or *staggered* alignment, where the bands favour electrons in InP but holes in $\text{In}_{0.52}\text{Al}_{0.48}\text{As}$. Thus a sandwich results in a potential well for one species but a barrier for the other.

The third possibility is that the two band gaps do not overlap at all, giving *type III* or *broken-gap* alignment. The classic example is InAs–GaSb shown in

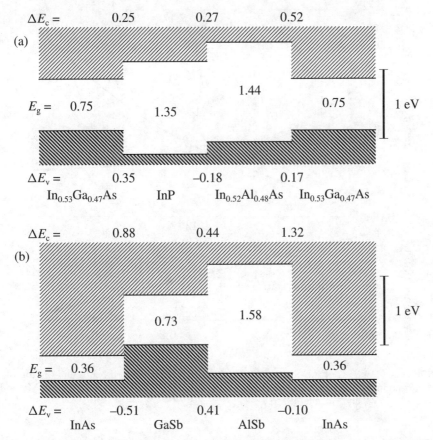

FIGURE 3.5. Alignment of conduction and valence bands in two lattice-matched systems of semi-conductors: $In_{0.53}Ga_{0.47}As$–$In_{0.52}Al_{0.48}As$–InP, which is widely used for high-speed electronic devices; and InAs–GaSb–AlSb, which shows types I, II, and III of matching. The convention is that ΔE_c, $\Delta E_v > 0$ for type I. [Redrawn from Yu, McCaldin, and McGill (1992) and Frensley (1994).]

Figure 3.5(b). In this case the conduction band of one material overlaps the valence band of the other, by 0.15 eV for this example. Electrons and holes therefore transfer spontaneously until they are restrained by the electronic field generated, much as in the depletion region of a p–n diode.

This pair of materials has another interesting feature, because the interface between InAs and GaSb may be a monolayer of either GaAs or InSb, depending on the precise order of growth. This may affect the alignment of the bands, and it has been suggested that the overlap of conduction and valence bands is 0.125 eV for a 'GaAs' interface and 0.160 eV for an 'InSb' interface. Less controlled growth leads to an average of these.

A further complication is that the components of a heterostructure may have minima in the bands at different points in the Brillouin zone. For example, the conduction band of GaAs has its minimum at Γ, whereas the lowest minima in AlAs are near X. The character of the lowest minimum therefore changes as the

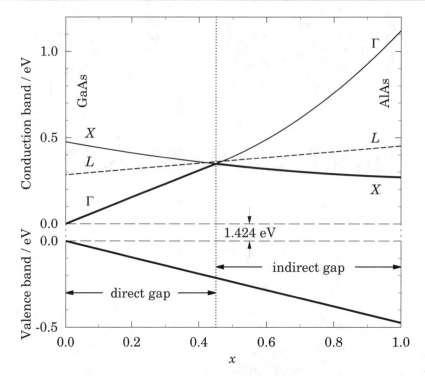

FIGURE 3.6. Energies of the bottoms of the three lowest conduction bands and the top of the valence band in $Al_xGa_{1-x}As$ as a function of the composition x. [Redrawn from Yu, McCaldin, and McGill (1992) and Adachi (1985).]

fraction x of Al in the alloy $Al_xGa_{1-x}As$ increases. Figure 3.6 shows the extrema as a function of x. The lowest minimum in the conduction band changes from Γ (direct gap) to X (indirect gap) at $x \approx 0.45$, where ΔE_c takes its maximum value of 0.35 eV. The effective mass and valley degeneracy vary abruptly at this crossover, and the change from a direct to an indirect energy gap gives a profound change in the optical properties. Note that the L minima are not far above the others near the crossover. Usually $Al_xGa_{1-x}As$ is grown with $x < 0.4$ to ensure that the lowest conduction band remains at Γ. The behaviour for $x > 0.45$ is less well established.

3.4 Layered Structures: Quantum Wells and Barriers

We have now seen how the conduction and valence bands can be engineered during growth to vary in one dimension. These layered structures are the building blocks of more complicated devices. We shall now briefly survey some of the simpler profiles that will be studied later. Most will be discussed in relation to electrons in the conduction band but work in a similar way for holes, subject to the complications of the valence band. Even the conduction band has complications when its nature changes in $Al_xGa_{1-x}As$ with $x > 0.45$, and this will be discussed at the end of the section.

3.4.1 TUNNELLING BARRIER

A simple example of a tunnelling barrier is provided by a layer of AlGaAs surrounded by GaAs (Figure 3.7(a)). The rectangular barrier is one of the elementary examples used to illustrate quantum mechanics in textbooks, and here it is in real life! Classically, an electron would not be able to pass the barrier unless it had sufficient kinetic energy to pass over the top, but in quantum mechanics it is able to tunnel through the barrier. Even a simple barrier such as this has practical uses, for example, as a 'throttle' to control the injection of electrons in some hot-electron transistors. The transmission coefficient increases rapidly with energy in the tunnelling regime, so the barrier selects electrons of higher energy.

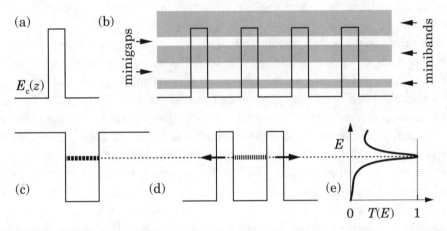

FIGURE 3.7. Profile of the conduction band $E_c(z)$ for various layered structures: (a) tunnelling barrier; (b) superlattice, showing the miniband structure; (c) quantum well, showing a bound state; (d) double barrier, where the state is now resonant rather than bound; (e) transmission coefficient as a function of energy for tunnelling through a double barrier, showing a peak at the energy of the resonant state.

3.4.2 QUANTUM WELL

The opposite of the barrier is a *quantum well*, such as a thin layer of GaAs sandwiched between two thick layers of AlGaAs (Figure 3.7(c)). We shall calculate in Section 4.2 the energies of these bound states as a function of the depth of the well, although the rough estimate for an infinitely deep well derived in Section 1.3 is often used.

There is a well in both conduction and valence bands if the heterojunction is of type I, which includes the important case of GaAs surrounded by AlGaAs. In this case the well traps both electrons and holes and the energy levels can be measured by optical experiments, as discussed in Section 1.3.1. This process forms the basis of most optoelectronic devices, and Chapter 10 is concerned with the details.

In fact one must remember that the material is three-dimensional, and Figure 3.7(c) shows only the variation along the direction of growth, conventionally taken as the z-direction. Electrons and holes remain free to move in the plane normal to growth, but they now have only two free dimensions rather than three. Optical absorption is governed by the density of states, which is stronger near the bottom of the band in two dimensions and helps to provide more-efficient optoelectronic devices.

3.4.3   TWO BARRIERS: RESONANT TUNNELLING

A quantum well has infinitely thick barriers on either side. New physics enters if the barriers are thin, as in Figure 3.7(d). Now an electron in the state can 'leak out', giving a *quasi-bound* or *resonant* state rather than a true bound state. This double-barrier structure provides the active region of a *resonant-tunnelling diode* and is a close analogue of an optical Fabry–Pérot etalon.

Consider electrons that impinge on the barrier from the left at an energy E. Their probability of passing through the barrier is the transmission coefficient T. In most cases this is roughly the product of the transmission coefficients of the two individual barriers: $T \approx T_L T_R$. Both T_L and T_R are small for typical barriers. The distinctive feature of *resonant tunnelling* is that T rises to much higher values than $T_L T_R$ for energies near the resonance, as shown in Figure 3.7(e), in the same way that multiple reflection leads to peaks in transmission through a Fabry–Pérot etalon. In a structure with identical barriers there is perfect transmission at the centre of the resonance, however small the individual values of T_L and T_R. The width of this peak is proportional to $T_L + T_R$, so in some sense there is less transmission through more opaque barriers. We shall calculate these results in detail in Section 5.5. Resonant-tunnelling diodes have been developed for use at extremely high frequencies.

3.4.4   SUPERLATTICE

A logical step beyond the double-barrier structure is to increase the number of barriers to give alternating layers of wells and barriers, as shown in Figure 3.7(b). Now an electron can tunnel from one well to its neighbours, rather than from a well to outside as in the double barrier. Again this destroys the sharp bound level associated with the well, but in this case the result is a *miniband*. The superlattice gives a periodic potential, so the general theory of one-dimensional crystals (Section 2.1) holds. Thus motion along the direction of growth is governed by Bloch's theorem and band structure.

The structure is called a *superlattice* because it is a second level of periodicity imposed on the first level, which is the crystalline nature of the semiconductors.

A major difference is that it affects motion in only one direction. The period is of course longer than that of the crystal and the periodic potential is weaker, with the result that the bands and gaps appear on a much smaller scale of energies, hence the name of minibands. The band structure can be tuned by varying the composition and thickness of the layers. Superlattices have been used to filter the energy of electrons, allowing only those within the minibands to pass or reflecting those in the minigaps, and absorption between the minibands can be used to detect infrared radiation.

The bands become narrower as the barriers between wells become thicker, and tunnelling between wells has less effect on the overall properties. The limit in which wells are practically isolated is called a *multiquantum well* (MQW). Such structures are widely used in optical applications to increase the active volume of the device.

In fact superlattices are widely used in an application that has nothing to do with their electronic properties. This is to improve the cleanliness of material during growth. It is common practice to grow a superlattice on top of the original substrate, then a buffer of GaAs, and finally the desired structure. The numerous interfaces trap many defects and impurities that would otherwise migrate up with growth and contaminate devices.

3.4.5 NATURE OF THE CONDUCTION BAND

Figure 3.6 showed the energies of the three lowest minima in the conduction band of $Al_xGa_{1-x}As$ as a function of x. The lowest minimum is at Γ for $x < 0.45$ but lies at X for larger x. The profiles of the three minima are shown in Figure 3.8 for a barrier of $Al_{0.3}Ga_{0.7}As$ and one of AlAs, both surrounded by GaAs. This illustrates the problems that arise when the two regions have different minima in the conduction band.

The Γ minimum is the lowest throughout the structure with the barrier of $Al_{0.3}Ga_{0.7}As$ and it is reasonable to neglect the higher bands, provided that the

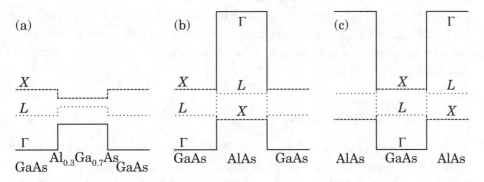

FIGURE 3.8. Barriers and wells in GaAs–$Al_xGa_{1-x}As$, showing the three lowest conduction bands. (a) Barrier of $Al_{0.3}Ga_{0.7}As$, where Γ is the lowest minimum throughout. (b) Barrier of AlAs, where X is the lowest minimum in the barrier. (c) Well of GaAs surrounded by AlAs.

energy of the incident electron is well below the top of the barrier. There is clearly a problem for the AlAs barrier, though. The sandwich appears as a barrier if we consider E_Γ, the lowest minimum outside, but is a well from the point of view of E_X, which lies lowest inside the barrier. It is also a barrier if we simply consider the lowest minimum in the conduction band, but what should we take for its height? This is not a simple problem to answer, and requires band-structure calculations. Although the barrier is lower if an incident electron transfers from Γ to X when it hits the barrier, the effective mass is higher in X, so the wave function may decay more rapidly. Thus the transmission coefficient may be dominated by E_Γ although it is not the lowest band. Transfer from Γ to X is also subject to restrictions in **k**-space, which will be discussed in Section 5.8.1. These problems are often avoided by keeping $x < 0.4$, although barriers of AlAs may be used to avoid alloy scattering, particularly in resonant tunnelling.

Similar problems afflict wells, such as that of GaAs between AlAs barriers shown in Figure 3.8(c). The lowest minimum in the well is Γ, but any mixing into an X-valley at the interface will allow states to leak out if $E > E_X^{AlAs}$, and this minimum barrier therefore sets the limit on true bound states in the well.

3.5   Doped Heterostructures

Devices such as the resonant-tunnelling diode discussed before rely on electrons or holes for conduction and these must be introduced by doping. The same is true in optoelectronics, where a quantum well might be embedded in a p–n junction to form a quantum-well laser. We must therefore consider the effect of doping on band diagrams. The general principles follow those for homostructures, but heterostructures give further control over the positions of impurities and the carriers that they release.

Also, the structures discussed already are *vertical*, in the sense that current flows along the direction of growth or normal to the interfaces. The opposite case is obviously *horizontal*, where current flows parallel to an interface. This is similar to the MOSFET, and a major application is to heterostructure field-effect transistors. These use a two-dimensional electron gas, which is the foundation for many electronic structures.

3.5.1   MODULATION DOPING

The obvious way of introducing carriers, used in classical devices, is to dope the regions where electrons or holes are desired. Unfortunately, charged donors or acceptors are left behind when electrons or holes are released, and scatter the carriers through their Coulomb interaction (ionized-impurity scattering). This spoils propagation within the carefully grown structure, blurring energy levels and disrupting

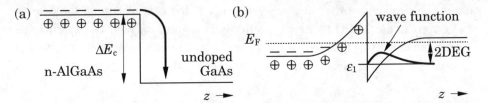

FIGURE 3.9. Conduction band around a heterojunction between n-AlGaAs and undoped GaAs, showing how electrons are separated from their donors to form a two-dimensional electron gas.

the interference between electron waves needed to see effects such as resonant tunnelling.

The solution is *remote* or *modulation doping*, where the doping is grown in one region but the carriers subsequently migrate to another. This is illustrated for a heterojunction between n-AlGaAs and undoped GaAs in Figure 3.9. The material is neutral everywhere and the bands are flat if the electrons sit on their donors in the n-AlGaAs (Figure 3.9(a)). The electrons travel around after being released and some cross into the GaAs. There they lose energy and become trapped because they cannot climb the barrier presented by ΔE_c. This motion separates the negatively charged electrons from their positively charged donors, which sets up an electrostatic potential $\phi(z)$ that tends to drive the electrons back into the AlGaAs.

The total energy for electrons is the sum of two terms. Their kinetic energy is given by the band structure. This does not vary with position inside each material but changes at the heterojunction, the main effect being the discontinuity ΔE_c. The second term is the potential energy due to the electrostatic potential, $-e\phi(z)$ for electrons. This must be added to the kinetic energy to give the total energy at the bottom of the conduction band $E_c(z)$, shown in Figure 3.9(b). The arguments are exactly the same as those for the band diagram of an ordinary p–n junction, except for the crucial extra ingredient of ΔE_c. This discontinuity has a major effect because it prevents the electric field from returning the electrons to their donors; the field can only squeeze the electrons against the interface, where they are trapped in a roughly triangular potential well. This well is typically about 10 nm wide at the energy of the electrons, and the energy levels for motion along z are quantized in a similar way to those in a square well. Often only the lowest level is occupied. All electrons then occupy the same state for motion in z, but remain free in the other two dimensions x and y. This is the *two-dimensional electron gas* or 2DEG. It is the basis for the majority of electronic devices in heterostructures, occupying much the same position as the quantum well for optical devices.

Thus modulation doping has achieved two benefits: it has separated electrons from their donors to reduce scattering by ionized impurities, and confined the electrons to two dimensions. A refinement made by leaving a spacer layer of undoped AlGaAs between the n-AlGaAs and GaAs increases the separation between electrons and donors. This reduces further the scattering, but at the cost of cutting the density of electrons in the 2DEG. A high mobility is vital in many physics experiments, whereas the density of electrons is unimportant, so a thick spacer is often used, but the requirements in engineering are usually the opposite.

3.5.2 CONSTRUCTION OF BAND DIAGRAMS

We shall now look a little more closely at the construction of the band diagram in Figure 3.9, and see how it may be generalized to other systems. Consider a junction between n-AlGaAs (material A) and p-GaAs (material B). Here is a recipe for sketching the band diagram, shown in Figure 3.10. It is very close to the method for a doped homostructure, but we need to take account of the discontinuities ΔE_c and ΔE_v.

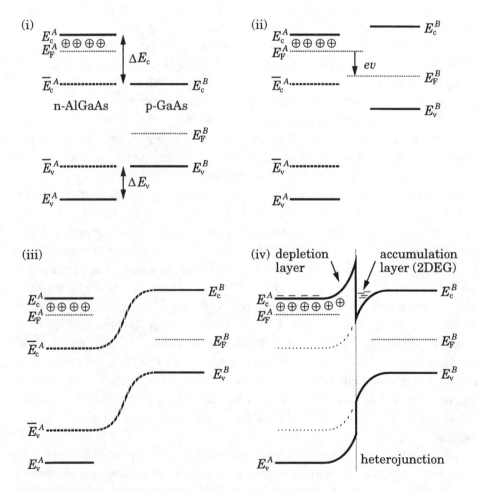

FIGURE 3.10. Steps in the construction of the band diagram for a doped heterojunction between material A, n-type AlGaAs, and material B, p-GaAs.

(i) Start with flat bands in each material, with the bands in their natural alignment (given by Anderson's rule or something superior) and the Fermi levels set by the doping on each side. This gives the position of the bands relative to the Fermi level far from the junction on each side. To cancel out the effect of the discontinuities temporarily, draw lines on side A at $\bar{E}_c^A = E_c^A - \Delta E_c$ and $\bar{E}_v^A = E_v^A + \Delta E_v$. Note that $\bar{E}_c^A - \bar{E}_v^A = E_c^B - E_v^B = E_g^B$, so the 'effective' band gap is the same on both sides.

(ii) Align the Fermi levels. The difference in Fermi level far from the junction is set by any applied voltage. Assume that there is a positive bias v applied to side B, so $E_F^A - E_F^B = ev$.

(iii) Join $\bar{E}_c^A$ to E_c^B, and $\bar{E}_v^A$ to E_v^B, with parallel curves due to the electrostatic potential. The precise form of these curves must generally be found numerically. For a qualitative picture, sketch an S-shaped curve whose curvature is set by the sign of the charge density. Usually this gives a point of inflexion at the junction.

(iv) Now restore E_c^A on side A as a line at $\bar{E}_c^A + \Delta E_c$ and E_v^A at $\bar{E}_v^A - \Delta E_v$, including the discontinuities in E_c and E_v at the junction. This completes the sketch of the band diagram.

The bands are bent near the heterojunction as they would be in a p–n diode, but the discontinuities in the bands add the new feature that carriers may be trapped in a potential well next to the junction. There is an accumulation layer of electrons on the p-type side in the preceding example, unlike the p–n homojunction where both sides are depleted.

Thought needs to be given to the form of the electrostatic potential. For example, the potential is parabolic in a typical depletion region with constant doping, as in an abrupt p–n diode. On the other hand, heterostructures often contain an undoped spacer in which there is negligible charge density; the electric field must be constant and the bands slope linearly as in a p–i–n diode. An accurate treatment requires a self-consistent numerical solution of Poisson's equation for the electrostatic potential and Schrödinger's equation for the wave functions and energy levels, with the Fermi–Dirac distribution to find the occupation of the levels. This is not a major problem by contemporary standards, and will be considered in detail in Section 9.3, but the qualitative procedure described already is often adequate.

3.5.3 MODULATION-DOPED FIELD-EFFECT TRANSISTOR

The density of a 2DEG can be controlled, as in a MOSFET, by forming a capacitor between the 2DEG and a metallic gate. We shall review this in detail in Section 9.1. Adding a source and drain completes a field-effect transistor (FET) called a modulation-doped field-effect transistor (MODFET). A simple structure

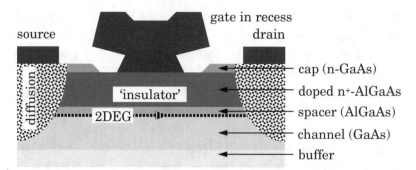

FIGURE 3.11. Simplified cross-section through a high-frequency GaAs–AlGaAs MODFET.

for a high-speed device is shown in Figure 3.11. The insulator between the gate and channel is AlGaAs in this structure. It presents a rather low barrier to electrons in the GaAs channel, and other combinations of materials are now preferred. The thin layer of GaAs on the surface prevents oxidation of the AlGaAs underneath. There is an undoped spacer of AlGaAs above the channel, kept very thin to raise the density of electrons; mobility is a secondary consideration. The T-shape of the gate is to reduce its resistance while keeping its contact with the channel short, and it sits in a recess that brings it closer to the channel.

Real devices have many refinements to control parasitic effects and improve ease of manufacture but these basic elements remain. MODFETs have found favour in high-frequency applications such as direct-broadcast satellite receivers, largely because of their low noise. Difficulties arise from the small barrier presented by ΔE_c in GaAs–AlGaAs. Although $In_{0.53}Ga_{0.47}As$–$In_{0.52}Al_{0.48}As$ on InP is better, performance can be further improved by abandoning the restriction to lattice-matched structures. We shall therefore consider strained active layers next.

3.6 Strained Layers

Most of the specific examples used in this book are taken from the GaAs–AlAs material system because it is the best developed and currently dominates physics experiments. Unfortunately its properties are inadequate for many practical applications. For example, the barrier retaining electrons in GaAs is typically only 0.25 eV, which is too small to confine energetic electrons in the channel of a short FET. In optical applications the band gap of 1.4 eV corresponds to a wavelength of about 0.9 μm, in the near infrared, which is unsuitable for driving long-haul optical fibres where wavelengths of 1.3 μm or 1.55 μm are desired.

A feature of GaAs and AlAs is that they have nearly the same lattice constant, so the microscopic structure of each material is not distorted in a heterostructure. There are few pairs of materials like this, and our choice is restricted to particular compositions of alloys if we insist that the lattice constants be identical. A practical example is $In_{0.53}Ga_{0.47}As$ and $In_{0.52}Al_{0.48}As$, which are matched to a substrate of InP. A much wider range of materials opens up if the restriction of equal lattice constant is relaxed. For example, an attraction of $In_{0.53}Ga_{0.47}As$–$In_{0.52}Al_{0.48}As$ for electronic systems is the large value of ΔE_c and small m^* for electrons in $In_{0.53}Ga_{0.47}As$. Raising the fraction of indium above 0.53 further improves both properties, at the cost of introducing strain. Adjusting the fraction of indium also allows the band gap to be matched to the needs of optical fibres. A further strained system is provided by $Si_{1-x}Ge_x$ alloys, the inclusion of silicon giving obvious commercial potential.

There are three main reasons for using strained layers:

- They broaden the range of materials available for tuning the properties of interest, such as band offsets and effective masses.
- Strain has strong effects on the valence band and therefore provides another tool for band engineering.
- They may allow the growth of desired layers on convenient but usually mismatched substrates.

All play a role in many devices. Concern has been voiced about the long-term stability of strained layers, and the growth of many materials remains beset by problems, but they have gained widespread use.

3.6.1 STRUCTURAL ASPECTS OF STRAINED LAYERS

Two extreme outcomes are possible if a material is grown on one with a different lattice constant, say, InGaAs on GaAs. In equilibrium InGaAs has a larger lattice constant (Figure 3.12(a)). Assume that the GaAs substrate is so thick that it cannot be distorted significantly. If the InGaAs is thin, it can strain to conform to the GaAs in the plane of the junction as in Figure 3.12(b). Thus its lattice constant in the plane is

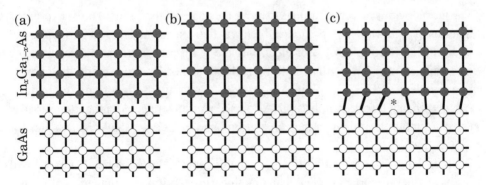

FIGURE 3.12. Growth of $In_xGa_{1-x}As$ on a GaAs substrate. (a) Separate layers at equilibrium. (b) Thin layer of InGaAs on GaAs. The InGaAs is strained to conform to the lattice constant of GaAs in the plane of the heterojunction. (c) Thicker layer, where the strain has relaxed due to a misfit dislocation at the heterointerface, shown by an asterisk.

reduced to that of GaAs, while the usual elastic response causes it to extend along the direction of growth. The InGaAs is severely distorted and elastic energy builds up. The stress is enormous in comparison with those tolerated by macroscopic samples (even GaAs–AlAs structures, normally regarded as 'lattice-matched', contain large stresses by usual standards). Such stress can be tolerated only in a thin film, and relaxation occurs in a thick film, so the InGaAs takes up its natural lattice constant. Perfect matching of the atoms at the heterojunction, a basic assumption of our previous discussion, is no longer possible. Instead the difference in lattice constants is taken up by misfit dislocations, shown in Figure 3.12(c).

Many bonds are broken when the dislocations form, and this requires an energy proportional to the area of the sample. The elastic energy of a strained layer, on the other hand, is also roughly proportional to its thickness. There is therefore a critical thickness beyond which it is energetically favourable for dislocations to appear. This balance of energies between strain and the creation of dislocations gives a simplified version of the *Matthews–Blakeslee* criterion for the stability of a strained layer.

It appears from Figure 3.12(c) that the defects associated with a relaxed layer are localized in the plane of the interface. This is not the case: dislocations cannot simply begin or end anywhere, and may turn upwards to become *threading* dislocations which terminate on the exposed surface of the sample. These are extremely damaging if they pass through the active region of a device. There is great current interest in growing thick relaxed buffer layers so that cheap and readily available substrates, notably Si or GaAs, can be used under materials of different lattice constant, which are selected for their superior electronic properties. The control of threading dislocations and other defects presents a major technological challenge in these *metamorphic* structures.

3.6.2 EFFECT OF STRAIN ON BAND STRUCTURE

The most obvious effect of strain is to move the edges of the bands. This is expressed in terms of *deformation potentials*, which we shall use to calculate electron–phonon scattering in Section 8.4.1. For electrons, the main use of strain is to control the discontinuity ΔE_c, which needs to be as large as possible to confine the carriers in a real device. This is exploited in *pseudomorphic* field-effect transistors with a strained channel, for example, InGaAs on a GaAs substrate. Complications arise if there are multiple valleys in the conduction band, which we shall discuss later for Si–Ge.

Strain has more dramatic effects on the degenerate valence bands near Γ and opens another important route to band engineering. A simple explanation of the top of the valence bands was given in Section 2.6.3 based on the picture of p_x, p_y, and p_z orbitals. This can be extended to include strain. Cubic symmetry demands that these orbitals, and their resulting bands, be degenerate in the unstrained crystals.

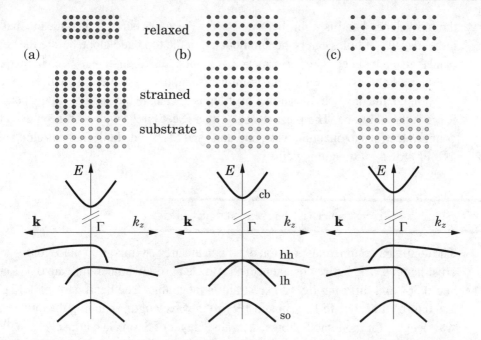

FIGURE 3.13. Effect of strain on the bands of a semiconductor, showing the splitting of the valence band; the band gap is much reduced. The active layer has a smaller lattice constant in (a), is unstrained in (b) to show the usual band structure, and has a larger lattice constant in (c), which applies, for example, to InGaAs on GaAs. The wave vector **k** is in the plane normal to growth, which takes place along z. [Redrawn from O'Reilly (1989).]

The distortion of the strained layer in Figure 3.12(b) reduces its symmetry to tetragonal (fourfold rotational symmetry about the axis of growth). The most significant electronic effect of the strain is to change the energy of the p_z orbital, aligned along the direction of growth, with respect to that of p_x and p_y, which remain degenerate.

The effect of compression of the lattice in the plane of the junction is illustrated in Figure 3.13(c). The energy of p_z falls, so the top valence band arises from p_x and p_y. This is heavy along k_z but has a light component for wave vectors **k** in the xy-plane (recall that $\mathbf{K} = (\mathbf{k}, k_z)$). The band is therefore anisotropic and motion in the plane of the junction is governed by a light mass, improving the mobility of holes. The band from p_z lies farther from the band gap; it is light for k_z and heavy for **k**. Unfortunately the picture is complicated by spin–orbit coupling but this explanation accounts for the main features of Figure 3.13. The opposite ordering of the bands is seen in Figure 3.13(a), where the active layer has a smaller natural lattice constant. An example is Si on Si–Ge.

The bands in a strained quantum well are affected both by the strain and by the confinement. They reduce the symmetry in the same way, and we shall see in Section 10.3 that confinement has much the same effect as the strain in Figure 3.13(c). The two effects add cooperatively in a strained layer of InGaAs between GaAs, and pull the top valence bands farther apart. This in turn extends the range of **k** for which the top valence band is light, before it meets the other band. It also means that a larger density of holes can be accommodated in the top band alone before the next band is also populated. The performance of semiconductor lasers has been improved substantially by using strained quantum wells and harnessing these changes.

Although the most obvious effect of strain is to change the energies of the bands, it is not the only one. The piezoelectric effect (Section 2.5) has also been used to confine electrons to quantum wires and dots, and to provide built-in fields along the axis of growth of quantum wells.

3.7 Silicon–Germanium Heterostructures

The dominant commercial position of silicon ensures an important place for its heterostructures. The lattice constants of Si and Ge at room temperature are 0.543 nm and 0.564 nm, differing by 4%, so strain is inevitable. The band gaps of 1.12 eV and 0.66 eV translate to 1.1 μm and 1.9 μm in wavelength, spanning the important wavelengths for monomode fibres. A strained layer of Si on Ge has an even smaller gap, further extending the range of wavelengths available. Unfortunately the band gap is indirect, so strong emission of light is unlikely but detection is possible, raising the potential of integrated optoelectronic devices.

A wide range of layers has been studied, which can generally be described in terms of a substrate of $Si_{1-y}Ge_y$, which sets the lattice constant in the plane, and an active layer of $Si_{1-x}Ge_x$. The active layer is strained unless $x = y$; both tension and compression are possible.

The effect of strain on the band structure is now even richer. The valence band is split as shown in Figure 3.13 with all three outcomes possible. Both Si and Ge have multiple valleys in the conduction band but these are at different points in **K**-space, X in Si and L in Ge. The minimum switches between these at $x \approx 0.85$ in the *unstrained* alloy $Si_{1-x}Ge_x$. The reduced symmetry of a strained layer breaks the degeneracy of the six X-valleys in silicon-rich alloys. The two valleys along the growth direction (assuming a (001)-substrate) form a pair $\Delta_{\perp}$ that splits from the other four valleys in the plane of the interface, denoted $\Delta_{\parallel}$, which remain degenerate. The four L-valleys in germanium-rich alloys stay equivalent.

Figure 3.14 shows calculations of the offsets in conduction and valence band for an active layer of $Si_{1-x}Ge_x$ that is strained to conform to a substrate of $Si_{1-y}Ge_y$. The offsets are defined by $E_c(x) - E_c(y)$ and $E_v(x) - E_v(y)$, so a positive offset in E_c tends to confine electrons in the substrate, whereas a positive E_v confines holes in the active layer. The discontinuities in the plot of ΔE_c arise from the change in nature of the lowest minima of the conduction band in the two materials; for the active layer it is $\Delta_{\perp}$ when $x < y$ and $\Delta_{\parallel}$ when $x > y$, except when both x and y are near 1, when the L minima enter play. The two discontinuities have the same sign over most of the plot, giving a type II or staggered alignment. Specific alignments

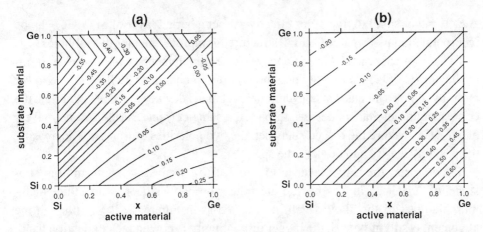

FIGURE 3.14. Offsets in (a) conduction band and (b) valence band between a substrate of $Si_{1-y}Ge_y$ and an active layer of $Si_{1-x}Ge_x$ that is strained to match the substrate. The offsets are defined by $E_c(x) - E_c(y)$ and $E_v(x) - E_v(y)$. [From Rieger and Vogl (1993) with correction.]

for two heterojunctions between Si and $Si_{0.7}Ge_{0.3}$ are shown in Figure 3.15. We shall look at these in a little more detail.

Silicon is the substrate in Figure 3.15(a) with a strained layer of $Si_{0.7}Ge_{0.3}$. Most of the offset is in the valence band, and the type II nature is clear. The natural lattice constant of the active layer is larger than that of the substrate so the valence bands are split according to Figure 3.13(c). This gives a light mass for motion in the plane of the interface. A two-dimensional hole gas (2DHG) can be trapped at this heterojunction, and a mobility of $2\,m^2\,V^{-1}\,s^{-1}$ has been measured. This approaches the peak value of $4\,m^2\,V^{-1}\,s^{-1}$ found for electrons in a MOSFET. Unfortunately

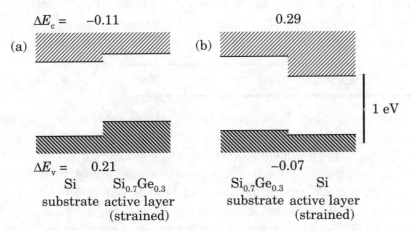

FIGURE 3.15. Band alignments at two heterojunctions between Si and $Si_{0.7}Ge_{0.3}$. (a) Substrate of silicon with strained active layer of $Si_{0.7}Ge_{0.3}$. (b) Substrate of $Si_{0.7}Ge_{0.3}$ with strained active layer of silicon; note the reduced band gap of the silicon. Here a positive discontinuity ΔE_c or ΔE_v gives confinement in the active layer.

ΔE_v is rather small to trap holes effectively, but a buried layer of $\mathrm{Si}_{1-x}\mathrm{Ge}_x$ can be incorporated in a normal MOSFET structure to improve the mobility of holes.

This is of great practical importance because complementary metal–oxide–semiconductor (CMOS) technology uses both n- and p-channel MOSFETs. The lower mobility of holes means that conventional p-channel devices must be larger than the corresponding n-channel devices to pass the same current. Buried strained layers of $\mathrm{Si}_{1-x}\mathrm{Ge}_x$ restore the balance between the two polarities of transistor, increasing the packing density and perhaps the speed.

Silicon forms the active layer in Figure 3.15(b), strained to conform to a 'substrate' of $\mathrm{Si}_{0.7}\mathrm{Ge}_{0.3}$. In fact the true substrate is probably silicon, with a thick metamorphic layer of the alloy which has relaxed as in Figure 3.12(c). Although the alignment is still of type II, the bigger discontinuity is now in the conduction band, and electrons can be trapped in the active layer to form a 2DEG. The lowest valleys in the conduction band are the $\Delta_\perp$ pair, which have the heavy longitudinal mass $m_\mathrm{L} \approx 0.98$ along the direction of growth and the light transverse mass $m_\mathrm{T} \approx 0.19$ in the plane of the interface. Thus motion in the 2DEG is governed by m_T, which is smaller than the average mass of 0.33 in unstrained silicon. This contributes to the mobility of $20\,\mathrm{m^2\,V^{-1}\,s^{-1}}$ found for electrons at this heterojunction, well above that in a MOSFET where it is limited by properties of the Si–$\mathrm{SiO_2}$ interface.

Another application of Si–Ge heterostructures is in heterojunction bipolar transistors. These employ an emitter with a wider band gap than the base to improve the emitter efficiency (see the exercises). This in turn permits the base to be more highly doped, decreasing its resistance and removing one of the limitations on high-frequency performance. Considerable work has also been directed at superlattices of Si–Ge. A particular aim is to modify the conduction-band structure to bring the minimum to Γ, giving a direct gap. The goal, not yet realized, is the efficient emission of light – an outcome that could put III–V materials out of business!

3.8   Wires and Dots

The structures described above have been layered, and either confine carriers to a plane or provide a one-dimensional potential through which they move. The next step in complexity is to pattern another dimension so that free motion is possible along only one direction. There are two approaches, broadly speaking: one can grow a layered structure and impose further structure on that, or form a two-dimensional pattern during growth. The former approach is more widely used, particularly for electronic applications, but cannot match the atomic sizes offered by controlled growth.

Layers are usually patterned in two steps. First the desired area is marked out in a resist by lithography, and the pattern is then transferred to the semiconductor

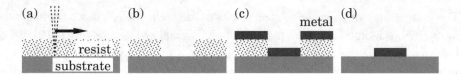

FIGURE 3.16. Production of a metal gate by electron-beam lithography and lift-off. (a) Exposure of resist by electron beam, (b) development of resist, (c) deposition of metal, and (d) lift-off to leave metal gate.

either by etching material away or by depositing another layer, usually metal, on its surface. A simple example, the production of a metal gate by electron-beam lithography and lift-off, is shown in Figure 3.16. The resist is a thin layer of poly-methylmethacrylate (PMMA, better known as Perspex or Plexiglas). An electron beam is scanned over the areas of resist to be removed in an instrument that closely resembles a scanning electron microscope. The advantages of this technique include the fine scale of patterns that can be drawn and its versatility; a severe disadvantage is that it is a serial process and therefore very slow. The electron beam breaks the polymer chains, which can then be more easily dissolved in a developer, leaving the areas that were not exposed. A thin layer of metal is next deposited over the whole surface. Finally the sample is immersed in a solvent that removes the remaining resist, and the metal on top of the resist lifts off, leaving only the metal deposited on the surface of the semiconductor. The edges of the resist need to be vertical or undercut for lift-off to be reliable, which often requires two- or even three-layer resists.

There is an enormous range of other processes in use. In particular, the remaining resist can be used as a mask to protect the area underneath from etching. Etching can be either wet, using chemicals in solution, or dry, where a plasma is employed. Wet etches produce little damage, and many selective processes have been developed to remove material at a rate that depends on its composition or even its crystallographic orientation. Dry etching produces sharper structures with vertical sidewalls, but usually at the cost of damage to the material under the etched surface.

Figure 3.17 shows three ways of confining the electrons in a 2DEG to a narrow strip, forming a *quantum wire* where free motion is possible only along the length.

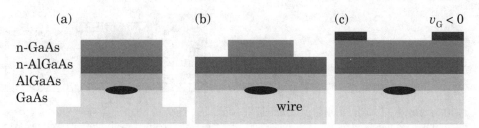

FIGURE 3.17. Restriction of a 2DEG to a wire by (a) deep etching, (b) shallow etching, and (c) negative bias on a split gate.

These structures are clearly related to electromagnetic wave guides. Perhaps the most obvious method is to etch the unwanted material away, removing the 2DEG to leave a narrow rib or mesa that contains the wire. A feature of GaAs, to be discussed at length in Section 9.1, is that there are surface states that absorb a large density of electrons or holes and create depletion regions underneath that contain no electrons. Thus the active width of the wire is much less than that of the rib in Figure 3.17(a). This feature can be turned to our advantage because it means that only a shallow etch is sufficient to deplete the 2DEG underneath, as in Figure 3.17(b). Any damage that occurs during etching is kept well away from the active region. The third method shown is to deposit a split gate on the surface. A negative bias v_g on this gate repels the electrons underneath, leaving a narrow channel under the gap which forms the wire. The width of the wire can be controlled through the bias on the gates, a significant advantage for experimental devices. This technique has proved to be extremely versatile and has been used to define channels far more complicated than a simple wire. An example is the narrow constriction used to demonstrate the quantized conductance of a one-dimensional system (Section 5.7.1).

A disadvantage of all three methods is that the potential seen by the electrons is rather loosely controlled. The transverse dimensions, such as the gap between the gates, are limited by processing and are usually above 50 nm. Also, the electrons are typically 50 nm or more below the surface. Although the electrostatic potential at the edge of a gate might be modelled as a sharp step, it broadens as it passes down to the 2DEG. More precise control of the wire can be obtained if the carriers are confined in both dimensions by heterojunctions. Great effort has therefore been expended on the direct growth of wires. Most methods require that the substrate be patterned in some way before growth, and Figure 3.18 shows one successful method.

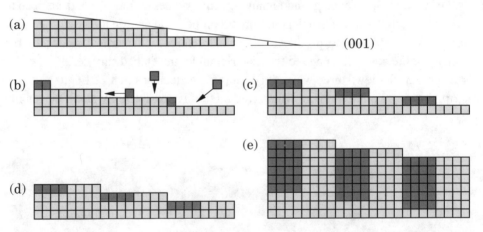

FIGURE 3.18. Growth of wires on a vicinal substrate. (a) Substrate before growth, showing steps and terraces on the vicinal (off-axis) surface. (b) Flux of AlAs applied, with motion of atoms over surface to steps where they stick. (c) Completion of a half-monolayer of AlAs. (d) Growth of a further half-monolayer of GaAs. (e) Growth of many layers, showing development of superlattice.

The substrate is deliberately prepared slightly off the (001)-surface usually used for growth. On the atomic scale, the surface of such a *vicinal* substrate consists of short terraces separated by steps. When a flux of Al and As is turned on, atoms land on the surface. Conditions are chosen so that these are mobile and can diffuse around the surface. It is energetically favourable for them to stick at the steps rather than the middle of the terraces. If sufficient reactants for half a monolayer are applied and the surface takes the state of lowest energy, the structure in Figure 3.18(c) results. Repeating the process to grow half a monolayer of GaAs restores the steps to their original positions in the plane, but a monolayer higher. The sequence of half-layers is repeated and the result is ideally as shown in Figure 3.18(e). Exquisite control of growth is essential but some remarkable structures have been grown, notably by MOCVD.

Many of the techniques described so far can readily be extended to confine electrons in all three dimensions, forming quantum dots. Again, there is interest in grown structures that do not need lithography or etching. A thin layer of InAs on GaAs tends naturally to cluster into dots and these *self-organized* structures currently receive great attention.

3.9 Optical Confinement

We have considered only the trapping of electrons and holes in the preceding sections, but the optical field must also be confined in an optoelectronic structure. Fortunately the problem is very similar in principle, although the vector nature of light makes calculations more difficult. The wave number and frequency of particles are related by $k^2 = 2m(\hbar\omega - V)/\hbar^2$, while for photons $k^2 = \tilde{\epsilon}_r \epsilon_0 \mu_0 \omega^2 = (\tilde{n}_r \omega/c)^2$. Here $\tilde{\epsilon}_r$ is the complex dielectric constant and $\tilde{n}_r = \tilde{\epsilon}_r^{1/2}$ is the refractive index; these optical constants will be discussed more fully in Section 8.5.1. The relations between wave number and frequency for particles and photons are similar, but there are some important differences. The difference $(\hbar\omega - V)$ is negative in a barrier, giving an imaginary wave number, whereas this is rarely true for $\tilde{\epsilon}_r$. On the other hand, the dielectric constant is complex, with an imaginary part that usually represents loss. This describes the absorption of photons, but electrons must be conserved and therefore cannot have a complex 'dielectric constant'.

Light tends to propagate in regions with a higher refractive index, in the same way that particles can be trapped in regions of lower potential energy. This can be viewed as total internal reflection of rays within the medium of higher index. A smaller band gap is usually associated with a larger refractive index, and a sandwich such as GaAs between AlGaAs will therefore tend to guide both particles and light waves in the GaAs. Indeed the theory of slab wave guides is very similar to that of a particle in a finite potential well (Section 4.2). The length scales are rather different, because the wavelength is around 1 μm for visible light but some 50 nm for electrons. Thus

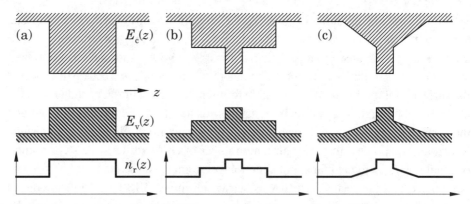

FIGURE 3.19. Profiles of conduction and valence bands and refractive index through layers designed both to guide light and trap carriers: (a) double heterostructure (DH), (b) separately confined heterostructure (SCH), (c) graded-index separately confined heterostructure (GRINSCH).

a double heterostructure (DH) such as Figure 3.19(a) guides light effectively only when it is so wide that quantization of the electronic states is unimportant. A narrow well with widely spaced energy levels gives very weak confinement of the optical mode. The solution is to confine light and particles separately, with a narrow inner well for the particles and a broader outer region for the light. Two common profiles are the separately confined heterostructure (SCH) with a stepped composition, or the graded-index separately confined heterostructure (GRINSCH) shown in Figures 3.19(b) and (c). The inner region may be a multiquantum well rather than a single layer.

Just as electrons can be further confined to wires, light can be confined in two dimensions like microwaves in a conventional metal wave guide. Many techniques are used, of which three are shown in Figure 3.20. The buried double heterostructure is complex to fabricate because a mesa must be etched to limit the active region and then followed by regrowth of a material with lower refractive index to guide the light parallel to the original interfaces. The advantage is that strong confinement is achieved. The ridge wave guide in Figure 3.20(b) is much easier to fabricate, requiring only etching of the surface, but guides the light more weakly. Finally, Figure 3.20(c) shows an optical application of a superlattice. Light travelling along the ridge wave guide experiences a periodic perturbation due to the transverse ridges etched into the wave-guide layer. These cause reflection according to the

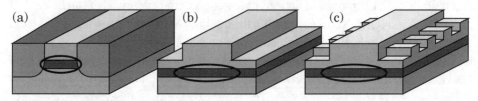

FIGURE 3.20. Cross-sections through three examples of laterally confined optical wave guides: (a) buried heterostructure, (b) rib wave guide, (c) distributed Bragg reflector on a rib wave guide.

Bragg condition and can be used to form an optical cavity in a laser. The method is known as distributed feedback (DFB) or a distributed Bragg reflector (DBR), depending on the configuration. Such lasers have a smaller linewidth than those using a conventional Fabry–Pérot cavity. Superlattices can also be formed by growth, as in Section 3.4.4, and used as distributed Bragg reflectors to form the cavity in vertical surface-emitting lasers.

A layered superlattice generates a band gap only for motion in one direction; there is no effect parallel to the layers. Further patterning to give a three-dimensional periodic potential, as in a solid crystal, may produce a range of frequencies or *photonic band gap* in which no modes of light can propagate. These devices are the subject of active research because of the control they offer over the optical properties, particularly spontaneous emission, which may lead to improved optoelectronic devices.

This concludes our review of the vast range of heterostructures in use. Before moving on to study their electronic structure, we shall take a diversion to explore the effective-mass approximation, a foundation of the theory of inhomogeneous semiconductors.

3.10 Effective-Mass Approximation

One of the simplifications that has been made throughout the foregoing survey is to assume that the band structure of the crystalline semiconductors can largely be ignored, retaining only the energy and effective mass at the extremity of each band. We shall now explore this *effective-mass approximation* to find its limitations and verify where it can be used safely. This section could be omitted on a first reading.

Suppose that we add a perturbation such as an impurity to a perfect crystal. More generally, the perturbation could be a quantum well, barrier, superlattice, or the potential from a patterned gate in a two-dimensional electron gas. The Schrödinger equation takes the form

$$[\hat{H}_{\text{per}} + V_{\text{imp}}(\mathbf{R})]\psi(\mathbf{R}) = E\psi(\mathbf{R}). \tag{3.2}$$

Here $\hat{H}_{\text{per}}$ is the Hamiltonian for the perfect crystal and $V_{\text{imp}}(\mathbf{R})$ is the additional contribution from the impurity. It is difficult to solve the Schrödinger equation for a perfect crystal, so equation (3.2) presents a formidable challenge.

On the other hand, elementary textbooks give simple results for a donor or acceptor in a semiconductor. A donor behaves as though it has an extra positive ionic charge compensated by an extra electron, with the signs reversed for acceptors. The impurity therefore behaves like a hydrogen atom, but the mass is rescaled to the effective mass of electrons or holes, and the charge is screened by the dielectric constant of the host. Both effects reduce the ionization energy substantially from 13.6 eV in hydrogen to only 5 meV for a donor in GaAs. This simple procedure can be justified by effective-mass (or effective-Hamiltonian) theory. It is attractive that the theory yields so straightforward a result, but it is also important to realize its limitations.

Consider a one-dimensional system to simplify the notation. Suppose that we have solved the Schrödinger equation for the perfect crystal

$$\hat{H}_{\text{per}}\phi_{nk}(x) = \varepsilon_n(k)\phi_{nk}(x). \tag{3.3}$$

These solutions form a complete set (Section 1.6). The wave function $\psi(x)$ of the system with the impurity can therefore be expanded in terms of $\phi_{nk}(x)$ as

$$\psi(x) = \sum_n \int_{-\pi/a}^{\pi/a} \tilde{\chi}_n(k)\phi_{nk}(x)\frac{dk}{2\pi}, \tag{3.4}$$

where $\tilde{\chi}_n(k)$ are expansion coefficients. This has both a summation over all bands n and an integral of k over the Brillouin zone to include all states. We could substitute the expansion (3.4) into the Schrödinger equation (3.2) and attempt to find $\tilde{\chi}_n(k)$. This is, of course, still an exact solution and is no easier to do in principle than solving the original equation directly. The wave function must be simplified to make further progress. We shall make a set of drastic approximations to get a straightforward result, and review afterwards how they may be relaxed.

The first step is to assume that wave functions from only one band play a significant part, so the summation over n can be dropped. Taking a donor in GaAs as an example, we expect the states to be drawn largely from the bottom conduction band (Γ_6). The valence and higher conduction bands should make only a small contribution. This has to be justified a posteriori: we find that the energy of the donor is close to the edge of the conduction band but a long way from the other bands, so the approximation is probably good.

The second step is to assume that states from only a small region of k-space contribute significantly to the integral. In the case of a donor, we expect that k lies around 0 as the bottom of the conduction band is at Γ. Now the Bloch functions can be written as $\phi_{nk}(x) = u_{nk}(x) \exp(ikx)$, where $u_{nk}(x)$ is a periodic function of x. Assume that most of the variation in $\phi_{nk}(x)$ with k comes from the plane wave and that $u_{nk}(x)$ can be treated as independent of k over a small region of k-space. Thus we write

$$\phi_{nk}(x) = u_{nk}(x)\,e^{ikx} \approx u_{n0}(x)\,e^{ikx} = \phi_{n0}(x)\,e^{ikx} \tag{3.5}$$

for small values of k.

With these two simplifications, the wave function (3.4) takes the form of an inverse Fourier transform,

$$\psi(x) \approx \phi_{n0}(x) \int_{-\pi/a}^{\pi/a} \tilde{\chi}(k) \exp(ikx)\frac{dk}{2\pi} = \phi_{n0}(x)\,\chi(x). \tag{3.6}$$

This is the first major result of effective-mass theory, illustrated in Figure 3.21. The wave function can be written approximately as the product of the Bloch function

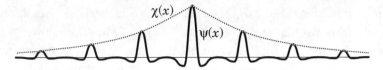

FIGURE 3.21. Wave function around an impurity, showing the envelope function $\chi(x)$ that modulates the Bloch function to give the full wave function $\psi(x)$.

of the local extremum of the host's energy bands and an *envelope function* $\chi(x)$. We assumed that $\tilde{\chi}(k)$ contains only a small range of wave numbers, which in turn means that $\chi(x)$ must be a slowly varying function in real space. Again this must be checked afterwards: the radius of a donor is about 10 nm in GaAs, which is large compared with the lattice constant of 0.5 nm.

This is an attractive step forwards, and we next need an equation for the envelope function. Substitute the expansion (3.4) into the Schrödinger equation (3.2). The main problem is the effect of $\hat{H}_{\text{per}}$. If we make the reduction to a single band, this operator gives

$$\hat{H}_{\text{per}} \psi(x) = \hat{H}_{\text{per}} \int_{-\pi/a}^{\pi/a} \tilde{\chi}(k)\, \phi_{nk}(x) \frac{dk}{2\pi} = \int_{-\pi/a}^{\pi/a} \tilde{\chi}(k)\, \varepsilon_n(k)\, \phi_{nk}(x) \frac{dk}{2\pi}$$

$$\approx \phi_{n0}(x) \int_{-\pi/a}^{\pi/a} \tilde{\chi}(k)\, \varepsilon_n(k)\, e^{ikx} \frac{dk}{2\pi}. \tag{3.7}$$

The Hamiltonian $\hat{H}_{\text{per}}$ acts on x and therefore affects only $\phi_{nk}(x)$. We know that this is an eigenfunction of the Schrödinger equation for the pure crystal, equation (3.3), so the operator can be replaced by its eigenvalue $\varepsilon_n(k)$. Finally, we make the approximation (3.5) for the Bloch functions.

Next, expand the energy band as a power series in k: $\varepsilon_n(k) = \sum_m a_m k^m$. This gives

$$\hat{H}_{\text{per}} \psi(x) \approx \phi_{n0}(x) \sum_m a_m \int_{-\pi/a}^{\pi/a} \tilde{\chi}(k)\, k^m\, e^{ikx} \frac{dk}{2\pi}. \tag{3.8}$$

Now we use the standard result for the Fourier transform of a derivative, obtained by integration by parts,

$$\int \frac{df(x)}{dx} e^{-ikx}\, dx = ik \int f(x) e^{-ikx}\, dx = ik \tilde{f}(k). \tag{3.9}$$

Thus the inverse Fourier transform of $k \tilde{f}(k)$ is $-i\, df(x)/dx$, and this can be generalized to show that the inverse transform of $k^m \tilde{f}(k)$ is $(-i\, d/dx)^m f(x)$. Thus equation (3.8) becomes

$$\hat{H}_{\text{per}} \psi(x) \approx \phi_{n0}(x) \sum_m a_m \left(-i\frac{d}{dx}\right)^m \chi(x) \equiv \phi_{n0}(x)\, \varepsilon_n\left(-i\frac{d}{dx}\right) \chi(x). \tag{3.10}$$

The weird expression $\varepsilon_n(-i\,d/dx)$ is no more than an abbreviation for the expansion on the left. It means that one should take $\varepsilon_n(k)$, expand it as a power series in k, and replace k with $-i\,d/dx$ in every term. In general there is an infinite number of terms so $\varepsilon_n(-i\,d/dx)$ is a nasty expression despite its neat appearance.

We have thus reduced the Schrödinger equation (3.2) to a remarkably simple form because the remaining terms $V_{\text{imp}}\psi$ and $E\psi$ simply multiply the wave function. The common factor of $\phi_{n0}(x)$ cancels to leave

$$\left[\varepsilon_n\left(-i\frac{d}{dx}\right) + V_{\text{imp}}(x)\right]\chi(x) = E\chi(x). \tag{3.11}$$

This is our final result: a Schrödinger equation for the envelope function alone, containing an *effective Hamiltonian*. The Bloch functions have vanished, as has the periodic potential of the host, in favour of a complicated kinetic energy operator that contains the band structure. In three dimensions, $\varepsilon_n(-i\,d/dx)$ is replaced by $\varepsilon_n(-i\nabla)$. The normalization of the envelope function is a minor problem, because the full wave function is a product of χ and ϕ_{n0}, and will be addressed in Section 10.4.

The effective Hamiltonian would still be extremely complicated if we retained the full band structure for $\varepsilon_n(\mathbf{K})$. However, we have already assumed that the wave function is drawn from only a small region of $\mathbf{K}$-space, and we can simplify $\varepsilon_n(\mathbf{K})$ to be consistent with this. For example, the bottom of the conduction band in GaAs is approximately

$$\varepsilon_n(\mathbf{K}) \approx E_c + \frac{\hbar^2 K^2}{2m_0 m_e}, \tag{3.12}$$

and the replacement $\mathbf{K} \to -i\nabla$ gives

$$\varepsilon_n(-i\nabla) \approx E_c - \frac{\hbar^2}{2m_0 m_e}\nabla^2. \tag{3.13}$$

Substituting this into the effective Schrödinger equation (3.11), and moving E_c onto the right-hand side, gives

$$\left[-\frac{\hbar^2}{2m_0 m_e}\nabla^2 + V_{\text{imp}}(\mathbf{R})\right]\chi(\mathbf{R}) = (E - E_c)\chi(\mathbf{R}). \tag{3.14}$$

This is exactly the result that we sought: a Schrödinger equation that resembles that for free electrons, except for the effective mass, with energy measured from the bottom of the conduction band.

In the case of a donor, with an attractive Coulomb potential, the Schrödinger equation becomes

$$\left(-\frac{\hbar^2}{2m_0 m_e}\nabla^2 - \frac{e^2}{4\pi\epsilon_0\epsilon_b R}\right)\chi(\mathbf{R}) = (E - E_c)\chi(\mathbf{R}). \tag{3.15}$$

This is exactly like a hydrogen atom except for the effective mass m_e, the offset in energy from the conduction band E_c, and the dielectric constant of the semiconductor ϵ_b (it seems natural that this should appear although we have not proved it). We can simply insert the constants into the results for hydrogen, given later in Section 4.7.5. This shows that the energy of the lowest state is $E = E_c - \mathcal{R}$, where $\mathcal{R}$ is the *Rydberg energy* (equation 4.66). The corresponding wave function is $\chi(R) = (\pi a_B^3)^{-1/2} \exp(-R/a_B)$, where a_B is the *Bohr radius* (equation 4.67). The Rydberg energy and Bohr radius depend on m_e and ϵ_b, with values of $\mathcal{R} \approx 5.2 \, \mathrm{meV}$ and $a_B \approx 10 \, \mathrm{nm}$ for electrons in GaAs.

Before passing on to consider effective-mass theory in heterostructures, we must review what has been lost in the approximations. The validity of the simplest form of Hamiltonian, such as that in equation (3.14), is limited by the parabolic approximation to the energy band. Even in the GaAs–AlGaAs system with its 'small' offsets, we often consider energies ±0.3 eV from the edge of a band. The parabolic approximation is dubious at so large an energy, and numerical results should not be taken too seriously.

The restriction to a single band is a serious approximation that must be relaxed in several applications, including the following:

(i) Materials such as Si have several equivalent valleys in the conduction band, all of which must be retained for electronic states.

(ii) The valence band contains light and heavy holes that are degenerate at the top, separated from a third band by only the weak spin–orbit splitting. Again we must retain a summation over all these bands in the wave function.

(iii) We must keep both the conduction and valence bands to describe interband effects such as optical absorption in a quantum well (Figure 1.4). This raises a further problem. The bands around the fundamental gap in a semiconductor can be described using the Kane model (Section 10.2), which shows that the small effective masses arise from mixing of the wave functions between the conduction and valence bands as **K** increases from zero. The approximation made in equation (3.5), that variation of the periodic part of the Bloch functions can be neglected, therefore needs scrutiny too.

When several bands are retained the effective Hamiltonian becomes a matrix of differential equations acting on a vector containing the components of the wave function in each band. An example will be given in Section 10.3.

Fortunately, investigation has shown that the effective-mass approximation is robust and generally successful, although particular care must be exercised when calculating optical matrix elements.

3.11 Effective-Mass Theory in Heterostructures

A little more attention is needed before applying effective-mass theory to heterostructures. The important features are the Bloch part of the wave function and the way in which the envelope functions are matched.

Remember that the total wave function is a product of the slowly varying envelope function and the Bloch function of the local extremum in the host's band structure (equation 3.6). The Bloch functions in the two materials on either side of a heterojunction must be similar for the effective-mass approximation to be valid. An obvious condition is that they must belong to the same point in $\varepsilon_n(\mathbf{K})$. This would fail for the conduction band at a junction between GaAs and AlAs, where the conduction bands are at Γ and X in the two materials. Even where the minima are the same, as in GaAs and $Al_{0.3}Ga_{0.7}As$, one might guess that the Bloch functions would have to be nearly identical. This is because their kinetic energy is large, some 10 eV from Figure 2.16, so differences in the Bloch functions could easily introduce errors that would swamp the energies of around 0.1 eV that we are normally concerned with. Happily, detailed calculations confirm that effective-mass theory remains acceptable despite the different Bloch functions.

The second point concerns the matching of the envelope functions at the interface. Consider a junction at $z = 0$ between neutral regions of materials A and B (GaAs and $Al_xGa_{1-x}As$ with $x < 0.45$, for example). The Schrödinger equation for the envelope function in the two regions, if we consider only one dimension for simplicity, is

$$\left(E_c^A - \frac{\hbar^2}{2m_A m_0} \frac{d^2}{dz^2} \right) \chi(z) = E\chi(z), \tag{3.16}$$

$$\left(E_c^B - \frac{\hbar^2}{2m_B m_0} \frac{d^2}{dz^2} \right) \chi(z) = E\chi(z). \tag{3.17}$$

The difference in the bottoms of the conduction bands behaves like a step potential with material B higher by $\Delta E_c = E_c^B - E_c^A$. If the materials were the same we would simply match the value and derivative of the wave function at the interface, giving the usual conditions

$$\chi(0_A) = \chi(0_B), \qquad \left.\frac{d\chi(z)}{dz}\right|_{z=0_A} = \left.\frac{d\chi(z)}{dz}\right|_{z=0_B}, \tag{3.18}$$

where 0_A means the side of the interface in material A and so on. This simple condition is not correct for a heterojunction where the two effective masses are different, and we shall see in Section 5.8 that equation (3.18) does not conserve current. A correct set of matching conditions is

$$\chi(0_A) = \chi(0_B), \qquad \frac{1}{m_A} \left.\frac{d\chi(z)}{dz}\right|_{z=0_A} = \frac{1}{m_B} \left.\frac{d\chi(z)}{dz}\right|_{z=0_B}. \tag{3.19}$$

The condition for matching the derivative now includes the effective mass. Since the derivative is essentially the momentum operator, equation (3.19) requires the velocity to be the same on both sides to conserve current. The envelope function gains a kink at the interface if $m_A \neq m_B$.

A more mathematical argument is that the erroneous matching condition (3.18) implicitly assumes that the Schrödinger equation takes the form

$$-\frac{\hbar^2}{2m_0 m(z)}\frac{d^2\chi}{dz^2} + V(z)\chi(z) = E\chi(z).\qquad (3.20)$$

This is not Hermitian (or of Sturm–Liouville form) if $m(z)$ varies, and many of the crucial properties of wave functions disappear as a result. The apparently trivial change of rewriting the equation in the form

$$-\frac{\hbar^2}{2m_0}\frac{d}{dz}\left[\frac{1}{m(z)}\frac{d\chi}{dz}\right] + V(z)\chi(z) = E\chi(z)\qquad (3.21)$$

restores the Hermitian nature and can be justified more thoroughly. Both versions reduce to the usual Schrödinger equation for a homostructure, but the second form ensures that the wave functions are orthogonal and current is conserved, and preserves other conditions that we hold dear. The penalty is the revised condition (3.19) for matching.

The kink in the envelope function at a heterointerface calls its validity into question, because 'real' wave functions must be smooth. Fortunately, thorough calculations show that the effective-mass approximation continues to be valid provided that the envelope function varies slowly on the atomic scale; this condition need *not* apply to the potential. Figure 3.22 shows an example of the full wave function (including the Bloch functions) for a 6 nm quantum well. This function is smooth, but the envelope function constructed roughly by joining the peaks has an abrupt

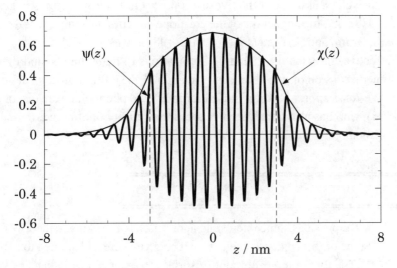

FIGURE 3.22. Wave function for the lowest state in a 6 nm quantum well in a heterostructure, including the Bloch functions. The thin curve is an approximate envelope function joining the peaks of the full wave function. [Redrawn from Burt (1994).]

change in slope at the heterointerfaces. Thus the kink in the envelope function imposed by the matching condition (3.19) correctly reflects a change in the full wave function on the atomic scale.

This concludes the complications introduced by heterojunctions for the present; we shall return to them again in Section 4.9. The next step is to put together the results of the first three chapters to see how electrons can be persuaded to behave as though they are confined to only two dimensions or fewer.

Further Reading

Stradling and Klipstein (1990) describe the growth and characterization of heterostructures. The review by Adachi (1985) is a useful source of data on GaAs, AlAs, and their alloys. It is reprinted in Blakemore (1987), which contains many other helpful papers.

Yu, McCaldin, and McGill (1992) give an extensive review of band offsets at heterojunctions, with a broad description of both theoretical and experimental approaches to this long-standing problem. Wolfe, Holonyak, and Stillman (1989) have a good discussion of band diagrams for heterostructures. The use of strain for valence-band engineering is reviewed by O'Reilly (1989).

Weisbuch and Vinter (1991), and Kelly (1995), describe many applications of heterostructures, including the practical difficulties. There are some useful reviews in Willardson and Beer (1966–). Applications to optoelectronics are described from many points of view; Chuang (1995) has a theoretical bias, whereas Gowar (1993) is more oriented towards engineering.

An extensive discussion of effective-mass theory in heterostructures is given by Bastard (1988), and further details of the envelope-function formalism are described by Bastard, Brum, and Ferreira (1991). Datta (1989) gives a straightforward derivation of effective-mass theory including applications to many bands, and there is a penetrating discussion of its validity in Anderson (1963). Problems that arise when the effective-mass approximation is used for interband phenomena are discussed by Burt (1995), who has also considered in detail the validity of effective-mass theory in heterostructures.

EXERCISES

3.1   MBE provides some interesting applications of the kinetic theory of gases. Some of the standard formulas will be given here and applied to MBE.

   The flux of molecules onto a surface (the number of molecules striking unit area in unit time) is given by $F = \frac{1}{4}n\bar{c}$, where n is the number density of molecules (number per unit volume) and $\bar{c}$ is their mean velocity. Pressure, number density, and temperature are related by the ideal gas law, which in microscopic terms reads $p = nk_B T$. The average velocity is related to the thermal energy by $\frac{1}{2}mc_{rms}^2 = \frac{3}{2}k_B T$, where c_{rms} is the root-mean-square velocity. This is related to the mean velocity by $\bar{c} = \sqrt{8/3\pi}\, c_{rms}$. Combining these equations gives the familiar result $p = \frac{1}{3}nmc_{rms}^2$.

Use these results to find out how long a surface in UHV remains clean, before a monolayer is deposited on it from the background gas. First, calculate the flux F_{bg} of background gas onto the substrate, assuming that H_2 is the dominant species. Only a fraction ξ, the sticking coefficient, of the incident atoms adhere to the surface. Estimate t_{bg}, the time taken to coat the surface with a monolayer. Assume that $\xi = 1$ (a very pessimistic assumption) and that the background gas is at a pressure of 5×10^{-11} mbar and room temperature $T_{bg} = 300$ K.

3.2   Estimate the growth rate, assuming that growth is limited by the supply of Ga, whose pressure in the K-cell is about 5×10^{-4} mbar at $T_{Ga} = 900°C$. The preceding formulas can be used to find the flux F_{Ga} of Ga atoms at the mouth of the K-cell, whose radius $A \approx 10$ mm. If the substrate is a distance $R \approx 200$ mm from the K-cell, the flux is spread over a hemisphere of area $2\pi R^2$ but is double the average value in the forward direction. Hence find the time t_{ML} to grow a monolayer. It is also instructive to convert this to the time to grow a thickness of 1 μm.

3.3   Estimate the degree of contamination in the grown material simply as t_{ML}/t_{bg} (although hydrogen is not a realistic material to consider as a contaminant). Good material needs this fraction of impurities to be well below 10^{-6}, so it is as well that the assumption $\xi = 1$ is pessimistic!

3.4   Verify that the mean free path L is much greater than the size of the apparatus. If the molecules have diameter d, they will hit all other molecules in a volume $\pi d^2 x$ while travelling a distance x. On average they hit one other molecule when travelling one mean free path, so $\pi d^2 L n = 1$. (The same argument will be used for the mean free path of an electron in a solid in Section 8.2.) Estimate L both for the background gas and for the molecular beams, using the foregoing conditions.

3.5   Use the data in Appendix 2 to find the compositions of $In_x Ga_{1-x} As$ and $In_x Al_{1-x}$ As that can be grown without strain on a substrate of InP.

3.6   Use Anderson's rule to estimate the discontinuities in the conduction and valence bands for two lattice-matched systems, GaAs–AlAs and InAs-GaSb. Band gaps and electron affinities are given in Appendix 2. How well do the results agree with the accepted values shown in Figure 3.5?

3.7   Figure 3.15 showed the alignment of the bands for the two pairings of Si and $Si_{0.7}Ge_{0.3}$. Repeat this for Ge and $Si_{0.2}Ge_{0.8}$. What types of alignment are found, and can you suggest applications for them?

3.8   Sketch the band diagram for a heterojunction between p-type AlGaAs and n-type GaAs. Show that this can trap a (two-dimensional) hole gas at the interface.

3.9   Interesting results can be achieved with type II (staggered) heterojunctions (Figure 3.5). Sketch the bands for both p–n and n–p junctions, and determine whether electrons or holes can be trapped at the interface.

What happens at an undoped type III heterojunction, such as InAs–GaSb?

3.10   Figure 3.5 shows that the bottom of the conduction band in InAs lies below the top of the valence band in GaSb. Thus electrons transfer spontaneously from GaSb to InAs. Suppose that a sequence of narrow layers is grown. Each layer of InAs now behaves as a quantum well for electrons, and the lowest state is raised above the bottom of the conduction band. The energy of the lowest state for holes in GaSb is also raised. For very thin layers the states will be raised so far in energy that they no longer overlap. Estimate the critical thickness for alternating layers of equal thickness, treating the wells as infinitely deep.

3.11   Sketch the band diagram for a homojunction n–p diode in GaAs, assuming that classical semiconductor statistics can be used (Section 1.8.3). Let $N_D = 2 \times 10^{21} \, \mathrm{m}^{-3}$ on the left and $N_A = 10^{23} \, \mathrm{m}^{-3}$ on the right. Calculate the position of the Fermi level in the two materials and use this to sketch the band diagram at equilibrium. Take the effective densities of states to be $N_c = 4.7 \times 10^{23} \, \mathrm{m}^{-3}$ and $N_v = 7.0 \times 10^{24} \, \mathrm{m}^{-3}$. Calculate also the built-in voltage, defined as the difference in Fermi level between the two sides when the bands are flat, so $eV_{bi} = E_F^{(n)} - E_F^{(p)}$.

3.12   The previous diode is replaced by a heterojunction between n-$Al_{0.3}Ga_{0.7}As$ and p-GaAs. The fraction of Al is graded rather than changed abruptly and the doping is unchanged. Calculate the new positions of the Fermi levels (very little work is needed!), sketch the band diagram, and calculate the built-in voltage. Show that

$$eV_{bi} = E_g^{(p)} + \Delta E_c - [E_c^{(n)} - E_F^{(n)}] - [E_F^{(p)} - E_v^{(p)}]. \qquad (E3.1)$$

The two differences in square brackets can be calculated simply from the doping. Deduce that the effect of changing from GaAs to AlGaAs on the n-side is simply to increase eV_{bi} by ΔE_c, assuming that the doping remains the same and that the effective densities of states do not change significantly. How does this follow from the band diagram? In a homojunction eV_{bi} can also be defined as the change in conduction or valence bands across the junction; how is this modified in a heterojunction?

3.13   The emitter efficiency of a bipolar junction transistor is governed by the ratio of current densities carried across the emitter–base junction by electrons and holes, which should be as large as possible. The dominant factor in this ratio is the ratio of *minority* carriers at equilibrium on the two sides. This can be controlled only through the doping in a homojunction, so bipolar transistors have a heavily doped emitter and lightly doped base. Calculate the ratio n_p/p_n for the homojunction, where n_p is the minority density of electrons in p-type material and similarly for p_n.

Show that the band discontinuities provide an additional degree of freedom in a heterojunction where the emitter has the wider band gap. Consider again the specific example of n-AlGaAs and p-GaAs, which is often used for the emitter–base junction of an npn heterojunction bipolar transistor (HBT). Show that the change from GaAs to AlGaAs in the emitter, keeping everything else fixed, decreases the density of minority holes there. What happens to the ratio n_p/p_n? This in turn reduces the current density carried by holes; the electrons are unaffected.

3.14 Electrons are trapped in a well whose effective mass is very different from that of the barriers. What happens to the boundary condition at the heterojunction? Consider both $m_W \ll m_B$ and $m_W \gg m_B$.

4

QUANTUM WELLS AND LOW-DIMENSIONAL SYSTEMS

Real electrons are three-dimensional but can be made to behave as though they are only free to move in fewer dimensions. This can be achieved by trapping them in a narrow potential well that restricts their motion in one dimension to discrete energy levels. If the separation between these energy levels is large enough, the electrons will appear to be frozen into the ground state and no motion will be possible in this dimension. The result is a two-dimensional electron gas (2DEG). The same effect can be achieved with a two-dimensional potential well, which leaves the electrons free to move in one dimension only – a quantum wire.

In the first part of this chapter we shall study some simple one-dimensional potential wells used to trap electrons. The infinitely deep square well cannot be made in practice, but its simplicity makes it a frequently used model. A well of finite depth provides a much better description of a real quantum well. Parabolic wells can be grown by changing the composition of the semiconductor continuously, but this potential proves to be most relevant for the study of magnetic fields. The final example is a triangular well, which can be used as a rough description of the two-dimensional electron gas formed at a doped heterojunction. Next we shall see how these potential wells make electrons behave as though they are two-dimensional. The final sections are concerned with further confinement to give one- or even zero-dimensional systems, and some details that change due to the different effective masses in heterostructures.

We shall be dealing with one-, two- and three-dimensional systems and it is important to have a clear notation to reflect this. Recall that z is the direction of growth, normal to the planes of a layered structure; the lower-case vector $\mathbf{r} = (x, y)$ is two-dimensional, normal to z; and a three-dimensional vector is denoted $\mathbf{R} = (\mathbf{r}, z) = (x, y, z)$.

4.1   Infinitely Deep Square Well

This is by far the simplest example of a quantum well. It has already been considered in Chapter 1, but the results will be repeated here as a reminder. For comparison with the well of finite depth, take the origin in the middle of the well so that $V(z) = 0$

in the region $-a/2 < z < a/2$ and infinitely high potential barriers prevent the trapped particle from straying beyond this region. We use the coordinate z rather than x in this chapter as it is conventionally used to denote the direction of growth in heterostructures. The wave functions are

$$\phi_n(z) = \begin{cases} \sqrt{\dfrac{2}{a}} \cos \dfrac{n\pi z}{a}, & n \text{ odd,} \\[2mm] \sqrt{\dfrac{2}{a}} \sin \dfrac{n\pi z}{a}, & n \text{ even,} \end{cases} \tag{4.1}$$

with energy

$$\varepsilon_n = \frac{\hbar^2}{2m}\left(\frac{n\pi}{a}\right)^2, \tag{4.2}$$

where $n = 1, 2, 3, \ldots$. The wave functions are even functions of z for odd n (even parity) and have odd parity for even n, with the lowest state being even. The results of this model are so simple that they are widely used despite its unrealistic nature.

4.2 Square Well of Finite Depth

Quantum wells made in the GaAs–AlGaAs system are far from the idealized case of infinite depth. The depth for electrons is set by the discontinuity in the conduction band ΔE_c, which is usually kept below about $0.3\,\text{eV}$ to avoid the problem of an indirect band gap in the AlGaAs. The discontinuity in the valence band ΔE_v is even smaller, although this is offset by the larger effective mass. These depths are rather small for many applications, particularly at room temperature, which has encouraged the use of other junctions where ΔE_c and ΔE_v are larger.

The square well with a depth V_0, shown in Figure 4.1, allows for the finite value of ΔE_c or ΔE_v. It is still a considerable simplification of a real system: the potential should be curved if the system is not electrically neutral everywhere, for example. The energy ε is measured from the bottom of the well to allow easy comparison with the results for the infinitely deep well. States with $\varepsilon < V_0$ are trapped inside the well, whereas those with $\varepsilon > V_0$ can propagate from $z = -\infty$ to $z = +\infty$. Bound states are often described by their binding energy B, the energy required to lift an electron from its bound state so that it can just escape from the well; clearly $B = V_0 - \varepsilon$.

Start by examining the bound states. The wave functions inside the well are similar to those for the infinitely deep well and have the same even/odd symmetry (equation 4.1). Write both possibilities as

$$\psi(z) = C \begin{Bmatrix} \cos \\ \sin \end{Bmatrix} kz, \tag{4.3}$$

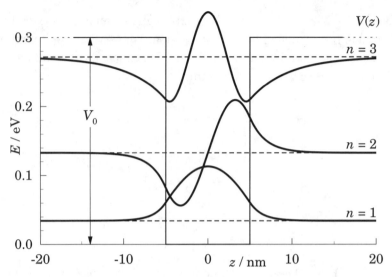

FIGURE 4.1. Finite square well in GaAs of depth $V_0 = 0.3\,\text{eV}$ and width $a = 10\,\text{nm}$, showing three bound states.

for $-a/2 < z < a/2$, with $\varepsilon = \hbar^2 k^2/2m$. Outside the well, $\psi(z)$ satisfies

$$-\frac{\hbar^2}{2m}\frac{d^2}{dz^2}\psi(z) + V_0\psi(z) = \varepsilon\psi(z) \tag{4.4}$$

with $\varepsilon < V_0$. The solutions are

$$\psi(z) = D\exp(\pm\kappa z) \tag{4.5}$$

with

$$\frac{\hbar^2\kappa^2}{2m} = V_0 - \varepsilon = B. \tag{4.6}$$

It must be possible to normalize the wave functions, so the positive exponential in equation (4.5) holds for $z < 0$ and the negative one for $z > 0$. As we know that the wave functions are either even or odd, we can concentrate on $z > 0$ with the negative exponential and use symmetry to find $\psi(z)$ for $z < 0$.

The wave functions (4.3) and (4.6) must now be matched at $z = \frac{1}{2}a$. Continuity of $\psi(z)$ requires

$$\psi(a/2) = C\begin{Bmatrix} \cos \\ \sin \end{Bmatrix}\left(\frac{ka}{2}\right) = D\exp\left(-\frac{1}{2}\kappa a\right). \tag{4.7}$$

Similarly, matching the derivatives gives

$$\frac{d\psi}{dz}\bigg|_{z=a/2} = Ck\begin{Bmatrix} -\sin \\ \cos \end{Bmatrix}\left(\frac{ka}{2}\right) = -D\kappa\exp\left(-\frac{1}{2}\kappa a\right). \tag{4.8}$$

This must be modified if the effective masses in the two materials are different, and will be described in Section 4.9.

There are three unknowns in (4.7) and (4.8): the normalization factors C and D, and the energy ε, which determines k and κ. We can eliminate C and D by dividing equation (4.7) into (4.8), giving

$$k \left\{ \begin{array}{c} -\tan \\ \cot \end{array} \right\} \left(\frac{ka}{2} \right) = -\kappa. \tag{4.9}$$

Effectively, we are matching the single function

$$\frac{1}{\psi} \frac{d\psi}{dz} = \frac{d \log \psi}{dz}, \tag{4.10}$$

known as the logarithmic derivative, from which the normalization cancels. Expanding equation (4.9) gives

$$\left\{ \begin{array}{c} \tan \\ -\cot \end{array} \right\} \left(\frac{ka}{2} \right) = \frac{\kappa}{k} = \frac{1}{k} \sqrt{\frac{2m}{\hbar^2} (V_0 - \varepsilon)} = \sqrt{\frac{2m V_0}{\hbar^2 k^2} - 1}. \tag{4.11}$$

The solution of this transcendental equation cannot be obtained exactly but is an easy numerical task; it can readily be programmed into a pocket calculator using functional iteration or something more fancy.

It is useful to get a qualitative idea of the solutions from a graph, which can also be used to guess an initial value for iteration. With the dimensionless variable $\theta = ka/2$, equation (4.11) becomes

$$\left\{ \begin{array}{c} \tan \\ -\cot \end{array} \right\} \theta = \sqrt{\frac{m V_0 a^2}{2\hbar^2} \frac{1}{\theta^2} - 1} \equiv \sqrt{\frac{\theta_0^2}{\theta^2} - 1}, \tag{4.12}$$

where

$$\theta_0^2 = \frac{m V_0 a^2}{2\hbar^2}. \tag{4.13}$$

All the physical input – the mass of the particle, the depth and width of the well – has been absorbed into the one dimensionless parameter θ_0^2. This parameter determines the allowed values of θ, and it is also simpler to solve the equation numerically in this dimensionless form. Both sides of equation (4.12) are plotted against θ in Figure 4.2 for electrons in a well in GaAs. This has $m = m_0 m_e$ with $m_e = 0.067$, $V_0 = 0.3\,\text{eV}$, and $a = 10\,\text{nm}$, which gives $\theta_0^2 = 13.2$. The square root on the right-hand side is always positive so we need only consider ranges of θ where each of the two left-hand sides is also positive. For $\tan \theta$ this means $(0, \frac{1}{2}\pi), (\pi, \frac{3}{2}\pi), \ldots$, while $-\cot \theta$ fills the gaps as it is positive for θ in the ranges $(\frac{1}{2}\pi, \pi), (\frac{3}{2}\pi, 2\pi), \ldots$. There are other ways of rewriting equation (4.12); for example, it can be reduced to

$$\left\{ \begin{array}{c} |\cos \theta| \\ |\sin \theta| \end{array} \right\} = \frac{\theta}{\theta_0}, \tag{4.14}$$

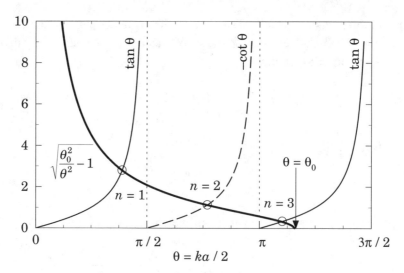

FIGURE 4.2. Graphical solution of equation (4.12) for a square well in GaAs with depth $V_0 = 0.3\,\mathrm{eV}$ and width $a = 10\,\mathrm{nm}$, giving $\theta_0^2 = 13.2$. There are three bound states.

but one must be careful to use the correct choice of the left-hand side in each range of θ, so we shall use the original expression. Solutions to (4.12) occur when the two lines intersect: there are three in this case, labelled by n. Several important results follow from this graph.

(i) The curve for the right-hand side of (4.12) intersects the θ-axis at θ_0, while the curves of $\tan \theta$ and $-\cot \theta$ intersect at $\theta = n\pi/2$ ($n = 0, 1, 2, \ldots$). The number of solutions is therefore given by

$$\frac{2}{\pi}\theta_0 = \frac{2}{\pi}\sqrt{\frac{m V_0 a^2}{2\hbar^2}} \tag{4.15}$$

rounded *up* to the nearest integer. This shows that there is always at least one solution to equation (4.12): a one-dimensional square well always has at least one bound state, however shallow or narrow the well (although the state may be very weakly bound). This result holds for any one-dimensional well, not just a square well. It also holds in two dimensions, although shallow wells have a very weakly bound state. It is not true for three-dimensional wells, which must exceed a critical radius or depth for a bound state to exist.

(ii) Consider a very shallow well, with only one bound state and small θ_0. Use the approximation $\tan \theta \approx \theta$ in equation (4.12), which becomes

$$\theta \approx \sqrt{\frac{\theta_0^2}{\theta^2} - 1}. \tag{4.16}$$

This can be turned into a quadratic equation, $\theta^4 + \theta^2 - \theta_0^2 = 0$, whose solution is

$$\theta^2 = \tfrac{1}{2}\left(-1 \pm \sqrt{1 + 4\theta_0^2}\right). \tag{4.17}$$

The '$-$' sign can be discounted since θ must be real. Expanding the square root with the binomial theorem to second order gives

$$\theta^2 \approx \theta_0^2 - \theta_0^4 = \frac{mV_0a^2}{2\hbar^2} - \left(\frac{mV_0a^2}{2\hbar^2}\right)^2. \tag{4.18}$$

The energy is therefore

$$\varepsilon = \frac{\hbar^2}{2m}\frac{4}{a^2}\theta^2 \approx V_0 - \frac{ma^2V_0^2}{2\hbar^2}. \tag{4.19}$$

The binding energy is more significant:

$$B = V_0 - \varepsilon \approx \frac{ma^2V_0^2}{2\hbar^2}. \tag{4.20}$$

This depends quadratically on V_0 and is therefore small for a shallow well.

(iii) If the well is very deep, $V_0 \to \infty$, the solutions lie on the steep parts of the tangent curves, and the intersections approach $\theta = \tfrac{1}{2}n\pi$ for $n = 1, 2, 3, \ldots$. Then $k = n\pi/a$ and the energies are given by $(\hbar^2/2m)(n\pi/a)^2$, the same as those for an infinitely deep well (equation 4.2).

(iv) Focus on a solution with a given n as a function of V_0. Equation (4.15) shows that this bound state appears when the intersections approach $\theta = \tfrac{1}{2}(n-1)\pi$, $k = (n-1)\pi/a$, and the depth of the well is $(\hbar^2/2m)[(n-1)\pi/a]^2$. The state is only weakly bound, and B is small, when V_0 is not much larger than this. The exponential decay constant κ in the wave function (4.5) outside the well is small, and the wave function penetrates a long way into the barriers. The probability of finding the particle inside the well is small, measured by

$$\int_{-a/2}^{a/2} |\psi(z)|^2 dz \bigg/ \int_{-\infty}^{\infty} |\psi(z)|^2 dz. \tag{4.21}$$

Making the well deeper causes ε, B, and κ all to rise, and the state becomes better bound.

(v) The character of the wave functions changes as the state becomes better bound. When the state has only just become bound in the well and $\theta = \tfrac{1}{2}(n-1)\pi$, the derivative is almost zero at the edge of the well, matching to a slowly decaying wave in the barriers. Increasing V_0 increases its kinetic energy to give $\theta \approx \tfrac{1}{2}n\pi$, at which point the amplitude of the wave function is almost zero at the boundaries, and the exponential tail has only a small amplitude.

Most of these results apply to any potential well.

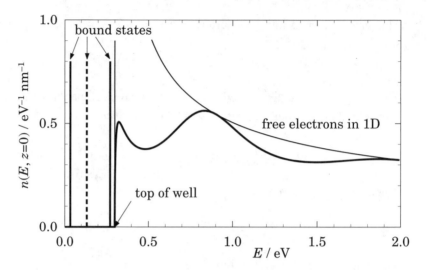

FIGURE 4.3. Local density of states $n(E, z = 0)$ in the middle of a square well in GaAs of width $10\,\text{nm}$ and depth $0.3\,\text{eV}$. The result for free electrons, proportional to $E^{-1/2}$, is shown for comparison.

States with $\varepsilon > V_0$ are not bound. They extend from $-\infty$ to $+\infty$ in z, and all energies are allowed, with two solutions for each energy. This does not mean that the wave functions are simple plane waves with uniform density everywhere; they are distorted as they pass over the well, particularly when their energy is not far above V_0. This can be calculated in the same way as the square barrier to be discussed in Section 5.2.1.

One way of showing the distortion is with the *local* density of states $n(\varepsilon, z)$. This was defined in equation (1.102) and weights the energy levels of the system with the density of their wave functions at z. It is plotted in Figure 4.3 for the centre of the well in Figure 4.1. Each of the three bound states contributes a δ-function. The second of these has a node at $z = 0$ so its local density of states vanishes at this point, and it is therefore shown as a broken line. The states are free for $\varepsilon > V_0$ but the local density of states is strongly distorted for low energies, reflecting a distortion of the plane waves by the well. In particular, $n(\varepsilon, 0)$ rises from zero like $(\varepsilon - V_0)^{1/2}$, rather than diverging like $(\varepsilon - V_0)^{-1/2}$ for free electrons. The distortion of the wave functions is important in processes where an electron is excited from a bound state to one only just above the top of the well, and leads to *final-state effects*; an example of optical absorption that reflects $n(\varepsilon, 0)$ is shown in Figure 10.7. The local density of states for $\varepsilon > V_0$ is lower than that for free electrons because part of its weight has gone into the bound states.

A useful theoretical simplification of the square well is to reduce it to a δ-function: $V(z) = -S\delta(z)$. The strength S has dimensions of (energy) × (length). To use our results for the finite well we take the limit $a \to 0$, $V_0 \to \infty$ while keeping $S = V_0 a$ constant. Thus $\theta_0 \to 0$ and we can use the result for a shallow well, equation (4.20).

This shows immediately that there is only one bound state, of binding energy

$$B = \frac{m S^2}{2\hbar^2}.$$ (4.22)

The direct solution for this potential is left as an exercise.

This has a potential energy given by

$$V(z) = \tfrac{1}{2} K z^2$$ (4.23)

and describes a harmonic oscillator. A simple physical realization is a mass on the end of a spring, in which case z is the displacement from equilibrium and K is the spring constant (measured in $N\,m^{-1}$ or $J\,m^{-2}$). Vibrations of the crystal lattice (phonons) can also be described by parabolic potentials. A further example is a region with uniform charge density where the solution to Poisson's equation is parabolic. It is also possible to grow parabolic wells by continuously varying the composition of an alloy. We shall see in Chapter 6 that a magnetic field also gives rise to a parabolic potential.

A classical particle of mass m moving in the potential (4.23) executes harmonic motion, $z = z_0 \cos \omega_0 t$, with angular frequency

$$\omega_0 = \sqrt{\frac{K}{m}}.$$ (4.24)

The key feature is that the frequency is independent of the amplitude z_0 (although this is not unique to the parabolic potential). This result is limited in practice because the potential (4.23) is usually an approximation that holds only for small z.

In quantum mechanics we must solve the time-independent Schrödinger equation

$$\left(-\frac{\hbar^2}{2m} \frac{d^2}{dz^2} + \tfrac{1}{2} m \omega_0^2 z^2 \right) \psi(z) = \varepsilon \psi(z),$$ (4.25)

where equation (4.24) has been used to eliminate K in favour of ω_0. The first step is to get rid of the physical quantities z and ε, and replace them with pure numbers (dimensionless quantities). The physical problem (4.25) is thereby reduced to a purely mathematical problem. We do this by defining a 'length scale' z_0 and an 'energy scale' ε_0, which can be done by inspection here. Multiplying equation (4.25) by $(2m/\hbar^2)$ gives

$$\left[-\frac{d^2}{dz^2} + \left(\frac{m \omega_0}{\hbar} \right)^2 z^2 \right] \psi(z) = \frac{2m \varepsilon}{\hbar^2} \psi(z).$$ (4.26)

The first term inside the square brackets has dimensions of $(\text{length})^{-2}$ while the second has that of $(\text{length})^{+2}$ from z^2, so the factor in front of it must have dimensions of $(\text{length})^{-4}$. This suggests that we should define the length scale z_0 to eliminate this constant by setting

$$\bar{z} = \frac{z}{z_0}, \qquad z_0 = \sqrt{\frac{\hbar}{m\omega_0}}. \tag{4.27}$$

Carrying this out leaves

$$\left[-\frac{d^2}{d\bar{z}^2} + \bar{z}^2 \right] \psi(\bar{z}) = 2\frac{\varepsilon}{\hbar\omega_0}\psi(\bar{z}). \tag{4.28}$$

It is easy to get rid of the remaining physical quantities by defining the energy scale ε_0:

$$\bar{\varepsilon} = \frac{\varepsilon}{\varepsilon_0}, \qquad \varepsilon_0 = \hbar\omega_0. \tag{4.29}$$

The result is the dimensionless Schrödinger equation

$$\frac{d^2}{d\bar{z}^2}\psi(\bar{z}) + (2\bar{\varepsilon} - \bar{z}^2)\psi(\bar{z}) = 0. \tag{4.30}$$

Remember that $\bar{z}$ and $\bar{\varepsilon}$ are pure numbers.

We have achieved something already: we expect that the size of the wave functions will be roughly z_0, and that the separation between energy levels will be roughly ε_0. The exact numbers cannot be found without solving the equations, but these estimates are valuable.

We must now solve equation (4.30). The lazy route is to look for this equation in Abramowitz and Stegun (1972, section 22), or a similar book, but the full method proceeds as follows. At large $\bar{z}$ the term $2\bar{\varepsilon}$ becomes negligible compared with $\bar{z}^2$ and $\psi'' \sim \bar{z}^2\psi$. This term can be killed off by substituting

$$\psi(\bar{z}) = \exp(-\tfrac{1}{2}\bar{z}^2)u(\bar{z}) \tag{4.31}$$

into equation (4.30). A positive exponential would also remove the $\bar{z}^2$ but the wave function could not be normalized. The result is Hermite's equation for $u(\bar{z})$,

$$u'' - 2\bar{z}u' + (2\bar{\varepsilon} - 1)u = 0. \tag{4.32}$$

This can be solved by expanding $u(\bar{z})$ as a power series (Appendix 4). It has acceptable (polynomial) solutions $u_n(\bar{z})$ only if $(2\bar{\varepsilon} - 1)$ is an even integer. Therefore $\bar{\varepsilon} = n - \tfrac{1}{2}$, that is,

$$\varepsilon_n = (n - \tfrac{1}{2})\hbar\omega_0, \qquad n = 1, 2, 3, \ldots. \tag{4.33}$$

This is the result for the energy levels of a harmonic oscillator in quantum mechanics: they are equally spaced by $\hbar\omega_0$ above a zero-point energy of $\tfrac{1}{2}\hbar\omega_0$. (The energy

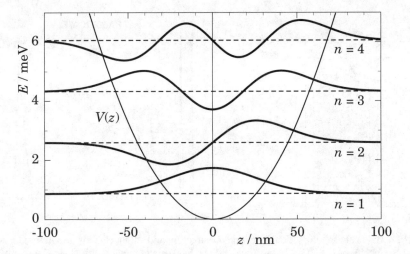

FIGURE 4.4. Potential well $V(z)$, energy levels, and wave functions of a harmonic oscillator. The potential is generated by a magnetic field of 1 T acting on electrons in GaAs.

levels of the harmonic oscillator are often counted from 0, but I have used 1 to be consistent with other potential wells.) The functions $u_n(\bar{z})$ are Hermite polynomials $H_{n-1}(\bar{z})$ apart from a factor for normalization; the first few are

$$H_0(t) = 1,$$
$$H_1(t) = 2t,$$
$$H_2(t) = 4t^2 - 2,$$
$$H_3(t) = 8t^3 - 12t, \tag{4.34}$$

using the notation of Abramowitz and Stegun (1972, section 22). The wave functions in terms of z, including normalization, are

$$\phi_{n+1}(z) = \left(\frac{1}{2^n n! \sqrt{\pi}}\right)^{1/2} \left(\frac{m\omega_0}{\hbar}\right)^{1/4} \exp\left(-\frac{m\omega_0 z^2}{2\hbar}\right) H_n\left[\left(\frac{m\omega_0}{\hbar}\right)^{1/2} z\right]. \tag{4.35}$$

The lowest few wave functions are plotted in Figure 4.4. They show the even–odd alternation seen in symmetric square wells. The wave function for $n = 1$ is a simple Gaussian function whose probability density is

$$|\phi_1(z)|^2 = \left(\frac{m\omega_0}{\pi\hbar}\right)^{1/2} \exp\left(-\frac{m\omega_0 z^2}{\hbar}\right). \tag{4.36}$$

The standard deviation of this density is

$$\Delta z = \sqrt{\frac{\hbar}{2m\omega_0}} = \frac{z_0}{\sqrt{2}}. \tag{4.37}$$

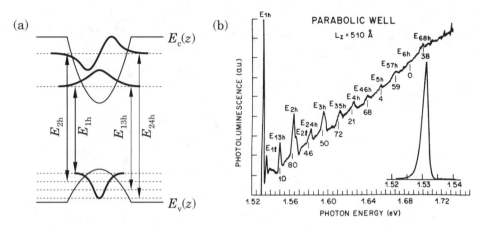

FIGURE 4.5. (a) Parabolic potential in both conduction and valence bands grown into GaAs by a graded composition of $Al_x Ga_{1-x} As$. The band gap has been reduced in this sketch, and only heavy holes are shown. (b) Photoluminescence in parabolic wells. [From Miller et al. (1984).]

These results are of great importance for a wide range of problems. The equal spacing of the energy levels is the analogue of the classical frequency being independent of the amplitude. It means that any wave packet made by superposing different states oscillates with the same frequency ω_0 (compare equation 1.55).

An example of a parabolic well grown by varying the composition of $Al_x Ga_{1-x} As$ is shown in Figure 4.5. The energy levels can be measured by optical transitions between states in the wells in the valence and conduction bands. A selection rule (Section 10.4) requires that both states have the same parity. The transitions are labelled by E_{mnh}, where m is the state of the electron, n is the state of the hole (omitted if it is the same), and 'h' refers to heavy holes, with 'l' for light holes. The composition of each well was graded from GaAs to $Al_{0.3}Ga_{0.7}As$ over a distance of 25.5 nm. Using the presently accepted value of $Q = \Delta E_c / \Delta E_g$ gives maximum values of $\Delta E_c = 0.23$ eV and $\Delta E_v = 0.14$ eV. The curvature of the conduction and valence bands, and therefore the energy levels, depends on ΔE_c and ΔE_v. This contrasts with a square well, where there is no dependence on ΔE_c and ΔE_v of the barriers within the simplest model, an infinitely deep well, and the dependence is weak for deep states in a finite well. Thus the experiment on parabolic wells is a sensitive test of the value of Q, which was long controversial (Section 3.3).

4.4 Triangular Well

The triangular well sketched in Figure 4.6 is useful because it is a simple description of the potential well at a doped heterojunction, to be studied in Chapter 9. There is an infinitely high barrier for $z < 0$ with a linear potential $V(z) = eFz$ for $z > 0$. It is convenient to write $V(z)$ in this way so that it describes a charge e in an electric

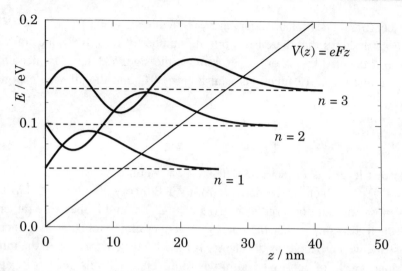

FIGURE 4.6. Triangular potential well $V(z) = eFz$, showing the energy levels and wave functions. The scales are for electrons in GaAs and a field of $5\,\mathrm{MV\,m^{-1}}$.

field F (the product eF is assumed to be positive). Note that F is used for the electric field rather than E to avoid confusion with the energy.

We must solve the Schrödinger equation

$$\left[-\frac{\hbar^2}{2m}\frac{d^2}{dz^2} + eFz \right] \psi(z) = \varepsilon \psi(z) \tag{4.38}$$

subject to the boundary condition $\psi(z=0) = 0$ imposed by the infinite barrier. Again we introduce dimensionless variables. Similar manipulation as for the harmonic oscillator shows that scales of distance and energy are

$$z_0 = \left(\frac{\hbar^2}{2meF} \right)^{1/3}, \qquad \varepsilon_0 = \left[\frac{(eF\hbar)^2}{2m} \right]^{1/3} = eFz_0. \tag{4.39}$$

The Schrödinger equation becomes

$$\frac{d^2\psi}{d\bar{z}^2} = (\bar{z} - \bar{\varepsilon})\psi(\bar{z}). \tag{4.40}$$

This can be simplified further by defining a new independent variable $s = \bar{z} - \bar{\varepsilon}$, and equation (4.40) reduces to the Stokes or Airy equation

$$\frac{d^2\psi}{ds^2} = s\psi. \tag{4.41}$$

This equation is discussed in Appendix 5. Its two independent solutions, the Airy functions $\mathrm{Ai}(s)$ and $\mathrm{Bi}(s)$, are sketched in Figure A5.1. We require a wave function that is well behaved as $z \to +\infty$, which is the same as $s \to +\infty$. This means that we can reject $\text{Bi}(s)$. The high boundary at $z = 0$ requires $\psi(z = 0) = \psi(s = -\bar{\varepsilon}) = 0$. Figure A5.1 shows that there is an infinite number of negative values of s where $\text{Ai}(s) = 0$ denoted by a_n, or $-c_n$ to remove the sign. We therefore need $\bar{\varepsilon} = c_n$ to ensure that the wave functions vanish at $z = 0$, and the allowed energies are given by

$$\varepsilon_n = c_n \left[\frac{(eF\hbar)^2}{2m} \right]^{1/3}, \qquad n = 1, 2, 3, \ldots. \tag{4.42}$$

The lowest level has $c_1 = 2.338$. There is a useful approximate formula $c_n \sim [\frac{3}{2}\pi(n - \frac{1}{4})]^{2/3}$, which can be derived from WKB theory (Section 7.4.2). Although this is most accurate for large n, it gives $c_1 \approx 2.320$ and is therefore rather good for all n. It also shows that the energy levels get closer together as n increases, because the well broadens as the energy is raised. This contrasts with the infinitely deep square well, of constant width, where the energy levels get farther apart as n increases. The parabolic well provides the middle case, with energy levels of constant separation.

The unnormalized wave functions are given by

$$\phi_n(z) = \text{Ai}(s) = \text{Ai}(\bar{z} - \bar{\varepsilon}) = \text{Ai}\left(\frac{eFz - \varepsilon}{\varepsilon_0} \right). \tag{4.43}$$

All wave functions have the same functional form and simply slide along z as the energy is changed: ϕ_1 contains one half-cycle, ϕ_2 contains two half-cycles, and so on. The wave functions lack the even or odd symmetry that we found in the wells considered previously because the triangular potential is not itself symmetric in z. Normalization is mentioned in Appendix 5.

4.5 Low-Dimensional Systems

We shall now use these results to see how three-dimensional electrons can be made to behave as though they are low-dimensional.

The starting point is the three-dimensional time-independent Schrödinger equation

$$\left[-\frac{\hbar^2}{2m} \nabla^2 + V(\mathbf{R}) \right] \psi(\mathbf{R}) = E\,\psi(\mathbf{R}). \tag{4.44}$$

There is no easy route to solving this equation if $V(\mathbf{R})$ is a general potential energy, but great simplifications occur for some forms of $V(\mathbf{R})$. In a layered structure the potential energy depends only on the coordinate z normal to the layers. This includes quantum wells made from alternating layers of GaAs and AlGaAs, and electrons trapped at a doped heterojunction. Thus $V(\mathbf{R}) = V(z)$ only, and the

Schrödinger equation (4.44) becomes

$$\left[-\frac{\hbar^2}{2m} \left(\frac{\partial^2}{\partial x^2} + \frac{\partial^2}{\partial y^2} + \frac{\partial^2}{\partial z^2} \right) + V(z) \right] \psi(x, y, z) = E \, \psi(x, y, z). \qquad (4.45)$$

The potential energy leaves the electrons free to move along x and y. The wave functions would be plane waves if there were no potential at all, which suggests that we should try plane waves for the motion along x and y. Write the wave function in the form

$$\psi(x, y, z) = \exp(i k_x x) \, \exp(i k_y y) \, u(z). \qquad (4.46)$$

We can then substitute this into the Schrödinger equation, check that it gives a correct solution for x and y, and find the equation for the unknown function $u(z)$. Equation (4.45) becomes

$$\left[-\frac{\hbar^2}{2m} \left(\frac{\partial^2}{\partial x^2} + \frac{\partial^2}{\partial y^2} + \frac{\partial^2}{\partial z^2} \right) + V(z) \right] \exp(i k_x x) \, \exp(i k_y y) \, u(z)$$

$$= \left[\frac{\hbar^2 k_x^2}{2m} + \frac{\hbar^2 k_y^2}{2m} - \frac{\hbar^2}{2m} \frac{\partial^2}{\partial z^2} + V(z) \right] \exp(i k_x x) \, \exp(i k_y y) \, u(z)$$

$$= E \, \exp(i k_x x) \, \exp(i k_y y) \, u(z). \qquad (4.47)$$

The exponential functions cancel from both sides of equation (4.47), confirming that the guess (4.46) is correct. Only functions of z remain:

$$\left[\frac{\hbar^2 k_x^2}{2m} + \frac{\hbar^2 k_y^2}{2m} - \frac{\hbar^2}{2m} \frac{d^2}{dz^2} + V(z) \right] u(z) = E \, u(z). \qquad (4.48)$$

The energy of the plane waves can be moved over to the right-hand side, giving

$$\left[-\frac{\hbar^2}{2m} \frac{d^2}{dz^2} + V(z) \right] u(z) = \left[E - \frac{\hbar^2 k_x^2}{2m} - \frac{\hbar^2 k_y^2}{2m} \right] u(z). \qquad (4.49)$$

A further substitution for the energy,

$$\varepsilon = E - \frac{\hbar^2 k_x^2}{2m} - \frac{\hbar^2 k_y^2}{2m}, \qquad (4.50)$$

reduces equation (4.49) to

$$\left[-\frac{\hbar^2}{2m} \frac{d^2}{dz^2} + V(z) \right] u(z) = \varepsilon \, u(z). \qquad (4.51)$$

This is a purely one-dimensional Schrödinger equation in z – the other two dimensions have been eliminated.

Suppose for now that we have solved this equation – it might be a square well, one of the examples considered in this chapter, or it might have to be done numerically – and that the wave functions are $u_n(z)$ with energy ε_n. Equations (4.46) and (4.50) show that the solution to the original three-dimensional problem is

$$\psi_{k_x,k_y,n}(x,\,y,\,z) = \exp(ik_x x)\,\exp(ik_y y)\,u_n(z), \qquad (4.52)$$

$$E_n(k_x,\,k_y) = \varepsilon_n + \frac{\hbar^2 k_x^2}{2m} + \frac{\hbar^2 k_y^2}{2m}. \qquad (4.53)$$

Three quantum numbers, k_x, k_y, and n, are needed to label the states because there are three spatial dimensions. Equations (4.52) and (4.53) can be written slightly more compactly by defining two-dimensional vectors for motion in the xy-plane, $\mathbf{r} = (x,\,y)$ and $\mathbf{k} = (k_x,\,k_y)$. This gives

$$\psi_{\mathbf{k},n}(\mathbf{r},\,z) = \exp(i\mathbf{k} \cdot \mathbf{r})\,u_n(z), \qquad (4.54)$$

$$E_n(\mathbf{k}) = \varepsilon_n + \frac{\hbar^2 \mathbf{k}^2}{2m}. \qquad (4.55)$$

The results are illustrated in Figure 4.7. The left-hand sketch shows the potential well $V(z)$ with its allowed energies ε_n and wave functions $u_n(z)$. The dispersion relation (4.55) is plotted in the centre. For a fixed value of n, this is simply the energy–wave-vector relation of a free two-dimensional electron gas with the bottom of the band shifted to ε_n. The relation for each n gives a parabola, called a *subband*

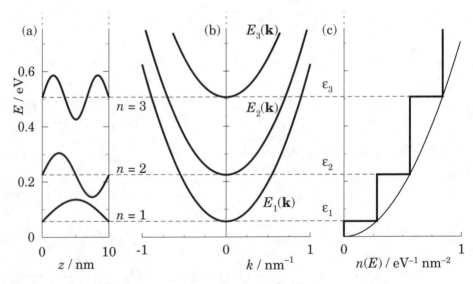

FIGURE 4.7. (a) Potential well with energy levels, (b) total energy including the transverse kinetic energy for each subband, and (c) steplike density of states of a quasi-two-dimensional system. The example is an infinitely deep square well in GaAs of width 10 nm. The thin curve in (c) is the parabolic density of states for unconfined three-dimensional electrons.

(*electric subband* for precision), starting at the energy $E_n(\mathbf{k} = 0) = \varepsilon_n$ when plotted against $|\mathbf{k}|$. There are no subbands for $0 < E < \varepsilon_1$, although there would have been allowed states here if there were no confinement. For $\varepsilon_1 < E < \varepsilon_2$ there are states only in the lowest subband. For $\varepsilon_2 < E < \varepsilon_3$ there are states in the two lowest subbands, with $n = 1$ and $n = 2$.

Energy is partitioned differently in the two subbands. The subband with $n = 2$ has a higher kinetic energy in the z-direction, ε_2 rather than ε_1, and consequently a lower kinetic energy and velocity in the transverse $\mathbf{k}$-plane. This separation of energy into different 'components', as if it were a vector, relies on the simple form of the kinetic energy operator in Schrödinger's equation and would not hold if we had to use a more complicated effective Hamiltonian, to be discussed in Section 4.9. There are more subbands to choose from at higher total energies, so electrons with the same total energy may have a number of different transverse wave vectors $\mathbf{k}$. This will be discussed further in Section 4.6, and is similar to the problem in electromagnetic theory of a wave guide with many allowed modes.

The subbands change the shape of the density of states $n(E)$. For a given subband (fixed n), the energy (4.55) is that of a two-dimensional electron gas with the bottom of the band at ε_n. The density of states is therefore that of a two-dimensional electron gas, a step function of height $m/\pi\hbar^2$, starting at ε_n. Each subband contributes a step, so the total density of states $n(E)$ looks like a staircase with jumps at the energies of the subbands. Note that this is a density of states per unit *area*, not volume. We shall see later that optical absorption measures $n(E)$, and Figure 8.4(b) shows the stepped density of states of a quantum well.

4.6 Occupation of Subbands

Now that we have calculated the allowed energy levels, we need to see what happens as we fill the system with electrons. The number of occupied subbands depends on the density of electrons and the temperature. The density of electrons per unit area $n_{2\mathrm{D}}$ can be found in the usual way by integrating the product of the density of states $n(E)$ and the Fermi–Dirac occupation function $f(E, E_\mathrm{F})$, where E_F is the Fermi energy:

$$n_{2\mathrm{D}} = \int_{-\infty}^{\infty} n(E)\, f(E, E_\mathrm{F})\, dE. \tag{4.56}$$

It is convenient to split this into subbands,

$$n_{2\mathrm{D}} = \sum_j n_j, \tag{4.57}$$

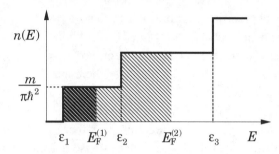

FIGURE 4.8. Occupation of steplike density of states for a quasi-two-dimensional system. Only one subband is occupied if the Fermi energy takes the lower value $E_F^{(1)}$, but two are occupied at the higher value $E_F^{(2)}$.

where n_j is the density of electrons in a two-dimensional band starting at ε_j. This is given by equation (1.114) in Section 1.8, which becomes

$$n_j = \frac{m}{\pi\hbar^2} \int_{\varepsilon_j}^{\infty} f(E, E_F)\, dE = \frac{m k_B T}{\pi\hbar^2} \ln\left[1 + \exp\left(\frac{E_F - \varepsilon_j}{k_B T} \right) \right]. \qquad (4.58)$$

This can be simplified in the limit of high or low temperature (as compared with $(E_F - \varepsilon_j)/k_B$) as before.

Suppose that we are in the limit of low temperatures where the electrons are degenerate. Then

$$n_{2D} = \sum_j n_j = \frac{m}{\pi\hbar^2} \sum_j (E_F - \varepsilon_j)\, \Theta(E_F - \varepsilon_j). \qquad (4.59)$$

This is illustrated in Figure 4.8 for two values of E_F. The lower value $E_F^{(1)}$ is less than ε_2 and lies only in the lowest subband. There is negligible occupation of the second and higher subbands provided that $\varepsilon_2 - E_F^{(1)} \gg k_B T$. The electrons behave as though they were in a real two-dimensional system with a single steplike density of states. All electrons are stuck in the same state $u_1(z)$ for the motion normal to the confining potential, and cannot move along z as this would require them to change their state. This limit can be achieved experimentally for a two-dimensional electron gas. The two-dimensional nature is somewhat delicate, and is easily lost if the temperature is raised or if the electrons gain energy from some external source such as an electric field and enter higher subbands.

The Fermi energy enters the second subband if the density of electrons is increased too far, as is the case for $E_F^{(2)}$. Electrons at the Fermi energy may now be in one of two subbands, each with a different velocity in the transverse plane. A limited degree of motion along z is possible by scattering between $u_1(z)$ and $u_2(z)$. This gives an experimental signature when more than one subband becomes occupied: scattering between subbands causes the mobility to drop (Figure 9.12). The system is only quasi-two-dimensional, although it remains far from the limiting case of free motion in all three dimensions. The maximum density of electrons that can occupy

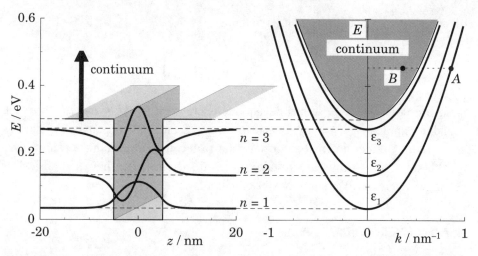

FIGURE 4.9. Quasi-two-dimensional system in a potential well of finite depth. Electrons with the same total energy can be bound in the well (A) or free (B).

the system before entering the second subband is given by $(m/\pi\hbar^2)(\varepsilon_2 - \varepsilon_1)$. The separation between the energy levels along z should be maximized to increase this density. A narrow well achieves this, until the energy levels get squeezed out of the top.

An interesting situation occurs if the energy levels are redrawn for a potential well of finite depth. Figure 4.9 shows three bound states in the well. All energies are allowed for motion along z above the top of the well. It is now possible for two electrons with the same total energy to be bound in the well with large $\mathbf{k}$ (A) or free along z (B), depending on how the energy is partitioned between z and the transverse plane. Electron A is rather precariously bound because even an elastic scattering event, one that preserves the total energy of the electron, can take it from A to B and allow it to escape from the well. This is called *real-space transfer* by analogy with $\mathbf{k}$-space transfer in materials such as GaAs where electrons scatter from the Γ-valley to the X-valleys at high electric fields. In both cases the reduction in mobility associated with transfer may give negative differential resistance.

4.7 Two- and Three-Dimensional Potential Wells

Many models of potential wells in two and three dimensions are used. The simplest is to extend the infinitely deep square well by multiplying together such a potential for each dimension. In two dimensions this leads to a rectangular potential for a box (with sides of length a and b) in the xy-plane. The wave function is a product of sine waves in each dimension, giving energies

$$\varepsilon_{n_x,n_y} = \frac{\hbar^2\pi^2}{2m}\left(\frac{n_x^2}{a^2} + \frac{n_y^2}{b^2}\right).$$

(4.60)

A square well with $a = b$ has degenerate energy levels; the symmetry between x and y requires $\varepsilon_{n_x,n_y} = \varepsilon_{n_y,n_x}$ and there are 'accidental' degeneracies such as $\varepsilon_{8,1} = \varepsilon_{4,7}$. A three-dimensional rectangular well can be treated in the same way.

Although these results are simple, practical systems often have cylindrical or spherical symmetry. We shall look briefly at a few examples.

4.7.1 CYLINDRICAL WELL

Start with free electrons in two dimensions, with $\mathbf{r} = (x, y)$. The most straightforward wave function has plane waves in both x and y, giving $\phi(\mathbf{r}) = \exp[i(k_x x + k_y y)] = \exp(i\mathbf{k} \cdot \mathbf{r})$. This is a plane wave moving in a direction set by $\mathbf{k}$ and with energy $\varepsilon_0(k) = \hbar^2 k^2/2m$. It can instead be written in polar coordinates $\mathbf{r} = (r, \theta)$ as $\phi(\mathbf{r}) = \exp(ikr\cos\theta)$, where θ is measured from the direction of $\mathbf{k}$. This is still a plane wave.

Instead we might wish to describe waves that radiate out in all directions from a point source, cylindrical rather than plane waves. To do this we must rewrite the Schrödinger equation for free electrons in two dimensions using polar coordinates:

$$-\frac{\hbar^2}{2m}\left(\frac{\partial^2}{\partial r^2} + \frac{1}{r}\frac{\partial}{\partial r} + \frac{1}{r^2}\frac{\partial^2}{\partial\theta^2}\right)\psi(r,\theta) = E\,\psi(r,\theta).$$

(4.61)

The angle θ appears only as a derivative, so a separable wave function of the form $\psi(r,\theta) = u(r)\exp(il\theta)$ is a solution. This resembles a plane wave in θ, reflecting the rotational symmetry of the system. Although it is a solution for any l, the wave function must be single valued: it should return to the same value if we encircle the origin and add 2π to θ. This restricts the angular momentum quantum number l to integral values, $l = 0, \pm1, \pm2, \ldots$. It is often written as m in two dimensions but can then be confused with the mass.

The radial function $u(r)$ obeys

$$\left[-\frac{\hbar^2}{2m}\left(\frac{d^2}{dr^2} + \frac{1}{r}\frac{d}{dr}\right) + \frac{\hbar^2 l^2}{2mr^2}\right]u(r) = E\,u(r).$$

(4.62)

The angular motion leaves a *centrifugal* term $\hbar^2 l^2/2mr^2$ in the potential energy which pushes states away from the origin as their angular momentum increases. Replacing E by $k = \sqrt{2mE}/\hbar$ for $E > 0$ gives

$$r^2\frac{d^2u}{dr^2} + r\frac{du}{dr} + [(kr)^2 - l^2]u = 0.$$

(4.63)

This is Bessel's equation with solutions $J_l(kr)$ and $Y_l(kr)$, Bessel functions of order l of the first and second kind (Abramowitz and Stegun 1972, chapter 9). The second kind Y_l diverge at the origin and therefore cannot be used over all space. These Bessel functions are standing waves but can be combined to give travelling waves. Their wavelike nature is clear in the asymptotic form for large arguments,

$$J_l(kr) \sim \sqrt{\frac{2}{\pi kr}} \cos\left(kr - \tfrac{1}{2}l\pi - \tfrac{1}{4}\pi\right); \tag{4.64}$$

Y_l has sine instead of cosine. The waves oscillate as expected, and the decay like $r^{-1/2}$ in the amplitude becomes r^{-1} in the intensity, which balances the increase in perimeter as the wave spreads out in the plane. If $E < 0$ the solutions are modified Bessel functions $I_l(\kappa r)$ and $K_l(\kappa r)$. These resemble real exponentials; I_l grows while K_l decreases from a divergence at the origin.

The solution for a cylindrical well with infinitely high walls follows from these results. This well has $V(r) = 0$ for $r < a$ and an impenetrable barrier for $r > a$. The wave function must vanish at $r = a$, which in turn requires $J_l(ka) = 0$. The Bessel function vanishes at zeros denoted by $j_{l,n}$ for $n = 1, 2, \ldots$. Thus the allowed values of the wave vector are $k = j_{l,n}/a$, and the wave functions and energies are

$$\phi_{nl}(\mathbf{r}) \propto J_l\left(\frac{j_{l,n}r}{a}\right) \exp(il\theta), \qquad \varepsilon_{nl} = \frac{\hbar^2 j_{l,n}^2}{2ma^2}. \tag{4.65}$$

The state of lowest energy has zero angular momentum, $l = 0$. The asymptotic expansion (4.64) shows that $j_{l,n} \sim (n + \tfrac{1}{2}|l| - \tfrac{1}{4})\pi$. This is accurate as $n \to \infty$ but is not far wrong even for $j_{0,1} \approx 2.405 = 0.765\pi$, compared with the asymptotic approximation of $\tfrac{3}{4}\pi$.

Figure 4.10 shows an experiment where iron atoms on the surface of gold were manipulated with the tip of a scanning tunnelling microscope to form a circular enclosure dubbed a *quantum corral*. The scanning tunnelling microscope was then used to image the states within the corral, a remarkable demonstration of cylindrical confinement. The measurements could be fitted using the particle-in-a-box model that we have just discussed, although later work showed that the iron atoms behave in a more complicated way than a simple hard wall.

4.7.2 TWO-DIMENSIONAL PARABOLIC WELL

There are two approaches to solving a parabolic potential $V(r) = \tfrac{1}{2}Kr^2$ in two dimensions. One is to add this potential to the radial Schrödinger equation (4.62) and find the allowed energies and wave functions. The resulting states have definite values of angular momentum l and are needed to treat a magnetic field using the 'symmetric gauge' in Section 6.4.2, where the full solution will be given. The energy levels of the oscillator are $\varepsilon_{nl} = (2n + |l| - 1)\hbar\omega_0$ where $\omega_0 = \sqrt{K/m}$ as before and $n = 1, 2, \ldots$. The lowest level has zero-point energy $\varepsilon_{1,0} = \hbar\omega_0$ and there are

(a)

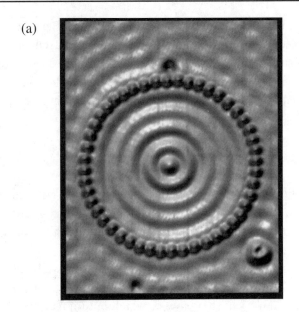

(b)

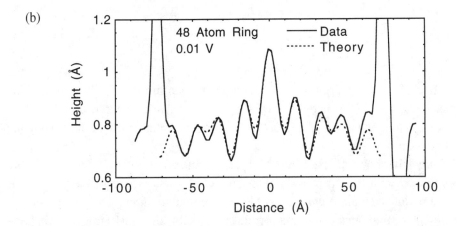

FIGURE 4.10. (a) Spatial image of the eigenstates of a 'quantum corral', defined by a ring of 48 iron atoms on a (111)-surface of copper. (b) Cross-section of the corral, fitted to a combination of states for a cylindrical box. [Reprinted with permission from Crommie, Lutz, and Eigler (1993). Copyright 1993 American Association for the Advancement of Science.]

N degenerate states with energy $N\hbar\omega_0$. The ladder of equally spaced energy levels, the most significant feature of the one-dimensional oscillator, is preserved.

The second approach is to note that the potential is separable, $\frac{1}{2}Kr^2 = \frac{1}{2}Kx^2 + \frac{1}{2}Ky^2$, so the Schrödinger equation in Cartesian coordinates can be reduced to separate equations for x and y. Each of these is just the one-dimensional problem that we solved in Section 4.3. Thus the total energy is $\varepsilon_{n_x,n_y} = [(n_x - \frac{1}{2}) + (n_y - \frac{1}{2})]\hbar\omega_0$, which is again a ladder rising from $\hbar\omega_0$. It is also clear that an energy $N\hbar\omega_0$ can be partitioned in N different ways between n_x and n_y, confirming the result for the degeneracy. Clearly this approach can be extended to less symmetric parabolic potentials of the form $\frac{1}{2}K_x x^2 + \frac{1}{2}K_y y^2$ with $K_x \neq K_y$, and three-dimensional parabolic potentials can be solved in the same way.

4.7.3 TWO-DIMENSIONAL COULOMB POTENTIAL

An attractive Coulomb potential $V(r) = -e^2/4\pi \epsilon r$ has an infinite number of bound states with energy $\varepsilon_n = -\mathcal{R}/(n - \frac{1}{2})^2$. The lowest wave function has a simple exponential decay, $\phi_1(r) \propto \exp(-2r/a_B)$. The radial Schrödinger equation can be solved in the same way as that for the parabolic well in Section 4.3, and the details will not be given here. The scales of energy and length, the *Rydberg energy* $\mathcal{R}$ and *Bohr radius* a_B, are given by

$$\mathcal{R} = \left(\frac{e^2}{4\pi\epsilon}\right)^2 \frac{m}{2\hbar^2} = \frac{\hbar^2}{2ma_B^2} = \frac{1}{2}\frac{e^2}{4\pi\epsilon a_B}, \tag{4.66}$$

$$a_B = \frac{4\pi\epsilon\hbar^2}{me^2}. \tag{4.67}$$

For hydrogen, where $m = m_0$ and $\epsilon = \epsilon_0$, the scales are $\mathcal{R} = 13.6\,\text{eV}$ and $a_B = 0.053\,\text{nm}$. In a semiconductor we replace m by $m_0 m^*$, where m^* is the effective mass of the carrier involved, and ϵ by $\epsilon_0\epsilon_b$, where ϵ_b is the background dielectric constant of the material. These substitutions change the scales drastically, giving $\mathcal{R} \approx 5\,\text{meV}$ and $a_B \approx 10\,\text{nm}$ for electrons in GaAs. The numbers are important because they set the natural scales for many processes in semiconductors. Indeed they are often used as the scales of energy and length in dimensionless atomic or Rydberg units to simplify calculations.

The lowest state has a binding energy of $4\mathcal{R}$, four times larger than the corresponding three-dimensional result. This is particularly important for an *exciton*, an electron, and a hole bound by their Coulomb attraction, as we shall see in Section 10.7.2.

4.7.4 SPHERICAL WELL

The starting point is again the solution for free three-dimensional motion in polar coordinates. There are now two angular momentum quantum numbers: $l = 0, 1, 2, \ldots$, which gives the total angular momentum, and $m = 0, \pm 1, \pm 2, \ldots, \pm l$, which gives its component along a particular axis, conventionally chosen to be z. The radial part of the solution $u(R)$ can be written as $w(R)/R$, where $w(R)$ obeys

$$\left[-\frac{\hbar^2}{2m}\frac{d^2}{dR^2} + \frac{\hbar^2}{2m}\frac{l(l+1)}{R^2} + V(R)\right]w(R) = E\,w(R), \tag{4.68}$$

very close to the usual one-dimensional Schrödinger equation. This includes a centrifugal potential as in the two-dimensional case, and a spherically symmetric potential energy $V(R)$ has been inserted.

The energy does not depend on the quantum number m. The solution of lowest energy is spherically symmetric, with $l = m = 0$, and in this case we have the familiar one-dimensional problem for $w(R)$. There is an important difference in the boundary conditions: the wave function $u(R)$ must not diverge at the origin, which requires $w(R{=}0) = 0$. This is the same condition obeyed by the *odd* states in a symmetric one-dimensional well and forbids the even solutions. Thus the lowest state in an infinitely deep spherical well of radius a has $w(R) = \sin(\pi R/a)$ with energy $\varepsilon_{100} = \hbar^2\pi^2/2ma^2$.

A finite well of radius a and depth V_0 can be solved by analogy with a one-dimensional well of full width $2a$. The lowest state in the spherical well corresponds to the *second* state in the one-dimensional well, given by the curve in Figure 4.2 for θ between $\pi/2$ and π; the region for $\theta \le \pi/2$ is forbidden as it gives an even state. This has the important consequence that a shallow spherical well has *no* bound states, in contrast with one dimension where at least one state is always bound. The spherical well requires $\theta_0 > \pi/2$ or $a^2V_0 > \pi^2\hbar^2/8m$ to bind a state. Other features of the solution follow from the one-dimensional case.

4.7.5   THREE-DIMENSIONAL COULOMB POTENTIAL

The three-dimensional Coulomb well, $V(R) = -e^2/4\pi\epsilon R$, has an infinite number of bound states with energies $\varepsilon_n = -\mathcal{R}/n^2$. The lowest state is an exponential:

$$\phi(R) = (\pi a_B^3)^{-1/2}\exp(-R/a_B). \tag{4.69}$$

The Rydberg energy and Bohr radius are defined by equations (4.66) and (4.67). These results control the binding of electrons on hydrogenic (ordinary) donors (Section 3.10) and three-dimensional excitons (Section 10.7.1).

4.8   Further Confinement Beyond Two Dimensions

In Section 4.5 we investigated electrons that were confined in a bound state along z and behaved as though they were two-dimensional. It is possible to confine them further and reduce their effective dimensionality to one or zero. If we take the confining potential to be a function of $\mathbf{r} = (x, y)$, the electrons remain free to move along z and the result is a *quantum wire*, closely analogous to an electromagnetic wave guide.

The analysis follows that for the two-dimensional electron gas. Start with the two-dimensional Schrödinger equation for the confining potential,

$$\left[-\frac{\hbar^2}{2m} \left(\frac{\partial^2}{\partial x^2} + \frac{\partial^2}{\partial y^2} \right) + V(\mathbf{r}) \right] u_{m,n}(\mathbf{r}) = \varepsilon_{m,n} u_{m,n}(\mathbf{r}). \qquad (4.70)$$

One of the simplified models considered in the previous section may be used for this, or it may require numerical methods, but assume that it has been solved. Then the total wave function and energy are given by

$$\psi_{m,n,k_z}(\mathbf{R}) = u_{m,n}(\mathbf{r}) \exp(ik_z z), \qquad (4.71)$$

$$E_{m,n}(k_z) = \varepsilon_{m,n} + \frac{\hbar^2 k_z^2}{2m}. \qquad (4.72)$$

These are the analogues of equations (4.52) and (4.53), and their interpretation is similar. Each value of $\varepsilon_{m,n}$ becomes the bottom of a one-dimensional subband, whose density of states goes like $E^{-1/2}$. The total density of states (per unit length) is

$$n(E) = \sum_{m,n} \frac{1}{\pi \hbar} \sqrt{\frac{2m}{E - \varepsilon_{m,n}}} \Theta(E - \varepsilon_{m,n}). \qquad (4.73)$$

This is sketched in Figure 4.11 with the parabola for free three-dimensional electrons for comparison. The density is distorted from the free case even more than in two dimensions. The bottoms of the subbands have become stronger features, divergences rather than steps, which is important in optical effects that reflect the density of states.

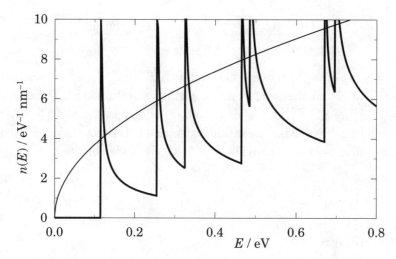

FIGURE 4.11. Density of states of a quasi-one-dimensional system. The curve was calculated for electrons in a 9×11 nm infinitely deep well in GaAs. The thin parabola is the density of states for unconfined three-dimensional electrons.

It is possible to go one stage further and confine electrons or holes in all three dimensions. Typically they are confined in one dimension by growth, a quantum well, or a doped heterojunction, and then restricted to a small area by etching or an electrostatic potential. The result is a *quantum dot*, essentially an artificial atom. The density of states is just a set of δ-functions as there is no free motion in any dimension.

4.9 Quantum Wells in Heterostructures

Although all the potentials described in the preceding sections need heterostructures to create them, we have just treated them as simple potential wells and ignored all the difficulties described in Section 3.11. We shall continue to assume that the bottom of the conduction band is at the same point in k-space (Γ) in all materials involved, and defer the much greater complications associated with valence bands to Section 10.3. The remaining issue is the different effective masses in the materials. This has two effects: the solution of the confining potential must take account of the different effective masses when matching the wave functions in different materials, and the reduction of the original three-dimensional problem to a one- or two-dimensional one becomes slightly less clean.

For simplicity consider electrons bound in a quantum well of GaAs sandwiched between two layers of AlGaAs. We shall treat first the effect of different masses on the one-dimensional problem, the finite well of width a and depth V_0 that we solved for a homostructure in Section 4.2. The wave numbers inside and outside are given by the modified expressions

$$k = \frac{\sqrt{2m_0 m_W (E - E_c^W)}}{\hbar}, \qquad \kappa = \frac{\sqrt{2m_0 m_B (E_c^B - E)}}{\hbar}. \qquad (4.74)$$

Here m_W is the effective mass in the well with E_c^W the bottom of the conduction band; m_B and E_c^B are the corresponding quantities in the barrier. The depth of the well $V_0 = E_c^B - E_c^W \equiv \Delta E_c$. The matching condition on the wave function at an interface must be changed, as we saw in Section 3.11. The straightforward matching of the derivative used in equation (4.8) is replaced by the condition

$$\frac{1}{m_W} \frac{d\psi}{dz}\bigg|_{z=a/2_-} = \frac{1}{m_B} \frac{d\psi}{dz}\bigg|_{z=a/2_+}. \qquad (4.75)$$

Thus the derivatives obey

$$\frac{Ck}{m_W} \left\{ \begin{matrix} -\sin \\ \cos \end{matrix} \right\} \left(\frac{ka}{2}\right) = -\frac{D\kappa}{m_B} \exp\left(\frac{\kappa a}{2}\right), \qquad (4.76)$$

TABLE 4.1 Dependence on the effective mass in the barriers m_B of the energies of the states bound in a well 5 nm wide and 1 eV deep, with effective mass $m_W = 0.067$ inside the well.

m_B	ε_1 (eV)	ε_2 (eV)	ε_3 (eV)
0.067	0.131	0.504	0.981
0.15	0.108	0.446	0.969

and dividing by the unchanged matching condition for $\psi(a/2)$ gives

$$\left\{\begin{array}{c} \tan \\ -\cot \end{array}\right\}\left(\frac{ka}{2}\right) = \frac{m_W \kappa}{m_B k} = \sqrt{\frac{m_W}{m_B}\left(\frac{2m_0 m_W V_0}{\hbar^2 k^2} - 1\right)}. \tag{4.77}$$

Again we define $\theta = ka/2$ and

$$\theta_0^2 = \frac{m_0 m_W V_0 a^2}{2\hbar^2}, \tag{4.78}$$

which depends only on the mass inside the well. The matching condition then becomes

$$\left\{\begin{array}{c} \tan \\ -\cot \end{array}\right\}\theta = \sqrt{\frac{m_W}{m_B}\left(\frac{\theta_0^2}{\theta^2} - 1\right)}. \tag{4.79}$$

This can be solved in exactly the same way as for the case of equal masses. The graphical solution in Figure 4.2 proceeds as before except that the square-root curve is scaled by a factor of $\sqrt{m_W/m_B}$. Suppose, for example, that m_B can be varied, holding m_W and V_0 constant. Raising m_B lowers the right-hand side of equation (4.79), so the energies of the bound states are all lowered. This is not surprising as we generally expect a higher mass to lead to lower energies. The number of bound states remains constant, however, because it depends on θ_0, which contains only properties of the well.

As an example, consider a well with 5 nm of GaAs sandwiched between AlAs, making the dubious assumptions that we need only consider the Γ-valley and that the bands are parabolic. The masses are $m_W = 0.067$ and $m_B = 0.15$, with $V_0 \approx 1$ eV. This is a rather more extreme example than the more usual GaAs–Al$_x$Ga$_{1-x}$As structure with $x \approx 0.3$, and was chosen to amplify the effect of the different mass. This well has $\theta_0 = 3.3$ and therefore three bound states, whose energies are listed in Table 4.1. All energies move down as m_B is increased, as expected, the middle one by over 50 meV. The top state is so weakly bound that its derivative at the boundary is almost zero, and a change in mass has little effect on this. The wave functions for $m_B = 0.15$ are plotted in Figure 4.12. The kink in the wave functions introduced by the matching condition (4.75) is obvious. Although the full wave function must

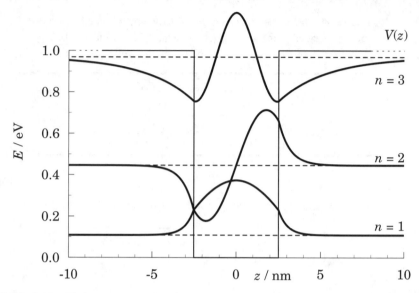

FIGURE 4.12. Finite square well of depth $V_0 = 1\,\mathrm{eV}$, width $a = 5\,\mathrm{nm}$ along z, and effective masses $m_\mathrm{W} = 0.067$ in the well and $m_\mathrm{B} = 0.15$ in the barrier.

be smooth, complete calculations confirm that the envelope function shows this behaviour (Figure 3.22).

The second issue is the reduction of the original three-dimensional Schrödinger equation to a one-dimensional equation. The Schrödinger equations in the two materials are

$$\left[E_\mathrm{c}^\mathrm{W} - \frac{\hbar^2}{2m_0 m_\mathrm{W}} \nabla^2 \right] \psi(\mathbf{r}) = E\,\psi(\mathbf{r}) \qquad \text{(well)}, \tag{4.80}$$

$$\left[E_\mathrm{c}^\mathrm{B} - \frac{\hbar^2}{2m_0 m_\mathrm{B}} \nabla^2 \right] \psi(\mathbf{r}) = E\,\psi(\mathbf{r}) \qquad \text{(barrier)}. \tag{4.81}$$

The effective potential from ΔE_c varies only along z, as in Section 4.5, so we can again write the wave function as

$$\psi_{\mathbf{k},n}(\mathbf{r}, z) = \exp(i\mathbf{k} \cdot \mathbf{r})\, u_n(z). \tag{4.82}$$

Substituting this wave function into the two three-dimensional Schrödinger equations yields a pair of one-dimensional equations

$$\left[E_\mathrm{c}^\mathrm{W} - \frac{\hbar^2}{2m_0 m_\mathrm{W}} \frac{d^2}{dz^2} + \frac{\hbar^2 \mathbf{k}^2}{2m_0 m_\mathrm{W}} \right] u_n(z) = E\,u_n(z), \tag{4.83}$$

$$\left[E_\mathrm{c}^\mathrm{B} - \frac{\hbar^2}{2m_0 m_\mathrm{B}} \frac{d^2}{dz^2} + \frac{\hbar^2 \mathbf{k}^2}{2m_0 m_\mathrm{B}} \right] u_n(z) = E\,u_n(z). \tag{4.84}$$

TABLE 4.2 Dependence on transverse wave vector $\mathbf{k}_\perp$ of the energies of the states bound in a well 5 nm wide and 1 eV deep, with effective mass $m_W = 0.067$ inside the well and $m_B = 0.15$ outside.

k	$\dfrac{\hbar^2 k^2}{2m_0 m_W}$	$\dfrac{\hbar^2 k^2}{2m_0 m_B}$	$V_0(k)$	ε_1	ε_2	ε_3	m_{eff}
(nm^{-1})	(eV)	(eV)	(eV)	(eV)	(eV)	(eV)	
0.0	0.000	0.000	1.000	0.108	0.446	0.969	0.067
0.5	0.142	0.064	0.921	0.106	0.435	0.919	0.069
1.0	0.570	0.254	0.685	0.096	0.397	—	0.076

The difference in energy between the two regions, which forms the well, now depends on $k = |\mathbf{k}|$ and is given by

$$
V_0(k) = \left(E_c^B + \frac{\hbar^2 k^2}{2m_0 m_B} \right) - \left(E_c^W + \frac{\hbar^2 k^2}{2m_0 m_W} \right)
$$

$$
= \Delta E_c + \frac{\hbar^2 k^2}{2m_0} \left(\frac{1}{m_B} - \frac{1}{m_W} \right). \tag{4.85}
$$

The correction is negative for GaAs–AlGaAs because $m_B > m_W$, so the potential well appears to become shallower as the transverse kinetic energy increases. Thus the total energy of an electron in a bound state is given by

$$
E_n(\mathbf{k}) = \frac{\hbar^2 k^2}{2m_0 m_W} + \varepsilon_n(k), \tag{4.86}
$$

where the energy of the bound state ε_n also depends on k through the variation in depth of the well.

Again take the 5 nm well of GaAs between AlAs as an example, with wave numbers $k = 0, 0.5$, and $1.0\,\text{nm}^{-1}$. The kinetic energies, effective depth, and energies of the bound states are listed in Table 4.2. The depth of the well is reduced so much at the highest wave number (admittedly a rather large value) that the third state is no longer bound. The energy of the second state falls by 49 mV, compared with its kinetic energy of 570 meV. We can account for this approximately by introducing yet another effective mass, such that $\hbar^2 k^2/2m_0 m_{\text{eff}}$ is the energy above that at $k = 0$, taking account of both the increased transverse kinetic energy and the decreased energy of the bound state along z. Thus

$$
E_n(\mathbf{k}) \approx \varepsilon_n(k=0) + \frac{\hbar^2 k^2}{2m_0 m_{\text{eff}}}. \tag{4.87}
$$

These new effective masses depend on the index n of the bound state and are tabulated. A more physical way of explaining their origin is to note that the electron spends part of its time in the barrier, not just the well, and therefore tends to acquire some of the barrier's characteristics. It can be shown that $m_{\text{eff}} \approx m_W P_W + m_B P_B$, where P_W is the probability of finding the electron in the well and P_B is that for finding it in the barrier.

Further considerations enter if the band structure in the two materials on either side of a heterojunction is qualitatively different. An obvious example is GaAs–AlAs, where the lowest conduction band is at Γ on one side of the junction and at X on the other. The issues are discussed briefly in connection with tunnelling in Section 5.8.1. The valence and conduction bands overlap in a type III junction, as between InAs and GaSb, which presents further problems which will not be addressed here.

This concludes the discussion of electrons bound in quantum wells. In the next chapter we shall consider the opposite situation of barriers rather than wells, and transport due to the tunnelling of electrons.

Further Reading

The general theory of low-dimensional structures is described by Bastard (1988), Weisbuch and Vinter (1991), and Kelly (1995); Bastard has a good treatment of the effect of heterostructures on the energy levels and effective mass.

Landau, Lifshitz, and Pitaevskii (1977) contains the solutions of many problems involving potential wells as examples. These often provide useful models.

EXERCISES

4.1   Calculate the number of bound states and the lowest energy level for electrons and light and heavy holes in a GaAs well 6 nm wide sandwiched between layers of $Al_{0.35}Ga_{0.65}As$. How good an approximation is an infinitely deep well?

       Hence recalculate the energy of the optical transition in the 6 nm well in the sample whose photoluminescence was shown in Figure 1.4. How much difference does the finite depth make? Does it improve agreement with experiment, or are there any signs of error (in the model or the growth)?

       How large a shift would occur if the thickness of the well were to fluctuate by a monolayer?

4.2   Calculate the probability of finding an electron in the lowest bound state inside the 4 nm well, using equation (4.21). This requires the ratio of coefficients D/C, which can be calculated from equation (4.7) or (4.8) after ε (and hence k and κ) has been found. Explain qualitatively how this fraction depends on the width of the well.

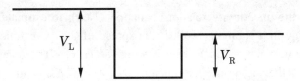

FIGURE 4.13. An asymmetric potential well with barriers of height V_L on the left and V_R on the right.

4.3   Plot a graph of the energy of the bound states in a GaAs well 0.3 eV deep as a function of width from 0 to 20 nm.

4.4   How would you go about finding the bound states in an asymmetric well, such as that shown in Figure 4.13? A detailed solution is not recommended. Remember that you will have to discard all the simplifications due to symmetry that we have made in this chapter! It is fortunate that most wells grown in practice are symmetric.

Suppose that V_L becomes infinitely large to make a hard wall. Show that the problem now becomes simple again using the results of Section 4.2.

4.5   Solve the δ-function well with $V(z) = -S\delta(z)$ directly. The wave function is a decaying exponential everywhere, $\phi(z) \propto \exp(-\kappa|z|)$, like the wave function outside a finite well. Its discontinuity in slope at $z = 0$ must be balanced by the δ-function in the potential. Integrate the Schrödinger equation over z from just below zero to just above to show that

$$\left.\frac{d\phi}{dz}\right|_{z=0_+} - \left.\frac{d\phi}{dz}\right|_{z=0_-} = -\frac{2mS}{\hbar^2}\phi(0) \qquad (E4.1)$$

(the right-hand side of the Schrödinger equation vanishes because it contains no singularities). Show that this gives $\kappa = mS/\hbar^2$ and $B = \hbar^2\kappa^2/2m$, in agreement with equation (4.22).

4.6   We shall see in Chapter 6 that a magnetic field B gives rise to a parabolic potential, with a 'spring constant' $K = e^2B^2/m = m\omega_c^2$, where e is the charge of the particle, m is its mass, and $\omega_c = eB/m$ is called the cyclotron frequency. Calculate the cyclotron energy and length scale z_0 (known as the magnetic length l_B) for an electron in GaAs in a field of 1 T.

4.7   What are the energies of the bound states in a potential that is parabolic for $x > 0$ with a hard wall at $x = 0$?

4.8   Estimate the spacing of the energy levels for electrons and holes (both light and heavy), and the optical transition energies, for the sample shown in Figure 4.5. Make the obvious approximation that the parabolic potential continues upwards, rather than stopping when it reaches the maximum imposed by the bands of the AlGaAs; this should be adequate for the lowest levels. Ignore also the variation in effective mass through the heterostructure.

4.9   Use the uncertainty relation, equation (1.63), to estimate the zero-point energy in a parabolic well. If the spread of the wave function is Δx, the potential energy can be estimated as $\frac{1}{2}K(\Delta x)^2$. The uncertainty relation shows that the spread in momentum $\Delta p \geq \hbar/(2\Delta x)$, and the kinetic energy is $(\Delta p)^2/2m$. Write the total energy ε in terms of Δx, find the value of Δx that minimizes it, and obtain the exact result $\varepsilon = \frac{1}{2}\hbar\omega_0$.

4.10 A rough value for the electric field confining electrons near a doped heterojunction is $5\,\mathrm{MV\,m^{-1}}$ (we shall see how to calculate it in Section 9.3.1). Calculate the first few energy levels in a triangular potential with this slope. Are the results useful for the GaAs–AlGaAs system, where the barrier at $z = 0$ is only about $0.3\,\mathrm{eV}$ high?

4.11 A symmetric triangular well, with $V(z) = |eFz|$, can be produced by depositing a sheet of donors in the middle of a semiconductor (δ-doping). Gauss's theorem shows that $F = eN_{\mathrm{D}}^{(2\mathrm{D})}/2\epsilon_0\epsilon_b$ if all donors are ionized, where $N_{\mathrm{D}}^{(2\mathrm{D})}$ is the areal density of donors and ϵ_b is the dielectric constant of the host semiconductor. Calculate the first few energy levels in GaAs if $N_{\mathrm{D}}^{(2\mathrm{D})} = 5 \times 10^{15}\,\mathrm{m^{-2}}$ and $\epsilon_b = 13$. You will need the zeros of $\mathrm{Ai}'(x)$, tabulated in Appendix 5.

4.12 Calculate $n(E)$ for an infinitely deep well of width a, and compare it with the results for free three-dimensional electrons (Figure 4.7(c)). There is a slight difficulty in comparing these two functions, because the steplike density of states for the well is a density per unit *area*, whereas the $E^{1/2}$ function for free electrons is a density per unit *volume*. We can allow for this by multiplying the three-dimensional density of states by the width a of the well, which turns it into a density of states per unit area that can be directly compared with the two-dimensional result (or, of course, both can be converted to three-dimensional units). Plot on the same axes the density of states for free electrons in 3D and the density of states for a GaAs well 10 nm wide and for a well 20 nm wide. Show that the top of each step just touches the parabola. What happens as the well gets wider?

It is desired to put as many electrons as possible in the lowest subband without occupying the second subband. Does it help to use a material with a different effective mass? Would the conclusion be different if a finite well were used?

4.13 How many electrons can be trapped at the heterojunction described by the triangular potential in the earlier problem with only the lowest subband occupied? Roughly how cold must the sample be for thermal occupation of higher subbands to be negligible?

4.14 A quantum dot is confined in two dimensions by a parabolic potential that rises 50 meV over a radius of 100 nm. Confinement is much stronger in the third dimension, and it can be assumed that all electrons remain in the lowest state for this dimension. How many states are occupied if the Fermi level lies 12 meV above the minimum? How would the results differ if the dot were instead modelled with a hard wall at 50 nm radius?

4.15 Is the condition for a bound state to exist likely to pose practical problems for a spherical well of GaAs surrounded by AlGaAs? Does this change if the well is InAs, with its lower effective mass? Take the barrier to be 1 eV high. There is considerable interest in 'self-organized' dots of InAs grown on GaAs or AlGaAs.

4.16 Calculate the density of states for electrons trapped in an infinitely deep two-dimensional 11×9 nm well. The energies of the subbands are given in equation (4.60). Compare it with the three-dimensional result (converted to a density per unit length).

4.17 How does the changing depth of a potential well in a heterostructure as a function of the transverse wave vector $\mathbf{k}$ affect the trapping of electrons in different subbands, shown in Figure 4.9 for a homostructure? Use a GaAs–AlAs structure for illustration.

4.18 A monolayer of InAs between GaAs can be modelled very roughly as a potential well for electrons of depth 1 eV and width 0.3 nm, with effective masses 0.025 inside and 0.067 outside. Calculate its bound state(s) for $\mathbf{k} = \mathbf{0}$. How far does the wave function extend? Is a δ-function a good approximation to the potential?

5 TUNNELLING TRANSPORT

In Chapter 4 we looked at how electrons could be trapped in various examples of potential wells and made to behave as though they were only two-dimensional (or less). In this chapter we shall look at free electrons that encounter barriers or other obstacles as they travel. Again, most of the potential profiles will be one-dimensional and we need only solve the Schrödinger equation in this dimension, although the other dimensions enter into the calculation of the current. We shall use the general tool of T-matrices, which can simply be multiplied together to yield the transmission coefficient for an arbitrary sequence of steps and plateaus. Two particular applications are to resonant tunnelling through a double barrier and to an infinite, regularly spaced sequence of barriers, a superlattice. Two barriers show a narrow peak in the transmission when the energy of the incident electron matches that of a *resonant* or *quasi-bound* state between the barriers (Section 5.5). This peak broadens into a band in the superlattice, and Section 5.6 shows how band structure and Bloch's theorem emerge for a specific example.

Many low-dimensional structures cannot simply be factorized into one-dimensional problems but have many leads, each with several propagating modes. These will be treated in Section 5.7 and we shall derive one of the famous results of low-dimensional systems, the quantized conductance. Finally we need to address the extra details introduced by heterostructures, particularly the different effective masses on either side of a heterojunction.

An important restriction that applies to most of these results is that the electrons must remain *coherent*. In other words, we shall treat the electrons as pure waves, such as electromagnetic waves travelling through media without absorption. This may not be realistic in practice and will be discussed later.

5.1 Potential Step

Suppose that an electron of energy E, travelling in the $+z$-direction, hits the upward potential step at $z = 0$ shown in Figure 5.1, where $V(z)$ jumps from 0 for $z < 0$ to a positive value V_0 for $z > 0$. Classically the electron would always be reflected if its

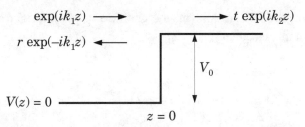

FIGURE 5.1. A step at $z = 0$ with potential energy $V = 0$ on the left and $V = V_0$ on the right. There is an incident wave from the left, giving a reflected wave on the left and a transmitted wave on the right.

energy $E < V_0$, whereas it always passes over the step if $E > V_0$. This is modified by the wavelike nature of electrons in quantum mechanics.

Let the incoming wave be $\exp(ik_1 z)$ in region 1 for $z < 0$ to the left of the step. The temporal factor of $\exp(-iEt/\hbar)$ is the same for all waves and will be ignored. There are two outgoing waves, a reflected wave $r \exp(-ik_1 z)$ in region 1 and a transmitted wave $t \exp(ik_2 z)$ in region 2. The coefficients r and t are *reflection* and *transmission amplitudes* and are complex in general. The wave numbers are given by $k_1^2 = 2mE/\hbar^2$ and $k_2^2 = 2m(E - V_0)/\hbar^2$. Assume first that $E > V_0$, so k_2 is real and all the waves propagate. There could be an incoming wave from the right as well, and it is no more difficult to solve the general problem illustrated schematically by Figure 5.2. Two of the amplitudes A, B, C, and D must be provided and the other two can then be deduced. The solutions to the Schrödinger equation are

$$\psi(z) = \begin{cases} A \exp(ik_1 z) + B \exp(-ik_1 z), & z < 0, \\ C \exp(ik_2 z) + D \exp(-ik_2 z), & z > 0. \end{cases} \tag{5.1}$$

As usual, the value and slope of the wave function must be continuous everywhere. Matching at the step at $z = 0$ gives

$$A + B = C + D, \qquad k_1(A - B) = k_2(C - D). \tag{5.2}$$

The next derivative, $d^2\psi/dz^2$, is not continuous because it is proportional to $V(z)$, which jumps at the step.

FIGURE 5.2. A general transmission problem with an unspecified 'barrier' separating region 1 on the left from region 2 on the right. There are incoming and outgoing waves on both sides. The potential is constant away from the barrier so the wave functions are plane waves but they may have different wave numbers k_1 and k_2.

If the waves on the left are known, the coefficients on the right are given by solving the simultaneous equations (5.2):

$$C = \tfrac{1}{2}(1 + k_1/k_2)A + \tfrac{1}{2}(1 - k_1/k_2)B,$$

$$D = \tfrac{1}{2}(1 - k_1/k_2)A + \tfrac{1}{2}(1 + k_1/k_2)B. \tag{5.3}$$

This can be related to the simpler situation of Figure 5.1 by setting $A = 1$, $B = r$, $C = t$, and $D = 0$, which gives

$$t = \frac{2k_1}{k_1 + k_2}, \qquad r = \frac{k_1 - k_2}{k_1 + k_2}. \tag{5.4}$$

Usually one wants to know the flux or current of electrons rather than the amplitude of the waves. A wave $F \exp(ikx)$ carries (number) current density $(\hbar k/m)|F|^2$, which depends on the wave number as well as the amplitude. The *flux transmission and reflection coefficients*, the ratios of the currents, are therefore

$$T = \frac{(\hbar k_2/m)|t|^2}{(\hbar k_1/m)} = \frac{k_2}{k_1}|t|^2 = \frac{4k_1 k_2}{(k_1 + k_2)^2}, \tag{5.5}$$

$$R = \frac{(\hbar k_1/m)|r|^2}{(\hbar k_1/m)} = |r|^2 = \left(\frac{k_1 - k_2}{k_1 + k_2}\right)^2. \tag{5.6}$$

An important check is that these coefficients obey the conservation of particles: every incident particle must be either reflected or transmitted, so

$$R + T = 1. \tag{5.7}$$

It is easy to verify that (5.5) and (5.6) satisfy this.

If $E < V_0$, the waves to the right of the step are evanescent (real, decaying exponentials) with a wave number κ_2 given by $\kappa_2^2 = 2m(V_0 - E)/\hbar^2$. The outgoing wave changes from $C \exp(ik_2 z)$ to the decaying exponential $C \exp(-\kappa_2 z)$, and the incoming one becomes the growing $D \exp(+\kappa_2 z)$. Thus k_2 is replaced by $i\kappa_2$. It is important that k_2 goes to $+i\kappa_2$ rather than $-i\kappa_2$; one way of remembering the sign is that it is the same sign that would be found for a forward-going wave in an absorbent medium (although absorption does *not* take place here). The waves can be matched at $z = 0$ again, or one can simply use the replacement $k_2 \to i\kappa_2$ to get

$$t = \frac{2k_1}{k_1 + i\kappa_2}, \qquad r = \frac{k_1 - i\kappa_2}{k_1 + i\kappa_2}. \tag{5.8}$$

These are complex. No flux is carried by the purely decaying wave $t \exp(-\kappa_2 z)$ (equation 1.36) so $T = 0$ and $R = |r|^2 = 1$. Thus there is perfect reflection of the flux, as found classically. An exponential tail of the wave function tunnels into the barrier but carries no current.

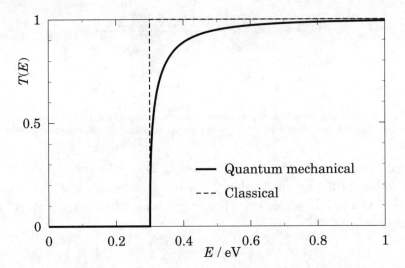

FIGURE 5.3. Transmission coefficient $T(E)$ as a function of the energy E of the incident electron for a step 0.3 eV high in GaAs. The broken line is the classical result.

The flux transmission coefficient $T(E)$ is plotted in Figure 5.3. Classically the transmission would be $T = 1$ for all energies above the step but the quantum-mechanical result only approaches this for high energies; in fact, $T \rightarrow 0$ like a square root as E decreases towards V_0.

Although steps are of importance in confining electrons, a more useful structure can be made by putting together an upward and a downward step to form a barrier. This can be handled by matching wave functions in the same way as the step, but a more interesting and general technique is to use T-matrices.

5.2 *T*-Matrices

The relation (5.3) that we found between the four waves at the step can be expressed as a matrix equation in a number of ways. The most fundamental is the S-matrix, which expresses 'what comes out' as a function of 'what goes in':

$$\begin{pmatrix} C \\ B \end{pmatrix} = \mathsf{S} \begin{pmatrix} A \\ D \end{pmatrix}. \tag{5.9}$$

We shall concentrate instead on the transfer or T-matrix, which gives the waves on the right (region 2 in Figure 5.2) as a function of those on the left (region 1):

$$\begin{pmatrix} C \\ D \end{pmatrix} = \mathsf{T}^{(21)} \begin{pmatrix} A \\ B \end{pmatrix} = \begin{pmatrix} T_{11}^{(21)} & T_{12}^{(21)} \\ T_{21}^{(21)} & T_{22}^{(21)} \end{pmatrix} \begin{pmatrix} A \\ B \end{pmatrix}. \tag{5.10}$$

Note the order of the superscripts on $\mathsf{T}^{(21)}$; the reason for this will soon become clear. The reason for choosing T-matrices is that they can easily be multiplied to

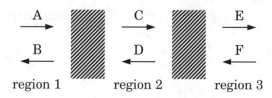

region 1   region 2   region 3

FIGURE 5.4. Waves in three regions, in each of which the potential energy is constant, separated by two barriers to show how T-matrices may be multiplied to get the overall transmission amplitudes.

build up complicated barriers in one dimension. This is illustrated in Figure 5.4 for three regions of flat potential separated by two features that scatter the electrons. The amplitudes of the waves in the three regions are related by T-matrices for the junctions:

$$\begin{pmatrix} C \\ D \end{pmatrix} = \mathsf{T}^{(21)} \begin{pmatrix} A \\ B \end{pmatrix}, \qquad \begin{pmatrix} E \\ F \end{pmatrix} = \mathsf{T}^{(32)} \begin{pmatrix} C \\ D \end{pmatrix}. \tag{5.11}$$

These can obviously be combined to give

$$\begin{pmatrix} E \\ F \end{pmatrix} = \mathsf{T}^{(32)} \mathsf{T}^{(21)} \begin{pmatrix} A \\ B \end{pmatrix} \equiv \mathsf{T}^{(31)} \begin{pmatrix} A \\ B \end{pmatrix}, \tag{5.12}$$

where $\mathsf{T}^{(31)} = \mathsf{T}^{(32)} \mathsf{T}^{(21)}$. This explains the order of the superscripts and can clearly be extended to an arbitrary number of features in the potential. Although the regions in space are conventionally numbered from left to right, the matrices must be written in the opposite order so that they act on the amplitudes of the waves in the correct sequence.

The reflection and transmission amplitudes can be recovered from the elements of T. The waves in the left and right regions are related by

$$\begin{pmatrix} t \\ 0 \end{pmatrix} = \mathsf{T} \begin{pmatrix} 1 \\ r \end{pmatrix} = \begin{pmatrix} T_{11} & T_{12} \\ T_{21} & T_{22} \end{pmatrix} \begin{pmatrix} 1 \\ r \end{pmatrix}, \tag{5.13}$$

whence

$$r = -\frac{T_{21}}{T_{22}}, \qquad t = \frac{T_{11}T_{22} - T_{12}T_{21}}{T_{22}}. \tag{5.14}$$

We shall see later that the numerator in the expression for t is often unity and $t = 1/T_{22}$.

The T-matrix for the step follows from equation (5.3):

$$\mathsf{T}^{(21)} = \frac{1}{2k_2} \begin{pmatrix} k_2 + k_1 & k_2 - k_1 \\ k_2 - k_1 & k_2 + k_1 \end{pmatrix} \equiv \mathsf{T}(k_2, k_1). \tag{5.15}$$

This is for $E > V_0$; set $k_2 = i\kappa_2$ if $E < V_0$.

The preceding matrix is for a step at the origin and it will be helpful to generalize it to a step at $z = d$. The only difference lies in the phase of the waves that hit the object. We can allow for these phases by constructing the new matrix $\mathsf{T}(d)$ from that at the origin $\mathsf{T}(0)$ in three steps.

(i) Translate the object from d to the origin by writing $z' = z - d$. The incoming wave on the left, $\exp(ik_1 z)$, becomes $\exp(ik_1 z') \exp(ik_1 d)$, so we need to multiply the corresponding amplitude by $\exp(ik_1 d)$. Similarly, the amplitude of the outgoing wave on the left needs to be multiplied by $\exp(-ik_1 d)$. These two factors can be written as a diagonal matrix multiplying the original vector of amplitudes.

(ii) The T-matrix calculated at the origin $\mathsf{T}(0)$ can now be used to give the amplitudes on the right.

(iii) The object must now be restored to d by reinstating $z = z' + d$, which again introduces phase factors. They have the opposite sign to those in step (i) and contain the wave number k_2 rather than k_1. Again they can be written as a diagonal matrix.

Thus we can write $\mathsf{T}(d)$ for an object at d in terms of that for the object at the origin:

$$\mathsf{T}(d) = \begin{pmatrix} e^{-ik_2 d} & 0 \\ 0 & e^{ik_2 d} \end{pmatrix} \mathsf{T}(0) \begin{pmatrix} e^{ik_1 d} & 0 \\ 0 & e^{-ik_1 d} \end{pmatrix}. \tag{5.16}$$

We shall make great use of this formula as it helps us to build more complicated potentials from a sequence of steps. If the potential and wave number are the same on both sides of the object (a square barrier rather than a step, for example), the translation formula takes a slightly simpler form because the two outside matrices become inverses of one another and $\mathsf{T}(d) = \mathsf{A}^{-1}(d)\mathsf{T}(0)\mathsf{A}(d)$, where A contains the phase factors. This is a similarity transformation of T and preserves properties such as its determinant.

5.2.1 SQUARE BARRIER

We can construct T for a rectangular potential barrier from the above ingredients. Let the barrier be centred on the origin with $V(z) = V_0$ for $|z| < a/2$ and $V(z) = 0$ elsewhere, as shown in Figure 5.5, and assume that $E > V_0$. The potentials on the left and right sides are both zero, so $k_3 = k_1$. The T-matrix can be constructed from those for two steps. The wave number changes from k_1 to k_2 at the first and we can modify equation (5.15) for the step at the origin using equation (5.16) to translate it to $z = -a/2$. At the second step the wave number changes in the opposite sense,

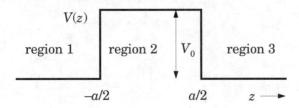

FIGURE 5.5. Potential barrier with $V(z) = V_0$ for $|z| < a/2$ and $V(z) = 0$ elsewhere.

from k_2 to $k_3 = k_1$, so we interchange k_1 and k_2 in equation (5.15) before translating it to $z = a/2$ with equation (5.16). The product of these two steps gives

$$T^{(31)} = \begin{pmatrix} e^{-ik_1a/2} & 0 \\ 0 & e^{ik_1a/2} \end{pmatrix} T(k_1, k_2) \begin{pmatrix} e^{ik_2a/2} & 0 \\ 0 & e^{-ik_2a/2} \end{pmatrix}$$

$$\times \begin{pmatrix} e^{ik_2a/2} & 0 \\ 0 & e^{-ik_2a/2} \end{pmatrix} T(k_2, k_1) \begin{pmatrix} e^{-ik_1a/2} & 0 \\ 0 & e^{ik_1a/2} \end{pmatrix}. \quad (5.17)$$

The middle pair can be multiplied simply to give a matrix with $\exp(\pm ik_2a)$ on the diagonal, reflecting the change in phase while travelling between the steps. Multiplying this between the two steps yields

$$T^{(31)} = \frac{1}{2k_1k_2} \begin{pmatrix} e^{-ik_1a/2} & 0 \\ 0 & e^{ik_1a/2} \end{pmatrix}$$

$$\times \begin{pmatrix} 2k_1k_2 \cos k_2a + i(k_1^2 + k_2^2) \sin k_2a & -i(k_1^2 - k_2^2) \sin k_2a \\ i(k_1^2 - k_2^2) \sin k_2a & 2k_1k_2 \cos k_2a - i(k_1^2 + k_2^2) \sin k_2a \end{pmatrix}$$

$$\times \begin{pmatrix} e^{-ik_1a/2} & 0 \\ 0 & e^{ik_1a/2} \end{pmatrix}. \quad (5.18)$$

The middle part is a function of the width of the barrier but not its absolute position. The location enters the phase factors on either side and would change if the barrier were between 0 and a, for example. After the final multiplications, the lower elements of $T^{(31)}$ are

$$T_{21}^{(31)} = \frac{i(k_1^2 - k_2^2) \sin k_2a}{2k_1k_2}, \quad (5.19)$$

$$T_{22}^{(31)} = \frac{2k_1k_2 \cos k_2a - i(k_1^2 + k_2^2) \sin k_2a}{2k_1k_2} e^{ik_1a}. \quad (5.20)$$

In the next section we shall derive the general results that the remaining two entries are given by $T_{11} = T_{22}^*$ and $T_{12} = T_{21}^*$, and that the determinant $\det |T| = 1$. We can then deduce the transmission amplitude

$$t = \frac{T_{11}T_{22} - T_{12}T_{21}}{T_{22}} = \frac{1}{T_{22}} = \frac{2k_1k_2e^{-ik_1a}}{2k_1k_2 \cos k_2a - i(k_1^2 + k_2^2) \sin k_2a}. \quad (5.21)$$

The flux coefficient $T = |t|^2$ is

$$T = \frac{4k_1^2k_2^2}{4k_1^2k_2^2 + (k_1^2 - k_2^2)^2 \sin^2 k_2a} = \left[1 + \frac{V_0^2}{4E(E - V_0)} \sin^2 k_2a\right]^{-1}, \quad (5.22)$$

where $k_2 = [2m(E - V_0)/\hbar^2]^{1/2}$. The reflection coefficient $R = 1 - T$. If $E < V_0$

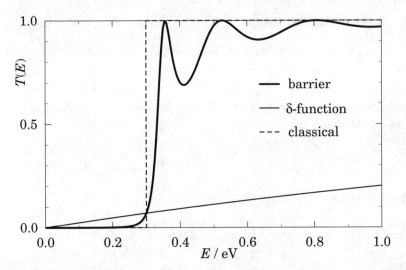

FIGURE 5.6. Transmission coefficient $T(E)$ as a function of energy E for a square potential barrier of height $V_0 = 0.3 \, \text{eV}$ and thickness $a = 10 \, \text{nm}$ in GaAs. The thin curve is for a δ-function barrier of the same strength $S = V_0 a$, and the broken curve is the classical result for a barrier of the same height.

we make the usual replacement $k_2 \to i\kappa_2$. Then $\sin k_2 a = \sin i\kappa_2 a = i \sinh \kappa_2 a$ and the transmission coefficient becomes

$$T = \frac{4k_1^2 \kappa_2^2}{4k_1^2 \kappa_2^2 + (k_1^2 + \kappa_2^2)^2 \sinh^2 \kappa_2 a} = \left[1 + \frac{V_0^2}{4E(V_0 - E)} \sinh^2 \kappa_2 a \right]^{-1} \quad (5.23)$$

where $\kappa_2 = [2m(V_0 - E)/\hbar^2]^{1/2}$. For the special case when the energy matches the top of the barrier, $E = V_0$, the transmission is

$$T(E{=}V_0) = \left[1 + \frac{ma^2 V_0}{2\hbar^2} \right]^{-1}. \quad (5.24)$$

The transmission coefficient is plotted in Figure 5.6. Classically $T = 0$ for $E < V_0$ and $T = 1$ for $E > V_0$. Quantum mechanics allows the electron to tunnel through the barrier for $E < V_0$, although the probability of transmission may be small. If $\kappa_2 a$ is large, equation (5.23) can be expanded approximately as

$$T \approx \frac{16E}{V_0} \exp(-2\kappa_2 a). \quad (5.25)$$

This is dominated by the exponential term, and $\exp(-2\kappa_2 a)$ is a simple estimate for the probability of tunnelling through any barrier.

When $E > V_0$, the transmission coefficient is unity only when $\sin k_2 a = 0$ and there is an exact number of half-wavelengths in the barrier. These 'over-the-barrier' resonances are a general phenomenon seen in other systems such as microwaves.

5.2.2 δ-FUNCTION BARRIER

A useful theoretical simplification of the rectangular barrier that we shall use later is the δ-function barrier. The height V_0 of the barrier goes to infinity while its length a is reduced to zero such that the product $S = V_0 a$ remains constant. This provides a measure of the strength of the barrier. In this limit $\kappa_2 \to \infty$ while $\kappa_2 a \to 0$, and the T-matrix becomes

$$\mathsf{T} = \frac{1}{2k_1}\begin{pmatrix} 2k_1 - i\kappa_2^2 a & -i\kappa_2^2 a \\ i\kappa_2^2 a & 2k_1 + i\kappa_2^2 a \end{pmatrix} \equiv \begin{pmatrix} 1 - iZ & -iZ \\ iZ & 1 + iZ \end{pmatrix}, \tag{5.26}$$

where all the entries depend on the single parameter

$$Z = \frac{\kappa_2^2 a}{2k_1} = \frac{mS}{\hbar^2 k_1} = S\sqrt{\frac{m}{2\hbar^2 E}}. \tag{5.27}$$

The flux transmission coefficient rises monotonically, with none of the structure seen in the finite barrier:

$$T = \frac{1}{1 + Z^2} = \left(1 + \frac{2m}{\hbar^2}\frac{S^2}{4E}\right)^{-1}. \tag{5.28}$$

This is plotted in Figure 5.6 for the same value of S as the thick barrier. The curves look quite different with these parameters, but become closer for thinner barriers.

5.3 More on T-Matrices

Although T-matrices are clearly useful, they appear to be too complicated. Wave functions are complex in general and so are the entries in T. It seems that we need four complex quantities, equivalent to eight real quantities, to specify T. In the elementary treatment we produced r and t only from matching the wave functions, just two complex or four real quantities. Moreover the conservation of current demands $R + T = 1$, which reduces the freedom to three real quantities. Fortunately, general principles show that the elements of T are related and reduce to only three independent real quantities. This relies on the conservation of current and *time-reversal invariance*, a general and important result that we shall now derive.

Take a wave function that satisfies the time-dependent Schrödinger equation

$$\hat{H}\Psi(t) = i\hbar \frac{\partial}{\partial t}\Psi(t). \tag{5.29}$$

The spatial arguments are irrelevant and have been hidden for clarity. It is assumed that $\hat{H}$ does not depend on t and is real – more on this later. First change the sign of t everywhere,

$$\hat{H}\Psi(-t) = -i\hbar \frac{\partial}{\partial t}\Psi(-t), \tag{5.30}$$

and then take the complex conjugate,

$$\hat{H}\Psi^*(-t) = i\hbar\frac{\partial}{\partial t}\Psi^*(-t). \tag{5.31}$$

The form of the final equation (5.31) is identical to the starting one (5.29). Thus if $\Psi(t)$ is a solution to the Schrödinger equation, so is $\Psi^*(-t)$. This is the statement of time-reversal invariance. The requirement that the Hamiltonian $\hat{H}$ be real is not trivial, and fails in the presence of a magnetic field **B**; in this case **B** must be reversed as well as the sign of time.

Time-reversal invariance has important consequences for the *T*-matrix. Assume for simplicity that the plateaus have the same level on either side of the obstacle, so the wave numbers are equal and describe propagating states. All the wave functions are built from plane waves such as

$$\Psi_k(z, t) = A\exp[i(kz - \omega t)]. \tag{5.32}$$

Applying the time-reversal operation to this yields

$$\Psi_k^*(z, -t) = A^*\exp[i(-kz - \omega t)]. \tag{5.33}$$

The sign of k, and therefore the direction of motion, has been reversed, as one might expect on reversing the direction of time. Less predictable is the change of the coefficient to its complex conjugate. The dependence on time remains $\exp(-i\omega t)$, so the energy of stationary states is not changed and can be dropped as usual. Thus if the waves in region 1 of Figure 5.2 are

$$A\exp(ikz) + B\exp(-ikz), \tag{5.34}$$

time-reversal invariance stipulates that

$$B^*\exp(ikz) + A^*\exp(-ikz) \tag{5.35}$$

must also be a solution to the Schrödinger equation.

The *T*-matrix was defined by equation (5.10), dropping the superscripts for clarity:

$$\begin{pmatrix} C \\ D \end{pmatrix} = \mathsf{T}\begin{pmatrix} A \\ B \end{pmatrix} = \begin{pmatrix} T_{11} & T_{12} \\ T_{21} & T_{22} \end{pmatrix}\begin{pmatrix} A \\ B \end{pmatrix} = \begin{pmatrix} T_{11}A + T_{12}B \\ T_{21}A + T_{22}B \end{pmatrix}. \tag{5.36}$$

Time-reversal invariance requires that another solution is

$$\begin{pmatrix} D^* \\ C^* \end{pmatrix} = \mathsf{T}\begin{pmatrix} B^* \\ A^* \end{pmatrix} = \begin{pmatrix} T_{11}B^* + T_{12}A^* \\ T_{21}B^* + T_{22}A^* \end{pmatrix}. \tag{5.37}$$

The physical system has not changed, so nor does T. Reordering these equations and taking their complex conjugate gives

$$\begin{pmatrix} C \\ D \end{pmatrix} = \begin{pmatrix} T_{22}^*A + T_{21}^*B \\ T_{12}^*A + T_{11}^*B \end{pmatrix} = \begin{pmatrix} T_{22}^* & T_{21}^* \\ T_{12}^* & T_{11}^* \end{pmatrix}\begin{pmatrix} A \\ B \end{pmatrix}. \tag{5.38}$$

Comparing equations (5.36) and (5.38) shows that they are consistent only if

$$T_{22} = T_{11}^*, \qquad T_{21} = T_{12}^*. \tag{5.39}$$

Thus only two of the four elements of the matrix are independent, and T can be written in the form

$$\mathsf{T} = \begin{pmatrix} T_{11} & T_{12} \\ T_{12}^* & T_{11}^* \end{pmatrix}. \tag{5.40}$$

A further relation between the two remaining entries comes from the conservation of current, which must be equal on the left and right of the obstacle. We have assumed that the two wave numbers are equal, so this requires

$$|A|^2 - |B|^2 = |C|^2 - |D|^2. \tag{5.41}$$

Multiplying this out, using the T-matrix to replace C and D, and eliminating elements that are already determined by equation (5.40) yields the condition

$$|T_{11}|^2 - |T_{12}|^2 = 1 = \det |\mathsf{T}|. \tag{5.42}$$

This constraint reduces the number of independent real quantities required to specify the T-matrix to three, as expected. It also shows that the previous expressions for r and t, equation (5.14), can be simplified under these conditions to

$$r = -\frac{T_{12}^*}{T_{11}^*}, \qquad t = \frac{1}{T_{11}^*}. \tag{5.43}$$

We can also write T in terms of r and t as

$$\mathsf{T} = \begin{pmatrix} 1/t^* & -r^*/t^* \\ -r/t & 1/t \end{pmatrix}, \tag{5.44}$$

subject to $|r|^2 + |t|^2 = 1$ for current conservation.

It is sometimes useful to have the T-matrix that gives the waves on the left in terms of the waves on the right. If $\mathsf{T}^{(21)}$ is given by the usual definition, equation (5.36), the 'reverse' matrix is $\mathsf{T}^{(12)}$, given by

$$\begin{pmatrix} B \\ A \end{pmatrix} = \mathsf{T}^{(12)} \begin{pmatrix} D \\ C \end{pmatrix}. \tag{5.45}$$

Note that $\mathsf{T}^{(12)}$ is *not* the usual inverse matrix of $\mathsf{T}^{(21)}$ because T-matrices are defined with the forward-going waves on top and this is reversed for the two matrices. To avoid a clutter of superscripts, put $\mathsf{T}^{(21)} = \mathsf{T}$ and $\mathsf{T}^{(12)} = \mathsf{T}'$. Comparing equations (5.36) and (5.45) shows that the reverse matrix T' is related to T by

$$\mathsf{T}' = \begin{pmatrix} T_{11}' & T_{12}' \\ T_{12}'^* & T_{11}'^* \end{pmatrix} = \begin{pmatrix} 1/t'^* & -r'^*/t'^* \\ -r'/t' & 1/t' \end{pmatrix}$$

$$= \begin{pmatrix} T_{11} & -T_{12}^* \\ -T_{12} & T_{11}^* \end{pmatrix} = \begin{pmatrix} 1/t^* & r/t \\ r^*/t^* & 1/t \end{pmatrix}. \tag{5.46}$$

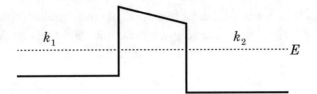

FIGURE 5.7. A barrier with the plateaus on either side at different energies, giving different wave numbers k_1 and k_2 for the electrons.

Thus $T'_{11} = T_{11}$ and $T'_{12} = -T^*_{12}$, and the t and r coefficients seen from the two sides are related by

$$t' = t, \qquad r' = -\frac{t}{t^*}r^*. \tag{5.47}$$

The transmission amplitudes seen from the two sides are identical; the reflection amplitudes differ only by a phase, so the flux reflection coefficient R is the same. These results apply to any barrier, however asymmetric the potential, provided that the plateaus are the same.

A barrier that is symmetric about the origin must have r and t identical from both sides (the symmetry is spoiled by phase factors if the barrier is put elsewhere). This puts further constraints on T. Combining equations (5.43) and (5.47) shows that T_{12} must be purely imaginary, which is indeed true for the rectangular barrier (equation 5.19).

We have normalized the plane waves in the usual way so that they have equal *densities* before being multiplied by the amplitude coefficients. This is adequate if the plateaus have the same height on either side of the obstacle. If this is not true, as in Figure 5.7, it is more convenient to normalize propagating states such that they carry equal *current*. We must cancel out the factor of $\hbar k/m$ in the current, and suitable wave functions on left and right are

$$A\sqrt{\frac{m}{\hbar k_1}}e^{ik_1 z} + B\sqrt{\frac{m}{\hbar k_1}}e^{-ik_1 z}, \qquad C\sqrt{\frac{m}{\hbar k_2}}e^{ik_2 z} + D\sqrt{\frac{m}{\hbar k_2}}e^{-ik_2 z}. \tag{5.48}$$

The flux transmission coefficient is now given by $T = |C|^2/|A|^2$ with no factors of k_1 or k_2. All the preceding results for equal plateaus can now be applied to the case where the levels on either side are different, although the prefactors in equation (5.48) must be included when the T-matrix is calculated.

The general theory of T-matrices applies to any barrier. Although we have treated only obstacles constructed from steps, a smoothly varying barrier can be handled by approximating its shape as a series of steps. The overall T-matrix is simply given by multiplying those for all the steps. However, a warning must be given! Although arbitrarily complicated systems can be treated in principle by multiplying T-matrices together, it is not numerically stable. The problem is the same one that arises when integrating the Schrödinger equation through a tunnelling barrier where there are exponentially growing and decaying solutions. Although the decaying solution is usually wanted, unavoidable numerical errors lead to the appearance of the growing exponential, which rapidly dominates the wave function. More accurate methods are available but lack the analytic simplicity of T-matrices.

5.4 Current and Conductance

The theory described in the previous sections allows us to calculate the transmission coefficient $T(E)$ as a function of energy E. The next job is to turn this into a more easily measured quantity, the current–voltage relation $I(V)$. A simple barrier surrounded by a Fermi sea of electrons is shown in Figure 5.8. A positive bias V applied to the right-hand side lowers the energies there by $-eV$. The distribution of electrons is given by a Fermi function in equilibrium but this no longer holds with a bias. Each side now has its own 'Fermi level', μ_L on the left and μ_R on the right. These differ by the applied bias giving $\mu_L - \mu_R = eV$. True Fermi levels exist only in equilibrium but a quasi-Fermi level can be defined if the system is not too strongly disturbed (Section 1.8.3).

A simple and successful procedure is to assume that the distribution of *incoming* electrons on each side is given by a Fermi distribution with the appropriate value of μ. Although this makes the calculation of the current into an almost trivial exercise, it makes strong demands on the leads, which we shall review later. For example, electrons must not be reflected after passing though the barrier or the incoming distribution will be perturbed. We shall apply this first to the simplest case of a purely one-dimensional system before including the integration over transverse degrees of freedom needed in higher dimensions. The 'leads' are defined as the regions to the left and right where the potentials have reached their plateaus, and

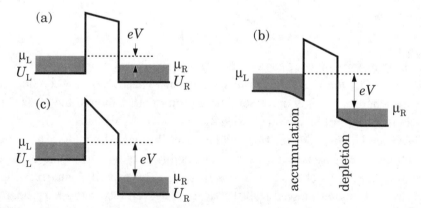

FIGURE 5.8. A barrier surrounded by a Fermi sea of electrons, with positive bias on the right. An idealized case with a small bias is shown in (a) with flat bands immediately outside the barrier. The more realistic diagram (b) shows curved bands, due to an accumulation layer on the left and a depletion layer on the right. Case (c) shows so large a bias that the electrons on the right make no contribution to the current.

the term 'barrier' will be used to encompass everything between; this may include depletion and accumulation regions as well as the true barrier itself (Figure 5.8(b)). The potential is constant throughout the leads, so their wave functions are plane waves, which is convenient for calculating the current.

5.4.1 CURRENT IN ONE DIMENSION

The method is first to calculate the current due to electrons impinging on the barrier from the left and then to add that due to electrons arriving from the right. The expressions for these are similar except for the Fermi levels.

The current due to electrons from the left is given by

$$I_L = 2e \int_0^\infty f[\varepsilon(k), \mu_L] \, v(k) \, T(k) \frac{dk}{2\pi}. \tag{5.49}$$

This integration should be carried out in the left lead, and the significance of each term is as follows:

(i) The factor of e converts number current into electrical current (the sign of the electronic charge $-e$ vanishes because conventional current flows from right to left in response to a positive bias on the right).

(ii) The integral is restricted to positive values of k because we include only electrons that impinge on the barrier from the left. The quotient $dk/2\pi$ is the usual one for counting k-states (Section 1.7), and the factor of 2 in front accounts for the two spins.

(iii) The Fermi function $f[\varepsilon(k), \mu_L]$ gives the probability that each state is occupied, governed by the Fermi level μ_L of the left lead.

(iv) The factor of velocity $v(k)$ turns the charge density into a current density as in the usual expression $J = nqv$.

(v) Finally, the (flux) transmission coefficient $T(k)$ gives the probability that an incident electron passes through the barrier and contributes to the current. If it is reflected it leaves the system to the left and makes no contribution.

It is usually more convenient to perform the integration over energy rather than wave number. This can be done by changing the variable of integration and using

$$dk = \frac{dk}{dE} \, dE = \frac{1}{\hbar v} \, dE. \tag{5.50}$$

Inserting this into equation (5.49), and denoting the bottom of the band in each lead by U_L and U_R, gives

$$I_L = 2e \int_{U_L}^\infty f(E, \mu_L) \, v \, T(E) \frac{dE}{2\pi \hbar v} = \frac{2e}{h} \int_{U_L}^\infty f(E, \mu_L) \, T(E) \, dE. \tag{5.51}$$

The velocity cancels in this expression, an important feature that underlies the quantized conductance (Section 5.7.1). Although the states at higher energy have a higher velocity, and might therefore be expected to carry more current, this is exactly cancelled by the reduction in their density of states.

The expression for the current due to electrons arriving from the right is almost identical. The only differences are the sign, as the electrons are travelling in the opposite direction, the Fermi level, and the lower limit of the integral over energy. Thus

$$I_R = -\frac{2e}{h} \int_{U_R}^{\infty} f(E, \mu_R) \, T(E) \, dE. \tag{5.52}$$

The transmission coefficient is the same from both sides of a barrier as we saw earlier (equation 5.47), so the same function $T(E)$ appears in I_L and I_R. It is clear that electrons in the range from U_R to U_L cannot contribute to the current because there are no propagating states with these energies on the left. Thus the lower limit on both integrals can be taken as U_L (or the higher of the two Us in general). Adding the two expressions gives the net current

$$I = I_L + I_R = \frac{2e}{h} \int_{U_L}^{\infty} [f(E, \mu_L) - f(E, \mu_R)] \, T(E) \, dE. \tag{5.53}$$

The current is not simply proportional to the bias; typically it is a complicated function and Ohm's law does not hold. An obvious check is that $I = 0$ when $V = 0$ and $\mu_L = \mu_R$.

This general result, due to Tsu and Esaki, can be simplified in a number of limits.

(i) When the bias is large (Figure 5.8(c)), *all* the incoming states on the right-hand side may be below the left-hand plateau U_L and therefore make no contribution to the current. In this case the occupation function $f(E, \mu_R)$ can be dropped from equation (5.53) and the right-hand electrons play no role at all.

(ii) At low temperature, where the electrons are highly degenerate, the Fermi occupation functions can be approximated by step functions. Only electrons with energies between μ_L and μ_R contribute to the current, which simplifies to

$$I = \frac{2e}{h} \int_{\mu_R}^{\mu_L} T(E) \, dE. \tag{5.54}$$

The lower limit should be replaced by U_L at large bias where $\mu_R < U_L$ and the right-hand electrons no longer contribute.

(iii) If the bias is very small the difference in Fermi functions can be expanded to lowest order in a Taylor series. Put $\mu_L = \mu + \frac{1}{2}eV$ and $\mu_R = \mu - \frac{1}{2}eV$ where μ is the Fermi level at equilibrium. Then

$$f(E, \mu_L) - f(E, \mu_R) \approx eV \frac{\partial f(E, \mu)}{\partial \mu} = -eV \frac{\partial f(E, \mu)}{\partial E}. \tag{5.55}$$

The last form follows because $f(E, \mu)$ is a function only of the difference $E - \mu$. Then

$$I = \frac{2e^2 V}{h} \int_{U_L}^{\infty} \left(-\frac{\partial f}{\partial E} \right) T(E) \, dE. \tag{5.56}$$

The current is directly proportional to the applied voltage in this limit, so Ohm's law holds. The conductance $G = I/V$ is given by

$$G = \frac{2e^2}{h} \int_{U_L}^{\infty} \left(-\frac{\partial f}{\partial E} \right) T(E) \, dE. \tag{5.57}$$

The integral itself is dimensionless and the prefactor provides the dimensions of conductance. The quotient e^2/h is often described as the quantum unit of conductance, with magnitude $38.7 \, \mu\text{S}$. The corresponding resistance $R_K = h/e^2 = 25.8 \, \text{k}\Omega$.

(iv) At very low temperature, where the Fermi function is much sharper than any features in $T(E)$, we can make the further replacement $-\partial f/\partial E = \delta(E-\mu)$ in the conductance. The integral collapses and we are left with the simple result

$$G = \frac{2e^2}{h} T(\mu). \tag{5.58}$$

This depends only on the transmission coefficient, not directly on any other parameters of the system such as the Fermi energy.

The factor of $-\partial f/\partial E$ emphasizes that conduction takes place near the Fermi level in a degenerate system, as the derivative is peaked about the Fermi level with a width of a few times $k_B T$ (Section 1.8). Equation (5.58) shows that the conductance is a perfect measure of the transmission coefficient at zero temperature. In general we must integrate equation (5.57), where $-\partial f/\partial E$ broadens with temperature and blurs sharp features in $T(E)$.

Equation (5.58) shows that a perfect wire containing no obstructions has $T = 1$ and therefore a conductance $G = 2e^2/h$, independently of its length. This is a curious result for a number of reasons. The conductance does not decrease inversely with length as it would classically, and one might ask why a perfect wire should have any resistance at all – should it not have $G = \infty$ like a superconductor? This question, which has generated much controversy over the years, will be addressed in Section 5.7.

5.4.2 CURRENT IN TWO AND THREE DIMENSIONS

Consider three dimensions; the result in two dimensions is very similar. Assume, as in Section 4.5, that the potential $V(\mathbf{R})$ is a function only of z so the system is translationally invariant in the xy-plane. The wave function can then be factored and the energy written as a sum. Use k_z to label the state, measured in (say) the left lead whose potential energy is U_L. Vectors are written with our usual notation $\mathbf{K} = (\mathbf{k}, k_z)$ with $\mathbf{k} = (k_x, k_y)$. Then

$$\psi_{\mathbf{k},k_z}(\mathbf{r}, z) = \exp(i\mathbf{k} \cdot \mathbf{r}) \, u_{k_z}(z), \tag{5.59}$$

$$\varepsilon(\mathbf{K}) = U_L + \frac{\hbar^2 \mathbf{k}^2}{2m} + \frac{\hbar^2 k_z^2}{2m}. \tag{5.60}$$

The transmission coefficient is a function only of k_z.

In the same way as before, the current density (current per unit cross-sectional area) due to electrons arriving from the left can be written as

$$J_L = 2e \int \frac{d^2\mathbf{k}}{(2\pi)^2} \int_0^\infty \frac{dk_z}{2\pi} f[\varepsilon(\mathbf{K}), \mu_L] \, v_z(\mathbf{K}) \, T(k_z). \tag{5.61}$$

The current flows along z so we need only the z-component of velocity $\hbar k_z/m$. The occupation function depends on the total energy, which can be split into the sum of transverse and z-components. Thus the current can be rewritten as

$$J_L = e \int_0^\infty \frac{dk_z}{2\pi} \frac{\hbar k_z}{m} T(k_z) \left[2 \int \frac{d^2\mathbf{k}}{(2\pi)^2} f\left(U_L + \frac{\hbar^2 k_z^2}{2m} + \frac{\hbar^2 \mathbf{k}^2}{2m}, \mu_L \right) \right]. \tag{5.62}$$

The expression in square brackets, with the factor of 2 for spin, counts the density of electrons in a two-dimensional electron gas in the xy-plane with the bottom of the band raised to $U_L + \hbar^2 k_z^2/2m$. It is the same as the expression for the density of electrons in a subband. Writing this density as $n_{2D}(\mu_L - U_L - \hbar^2 k_z^2/2m)$ we can use the previous result (equation 1.114) to find

$$n_{2D}(\mu) = \frac{m k_B T}{\pi \hbar^2} \ln \left(1 + e^{\mu/k_B T} \right). \tag{5.63}$$

The current can now be written as

$$J_L = e \int_0^\infty \frac{dk_z}{2\pi} \frac{\hbar k_z}{m} T(k_z) \, n_{2D} \left(\mu_L - U_L - \frac{\hbar^2 k_z^2}{2m} \right). \tag{5.64}$$

Now reintroduce $E = U_L + \hbar^2 k_z^2/2m$, which is the 'longitudinal component' of the total energy. This is illustrated in Figure 5.9, which shows a slice of constant longitudinal energy through a filled Fermi sphere. Then

$$J_L = \frac{e}{h} \int_{U_L}^\infty n_{2D}(\mu_L - E) \, T(E) \, dE. \tag{5.65}$$

Adding J_R gives the total current density

$$J = \frac{e}{h} \int_{U_L}^\infty [n_{2D}(\mu_L - E) - n_{2D}(\mu_R - E)] \, T(E) \, dE. \tag{5.66}$$

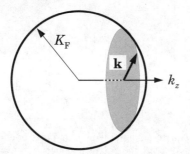

FIGURE 5.9. Decomposition of the wave vector **K** and energy into longitudinal (k_z) and transverse (**k**) components to calculate the tunnelling current in three dimensions.

The two-dimensional result is identical but with n_{1D}. The structure is very close to the one-dimensional Tsu–Esaki result (5.53).

Again this can be simplified in a number of limits. The contribution from the right can be ignored if the bias is large, and the Fermi function can be replaced by a step at low temperature. Then $n_{2D}(\mu) = (m/\pi\hbar^2)\mu\,\Theta(\mu)$ and the current becomes

$$J = \frac{e}{h}\frac{m}{\pi\hbar^2}\int_{U_L}^{\mu_L}(\mu_L - E)\,T(E)\,dE. \qquad (5.67)$$

The integral is over the whole range of energies impinging from the left, as expected. The factor of $\mu_L - E$ weights low energies more, which is slightly surprising. The reason is that a slice through the Fermi sphere captures more electrons in total for low longitudinal energies, as shown by Figure 5.9. This factor is important to explain the form of $J(V)$ in resonant tunnelling.

In the opposite limit of low bias, the conductance per unit area is

$$\bar{G} = \frac{J}{V} = \frac{e^2}{h}\frac{m}{\pi\hbar^2}\int_U^{\infty} f(E,\mu)\,T(E)\,dE \approx \frac{e^2}{h}\frac{m}{\pi\hbar^2}\int_U^{\mu} T(E)\,dE, \qquad (5.68)$$

where the second form holds at low temperature. Now there is an integral over energy even for the conductance. Although all active electrons have their *total* energy on the Fermi surface, their longitudinal energy can range from U to μ. In fact the current will be dominated by the electrons with the highest longitudinal energy in the case of a single barrier, because the transmission probability increases exponentially with energy (equation 5.25). From another point of view, electrons at the Fermi level impinge on the barrier from all angles, but those near normal incidence have a higher probability of transmission and the outgoing beam is collimated.

5.5 Resonant Tunnelling

We saw in Section 3.4.3 how a resonant state is formed in a well between two barriers. For example, the potential in Figure 5.10(a) is just a finite square well whose bound

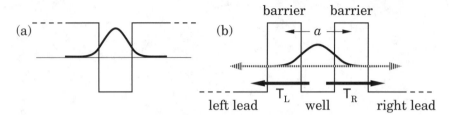

FIGURE 5.10. (a) A finite square potential well with a true bound state. (b) The same well but with barriers of finite thickness, where the bound state becomes resonant or quasi-bound.

states we studied in detail in Section 4.2. The potential in Figure 5.10(b) is identical except that the barriers that confine the electron in the well have finite thickness. There are no longer true bound states because an electron can tunnel through one of the barriers and escape from the well. However, the electron may remain in the well for a long time if the barriers are thick enough, and a remnant of the bound state persists as a *resonant* or *quasi-bound* state. The energy of this state cannot be precisely defined but is spread into a range $\hbar/\tau$, where τ is the lifetime of an electron in the well before it tunnels away.

Resonant states have a clear signature in the transmission coefficient, described in Section 3.4.3. In general, the transmission probability T of two barriers is roughly the product of the values for the two individual barriers. Near a resonance, however, T rises dramatically above the product and reaches its maximum value of unity if the structure is symmetric: there is perfect transmission through the double barrier, however opaque the individual barriers. This is *resonant tunnelling*. It is a widely observed effect, seen also in microwaves and in light where it is used in the Fabry–Pérot etalon. We shall first calculate the form of the transmission coefficient for a one-dimensional system and then use the results in the previous section to deduce the $I(V)$ characteristic of a resonant-tunnelling diode.

5.5.1 TRANSMISSION COEFFICIENT IN ONE DIMENSION

The general results do not depend on the details of the barriers that define the central well, so we shall not specify them. Let the right-hand barrier have transmission and reflection amplitudes t_R and r_R (both of which depend on k) when centred on the origin, so its T-matrix is

$$\mathsf{T}_R = \begin{pmatrix} 1/t_R^* & -r_R^*/t_R^* \\ -r_R/t_R & 1/t_R \end{pmatrix}. \tag{5.69}$$

The results are clearer if the equations are written to appear symmetric from the point of view of an electron trapped between the two barriers. Such an electron is travelling in the usual direction if it tries to escape to the right, but it is going in the opposite direction if it tries to escape to the left. It is therefore better to use the outward-going transmission and reflection amplitudes for the left-hand barrier, which are those seen by an electron hitting it from the right. The usual T-matrix applies to an electron impinging from the left, and equation (5.46) shows that it is given in terms of the *left-going* amplitudes by

$$\mathsf{T}_{\mathrm{L}} = \begin{pmatrix} 1/t_{\mathrm{L}}^* & r_{\mathrm{L}}/t_{\mathrm{L}} \\ r_{\mathrm{L}}^*/t_{\mathrm{L}}^* & 1/t_{\mathrm{L}} \end{pmatrix}. \tag{5.70}$$

We can now assemble the matrix $\mathsf{T} \equiv \mathsf{T}^{(31)}$ for the resonant-tunnelling structure in the same way that we treated the square barrier in Section 5.2.1. Shift the left barrier to $-\frac{1}{2}a$ and the right barrier to $\frac{1}{2}a$ using equation (5.16), which gives

$$\mathsf{T} = \begin{pmatrix} e^{-ika/2} & 0 \\ 0 & e^{ika/2} \end{pmatrix} \begin{pmatrix} 1/t_{\mathrm{R}}^* & -r_{\mathrm{R}}^*/t_{\mathrm{R}}^* \\ -r_{\mathrm{R}}/t_{\mathrm{R}} & 1/t_{\mathrm{R}} \end{pmatrix} \begin{pmatrix} e^{ika/2} & 0 \\ 0 & e^{-ika/2} \end{pmatrix}$$

$$\times \begin{pmatrix} e^{ika/2} & 0 \\ 0 & e^{-ika/2} \end{pmatrix} \begin{pmatrix} 1/t_{\mathrm{L}}^* & r_{\mathrm{L}}/t_{\mathrm{L}} \\ r_{\mathrm{L}}^*/t_{\mathrm{L}}^* & 1/t_{\mathrm{L}} \end{pmatrix} \begin{pmatrix} e^{-ika/2} & 0 \\ 0 & e^{ika/2} \end{pmatrix}$$

$$= \begin{pmatrix} (1 - r_{\mathrm{L}}^* r_{\mathrm{R}}^* e^{-2ika})/t_{\mathrm{L}}^* t_{\mathrm{R}}^* & (r_{\mathrm{L}} e^{ika} - r_{\mathrm{R}}^* e^{-ika})/t_{\mathrm{L}} t_{\mathrm{R}}^* \\ (r_{\mathrm{L}}^* e^{-ika} - r_{\mathrm{R}} e^{ika})/t_{\mathrm{L}}^* t_{\mathrm{R}} & (1 - r_{\mathrm{L}} r_{\mathrm{R}} e^{2ika})/t_{\mathrm{L}} t_{\mathrm{R}} \end{pmatrix}. \tag{5.71}$$

The transmission amplitude follows immediately from the bottom right entry:

$$t = \frac{t_{\mathrm{L}} t_{\mathrm{R}}}{1 - r_{\mathrm{L}} r_{\mathrm{R}} \exp(2ika)}. \tag{5.72}$$

The behaviour of t is clearer if the complex reflection amplitudes are written in polar form such as $r_{\mathrm{L}} = |r_{\mathrm{L}}| \exp(i\rho_{\mathrm{L}})$. The flux transmission coefficient becomes

$$T = |t|^2 = \frac{T_{\mathrm{L}} T_{\mathrm{R}}}{1 + R_{\mathrm{L}} R_{\mathrm{R}} - 2\sqrt{R_{\mathrm{L}} R_{\mathrm{R}}} \cos(2ka + \rho_{\mathrm{L}} + \rho_{\mathrm{R}})}$$

$$= \frac{T_{\mathrm{L}} T_{\mathrm{R}}}{\left(1 - \sqrt{R_{\mathrm{L}} R_{\mathrm{R}}}\right)^2 + 4\sqrt{R_{\mathrm{L}} R_{\mathrm{R}}} \sin^2 \frac{1}{2}\phi}, \tag{5.73}$$

where the phase $\phi = 2ka + \rho_{\mathrm{L}} + \rho_{\mathrm{R}}$.

We want to investigate this as a function of energy, and unfortunately every term varies. Usually the most rapid variation near a resonance is due to the change in phase of the wave between the barriers, $2ka$, and we can assume that the other terms vary slowly with respect to this. Then T has peaks when the sine in the denominator vanishes. This requires $\phi = 2n\pi$, which is therefore the condition for resonant states. At these points

$$T = T_{\mathrm{pk}} = \frac{T_{\mathrm{L}} T_{\mathrm{R}}}{\left(1 - \sqrt{R_{\mathrm{L}} R_{\mathrm{R}}}\right)^2} \approx \frac{4 T_{\mathrm{L}} T_{\mathrm{R}}}{(T_{\mathrm{L}} + T_{\mathrm{R}})^2}. \tag{5.74}$$

The second form follows by assuming that the individual transmission coefficients T_L and T_R are small (which is the usual case) and by expanding $\sqrt{R_L}$ and $\sqrt{R_R}$ with the binomial theorem.

This condition for resonance, $\phi = 2ka + \rho_L + \rho_R = 2n\pi$, is the requirement for constructive interference within the well. Consider an electron bouncing back and forth between the barriers. It picks up a phase of ka in each direction, with additional phases of ρ_L and ρ_R when it reflects from the barriers. Exactly the same condition applies to true bound states in a well if the problem is formulated in terms of T-matrices.

If the barriers have identical transmission coefficients, equation (5.74) shows the remarkable result that transmission at the peak is perfect, that is, $T_{pk} = 1$. If they are very different, the approximate form gives $T_{pk} \approx 4T_</T_>$, where $T_>$ and $T_<$ are the greater and lesser of T_L and T_R. Thus the transmission is limited by the more opaque barrier, which seems reasonable.

Assuming that the individual transmission coefficients are small, we can write the overall transmission coefficient as

$$T \approx \frac{T_L T_R}{\frac{1}{4}(T_L + T_R)^2 + 4\sin^2 \frac{1}{2}\phi} = T_{pk}\left[1 + \frac{16}{(T_L + T_R)^2}\sin^2\frac{\phi}{2}\right]^{-1}. \qquad (5.75)$$

The term with $\sin^2 \frac{1}{2}\phi$ usually dominates because of its large prefactor. Put $\sin^2 \frac{1}{2}\phi \approx \frac{1}{2}$ as a typical value, which gives $T \approx \frac{1}{2}T_L T_R$. Thus the overall transmission coefficient is typically the product of those for the two barriers, as might be expected.

Strong deviations from this occur when the sine vanishes. Put $\phi = 2n\pi + \delta\phi$ and expand the sine to first order, which gives

$$T \approx T_{pk}\left[1 + \frac{4(\delta\phi)^2}{(T_L + T_R)^2}\right]^{-1} = \frac{T_{pk}}{1 + (\delta\phi/\frac{1}{2}\phi_0)^2} \qquad (5.76)$$

with $\phi_0 = T_L + T_R$. The resonant peak has a Lorentzian shape. It falls to half its peak value at $\delta\phi = \pm\frac{1}{2}\phi_0$, showing that ϕ_0 is the full width at half-maximum (FWHM). This can be translated into a width in energy,

$$\Gamma = \frac{dE}{dk}\frac{dk}{d\phi}\phi_0 = \frac{\hbar v}{2a}(T_L + T_R), \qquad (5.77)$$

where v is the velocity of the electron between the barriers, and it is assumed that the variation of ϕ is dominated by $2ka$. Then the transmission coefficient as a function of energy is

$$T(E) \approx T_{pk}\left[1 + \left(\frac{E - E_{pk}}{\frac{1}{2}\Gamma}\right)^2\right]^{-1}, \qquad (5.78)$$

where the resonance is centred on E_{pk}. This Lorentzian shape with a FWHM of Γ is typical of resonant phenomena. It is known as the *Breit–Wigner* formula from nuclear physics, and is also familiar from the Fabry–Pérot etalon in optics. The shape changes if the flux transmission coefficients cannot be taken as constant over the width of the resonance. This might occur, for instance, in a resonant state near the top of the barriers, where T changes rapidly.

The width Γ can also be derived from elementary considerations. The velocity of an electron in the resonance is v and the distance of a round trip is $2a$, so it bumps against the left barrier $v/2a$ times per second. The probability of escaping is T_L on each occasion, so the average escape rate through the left barrier is $vT_L/2a$. Including the right barrier gives the total rate, which can be converted into an uncertainty in energy by multiplying by $\hbar$. The result is $\hbar v(T_L + T_R)/2a = \Gamma$, as found in the analysis. Alternatively, the lifetime $\tau = \hbar/\Gamma$. Definitions vary: often Γ is defined as the half-width rather than the full width.

This discussion has focussed on the flux transmission coefficient, but the transmission amplitude is complex and its phase also shows a signature of resonance. There is a rapid change through π as the energy passes through the resonance, superposed on a slowly varying background due to the barriers. The same change of phase is seen in classical resonant systems ranging from masses on springs to RLC circuits and is another general characteristic of resonant behaviour. The change in phase of $t(E)$ may be useful in searching for resonances, as the peak in $T(E)$ is often so narrow that it is easily missed.

An example of the transmission coefficient of a double-barrier structure is shown in Figure 5.11. The barriers were chosen to be δ-functions because of their monotonic transmission coefficients (equation 5.26), so any structure in $T(E)$ can be

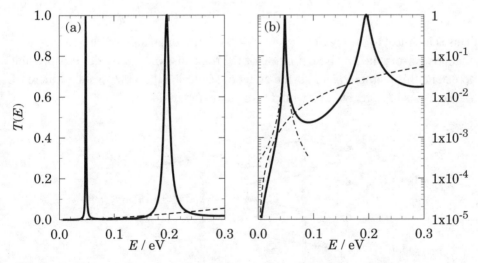

FIGURE 5.11. Transmission coefficient of a resonant-tunnelling structure on (a) linear and (b) logarithmic scales. The barriers are δ-functions of strength $0.3\,\mathrm{eV} \times 5\,\mathrm{nm}$ separated by $10\,\mathrm{nm}$. The solid curve is $T(E)$ for the whole structure, the dashed curve shows the square of $T(E)$ for a single barrier and would apply to the double-barrier structure if there were no resonance, and the chain curve is the Lorentzian approximation to the lowest resonance.

ascribed to interference between the two barriers. There is 10 nm between the two barriers, each of strength $0.3\,\text{eV} \times 5\,\text{nm}$. The plot of $T(E)$ shows two distinct peaks in this range of energies, each rising to 1. This is much bigger than the product of the two barriers' transmission coefficients, drawn as the broken line. The peak at higher energy is broader because the barriers are more transparent, increasing Γ (equation 5.77). The Lorentzian approximation (5.78) is shown as the chain curve using a width $\Gamma = 1.6\,\text{meV}$ calculated from equation (5.77) and is clearly a good fit near the peak.

5.5.2 PARTIAL WAVES

Another instructive way of deriving the transmission coefficient, often used in optics, is to sum 'partial waves', as shown in Figure 5.12. The incident wave is partly reflected at the first barrier, then bounces back and forth between the barriers, losing some of its amplitude by transmission through a barrier at each bounce. Use r_L' and t_L' to denote the amplitudes seen from the left-hand side of the left-hand barrier. Summing the contributions to the transmitted beam gives

$$t = t_L' e^{ika} t_R + t_L' e^{ika} r_R e^{ika} r_L e^{ika} t_R$$

$$+ t_L' e^{ika} r_R e^{ika} r_L e^{ika} r_R e^{ika} r_L e^{ika} t_R + \cdots . \tag{5.79}$$

Summing the geometric progression and using $t_L' = t_L$ gives

$$t = \frac{t_L t_R\, e^{ika}}{1 - r_L r_R \exp(2ika)}. \tag{5.80}$$

This is identical to the previous result (5.72) except for an unimportant phase factor that arises from the precise definition of the transmission amplitude. It is possible to extend this picture to include the effect of scattering that destroys the coherence of the resonant state, but that would take us too far afield.

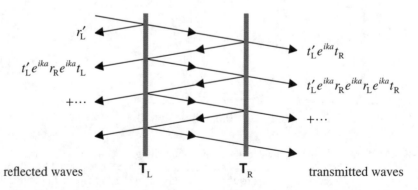

FIGURE 5.12. The partial waves that contribute to the transmission through a double-barrier structure.

The analysis using partial waves resolves a paradox that arises from the physical picture of resonant scattering. We imagine that an electron enters the resonant state and remains there for a long time, bouncing back and forth many times between the two barriers. The probabilities of tunnelling through the barriers are given by T_L and T_R, so one might guess that the transmitted and reflected currents would be proportional to these numbers. Unfortunately this argument predicts the overall transmission to be only $\frac{1}{2}$ for a symmetric structure instead of 1. The problem is that it neglects the directly reflected wave of amplitude r'_L caused by electrons that never enter the resonance (Figure 5.12). Perfect transmission occurs when this wave is precisely cancelled by the contributions from electrons that enter the well but return backwards after multiple reflections; see equation (E5.3) in the exercises.

Resonances are a characteristic feature of one-dimensional systems. Any two scattering centres can be described by T-matrices in the same way as the two barriers of the resonant-tunnelling structure calculated in this section. Resonances can *always* build up between them and there will be energies where the overall transmission and reflection coefficients are very different from the simple product of the two barriers. Thus it is *never* possible to treat scattering events in one dimension as though they are independent of one another. Fortunately this problem is less severe in two and three dimensions (Section 8.2). Instead an effect called *resonant scattering* may be seen, giving a peak in the scattering rate when the energy of an incident electron matches that of a quasi-bound state in the potential.

5.5.3 CURRENT THROUGH A RESONANT-TUNNELLING DIODE

Although there is always a background current due to electrons tunnelling at energies away from the resonant peaks, these peaks will dominate the transmission coefficient in a well-designed device. Assume that there is only one peak at E_{pk} in the region of interest and use the Lorentzian approximation (5.78) for $T(E)$ to calculate the current.

The effect of bias on a resonant-tunnelling structure is shown in Figure 5.13. The bias across the diode is small in (a), E_{pk} is above the sea of incoming electrons, and little current flows. In (b) the bias has brought down the energy of the resonant level, so the sea of electrons on the left can pass through it and a larger current flows. In three dimensions the current increases linearly as E_{pk} approaches the bottom of the sea of electrons on the left, as in (c). A further increase of the bias, as in (d), pulls the energy of the resonant level down so far that it is no longer available to electrons and the current decreases abruptly. The result is a current–voltage characteristic that shows negative differential conductivity, as in (e), which can be utilized in an amplifier or oscillator.

Figure 5.13 shows that the bias has at least three major effects on the electronic structure: it changes the Fermi levels, shifts the energy of the resonant state, and

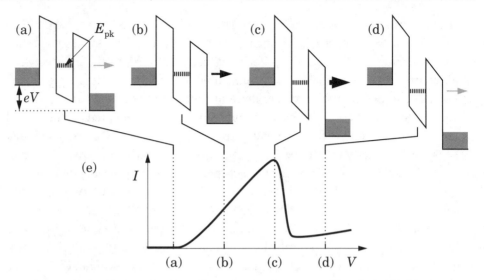

FIGURE 5.13. Profile through a three-dimensional resonant-tunnelling diode. The bias increases from (a) to (d), giving rise to the $I(V)$ characteristic shown in (e). The shaded areas on the left and right are the Fermi seas of electrons.

alters the profiles and transmission properties of the barriers. The first two are essential, but for simplicity we shall neglect the changes in T_L and T_R although this approximation can rarely be justified in practice.

Consider one dimension first. There are two extreme conditions under which the current through a resonant-tunnelling structure can be measured. The first, with a very small bias, gives the conductance (equation 5.57). Figure 5.14 shows such a measurement, the conductance of a quantum dot. This can be viewed as a one-dimensional resonant-tunnelling system with many resonant levels whose energy can be moved through the Fermi level by the gate voltage V_g. (In fact the energy levels are dominated by the electrostatic energy required to add extra electrons to the dot, a regime called the *Coulomb blockade*.) The integral for the conductance contains two peaked functions, the derivative of the Fermi function and the transmission coefficient of the device. The Fermi function is sharper if $k_B T \ll \Gamma$ and equation (5.58) shows that the conductance reflects $T(\mu)$. An example is provided by the peak around $V_g = 291.6\,\mathrm{mV}$, which can be fit by a Lorentzian function. The peaks at lower voltage are in the opposite limit where the width of the Fermi function is larger and the shape of $G(V_g)$ is due to $-\partial f / \partial E$ (Figure 5.14(b)). The shape of $T(E)$ affects the conductance only through the area of the peak, as discussed shortly for strong bias.

Next we consider the opposite case where the bias is so large that electrons from the right cannot contribute (Figure 5.8(c)). Measure energies with respect to the left-hand lead. The bias V pulls down the resonant state by about $\frac{1}{2}V$ if the structure is symmetric, so $E_{pk}(V) \approx E_{pk}(0) - \frac{1}{2}eV$. Hence the resonance is pulled through the range of incident energies by the bias. The current can be obtained from equation

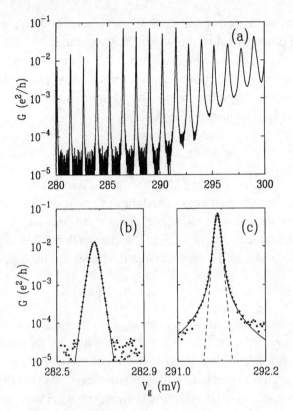

FIGURE 5.14. (a) Conductance of a quantum dot at 60 mK and $B = 2.53$ T as a function of gate voltage V_g, which changes the energy of the resonant level. (b) Enlargement of peak around $V_g = 282.7$ mV, fitted by thermal broadening. (c) Enlargement of peak around $V_g = 291.6$ mV, fitted by thermal broadening alone (dashed line) and a thermally broadened Lorentzian function (solid line). [From Foxman et al. (1993).]

(5.54), assuming that the temperature is low. Usually the right-hand electrons all have energies below the resonance and can be neglected. Very little current flows when the peak lies outside the range of integration, $E_{pk} < U_L$ or $E_{pk} > \mu_L$, as in Figures 5.13(a) and (d). Between these limits, the resonance lies well inside the range of integration and the only significant contribution is around the peak. It is then a good approximation to extend the integral over E from (U_L, μ_L) to $(-\infty, +\infty)$, and

$$I \approx \frac{2e}{h} \int_{-\infty}^{\infty} T(E) \, dE = \frac{2e}{h} T_{pk} \int_{-\infty}^{\infty} \left[1 + \left(\frac{E - E_{pk}}{\frac{1}{2}\Gamma} \right)^2 \right]^{-1} dE$$

$$= \frac{2e}{h} \frac{\pi}{2} \Gamma T_{pk}. \tag{5.81}$$

The current remains constant at this value while the peak remains well inside the range of incoming energies, and it falls to zero over a width of about Γ as E_{pk} passes outside this range.

It is important that the current depends on the *integral* of $T(E)$, not just its peak, and is therefore proportional to the width Γ. Although $T_{pk} = 1$ for a symmetric structure even if the barriers are highly opaque, the width is extremely small in this case and only a small current flows. Thus a device with opaque barriers passes a small current, as seems physically reasonable, despite its high peak in transmission. Expanding Γ and T_{pk} in equation (5.81) gives

$$I = 2\frac{ev}{2a}\frac{T_L T_R}{T_L + T_R} \approx \frac{ev T_<}{a}, \tag{5.82}$$

which confirms that current is limited by the more opaque barrier.

The same arguments can be used to find the current through a three-dimensional device. Consider only the case of large bias and low temperature, (5.67). In this case there is the additional factor of $\mu_L - E$ in the integrand, but it is slowly varying so we can evaluate it at E_{pk} and remove it from the integral. This gives

$$J = \frac{e}{h}\frac{m}{\pi\hbar^2}(\mu_L - E_{pk})\frac{\pi}{2}\Gamma T_{pk}, \tag{5.83}$$

assuming that the resonance lies within the range of incoming energies, $U_L < E_{pk} < \mu_L$. The current now depends on the applied bias through the factor of $(\mu_L - E_{pk}) \approx [\mu_L - E_{pk}(0) + \frac{1}{2}eV]$. This causes the characteristic triangular shape of $I(V)$ seen in Figure 5.13(e). The current is largest when the resonance is at the bottom of the range of incoming energies because this is when the maximum number of states have this value of longitudinal energy (Figure 5.9). The drop from this maximum current to a very small 'valley' current over a range of about Γ gives the negative differential resistance that makes the device useful.

We have assumed that the shape of the barriers does not change as a function of bias. Figure 5.13 shows that the bias reduces the heights of the barriers, the right-hand one in particular, as well as the energy of the resonance. This means that a structure that has identical barriers at equilibrium, and therefore $T_{pk} = 1$, will be far from symmetric under the large bias usually needed to make the resonance active in $I(V)$. Devices must be designed with thicker or higher right-hand barriers at equilibrium if they are to be reasonably symmetric, with high peaks in $T(E)$, when under bias. Also, the high density of electrons in the resonant state modifies the potential through Poisson's equation. Both effects must be included in a realistic calculation of $I(V)$, and a better description of the leads may also be necessary.

Figure 5.15 shows characteristics of real devices fabricated from three material systems. These were measured at room temperature, where many of the simplifications that we used to calculate $I(V)$ do not hold for GaAs–AlGaAs structures. The low-temperature limit is less stringent in InGaAs–AlAs because of its larger barriers, and $I(V)$ follows the triangular shape that we predict. A common figure of merit is the peak-to-valley ratio of the current, about 12 for this device.

The valley current that follows the sharp drop in $I(V)$ is always much larger than that predicted by simple theory. A realistic calculation must include direct

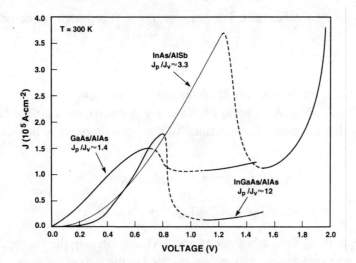

FIGURE 5.15. Characteristics of resonant-tunnelling diodes in three material systems measured at room temperature. [From Brown (1994).]

tunnelling, scattering by impurities and phonons, and the effect of randomness in the barriers if they are alloys (although AlAs is often used to avoid this). The third device has InAs wells with GaSb barriers and is interesting because this is a type III junction (Section 3.3) where the band gaps do not overlap. Although its peak-to-valley ratio is lower, the higher current density can be a practical advantage.

5.6 Superlattices and Minibands

We have now treated, at great length, tunnelling through a single barrier and through two barriers. The next step is to consider an infinite number of barriers, regularly positioned to form a superlattice. A simple example with alternating square wells and barriers, known as the *Kronig–Penney model*, is shown in Figure 5.16.

The T-matrix for the superlattice can be built up from those for each cell. Let the matrix for the cell at the origin be T_0, which is just that for a barrier in Figure 5.16. We can then use equation (5.16) for an obstacle translated along the

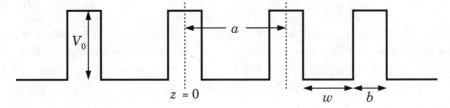

FIGURE 5.16. The Kronig–Penney model, a simple superlattice, showing wells of width w alternating with barriers of thickness b and height V_0. The (super)lattice constant is $a = b + w$.

axis to deduce T for other cells. For example, the next cell to the right has

$$\mathsf{T}_1 = \begin{pmatrix} e^{-ik_1a} & 0 \\ 0 & e^{ik_1a} \end{pmatrix} \mathsf{T}_0 \begin{pmatrix} e^{ik_1a} & 0 \\ 0 & e^{-ik_1a} \end{pmatrix} \equiv \mathsf{A}^{-1}\mathsf{T}_0\mathsf{A}. \tag{5.84}$$

Here k_1 is the wave number in the plateau outside the potential in each unit cell. Similarly, $\mathsf{T}_2 = \mathsf{A}^{-2}\mathsf{T}_0\mathsf{A}^2$, and so on. Thus the T-matrix for the superlattice, remembering that the matrices are written in the opposite order to the cells in real space, is

$$\mathsf{T} = \cdots (\mathsf{A}^{-2}\mathsf{T}_0\mathsf{A}^2)(\mathsf{A}^{-1}\mathsf{T}_0\mathsf{A})(\mathsf{T}_0)(\mathsf{A}\mathsf{T}_0\mathsf{A}^{-1})(\mathsf{A}^2\mathsf{T}_0\mathsf{A}^{-2}) \cdots$$

$$= \cdots \mathsf{A}\mathsf{T}_0\mathsf{A}\mathsf{T}_0\mathsf{A}\mathsf{T}_0\mathsf{A}\mathsf{T}_0\mathsf{A}\mathsf{T}_0\mathsf{A} \cdots. \tag{5.85}$$

Each unit cell is represented by the product $\mathsf{A}\mathsf{T}_0$, combining the scattering within each cell with the change in phase from one cell to the next.

The regularity of the structure means that the results derived in Chapter 2 for energy bands in crystals apply. In particular, Bloch's theorem tells us that the wave functions in one unit cell are related to those in the previous one simply by a phase factor of $\exp(ika)$. It is vital to distinguish between the two wave numbers involved.

    (i) The Bloch wave number k gives the change in phase of the wave function from one unit cell to the next.

    (ii) The wave number k_1 describes the wave function of the electron at a particular point in each cell and gives the energy through $E = \hbar^2 k_1^2/2m$.

Write the forward- and backward-going coefficients of the wave function in cell n as a_n and b_n. Combining the T-matrix and Bloch's theorem shows that

$$\begin{pmatrix} a_{n+1} \\ b_{n+1} \end{pmatrix} = \mathsf{A}\mathsf{T}_0 \begin{pmatrix} a_n \\ b_n \end{pmatrix} = \exp(ika) \begin{pmatrix} a_n \\ b_n \end{pmatrix}. \tag{5.86}$$

Thus $\exp(ika)$ is an eigenvalue of $\mathsf{A}\mathsf{T}_0$. Note that T-matrices are not Hermitian, so their eigenvalues may be complex. Writing T_0 in terms of reflection and transmission coefficients (equation 5.44), we need the eigenvalues of the product

$$\mathsf{A}\mathsf{T}_0 = \begin{pmatrix} e^{ik_1a} & 0 \\ 0 & e^{-ik_1a} \end{pmatrix} \begin{pmatrix} 1/t^* & -r^*/t^* \\ -r/t & 1/t \end{pmatrix} = \begin{pmatrix} e^{ik_1a}/t^* & -e^{ik_1a}r^*/t^* \\ -e^{-ik_1a}r/t & e^{-ik_1a}/t \end{pmatrix}.$$

$$\tag{5.87}$$

In general the product of the eigenvalues of a matrix is given by its determinant, which we know to be unity in this case (equation 5.42). Thus the eigenvalues are of the form $\exp(\pm ika)$ although k need not be real. The sum of the eigenvalues, $2\cos ka$, is given by the trace of the matrix (or one can expand the secular equation), whence

$$\cos ka = \operatorname{Re}\left\{\frac{1}{t\exp(ik_1a)}\right\} = \frac{1}{|t(k_1)|}\cos[k_1a + \tau(k_1)], \tag{5.88}$$

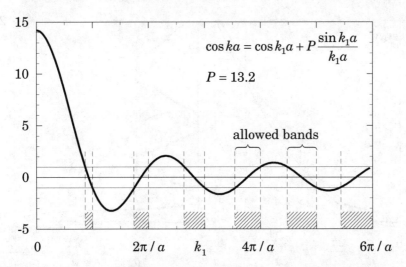

FIGURE 5.17. Solution of the Kronig–Penney model with δ-function barriers and $P = 13.2$, equation (5.91). Propagating solutions are permitted only in the shaded regions of k_1, where $|\cos ka| \leq 1$.

where $|t|$ and τ are the modulus and phase of t. Usually we specify a state and then find its energy, but here we first choose the energy, which gives k_1, then calculate the right-hand side of equation (5.88), and finally deduce the Bloch wave number k, which labels the state.

An example of this equation is plotted as a function of k_1 in Figure 5.17. The right-hand side oscillates and decays, but its amplitude is always greater than unity. The Bloch wave number k is real for propagating states, which implies $|\cos ka| \leq 1$ and in turn requires $|\cos(k_1 a + \tau)| \leq |t|$. Now $|t| < 1$ in general, so there are ranges of k_1 in which the inequality for $\cos(k_1 a + \tau)$ is not satisfied and propagating states are forbidden. These are the band gaps.

Take the simple Kronig–Penney model of a superlattice shown in Figure 5.16 as an example. The potential in a unit cell is a square barrier of height V_0 and thickness b. Equation (5.21) gives the transmission coefficient for this (with the trivial change $a \to b$), so the equation for $\cos ka$ becomes

$$\cos ka = \mathrm{Re} \left\{ \frac{2k_1 k_2 \cos k_2 b - i(k_1^2 + k_2^2) \sin k_2 b}{2k_1 k_2 e^{-ik_1 b} e^{ik_1 a}} \right\}$$

$$= \cos k_1 w \cos k_2 b - \frac{k_1^2 + k_2^2}{2k_1 k_2} \sin k_1 w \sin k_2 b, \qquad (5.89)$$

where $w = a - b$ is the width of the well between barriers, k_1 the wave number in the wells, and k_2 that in the barriers. These expressions are for $E > V_0$ with all waves propagating; the usual substitution $k_2 \to i\kappa_2$ for $E < V_0$ gives

$$\cos ka = \cos k_1 w \cosh \kappa_2 b - \frac{k_1^2 - \kappa_2^2}{2k_1 \kappa_2} \sin k_1 w \sinh \kappa_2 b. \qquad (5.90)$$

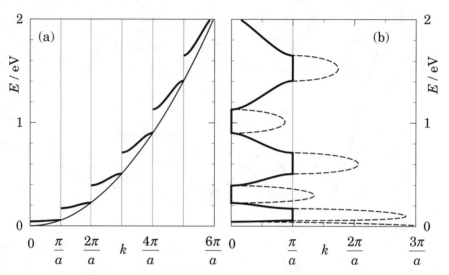

FIGURE 5.18. Energy bands for the Kronig–Penney model with δ-functions in (a) the extended and (b) the reduced zone schemes. The thick line is the solution of the Kronig-Penney model for propagating states, the thin line is the parabola for free electrons, and the broken line in (b) shows the imaginary part of the wave number in the band gaps.

A further simplification that can be made is to reduce the barrier to a δ-function (Section 5.2.2). We keep $V_0 b = S$ constant while letting $V_0 \to \infty$ and $b \to 0$. Then equation (5.90) reduces to

$$\cos ka = \cos k_1 a + \left(\frac{maS}{\hbar^2}\right)\frac{\sin k_1 a}{k_1 a}. \qquad (5.91)$$

For a numerical example, put $S = 0.3\,\mathrm{eV} \times 5\,\mathrm{nm}$ as in the resonant-tunnelling structure, $a = 10\,\mathrm{nm}$, and $m = 0.067\,m_0$, which gives $P = maS/\hbar^2 = 13.2$. Equation (5.91) is plotted as a function of k_1 in Figure 5.17. The right-hand side starts at $1 + P = 14.2$ when $k_1 = 0$ and oscillates with decreasing amplitude. Propagating bands occur when the function lies between ± 1 with band gaps between.

The energy $E = \hbar^2 k_1^2/2m$ is plotted against the Bloch wave number k, found from the solution of equation (5.91), in Figure 5.18. Band gaps occur at the zone boundaries $k = n\pi/a$. The lowest band is very narrow (only about 15 meV in this example) because the barriers are opaque at low energies. The cosine approximation (equation 2.9) would be good here. The bands become broader, and their effective mass decreases, at higher energies where $T(E)$ for each barrier rises. Note that the energies of the bands coincide with those of the peaks in $T(E)$ found for the resonant-tunnelling diode with the same barriers and separation (Figure 5.11): the peaks in transmission due to resonant states evolve into energy bands.

A feature peculiar to the model with δ-functions is that the bands touch the parabola at the bottom of the band gaps. The reason for this was illustrated in Figure 2.4. The wave functions at the zone boundary are standing waves such as sine and cosine. The sine waves have nodes at $z = na$, which coincide with the

δ-functions of $V(z)$, so they do not see the potential and their energy is unaffected. The cosines, on the other hand, have their antinodes at these points and receive the maximum increase in energy from the repulsive potentials. Band gaps usually get smaller as the energy increases but this does not apply to Figure 5.18, again because of the δ-function potentials.

Although we usually concentrate on propagating states, we can also investigate those in the band gaps. These have $|\cos ka| > 1$, which can be satisfied only if k is complex. The choice $k = n\pi/a + i\kappa$ gives $\cos ka = \cosh \kappa a > 1$ for even n and $\cos ka = -\cosh \kappa a < -1$ for odd n. Thus we can find solutions at all energies by adding an imaginary part to k in the band gaps. This is called the *complex band structure* and is shown as the broken line in Figure 5.18(b). You should imagine that the broken lines are in a plane perpendicular to the page for Im k. The states in the band gap decay or grow exponentially like $\exp(\pm \kappa z)$. They also change sign in adjacent cells for odd n like the standing waves on either side of the gap.

These states cannot be normalized over all space and are therefore unacceptable solutions for a pure infinite crystal. This does not mean that they can be neglected. For instance, they are important to describe the wave function of an impurity that causes a bound state in a band gap. They are also vital to describe tunnelling through a finite region of the superlattice. Figure 5.18(b) shows how the decay constant κ increases as the energy moves away from a band edge E_{edge} into the gap. It has the same dependence as for free electrons, $\kappa^2 = 2m|E - E_{\text{edge}}|/\hbar^2$, except for an effective mass that is the same as for propagating electrons near the edge of the band. The decay constant cannot increase without limit but reaches a maximum near the middle of the gap before reducing again as the edge of the next band is approached.

This behaviour of Im k applies to any band gap, in particular to that of AlGaAs, which is often used as a barrier in GaAs. The simple parabolic approximation for calculating the tunnelling decay constant κ holds only for small energies below the edge of the band, just as the parabolic approximation $\varepsilon(K) \approx \hbar^2 K^2 / 2m_0 m_e$ is accurate only for small energies inside the band (Section 2.6.4). Complex band structure should be used to find κ when tunnelling takes place far below the top of any barrier. This probably includes the usual case of $Al_{0.3}Ga_{0.7}As$ but the parabolic approximation is far too convenient!

Figure 5.19 shows the edges of the bands in a square-wave superlattice (not δ-functions) as a function of the thickness of the barriers. The calculation is for GaAs with 5 nm wells and barriers of 0.3 eV, and the energies were deduced from equations (5.89) and (5.90). The two lowest bands narrow into bound states in the well as the barriers become thicker (although the upper state is only weakly bound). All the gaps shrink to zero as the barriers vanish.

Nothing dramatic happens in Figure 5.19 at 0.3 eV, the top of the barriers, and there are distinct band gaps above this energy even though the states propagate everywhere. This is fortunate because most optical superlattices are constructed

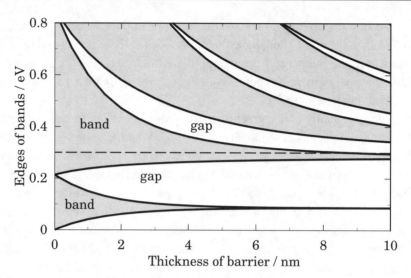

FIGURE 5.19. Bands of a superlattice in GaAs as a function of the thickness of the barriers. The wells remain 5 nm wide and 0.3 eV deep.

from materials in which light can propagate everywhere, and the only difference is the wavelength. The periodic nature alone is sufficient to cause coherent Bragg reflection and the appearance of gaps in the spectrum. The same phenomena are also seen in filter networks made up of lumped LC components in electronics, and can be described with similar matrices.

The density of states for a superlattice can be estimated using the same ideas as in Section 4.5. Each state of energy ε in a single quantum well becomes the bottom of a two-dimensional subband $n_{2D}(E - \varepsilon)$ in the density of states. In the same way, each Bloch state in a superlattice behaves as the bottom of a subband. Let the density of states of the one-dimensional superlattice be $n_{1D}^{(SL)}$. Then the three-dimensional density of states is given by the integral

$$n_{3D}(E) = \tfrac{1}{2} \int_{-\infty}^{\infty} n_{1D}^{(SL)}(\varepsilon) n_{2D}(E - \varepsilon)\, d\varepsilon = \frac{m}{2\pi\hbar^2} \int_{-\infty}^{E} n_{1D}^{(SL)}(\varepsilon)\, d\varepsilon. \qquad (5.92)$$

The factor of $\tfrac{1}{2}$ avoids double-counting the spin, which conventionally enters both densities of states. The cosine approximation for a single band of width W (equation 2.17) gives

$$n_{3D}(E) = \frac{m}{\pi a \hbar^2} \left(\frac{1}{2} + \frac{1}{\pi} \arcsin \frac{E - \tfrac{1}{2}W}{\tfrac{1}{2}W} \right), \qquad 0 \leq E \leq W. \qquad (5.93)$$

The density of states is constant at $m/\pi a \hbar^2$ above the top of the one-dimensional band. The same value is found in a multiquantum-well structure of period a, where there is no communication between wells. This is plotted for two minibands in Figure 5.20. The inclusion of tunnelling between the wells, which converts the

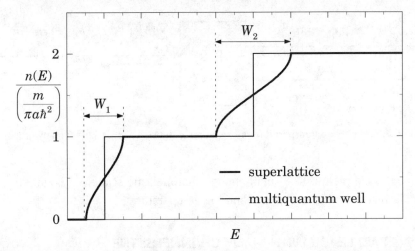

FIGURE 5.20. Density of states for a superlattice (thick line) and a multiquantum well (thin line) scaled to the same value in the plateau, $m/\pi a\hbar^2$. The two lowest bands have widths W_1 and W_2.

multiquantum well into a superlattice, broadens the sharp step in the density of states into an arcsine of width W. The bottom of each subband has a parabolic density of states, the general result for a three-dimensional system, but the superlattice is anisotropic. The effective mass for motion perpendicular to the direction of growth, m in the preceding formulas, is that for freely propagating electrons in the host, while that for motion along the superlattice depends on W.

5.7 Coherent Transport with Many Channels

The one-dimensional systems studied earlier in this chapter were purely one-dimensional in the sense that only one subband was occupied. Such systems are analogous to electromagnetic wave guides in which only one mode can propagate. Further subbands become occupied if the Fermi level of a quasi-one-dimensional system is raised, just as further modes can propagate in a wave guide as the frequency of operation is raised. In this section we shall first extend the theory of conduction to describe a scattering centre or 'sample' between two such quasi-one-dimensional systems or 'leads'. An example is a narrow constriction between two wider leads, which shows a quantized conductance. This is more complicated to describe than a purely one-dimensional system because reflection and transmission can now occur between different subbands or modes. The next step is to study samples that have more than two leads attached to them. The sample can now be a much more general object, and the most significant application will be to the quantum Hall effect in a sample with many leads. The theory remains restricted to coherent transport, so in

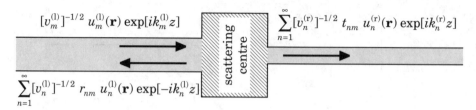

FIGURE 5.21. Coherent transport through a system with two leads, each with many propagating states.

practice the sample must be small enough that inelastic scattering is unlikely, and we shall treat only linear response and low temperature.

5.7.1 TWO LEADS WITH MANY CHANNELS: THE QUANTIZED CONDUCTANCE

The first extension beyond a strictly one-dimensional system is shown in Figure 5.21. There are two leads with a scattering centre between them, as in earlier sections, but each lead now has several subbands arising from the transverse states (Section 4.8). These subbands are also called *modes* or *channels*. It is essential that the transverse potential of each lead remains constant along its length, although its precise form is unimportant. This is part of the definition of a perfect lead, and any regions where the potential changes must be included as part of the scattering centre. The wave function within a lead takes the form

$$\psi(\mathbf{R}) = \sum_n v_n^{-1/2}[A_n \exp(ik_n z) + B_n \exp(-ik_n z)]u_n(\mathbf{r}). \qquad (5.94)$$

This generalizes equation (5.1). Transverse states are labelled by n (which should really be two labels), with wave functions $u_n(\mathbf{r})$ and energies ε_n. The total energy $E = \varepsilon_n + \hbar^2 k_n^2/2m$, so states propagate if $E > \varepsilon_n$ and decay otherwise. The factor with the velocity v_n of each mode normalizes the states by flux rather than density as in equation (5.48). The two leads are not in general identical, so the energies of the one-dimensional subbands are different and so are the number of propagating states in each, N_{left} and N_{right}.

Inject a wave from the left purely in mode m. The scattering centre mixes the different modes so the scattered wave has contributions from all outgoing modes on both sides. The wave functions in the left and right leads take the form

$$\psi_{\text{left}}(\mathbf{R}) = [v_m^{(l)}]^{-1/2}u_m^{(l)}(\mathbf{r})\exp(ik_m^{(l)}z)$$

$$+ \sum_{n=1}^{\infty}[v_n^{(l)}]^{-1/2}r_{nm}u_n^{(l)}(\mathbf{r})\exp(-ik_n^{(l)}z),$$

$$\psi_{\text{right}}(\mathbf{R}) = \sum_{n=1}^{\infty}[v_n^{(r)}]^{-1/2}t_{nm}u_n^{(r)}(\mathbf{r})\exp(ik_n^{(r)}z). \qquad (5.95)$$

The sums run over *all* values of n, not just the propagating states; the decaying states are essential to complete the wave function. There are now arrays of reflection and transmission coefficients rather than the simple numbers in the strictly one-dimensional case.

The S- and T-matrices defined in Section 5.2 can be extended to many modes. They are essential for detailed calculations but contain much information that may not be needed afterwards. A simpler matrix t can be constructed from the coefficients t_{nm} in equation (5.95), giving the transmission amplitude for an electron incident from the left in mode m to be transmitted on the right in mode n. We restrict t to propagating states, giving dimensions of $N_{\text{right}} \times N_{\text{left}}$.

One reason for using t is that it contains sufficient information to find the conductance. The derivation follows equation (5.58), which gave $G = (2e^2/h)T = (2e^2/h)|t|^2$ for one dimension. Consider electrons injected in a given mode m. Those that emerge in mode n make a contribution $(2e^2/h)|t_{nm}|^2$ to the conductance. The velocity of different modes is taken into account by the normalization and does not clutter this result. The total conductance is found by summing over all input and output modes:

$$G = \frac{2e^2}{h} \sum_m \sum_n |t_{nm}|^2. \tag{5.96}$$

A vital feature is that the sum is over *intensities* rather than *amplitudes*: it is assumed that there is no phase coherence between electrons injected in different modes, so that we can simply add the contributions to the current and not worry about interference between them. It is also implicit that each mode has the same Fermi level, yet another demand on the leads. The conductance can be written in a more compact form by using the Hermitian-conjugate matrix of t, defined by $(t^\dagger)_{mn} = (t_{nm})^*$. Then

$$G = \frac{2e^2}{h} \sum_{m,n} t_{nm} t_{nm}^* = \frac{2e^2}{h} \sum_{m,n} (\mathrm{t})_{nm} (\mathrm{t}^\dagger)_{mn}$$

$$= \frac{2e^2}{h} \sum_n (\mathrm{tt}^\dagger)_{nn} = \frac{2e^2}{h} \mathrm{Tr}\,(\mathrm{tt}^\dagger) = \frac{2e^2}{h} \mathrm{Tr}\,(\mathrm{t}^\dagger \mathrm{t}). \tag{5.97}$$

This is the form in which the result is usually quoted, where 'Tr' is the trace of the matrix (the sum of its diagonal elements). Note that the product $\mathrm{tt}^\dagger$ is square, although neither t nor $\mathrm{t}^\dagger$ need be, and that the two expressions for the trace are equal, although $\mathrm{tt}^\dagger$ and $\mathrm{t}^\dagger\mathrm{t}$ may not be the same size.

This result can be used to calculate the conductance of a short wire or constriction, the *quantum point contact*. A typical structure is illustrated in Figure 5.22(a). Two gates shaped like opposed fingers on the surface of a heterostructure are negatively biased to deplete the 2DEG underneath them. The remaining electrons are forced to travel through the gap between the gates, which behaves like a short quasi-one-dimensional system. The broad regions of 2DEG on either side act as the 'leads'.

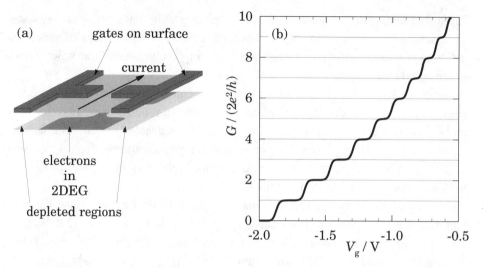

FIGURE 5.22. (a) Layout of a typical quantum point contact, a short constriction defined by patterned metal gates on the surface of a heterostructure containing a 2DEG. (b) Calculated conductance $G(V_g)$ as a function of gate voltage V_g. [From Nixon, Davies, and Baranger (1991).]

If the gates are short, the potential underneath them looks like the saddle shown in Figure 5.23(a). Because the potential varies so smoothly, it is possible to use the *adiabatic approximation*. The idea is to write the wave function of each mode in the separable form

$$\phi_n(\mathbf{R}) \approx u_n(\mathbf{r}; z)[v_n(z)]^{-1/2}\{A_n(z)\exp[ik_n(z)z] + B_n(z)\exp[-ik_n(z)z]\}. \tag{5.98}$$

The wave function and energy in the transverse potential at z, $u_n(\mathbf{r}; z)$ and $\varepsilon_n(z)$, are calculated as though this potential were constant along the wire. The wave number follows from $E = \varepsilon_n(z) + \hbar^2 k_n^2(z)/2m$. A mode may be propagating in one region and decaying in another. This approximation is related to the WKB method (Section 7.4) and is applicable only if the transverse potential changes slowly along z. It may also be possible to neglect scattering from one mode to another, in which case the amplitudes $A_n(z)$ and $B_n(z)$ may be calculated independently for each mode. The matrix t then becomes diagonal.

The energy $\varepsilon_n(z)$ of each subband varies with longitudinal position z through the constriction, rising to a broad peak in the middle (Figure 5.23(b)). Many states propagate while they are far from the constriction, but their wave number becomes imaginary as they approach the saddle point and see an apparent barrier when $\varepsilon_n(z) > E$. Such an electron (modes 2 and 3 in Figure 5.23(b)) may tunnel through the barrier but most of the amplitude is reflected unless the apparent barrier is low. Only electrons in the lowest N_{trans} modes propagate throughout the constriction (mode 1 in Figure 5.23(b)). Thus t has diagonal entries of nearly unity for the lowest N_{trans} modes, which are transmitted, and small values for the others. Equation (5.97)

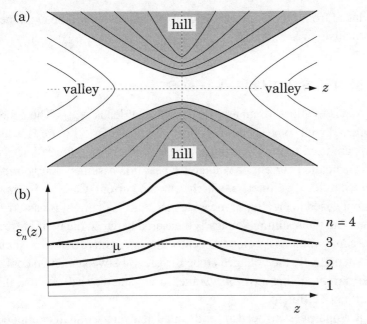

FIGURE 5.23. (a) Contours of a smooth saddle-point potential used to demonstrate the quantized conductance. The thicker line shows where the Fermi level meets the potential energy and the shaded region has high potential energy. (b) Energies $\varepsilon_n(z)$ of the transverse modes as a function of their position z.

shows that the conductance is given by

$$G = \frac{2e^2}{h}\mathrm{Tr}\,(\mathrm{t}^\dagger\mathrm{t}) \approx \frac{2e^2}{h}N_{\mathrm{trans}}\,. \tag{5.99}$$

This is the quantized conductance. The value of N_{trans} can be changed by altering the width and depth of the constriction, usually by varying the bias V_{g} on the gates. Thus a plot of $G(V_{\mathrm{g}})$ should give a steplike curve, with G jumping by $2e^2/h$ whenever another mode is allowed to propagate through the saddle point. An example is shown in Figure 5.22; this is a simulation but some experiments are even better! The rounding of the steps is due to tunnelling through the saddle point. Raising the temperature has a similar effect.

Adiabatic propagation of electrons is not a necessary condition for the quantized conductance. Scattering between modes will have no effect provided that only forward scattering occurs so the direction of the electron is preserved (although the magnitude of its velocity must change). Any backward scattering, on the other hand, will be disastrous and the constriction must be kept short to avoid this, typically below $1\,\mu$m. Sharp features in the potential also produce structure in $G(V_{\mathrm{g}})$. This sensitivity to the details of the potential means that the quantization of the conductance is not particularly accurate, and a result within 10% of $(2e^2/h)N_{\mathrm{trans}}$ is good. This contrasts strongly with the quantum Hall effect, where the Hall conductance takes the same value but in a good sample is in perfect agreement with the value of

$2e^2/h$ deduced from other high-precision experiments. It will be explained with a similar formalism in Section 6.6.

5.7.2 SYSTEMS WITH MANY LEADS

The next step is to allow for an arbitrary number of leads N_{leads}. The general form of the geometry, with some specific applications, is shown in Figure 5.24. Typically some leads are used to inject currents, whereas others measure voltages, and the leads are often called current or voltage probes. It is assumed that *voltage probes* are connected to ideal voltmeters, which draw no current. Case (c) is important as it is commonly used in practice to measure resistance. A current is passed along the straight path and the resulting voltage is measured between the two side arms. This *four-probe* configuration is preferred because it is insensitive to the resistance of the contacts. Alternatively the voltage can be measured between the two contacts used to pass the current, giving a *two-probe* measurement. We shall see that the results can be startlingly different.

Although S-matrices are needed for detailed analysis, we shall continue with the t-matrix defined in the previous section. Use m and n to label the leads and α and β to label the propagating modes within each lead; N_m modes propagate in lead m. As usual the leads must have constant cross-section, and states should be normalized to constant flux. An inward-flowing current in a lead is defined to be positive.

Consider a particular lead and mode, (m, α), say. We are interested only in deviations from equilibrium so the excess current that impinges on the sample from this lead is given, as in Section 5.4.1, by $I^{inc}_{m\alpha} = (-2e/h)\delta\mu_m$. The change in Fermi energy $\delta\mu_m = -eV_m$, where V_m is the applied voltage, so $I^{inc}_{m\alpha} = (2e^2/h)V_m$. This is the same for all N_m modes in this lead.

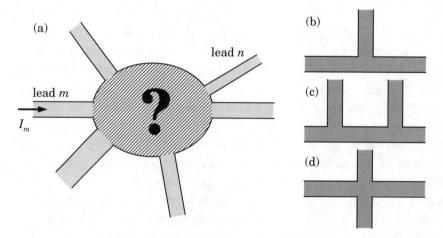

FIGURE 5.24. Geometry of a sample for coherent transport with many leads. The general case is shown in (a) with specific examples of (b) a T-junction, (c) four-terminal measurement of longitudinal resistance, and (d) a microscopic Hall bar.

Of the electrons incident from this mode and lead, those entering into mode β of lead n have amplitude $t_{n\beta,m\alpha}$. This contributes a current $-I_{m\alpha}^{\text{inc}}|t_{n\beta,m\alpha}|^2$, which is negative because it flows outwards. Some electrons are reflected back down lead m, not necessarily in their original mode, with reflection amplitudes $r_{m\beta,m\alpha}$. Since $t_{n\beta,m\alpha}$ is defined only for $n \neq m$ and $r_{n\beta,m\alpha}$ exists only for $n = m$, one often defines $t_{m\beta,m\alpha} = r_{m\beta,m\alpha}$.

The total current in lead n due to electrons injected from a different lead m is given by summing over all modes in the two leads, as in equation (5.96). Thus

$$I_{nm} = -\frac{2e^2}{h}V_m \sum_{\beta=1}^{N_n} \sum_{\alpha=1}^{N_m} |t_{n\beta,m\alpha}|^2. \qquad (5.100)$$

All expressions for currents involve sums such as this over the modes in both leads. It is like the trace in equation (5.97), although there are now extra subscripts on the transmission coefficients to label the leads. We can clarify the notation by introducing still more transmission and reflection coefficients which absorb these traces:

$$T_{nm} = \sum_{\beta=1}^{N_n} \sum_{\alpha=1}^{N_m} |t_{n\beta,m\alpha}|^2, \qquad R_m = \sum_{\beta=1}^{N_m} \sum_{\alpha=1}^{N_m} |r_{m\beta,m\alpha}|^2. \qquad (5.101)$$

These coefficients can be greater than unity, unlike the simple coefficients in one-dimensional systems. For example, T_{nm} reaches the smaller of N_m and N_n for perfect transmission. Again, one often sets $T_{mm} = R_m$ to treat the transmission and reflection coefficients on an equal footing.

The net current injected into lead m is given by the incident current less the reflected current. The total incident current, summed over all propagating modes, is $(2e^2/h)N_m V_m$, so the net current is

$$I_{mm} = (2e^2/h)(N_m - R_m)V_m. \qquad (5.102)$$

Conservation of current requires that this be equal to the total current injected from lead m that leaves the sample through other leads, that is, $I_{mm} = \sum_{n,n\neq m} I_{nm}$. This leads to a sum rule on the transmission coefficients,

$$R_m + \sum_{n,n\neq m} T_{nm} = N_m. \qquad (5.103)$$

This is an obvious generalization of $R + T = 1$.

We have now calculated the currents due to electrons injected from lead m. All that remains is to sum the contributions from all leads. Lead n causes a current $-(2e^2/h)T_{mn}V_n$ in lead m, so the total current in lead m is

$$I_m = \frac{2e^2}{h}\left[(N_m - R_m)V_m - \sum_{n,n\neq m} T_{mn}V_n \right]. \qquad (5.104)$$

This the Landauer–Büttiker formula for the conductance of a system with many leads. It can also be written in terms of a square conductance matrix whose dimension is given by the number of leads:

$$I_m = \sum_n G_{mn} V_n , \qquad G_{mn} = \frac{2e^2}{h}[(N_m - R_m)\delta_{mn} - T_{mn}]. \qquad (5.105)$$

This matrix must be treated with some care. The condition (5.103) for current conservation means that each column n of G_{mn} sums to zero. This shows immediately that the determinant vanishes and the matrix is singular. Another condition comes from the requirement that no current should flow if all voltages are equal, which means that

$$\sum_n G_{mn} = 0 = (N_m - R_m) - \sum_{n, n \neq m} T_{mn}. \qquad (5.106)$$

Thus each row m of the conductance matrix must also sum to zero. The fact that both the rows and columns sum to zero leads to the relation

$$\sum_n T_{mn} = \sum_n T_{nm}. \qquad (5.107)$$

These conditions can be used to rewrite the current, equation (5.104), in a number of ways. Replacing $(N_m - R_m)$ using the condition (5.103) for current conservation leads to

$$I_m = \frac{2e^2}{h} \sum_{n, n \neq m} (T_{nm} V_m - T_{mn} V_n). \qquad (5.108)$$

The diagonal term T_{nn} is not needed. Replacing the same term using the 'row' sum rule (5.106) gives

$$I_m = \frac{2e^2}{h} \sum_{n, n \neq m} (T_{mn} V_m - T_{mn} V_n) = \frac{2e^2}{h} \sum_{n, n \neq m} T_{mn}(V_m - V_n). \qquad (5.109)$$

This shows explicitly that only differences between applied voltages are significant.

Note that equation (5.107) does not imply that the conductance matrix must be symmetric, except in the case of only two leads. However, time-reversal symmetry makes the matrix symmetric in the absence of a magnetic field. This symmetry is broken by a magnetic field, which is important in the analysis of the Hall effect.

It is straightforward to verify that equation (5.104) agrees with our earlier results for systems with two probes. The next most complicated case is a sample with three leads, as in Figure 5.25(a). Let lead 3 be a voltage probe connected to an ideal voltmeter, which draws no current, so $I_3 = 0$. A current I flows into lead 2 and out of lead 1, so $I_1 = -I$ and $I_2 = I$. Finally, set $V_1 = 0$ as the reference

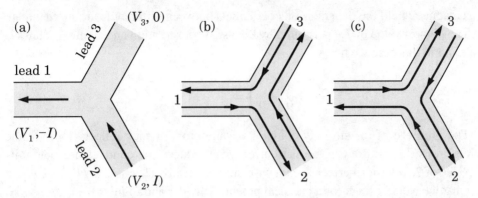

FIGURE 5.25. A sample with three leads to illustrate the multiprobe formula. A current I flows into lead 3 and out of lead 1; no current passes down lead 2, which is used purely to measure voltage. The two figures on the right show the transmission coefficients in large positive and negative magnetic fields when the device acts as a 'circulator'.

voltage. Equation (5.108) becomes

$$I_1 = -I = \frac{2e^2}{h}[-T_{12}V_2 - T_{13}V_3]$$

$$I_2 = I = \frac{2e^2}{h}[(T_{12} + T_{32})V_2 - T_{23}V_3]$$

$$I_3 = 0 = \frac{2e^2}{h}[-T_{32}V_2 + (T_{13} + T_{23})V_3]. \qquad (5.110)$$

Adding the three equations gives zero on both sides, confirming that one is redundant. Solution of these equations shows that the two-probe conductance of the system between leads 1 and 2 is

$$\frac{I}{V_2} = \frac{2e^2}{h}\left(T_{12} + \frac{T_{13}T_{32}}{T_{13} + T_{23}}\right). \qquad (5.111)$$

There are two contributions to the conductance. The first is due to those electrons that go directly from lead 1 to lead 2, as expected. The second is indirect and arises from electrons that go from lead 1 into 3. This is a voltage probe and carries no net current, so the flow must be balanced by an equal and opposite current that divides between leads 1 and 2 in the ratio of their transmission coefficients. Another useful result is the potential measured in lead 3:

$$\frac{V_3}{V_2} = \frac{T_{32}}{T_{13} + T_{23}} = \frac{T_{32}}{T_{31} + T_{32}}. \qquad (5.112)$$

This closely resembles a potential divider. The second form of the denominator follows from the rows-and-columns sum rule (5.107).

To check that these make sense, suppose that the leads each support only one transverse mode and that the structure has threefold symmetry. In the absence of a magnetic field the transmission coefficients between different leads are identical. The highest value of T_{12} permitted by the usual conservation and symmetry laws is $\frac{4}{9}$ and in this case we find

$$\frac{I}{V_2} = \frac{2e^2}{h}\left(\frac{4}{9} + \frac{2}{9}\right) = \frac{2}{3}\frac{2e^2}{h}, \qquad \frac{V_3}{V_2} = \frac{1}{2}. \tag{5.113}$$

The presence of the strongly coupled voltage probe, which reflects some of the electrons despite drawing no net current, has reduced the conductance below its value of $2e^2/h$ for a perfect system with only two leads. The ratio $V_3/V_2 = \frac{1}{2}$ is just what we would expect for a classical potential divider and follows from symmetry.

Now apply a large magnetic field. It is possible, as we shall see in Section 6.5, to arrange that the electrons are all forced to go down the lead to their right. In microwaves this would be a 'circulator'. Then $T_{12} = T_{23} = T_{31} = 1$, the others vanish, and we get

$$\frac{I}{V_2} = \frac{2e^2}{h}(1+0) = \frac{2e^2}{h}, \qquad \frac{V_2}{V_1} = 0. \tag{5.114}$$

Reversing the field gives $T_{21} = T_{32} = T_{13} = 1$ and

$$\frac{I}{V_2} = \frac{2e^2}{h}(0+1) = \frac{2e^2}{h}, \qquad \frac{V_2}{V_1} = 1. \tag{5.115}$$

The behaviour of the voltage probe is quite different for the two directions of magnetic field, although the conductance measured between two probes is not affected by the direction of the magnetic field (a general result for a two-probe measurement, mentioned earlier).

An important feature is that the indirect and direct currents are not coherent with each other, because the indirect current involves electrons that emerge from a different lead (the voltage probe 3). This can provide a useful theoretical trick to simulate lack of full coherence in tunnelling, which is very difficult to treat more formally. One just couples an additional voltage probe to the sample where the incoherence is supposed to originate. This picture can be verified by suppressing direct transmission from 1 to 2 so that all current is forced to take the indirect path. Then equation (5.111) becomes

$$\frac{I}{V_2} = \frac{2e^2}{h}\frac{T_{13}T_{32}}{T_{13} + T_{23}}. \tag{5.116}$$

The symmetry $T_{23} = T_{32}$ reduces this to the classical formula for two resistors in series, contact 3 acting as the joint between them.

Finally, assume that the coupling to lead 3 is very weak. This might be the case in practice because we would like the voltage probe to disturb the system as little as possible. Then the direct current dominates equation (5.111), which depends only

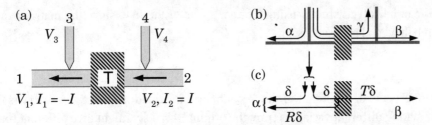

FIGURE 5.26. Two- and four-terminal measurements of the resistance of a tunnelling barrier. (a) Current is passed between probes 1 and 2; voltage can be measured either between these or between the weakly coupled probes 3 and 4. (b) Definition of the transmission coefficients coupling the voltage probes to the sample. (c) Derivation of the relation between the transmission coefficients α and β.

on T_{12} as we would hope. The voltage on probe 3, equation (5.112), unfortunately depends strongly on the ratio $T_{31} : T_{32}$ of the couplings to the two other leads. We would expect $V_3 = \frac{1}{2}V_2$ for a perfect structure (in the absence of a magnetic field). This requires the couplings to the two current leads to be equal, not very surprisingly. Any imperfections that break this symmetry will affect V_3.

Our final task is to derive a general formula for a four-probe resistance. The configuration is shown in Figure 5.26(a). Let current I enter through probe 2 and leave through probe 1; we wish to find the voltage between probes 4 and 3, which draw no current. There is a general notation $R_{mn,pq} = V_{pq}/I_{mn}$ for such quantities, where V_{pq} is the potential difference that appears between contacts p and q in response to a current between m and n. Thus we want the four-terminal resistance $R_{21,43}$. The full set of equations (5.109) is

$$
\begin{pmatrix}
T_{12} + T_{13} + T_{14} & -T_{12} & -T_{13} & -T_{14} \\
-T_{21} & T_{21} + T_{23} + T_{24} & -T_{23} & -T_{24} \\
-T_{31} & -T_{32} & T_{31} + T_{32} + T_{34} & -T_{34} \\
-T_{41} & -T_{42} & -T_{43} & T_{41} + T_{42} + T_{43}
\end{pmatrix}
$$

$$
\times
\begin{pmatrix} V_1 \\ V_2 \\ V_3 \\ V_4 \end{pmatrix}
= \frac{h}{2e^2}
\begin{pmatrix} I_1 \\ I_2 \\ I_3 \\ I_4 \end{pmatrix}
= \frac{h}{2e^2}
\begin{pmatrix} -I \\ I \\ 0 \\ 0 \end{pmatrix}.
\tag{5.117}
$$

We know that one of these equations is redundant, so drop that for I_1. We also know that only differences of voltages are significant, so we can set one to zero; V_3 is a convenient choice as we want V_{43}. We are then left with a 3×3 set of equations,

$$
\begin{pmatrix}
-T_{21} & T_{21} + T_{23} + T_{24} & -T_{24} \\
-T_{31} & -T_{32} & -T_{34} \\
-T_{41} & -T_{42} & T_{41} + T_{42} + T_{43}
\end{pmatrix}
\begin{pmatrix} V_1 \\ V_2 \\ V_4 \end{pmatrix}
= \frac{h}{2e^2}
\begin{pmatrix} I \\ 0 \\ 0 \end{pmatrix}.
\tag{5.118}
$$

This can be solved simply using Cramer's rule or the like, since the matrix is now well behaved. The result is

$$R_{21,43} = \frac{h}{2e^2} \frac{T_{42}T_{31} - T_{41}T_{32}}{S}, \tag{5.119}$$

where S is the determinant of the 3×3 matrix in equation (5.118). A worry is that eliminating different variables from the original 4×4 set might give different forms of this result. Fortunately the numerator does not change, and the sum rule that requires that the rows and columns all sum to zero means that any 3×3 submatrix of the original matrix has the same determinant (although one has to watch the sign!). Another useful result is the two-probe resistance, deduced from the voltage between the current probes,

$$R_{21,21} = \frac{h}{2e^2} \frac{(T_{31} + T_{32} + T_{34})(T_{41} + T_{42} + T_{43}) - T_{34}T_{43}}{S}. \tag{5.120}$$

These results will be used in Section 6.6.1 to study the propagation of edge states in the quantum Hall effect.

An interesting application, of great historical importance, is to compare the resistance of a tunnelling barrier measured using two or four probes. The system is shown in Figure 5.26(a), where the sample is the barrier in the middle with transmission coefficient T. The voltage probes are assumed to be identical and very weakly coupled to the structure to cause minimum disturbance. Assume for simplicity that only one mode propagates through the structure.

We need the transmission coefficients, of which there are six assuming time-reversal invariance. The largest is $T_{12} = T_{21} = T$ due to the barrier. All others involve coupling to the voltage probes 3 and 4 and are small by assumption. The transmission coefficients from voltage probe 3 to the other probes are shown in Figure 5.26(b). Let $T_{31} = \alpha$ and $T_{32} = \beta$, which are both of order δ, say. The third coefficient $T_{34} = \gamma$ will be of order δ^2, as it describes propagation through both of the weakly transmitting voltage contacts. The coefficients for probe 4 are the same but for reflection symmetry. The determinant of the matrix in equation (5.118) is

$$S \approx \det \begin{vmatrix} -T & T & -\alpha \\ -\alpha & -\beta & -\gamma \\ -\beta & -\alpha & \alpha + \beta \end{vmatrix} \approx T(\alpha + \beta)^2, \tag{5.121}$$

retaining only terms to lowest order. The two-probe resistance (equation 5.120) becomes

$$R_{21,21} \approx \frac{h}{2e^2} \frac{(\alpha + \beta)^2 - \gamma^2}{S} \approx \frac{h}{2e^2} \frac{1}{T}; \qquad G_{\text{2-probe}} = \frac{2e^2}{h} T. \tag{5.122}$$

This is a familiar result. The four-probe resistance (equation 5.119) is

$$R_{21,43} \approx \frac{h}{2e^2} \frac{\alpha^2 - \beta^2}{S} \approx \frac{h}{2e^2} \left(\frac{\alpha - \beta}{\alpha + \beta}\right) \frac{1}{T}. \tag{5.123}$$

As in the three-lead example, the voltage between the weakly coupled probes depends on the ratio of their transmission coefficients in the two directions. Figure 5.26(c) shows how to determine this ratio. Current from probe 3 divides equally in the two directions when it leaves the voltage probe, with transmission coefficients δ. One branch of this current flows into lead 1 without impediment but the other branch encounters the barrier in the middle of the device. A fraction R is reflected by this and only a fraction T passes through to reach probe 2; we can ignore the small effect of probe 4, which it passes on the way. Thus $\alpha = (1 + R)\delta$ and $\beta = T\delta$. This finally gives

$$R_{21,43} \approx \frac{h}{2e^2}\frac{R}{T}; \qquad G_{\text{4-probe}} = \frac{2e^2}{h}\frac{T}{R} = \frac{2e^2}{h}\frac{T}{1 - T}. \qquad (5.124)$$

This result is due to Landauer, and the difference between the two-probe and four-probe results was long a source of controversy. There is little difference between $G_{\text{2-probe}}$ and $G_{\text{4-probe}}$ for a weakly transmitting barrier but they disagree strongly as the barrier becomes more transparent. In the limiting case of $T = 1$ we have $G_{\text{2-probe}} = 2e^2/h$ but $G_{\text{4-probe}} = \infty$. What is the source of the difference?

If there is no barrier at all and $T = 1$, it seems clear that the distribution of electrons should be the same at all points within the wire and that a probe used to measure the voltage should return the same value at any point, giving $G_{\text{4-probe}} = \infty$. The difference with $G_{\text{2-probe}}$ is that the voltages reflect the Fermi levels of the *reservoirs*. There must be a difference between these Fermi levels in order to drive a current. The current that leaves a reservoir is proportional to the product of the density of states at the Fermi level, the Fermi velocity, and the difference in Fermi levels, and this product is finite. Thus a non-zero voltage must be applied simply to generate the non-equilibrium distribution of electrons needed to pass a current, even if that current is then transmitted perfectly to the other end of the sample. This extra voltage appears to be due to an extra contact resistance of $h/2e^2$ in series with the sample. It also reminds us that energy must be supplied to support conduction, even through a perfect wire, but leaves open the question of how this energy is dissipated.

5.8 Tunnelling in Heterostructures

We have assumed throughout the previous sections that the structures consist of a single material with a superposed potential. This is not strictly applicable to a heterostructure such as a barrier of AlGaAs surrounded by GaAs. In simple cases the changes needed are very similar to those for quantum wells, discussed in Section 4.9. For example, the wave function of a layered structure still separates into a transverse plane wave and a longitudinal part that can be treated with T-matrices, but the energy no longer separates totally and the effective height of a barrier depends on the transverse wave vector $\mathbf{k}$. The matching conditions at a heterointerface must again use the derivative divided by the effective mass, not just the derivative by itself.

This is a good place to demonstrate why the condition for matching the derivative of the wave function at a heterojunction needs to be modified. Consider the simple step in Section 5.1, for which we found $t = 2k_1/(k_1+k_2)$ and $r = (k_1-k_2)/(k_1+k_2)$ (equation 5.4). Applying these without modification to a heterojunction with $m_1 \neq m_2$ gives currents as follows:

$$\text{(incident)} \qquad I_{\text{inc}} = \frac{\hbar k_1}{m_0 m_1};$$

$$\text{(reflected)} \qquad I_{\text{ref}} = \frac{\hbar k_1}{m_0 m_1}|r|^2 = \frac{\hbar k_1}{m_0 m_1}\frac{(k_1 - k_2)^2}{(k_1 + k_2)^2};$$

$$\text{(transmitted)} \qquad I_{\text{trans}} = \frac{\hbar k_2}{m_0 m_2}|t|^2 = \frac{\hbar k_2}{m_0 m_2}\frac{4k_1^2}{(k_1 + k_2)^2}. \qquad (5.125)$$

Summing the outgoing currents gives

$$I_{\text{ref}} + I_{\text{trans}} = \frac{\hbar k_1}{m_1 m_0}\left[1 + \frac{4k_1 k_2}{(k_1 + k_2)^2}\left(\frac{m_1}{m_2} - 1\right)\right]. \qquad (5.126)$$

Conservation of current requires that this be equal to I_{inc} and is violated unless $m_1 = m_2$.

If we use the modified matching condition for a heterojunction at $z = 0$ between materials A and B,

$$\frac{1}{m_A}\frac{d\psi(z)}{dz}\bigg|_{z=0_A} = \frac{1}{m_B}\frac{d\psi(z)}{dz}\bigg|_{z=0_B}, \qquad (5.127)$$

rather than the derivative by itself, equation (5.2) is replaced by

$$A + B = C + D, \qquad \frac{k_1}{m_1}(A - B) = \frac{k_2}{m_2}(C - D). \qquad (5.128)$$

This leads to transmission and reflection amplitudes with k replaced by (k/m):

$$t = \frac{2k_1/m_1}{k_1/m_1 + k_2/m_2}, \qquad r = \frac{k_1/m_1 - k_2/m_2}{k_1/m_1 + k_2/m_2}. \qquad (5.129)$$

Now $\hbar k/m$ is the velocity so it is not surprising that such expressions should appear. The reflected and transmitted currents become

$$I_{\text{ref}} = \frac{\hbar k_1}{m_0 m_1}\frac{(k_1/m_1 - k_2/m_2)^2}{(k_1/m_1 + k_2/m_2)^2},$$

$$I_{\text{trans}} = \frac{\hbar k_1}{m_0 m_1}\frac{4(k_1/m_1)(k_2/m_2)}{(k_1/m_1 + k_2/m_2)^2}. \qquad (5.130)$$

Now we always have $I_{\text{ref}} + I_{\text{trans}} = I_{\text{inc}}$.

It is useful to redefine the coefficients of the wave function such that the flux transmission coefficient is given just by $|t|^2$. We modified our original description before to account for different plateaus on either side of a barrier (equation 5.48), and the obvious extension to account for different masses is to define the wave functions on the left and right as

$$\frac{A}{\sqrt{v_1(k_1)}}e^{ik_1z} + \frac{B}{\sqrt{v_1(k_1)}}e^{-ik_1z}, \qquad \frac{C}{\sqrt{v_2(k_2)}}e^{ik_2z} + \frac{D}{\sqrt{v_2(k_2)}}e^{-ik_2z}. \qquad (5.131)$$

Here $v_1(k) = \hbar k/m_0 m_1$, the velocity in material 1, and the particle current is simply $|A|^2 - |B|^2$ with no other factors. The transmission and reflection amplitudes for a step can be written in terms of velocities as

$$t = \frac{2\sqrt{v_1 v_2}}{v_1 + v_2}, \qquad r = \frac{v_1 - v_2}{v_1 + v_2}. \qquad (5.132)$$

These amplitudes relate the coefficients defined by equation (5.131). Current conservation now requires $|r|^2 + |t|^2 = 1$, which is clearly satisfied, and both r and t have an attractive symmetry between the two sides of the step. We can use r, t, and all the apparatus of T-matrices despite having different plateaus or effective masses on the two sides of a barrier. For example, the dispersion relation of a square-wave superlattice, equation (5.89), becomes

$$\cos ka = \cos k_1 w \cos k_2 b - \frac{(k_1/m_1)^2 + (k_2/m_2)^2}{2(k_1/m_1)(k_2/m_2)} \sin k_1 w \sin k_2 b, \qquad (5.133)$$

again with k replaced by k/m.

5.8.1 INTERVALLEY TRANSFER

Further interesting effects occur at a heterojunction where more than one valley is involved. Figure 3.8(b) showed the conduction bands for a barrier of AlAs, whose lowest minimum is at X, surrounded by GaAs, whose lowest minimum is at Γ. The lowest tunnelling barrier is obtained if an incident electron in the Γ-valley transfers to X inside the barrier. Moreover, such transfer enables it to pass over the barrier if its energy is above E_X in the AlAs even though its energy is well below E_Γ there.

The strength of intervalley transfer depends on the band structure and is not easy to estimate, but some important qualitative results follow from symmetry. Assuming that the interface and materials are perfectly ordered, the Bloch wave vector $\mathbf{k}$ in the plane of the interface is conserved. A relatively energetic electron on our scale of physics has $E \approx 0.3\,\text{eV}$, which gives $K \approx 0.7\,\text{nm}^{-1}$. This is small on the scale of the Brillouin zone, which is roughly $\pi/a \approx 6\,\text{nm}^{-1}$. Suppose that the heterojunction lies in the (001)-plane, the usual case. There are six X-valleys in AlAs, of which two lie along the $\pm[001]$-directions and are centred on $\mathbf{k} = \mathbf{0}$. An electron incident in the Γ-valley can transfer to either of these while conserving $\mathbf{k}$. There will be a

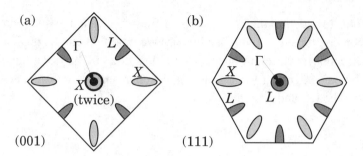

FIGURE 5.27. Surface Brillouin zone in **k** and lowest valleys in the conduction band of the usual semiconductors on (a) (001)- and (b) (111)-planes.

large change in k_z, from a small value to near the centre of the X-valley, but this is not forbidden (although the amplitude of this process may be small). The other four valleys, in contrast, lie out towards the zone boundaries along [100] or [010] and cannot be reached if **k** is conserved. For similar reasons it is impossible for the electron to transfer to the L-valleys, which lie at the zone boundary along [111]-directions. These selection rules can be illustrated by projecting the band structure onto the (001)-plane as in Figure 5.27(a).

Different selection rules apply if the heterointerface lies in the (111)-plane. In this case the projected band structure in Figure 5.27(b) shows that all the X-valleys lie at large values of **k** and cannot be reached from Γ. The L-valley along [111] is accessible but the other three are far away.

The projected band structures are particularly important when dealing with materials whose lowest minimum is not at Γ, such as Si and Ge. Consider the silicon inversion layer, for example, which is trapped at a heterojunction between Si and SiO_2. We can use Figure 5.27(a) if the silicon surface is (001), remembering that the X-valleys are lowest in Si. Two of these lie at the centre Γ' of the surface Brillouin zone. Their mass normal to the interface is the heavy longitudinal mass of the valley m_L. This governs the binding in the potential well at the interface, so the bound state has particularly low energy. The mass for motion parallel to the interface is isotropic, given by the lighter transverse mass m_T.

The other four valleys are situated near the boundary of the zone. Their mass normal to the interface is m_T, which is smaller, so the states are less deeply bound. They have an anisotropic mass for motion parallel to the interface, m_T along one principal axis and m_L along the other, but this anisotropy is lost on averaging over all four valleys. Thus the inversion layer has two series of subbands, a doubly degenerate series from the valleys at Γ' with deeper binding energies and low effective mass in the plane, and a fourfold degenerate series from the outer valleys with shallower binding energies and higher effective mass in the plane. Only the former set is occupied at low density and temperature but this changes as both are raised. Similar considerations apply to heterojunctions between Si, Ge, and their alloys.

5.9 What Has Been Brushed Under the Carpet?

It was mentioned earlier that transport, especially beyond the linear or ohmic region, is an extremely complex process that requires the solution of complicated kinetic equations. Despite this, we seem to have 'solved' a wide range of problems without vast effort, even highly nonlinear devices such as the resonant-tunnelling diode. You might reasonably assume that many problems have been 'brushed under the carpet', and you would be quite right. Here are a few issues, some of which are topics of current research.

5.9.1 POWER DISSIPATION

Where is power dissipated? We usually think of transport as a dissipative process (Joule heating), yet this has never been mentioned. A glib answer is to note that power is given by $I^2 R$ and is therefore second-order in the current, so we can ignore it within linear response. This is hardly illuminating. As an example, no dissipation occurs in coherent transport through a tunnelling barrier: the barrier throttles the flow of electrons but does not absorb their energy. This is something like water squirting from a hose: the nozzle controls the flow of water, but most of the energy is dissipated when the water splashes to the ground. Electrons that pass through the barrier travel into the far lead with a highly non-equilibrium distribution, particularly in something like a resonant-tunnelling diode. Processes within the leads restore the distribution close to equilibrium. We hope that these processes have a negligible effect on the conductance.

5.9.2 INCOMING DISTRIBUTION OF ELECTRONS

Do the incoming electrons really have an unperturbed Fermi distribution characteristic of the lead? It was argued in the preceding section that all energy is dissipated in the leads, and the scattering processes could distort this distribution. Another effect is shown in Figure 5.8(b), with an accumulation layer on the left of a tunnelling barrier. The triangular potential well produced by the accumulation layer is likely to be rather narrow, so states in it will be quantized like those at a doped hetero-junction and form a two-dimensional electron gas. Are electrons injected from the discrete energy levels of a 2DEG, rather than the three-dimensional continuum that we have assumed? The effect on $I(V)$ would be dramatic, particularly in a resonant-tunnelling diode, where the triangular shape arose from the density of states in the three-dimensional leads. The calculation of $I(V)$ would also become more complicated because one would have to consider how electrons enter the 2DEG from the leads.

5.9.3   INELASTIC SCATTERING

One of our key assumptions has been that transport within the active region of the device is coherent, with no inelastic scattering. What happens if this is not true? For example, a resonant-tunnelling diode under large bias has its resonance pulled below the range of incoming electrons (Figure 5.13(d)). Electrons cannot enter the resonant elastically, but can do so if they first emit a phonon. This gives rise to a satellite peak in $I(V)$ at higher bias than the elastic peak. This process is well understood but others are not.

5.9.4   IMPERFECTIONS

We have assumed that the structures are flawless, with no imperfections such as rough interfaces or impurities. Although neither of these cause inelastic scattering and destroy coherence, they can nevertheless have a drastic effect on tunnelling because they change the direction of an electron. This ruins the separation of momentum and energy into longitudinal and transverse components that was crucial to our calculation of the current in Section 5.4.2. Again, consider the resonant-tunnelling diode at so large a bias that elastic current through the resonance has ceased. If an incoming electron scatters from an impurity inside the double barrier, it can transfer so much of its energy to the transverse plane that its longitudinal energy falls into the resonance. This is rather like the real space transfer of Section 4.6 in reverse. The point is that the tunnelling process depends only on one component of the electron's momentum, and the change in direction caused by elastic scattering changes the energy of the electron as far as this one component is concerned. Unavoidable fluctuations in alloy barriers have the same effect, and resonant-tunnelling diodes based on GaAs often have barriers of AlAs to avoid this problem.

This list could continue for pages, but it is time to halt. We have now looked at the dynamics of electrons normal to heterostructures, both trapped and free; the next chapter looks at free electrons again, under the influence of uniform electric and magnetic fields.

Further Reading

Weisbuch and Vinter (1991) and Kelly (1995) describe a range of applications of tunnelling. Practical aspects of resonant-tunnelling diodes are reviewed by Brown (1994). Tiwari (1992) also addresses these issues.

The Landauer–Büttiker formalism is of great importance in mesoscopic systems, which by definition are small enough that the assumption of coherent transport holds. Büttiker (1988) gives a clear account of this theory, and it is discussed in the book by Datta (1995). There is a good chapter by Geerligs (1992) on Coulomb blockade. This and many other aspects of transport in two-dimensional systems are described by Beenakker and van Houten (1991).

5.1   Extend the calculation of transmission at an upward step to a downward step, $V_0 < 0$, and plot the result. Classical physics gives $T = 1$ for all incident energies; what about quantum mechanics?

5.2   Show by direct calculation that the T-matrix for a potential step situated at $z = d$ rather than the origin is given by

$$T^{(21)} = \frac{1}{2k_2} \begin{pmatrix} (k_2 + k_1)e^{i(k_1-k_2)d} & (k_2 - k_1)e^{-i(k_1+k_2)d} \\ (k_2 - k_1)e^{i(k_1+k_2)d} & (k_2 + k_1)e^{-i(k_1-k_2)d} \end{pmatrix}. \qquad \text{(E5.1)}$$

Show also that this can be rewritten as the product

$$T^{(21)}(d) = \begin{pmatrix} e^{-ik_2d} & 0 \\ 0 & e^{ik_2d} \end{pmatrix} T(k_2, k_1) \begin{pmatrix} e^{ik_1d} & 0 \\ 0 & e^{-ik_1d} \end{pmatrix}, \qquad \text{(E5.2)}$$

confirming the general rule (equation 5.16). (The states are normalized as in Section 5.2, without a factor to account for the different plateaus.)

5.3   How are t and r changed when a barrier is translated through d, as in the previous exercise, and has this any observable consequences?

5.4   As we build the T-matrices for complicated barriers out of those for simple components, it is important that the product of two T-matrices is itself a T-matrix and has the properties that we have just derived. Show that this is so. (This is one of the properties of a mathematical group, and T-matrices provide a representation of the group $SU(1, 1)$.)

5.5   It may be surprising to find that T-matrices can be used to treat bound states as well as propagating ones. A bound state, such as one in a finite square well, has decaying waves leaving it in both directions and the energy of these waves, measured from the plateau outside the well, is negative. These both correspond to outgoing waves, but there are no incoming waves which would grow in magnitude away from the well. The presence of outgoing waves and the absence of incoming waves means that both r and t are infinite at bound states.

Consider the δ-function potential as an example, with $S < 0$ for an attractive well. Equation (5.28) shows that there is a single energy at which the transmission goes to infinity, given by $E = -mS^2/2\hbar^2$, and this is the energy of the bound state. Repeat this for a finite square well, and show that it leads to the same condition for bound states as in Section 4.2.

5.6   The refraction of light at a dielectric boundary is summarized by Snell's law, $n_1 \sin \theta_1 = n_2 \sin \theta_2$. What is the corresponding result for electrons at a step?

5.7   A barrier of AlGaAs separates two regions of GaAs doped to $N_D = 3 \times 10^{24} \, \text{m}^{-3}$. What is the conductance of a sample $1 \, \mu\text{m}$ square if the barrier is $10 \, \text{nm}$ thick and $0.3 \, \text{eV}$ high? Use simple approximations and ignore the difference in effective mass. How sensitive is this result to monolayer fluctuations in the thickness of the barrier?

5.8   Rewrite equation (5.68) for the conductance of a barrier in three dimensions with the angle of the electrons (measured from z) as the variable of integration. Estimate the angular distribution of current for the previous example. Qualitatively, what effect would a large bias across the barrier have on this distribution?

5.9   Estimate the width of the lowest peak in the example of resonant tunnelling plotted in Figure 5.11, using the attempt frequency and transmission probabilities of the barriers.

5.10   Derive an approximate formula for the phase of the transmission amplitude near a peak in resonant tunnelling. How does it compare with classical resonant phenomena?

5.11   Use partial waves to show that the reflection amplitude of a double-barrier structure is

$$r = r'_L + \frac{t_L t'_L r_R \exp(2ika)}{1 - r_L r_R \exp(2ika)} = \frac{t_L}{t_L^* r_L} \left[\frac{|t_L|^2}{1 - r_L r_R \exp(2ika)} - 1 \right].$$

(E5.3)

You will need the relation between r_L and r'_L, the reflection amplitudes from the two sides of the barrier, given in equation (5.47).

5.12   It was shown in Section 5.5.3 that the current through a one-dimensional resonant-tunnelling diode should be constant while the resonance is within the range of incoming energies, and zero otherwise. What sets the width of the transitions? In particular, what is the effect of temperature, and is it the same for both transitions?

5.13   Consider a well of width $a = 10 \, \text{nm}$ confined between barriers of thickness $b = 3 \, \text{nm}$ and height $V_0 = 0.3 \, \text{eV}$. The energy E_{pk} of the resonant state can be estimated from the result for the true well (Section 4.2). Next estimate the transmission coefficient of each barrier at E_{pk} and use this to find the lifetime and width of the resonance.

5.14   Calculate the peak current density through the lowest resonance in the device analyzed in Figure 5.11, and sketch the form of $I(V)$. Assume that the left and right leads are n-GaAs doped to 5×10^{23} m^{-3}.

5.15   What shape would be expected for $I(V)$ in a two-dimensional resonant-tunnelling device, fabricated by making a double barrier in a 2DEG?

5.16   An impurity in a single barrier can sometimes be modelled as a one-dimensional resonant-tunnelling problem. The transmission coefficients T_L and T_R decay exponentially with the distance from the impurity to the edges of the barrier. Estimate how the peak and area of $T(E)$ depend on the position of the impurity within a barrier of fixed thickness, and show that resonant tunnelling is most effective when the impurity is near the centre.

5.17   Extend the calculation of band structure in a superlattice to rectangular rather than δ-function barriers. Use the same values, $V_0 = 0.3$ eV, $b = 5$ nm, $a = 10$ nm, and $m = 0.067 m_0$. It is easy to calculate and plot on a spreadsheet. Do you see any signature in the energy bands of the change in nature of the states when $E = V_0$? How good an approximation are δ-function barriers?

5.18   Calculate the edges of the lowest two or three bands using the same well but varying the width of the barriers b (Figure 5.19). This is useful as a demonstration of how energy bands form when atoms are brought closer together to form a solid (Section 7.7).

5.19   A superlattice can be used as an energy filter to reflect electrons whose energy lies in the band gaps. As a practical example, it has been suggested that a superlattice whose repeating unit is four monolayers of AlAs and twelve monolayers of GaAs should be a more effective barrier than the alloy Al$_{0.25}$Ga$_{0.75}$As, which has the same average composition [I. G. Thayne et al., *IEEE Transactions on Electron Devices*, **42** (1995): 2047–55]. Estimate the band structure to test this hypothesis using the model of a square-wave superlattice. Incident electrons may have energies up to 1 eV. Neglect complications such as higher valleys and non-parabolic bands, which ought really to be included.

The superlattice can be of only finite length; a few times the decay length is an obvious criterion. How many periods of the foregoing superlattice are needed to use the gap between the first and second bands as a filter?

5.20   What is the shape of the density of states for a lateral surface superlattice, a 2DEG with a one-dimensional periodic potential? (The exact expression for a cosine band involves an elliptic integral but the behaviour at the bottom of the band and at large energies can be estimated.)

5.21   Derive equation (5.120) for the two-probe resistance of a four-probe system. Show also that it reduces to the two-probe result $R = (h/2e^2)(1/T)$ if the voltage probes are very weakly coupled ($T_{12} = T_{21} = T$, all other Ts much smaller).

5.22   Calculate the transmission properties of a sandwich where the middle material has its conduction band at the same energy as the outer layers but has a different effective mass. The conduction band can be like this at heterojunctions between Si_xGe_{1-x}.

   Is it possible for electrons to be trapped in a sandwich like this?

5.23   How significant is the different effective mass for a typical barrier of AlGaAs in GaAs, with a height of about 0.3 eV?

5.24   Consider a superlattice consisting of alternating layers of equal thickness of GaAs and AlAs. The potential is a square wave with the minima in GaAs for the Γ-valley, but for the X-valleys the minima are in AlAs (Figure 3.8). The lowest parts of the conduction band are the Γ minima in GaAs, so it seems obvious that this is where electrons would migrate. However, this does not take account of the zero-point energy in the quantum wells, which is strongly affected by the difference in mass. The lowest state in the X-valleys arises from the high longitudinal mass, $m_L \approx 1.1$ in AlAs (Section 5.8.1). Show that the lowest state for electrons moves from GaAs to AlAs as the period of the superlattice is reduced, and find the critical value. Use an infinitely deep well for a rough estimate, or a finite well for a more accurate result.

   Are the holes affected in the same way? Is it possible to make a type II superlattice, where the electrons are confined in one material with the holes in the other?

5.25   Show that the matrix of transmission coefficients T_{mn} defined in Section 5.7.2 is symmetric for a sample with only two leads, even in the presence of a magnetic field B. This in turn implies that the two-terminal conductance of a sample must be an even function of B. The conductances of the 'circulators' shown in Figure 5.25(b) and (c) obey this relation (equations 5.114 and 5.115), but it looks rather mysterious because the path taken by the electrons is so different in the two cases. A more physical picture would be desirable.

   Consider a sample with two leads, labelled 1 and 2, with a conventional current flowing from 2 to 1 (so there is a net flow of electrons from 1 to 2). Our picture in Section 5.7.2 was that a current was injected by raising the chemical potential of lead 1 with a negative bias. This caused an excess flow of electrons into the sample from lead 1. Provided that the response is linear it is equally valid to apply a positive bias to the other lead (2), which *reduces* the flow of electrons into the sample from lead 2. The same net current flows in both cases. The second picture can be viewed as the injection of holes, although their charge and mass need to be treated carefully.

Now consider the two-terminal conductance between leads 1 and 2 in the two samples shown in Figure 5.25(b) and (c) within the picture of a positive bias on lead 2. What paths does the current take now? Show that a combination of these two pictures explains why the two samples have the same conductance. It is often useful to consider both pictures in the study of edge states and the quantum Hall effect (Section 6.6).

6 ELECTRIC AND MAGNETIC FIELDS

Electric and magnetic fields are among the most valuable probes of an electronic system. An obvious use of an electric field is to drive a current through a conductor; we studied conduction due to tunnelling in the previous chapter and will consider the opposite case of freely propagating electrons weakly scattered by impurities or phonons in later chapters. It is more surprising that useful information or practical applications can be obtained by applying an electric field to an insulator. An example of this is a change in optical absorption near a band edge caused by a strong electric field, the Franz–Keldysh effect, which we shall calculate in Section 6.2.1. This becomes even more useful when the electrons and holes are confined in a quantum well, and is used as an optoelectronic modulator.

A magnetic field has remarkable effects on a low-dimensional system. For example, the continuous density of states of a two-dimensional electron gas splits into a discrete set of δ-functions called *Landau levels*. This is reflected in the longitudinal conductivity as the Shubnikov–de Haas effect, giving a distinct signature of two-dimensional behaviour. The Hall effect is a widely used tool in semiconductors, and the combination with Landau levels in a two-dimensional electron gas gives the *integer quantum Hall effect*, where the Hall conductance is an exact multiple of e^2/h. This is now used as a fundamental standard. Strange values of the quantized Hall effect are found in samples with many leads and can be understood using the formalism for coherent transport developed in Section 5.7.2. Further modifications appear in a quasi-one-dimensional system and the name magnetic depopulation has been coined. Samples of the highest mobility show yet another behaviour, the fractional quantum Hall effect.

6.1 The Schrödinger Equation with Electric and Magnetic Fields

An electric field **F** or a magnetic field **B** must usually be introduced into quantum mechanics through the scalar potential ϕ and vector potential **A**. The symbol **F** will be used for the electric field to avoid confusion with the energy E, and **B** is measured in teslas (T) in SI units.

The fields are derived from the potentials using

$$\mathbf{F} = -\text{grad}\,\phi - \frac{d\mathbf{A}}{dt}, \qquad \mathbf{B} = \text{curl}\,\mathbf{A}. \tag{6.1}$$

An important point, to which we shall return several times, is that there is considerable freedom in the selection of potentials, especially $\mathbf{A}$. This is described as a choice of *gauge*.

Consider first a static electric field $\mathbf{F}$. This is most commonly described by a scalar electrostatic potential $\phi = -\mathbf{F} \cdot \mathbf{r}$. Even this has an arbitrary element because we can add any constant to ϕ without affecting the field. This in turn means that the absolute potential (or potential energy) has no significance and that physical results should depend only on differences.

Although this is the most common choice, we could instead derive the electric field from a vector potential $\mathbf{A} = -\mathbf{F}t$. There is more freedom here because we can add any function to $\mathbf{A}$ that does not depend on time without changing $\mathbf{F}$. We could also use a mixture of scalar and vector potentials. No choice is ideal because the solutions to Schrödinger's equation depend strongly on the choice of gauge. The electric field is the same at all points in space and it would be pretty if the potentials reflected this property, but the scalar potential does not; the vector potential respects spatial invariance but is a function of time, which means that there will be no stationary states. In practice one chooses the potential that simplifies the problem as far as possible, as we shall see in the next section.

The choice of gauge is greater for a magnetic field. Consider a field of magnitude B along z. The curl gives $B_z = \partial A_y/\partial x - \partial A_x/\partial y$. Two obvious choices are $A_y = Bx$ or $A_x = -By$. These are called the Landau gauge and have the advantage that only one component of vector potential is needed, which often simplifies calculations. The disadvantage is that they pick out a special direction in the xy-plane, which ought to be isotropic. Another choice is the symmetric gauge, $\mathbf{A} = \frac{1}{2}B(-y, x, 0) = \frac{1}{2}\mathbf{B} \times \mathbf{R}$. This preserves the isotropy of the plane transverse to $\mathbf{B}$ but the wave functions are more complicated.

Any function with zero curl can be added to $\mathbf{A}$ without affecting the value of $\mathbf{B}$. Since curl grad $\chi = 0$ for any function χ, adding grad χ to $\mathbf{A}$ will not change $\mathbf{B}$. It will change $\mathbf{F}$, however, and we must make a compensating change to the scalar potential ϕ to keep $\mathbf{F}$ constant. Thus a gauge transformation that leaves the fields unchanged is

$$\mathbf{A} \to \mathbf{A} + \text{grad}\,\chi, \qquad \phi \to \phi - \frac{d\chi}{dt}. \tag{6.2}$$

For example, $\chi = \pm\mathbf{F} \cdot \mathbf{R}t$ takes us between the scalar and vector potentials used to describe a uniform electric field. There are several special choices of gauge that are useful to simplify the treatment of time-dependent electromagnetic fields, but we will not need these.

Now that we have the potentials, the Schrödinger equation for a particle of charge q in an electromagnetic field is

$$\left\{\frac{1}{2m}[\hat{\mathbf{p}} - q\mathbf{A}(\mathbf{R}, \mathbf{t})]^2 + q\phi(\mathbf{R}, t)\right\}\Psi(\mathbf{R}, t) = i\hbar\frac{d}{dt}\Psi(\mathbf{R}, t). \quad (6.3)$$

An important feature is that there are now two momenta in the equation. One is $\hat{\mathbf{p}}$, called the *canonical* momentum, which is replaced by the operator $-i\hbar\nabla$. The second, $\hat{\mathbf{p}} - q\mathbf{A}$, is the expression that comes into the kinetic energy ('momentum squared over $2m$') and is called the *mechanical* or *kinematical* momentum. The same distinction is made in classical mechanics. Another feature, not present classically, arises because the Schrödinger equation (6.3) contains the potentials rather than the fields. This means that it is possible for electrons to be affected by the potentials even in regions where the fields themselves are absent. The Aharonov–Bohm effect is a consequence of this and will be described in Section 6.4.9.

The expression for the current density is modified in the presence of a vector potential. The original equation (1.32) becomes

$$\mathbf{J}(\mathbf{R}, t) = q\left[\frac{\hbar}{2im}\left(\Psi^*\nabla\Psi - \Psi\nabla\Psi^*\right) - \frac{q}{m}|\Psi|^2\mathbf{A}(\mathbf{R}, t)\right]. \quad (6.4)$$

The additional term is called the diamagnetic current. Its origin becomes a little clearer if the current is rewritten like equation (1.42) as

$$\mathbf{J}(\mathbf{R}, t) = \frac{q}{2}\left[\Psi^*\left(\frac{\hat{\mathbf{p}} - q\mathbf{A}}{m}\Psi\right) + \left(\frac{\hat{\mathbf{p}} - q\mathbf{A}}{m}\Psi\right)^*\Psi\right]. \quad (6.5)$$

This shows that the velocity operator is really $(\hat{\mathbf{p}} - q\mathbf{A})/m$, with the mechanical rather than the canonical momentum.

6.2 Uniform Electric Field

The classical behaviour of a charge q in an electric field $\mathbf{F}$ is simple: it accelerates uniformly at a rate $q\mathbf{F}/m$ and its motion normal to $\mathbf{F}$ is unaffected. Unfortunately the picture in quantum mechanics is not so straightforward, mainly because we have to use potentials rather than the field.

Consider a charge $q = -e$ in a uniform electric field F along z, so the potential energy is $q\phi = eFz$, which increases with z if $F > 0$. As the potential depends only on z the Schrödinger equation can be separated in the usual way (Section 4.5) and we can concentrate on the one-dimensional problem. The potential is constant in time so we seek stationary states of the Schrödinger equation

$$\left[-\frac{\hbar^2}{2m}\frac{d^2}{dz^2} + eFz\right]\psi(z) = \varepsilon\psi(z). \quad (6.6)$$

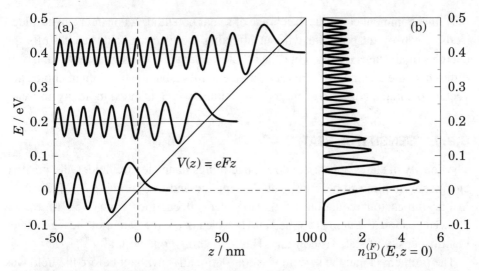

FIGURE 6.1. (a) Potential energy eFz, three wave functions, and energies for electrons in GaAs in a uniform electric field of $5\,\mathrm{MV\,m^{-1}}$. (b) Local density of states at $z = 0$, showing how the features correspond to the wave functions.

This is easy because we have solved it already for the triangular well in Section 4.4. The unnormalized wave functions with energy ε are

$$\phi(z, \varepsilon) = \mathrm{Ai}\!\left(\frac{z - \varepsilon/eF}{z_0}\right) = \mathrm{Ai}\!\left(\frac{eFz - \varepsilon}{\varepsilon_0}\right), \qquad (6.7)$$

where the length and energy scales are

$$z_0 = \left(\frac{\hbar^2}{2meF}\right)^{1/3}, \qquad \varepsilon_0 = \left[\frac{(eF\hbar)^2}{2m}\right]^{1/3} = eFz_0. \qquad (6.8)$$

This is easier than the triangular well because the electron can explore all space and the energy can take any value.

A few wave functions are plotted in Figure 6.1. Different states simply slide along in space and up in energy, retaining the same functional form. They tunnel into the potential for $z > \varepsilon/eF$, decaying more rapidly as the potential increases. The precise form follows from the asymptotic behaviour of the Airy function as $x \to \infty$ (equation A5.2). The wave functions oscillate for $z < \varepsilon/eF$, undulating more rapidly as $(z - \varepsilon/eF)$ becomes more negative and the kinetic energy increases. Their amplitude decreases at the same time to conserve current. Again, the functional form can be found from the asymptotic behaviour (A5.3).

These waves are standing rather than propagating. It is easy to see why this must be so: an electron with constant energy travels in the $+z$-direction until it hits the potential at ε/eF. It then reflects completely and returns along $-z$. Interference between the two waves of equal intensity sets up a standing wave. This contradicts our picture that the electron should accelerate uniformly in an electric field, and an alternative view is found by using a vector potential rather than a scalar potential.

A final comment concerns the choice of scalar potential. A constant can be added to the electrostatic potential without affecting the electric field, but it shifts the energies of all states rigidly. Thus the absolute energy has no significance, and only differences are meaningful. This is of course familiar, but the freedom of choice in a vector potential leads to an arbitrary momentum, which is much more mysterious!

6.2.1 DENSITY OF STATES

The density of states appears to do strange things in an electric field. Only positive energies were allowed before the field is applied and $n(E) = (1/\pi\hbar)(E/2m)^{-1/2}$ in one dimension (equation 1.89). However, all energies from $-\infty$ to ∞ are allowed as soon as a field is applied and it is clear from the nature of the eigenstates that the density of states is constant. Has something gone wrong?

The problem is that the system was translationally invariant before the field was applied, so we could measure the density of states at any point and find the same answer. This is clearly not true with the linear potential, which takes values from $-\infty$ to ∞ however weak the field. However, if we look at a particular point, we know that the majority of the wave functions there will have $\varepsilon > eFz$, with only the tails due to tunnelling at lower energies. The division between propagating and tunnelling states will occur at different energies at different places, so this feature is lost if we average the density of states over the whole system. The solution is to concentrate on one point and use the *local* density of states. This was defined in equation (1.102) as

$$n(E, z) = \sum_k |\phi_k(z)|^2 \, \delta(E - \varepsilon_k). \tag{6.9}$$

The sum is over all eigenstates, labelled by k. There is a small problem because our wave functions are not normalized but we shall ignore this and adjust the prefactor of the final result. The wave functions in equation (6.7) are labelled by their energy ε; there is no quantum number like the wave number for plane waves. There is also a continuous range of ε unlike the discrete set of k assumed in equation (6.9), so the sum is replaced by an integral (again neglecting a constant). Thus the density of states in an electric field in one dimension is

$$n_{1D}^{(F)}(E, z) \propto \int_{-\infty}^{\infty} \mathrm{Ai}^2\left(\frac{eFz - \varepsilon}{\varepsilon_0}\right) \delta(E - \varepsilon) \, d\varepsilon. \tag{6.10}$$

The integral is trivial and we obtain

$$n_{1D}^{(F)}(E, z) = C \, \mathrm{Ai}^2\left(-\frac{E - eFz}{\varepsilon_0}\right), \tag{6.11}$$

where C is the unknown prefactor. Note that $n_{1D}^{(F)}(E, z)$ depends, as we expect, only on $E - eFz$, which is the difference between the total energy and the potential energy at the point of observation. Classically this would simply be the kinetic energy.

To fix C, we might expect that the density of states at high kinetic energies will become similar to that for free electrons in the absence of a field,

$$n_{1D}^{FE}(E, z) \sim \frac{1}{\pi \hbar} \sqrt{\frac{2m}{E - eFz}}. \tag{6.12}$$

Equation (6.11) and the asymptotic form (A5.3) give

$$n_{1D}^{(F)}(E, z) \sim \frac{C}{\pi} \sqrt{\frac{\varepsilon_0}{E - eFz}} \cos^2 \left[\frac{2}{3} \left(\frac{E - eFz}{\varepsilon_0} \right)^{3/2} - \frac{\pi}{4} \right]. \tag{6.13}$$

The prefactor has the correct dependence on energy, but the $\cos^2$ function always oscillates between 0 and 1. Taking its average value of $\frac{1}{2}$ and equating the prefactors fixes C, and we finally get

$$n_{1D}^{(F)}(E, z) = \frac{2}{\hbar} \sqrt{\frac{2m}{\varepsilon_0}} \, \mathrm{Ai}^2 \left(-\frac{E - eFz}{\varepsilon_0} \right) \tag{6.14}$$

for the local density of states of a one-dimensional system in an electric field. This is plotted in Figure 6.2(a). The inverse-square-root singularity at the bottom of the band has been smeared out, with an exponentially decaying tail to negative energies. This corresponds to the tail of the wave functions that tunnel into the classically forbidden region where the kinetic energy is negative. The density of states oscillates for positive kinetic energies because it depends on the density of the wave function,

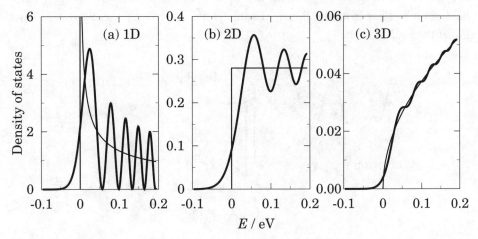

FIGURE 6.2. Local density of states $n^{(F)}(E, z)$ for electrons in GaAs in an electric field of $5 \, \mathrm{MV \, m^{-1}}$ as a function of local kinetic energy, $\varepsilon = E - eFz$. The thin curves are the results for free electrons. The units of $n(E, z)$ are $\mathrm{eV^{-1} nm^{-d}}$ in d dimensions.

which contains nodes from the standing wave. The correspondence is shown in Figure 6.1.

The density of states can be extended to three-dimensional systems as we did for the superlattice (equation 5.92). The result is

$$n_{3D}^{(F)}(E, z) = \frac{m}{\pi \hbar^3} \sqrt{\frac{2m}{\varepsilon_0}} \int_{-\infty}^{E} \mathrm{Ai}^2\left(-\frac{\varepsilon - eFz}{\varepsilon_0}\right) d\varepsilon \qquad (6.15)$$

$$= \frac{m}{\pi \hbar^3} \sqrt{2m\varepsilon_0}\{[\mathrm{Ai}'(s)]^2 - s[\mathrm{Ai}(s)]^2\}, \qquad s = -\frac{E - eFz}{\varepsilon_0}.$$

The integral is given in equation (A5.6). This is plotted with the corresponding result for two dimensions in Figure 6.2. The general features are the same as in one dimension but the oscillations are washed out by the convolution, and the underlying square-root behaviour is obvious.

The density of states can be measured by optical absorption (Section 8.6). The changes induced by an electric field are called the *Franz–Keldysh effect*, illustrated in Figure 6.3. An optical transition at frequency ω can occur only if two states can be found with separation $\Delta E = \hbar\omega$ and provided that the states overlap in space. Usually absorption in a semiconductor is impossible if $\hbar\omega < E_g$ because there are no states available. In an electric field, however, there are states in both valence and conduction bands at all energies. Their overlap in space depends on the difference in their energies. It is strong if $\Delta E > E_g$, when the oscillating part of the wave functions overlap. Only the tails overlap if $\Delta E < E_g$ and this decays rapidly with $E_g - \Delta E$. Thus the absorption edge gains a tail tunnelling into the previously forbidden gap, as well as structure due to changes in the wave functions, as shown in Figure 6.2. It is often said that the absorption edge is shifted down by an electric field but it is clear from the figure that this is not really so: the edge is broadened rather than shifted.

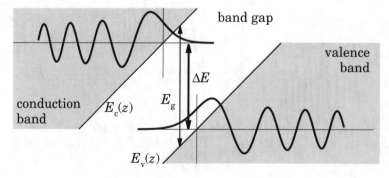

FIGURE 6.3. The Franz–Keldysh effect on interband absorption. The states shown in the valence and conduction bands are separated by $\Delta E < E_g$ but overlap because of the tail that tunnels into the band gap.

6.2.2 ELECTRIC FIELD FROM A VECTOR POTENTIAL

The quantum-mechanical description of an electron in a scalar potential has some undesirable features: the eigenstate is a standing wave, not an accelerating one as we might prefer, and the homogeneity of space is broken. These problems disappear if a vector potential $\mathbf{A} = -\mathbf{F}t$ is used instead; unfortunately other difficulties replace them.

The Schrödinger equation for an electron in three dimensions becomes

$$\frac{1}{2m}(\hat{\mathbf{p}} - e\mathbf{F}t)^2 \Psi(\mathbf{R}, t) = i\hbar \frac{\partial}{\partial t} \Psi(\mathbf{R}, t). \tag{6.16}$$

The Hamiltonian is a function of time so there are no stationary states. However, the potential does not depend on position so we can try a plane wave for the spatial part of the wave function, that is, $\Psi_{\mathbf{K}}(\mathbf{R}, t) = \exp(i\mathbf{K} \cdot \mathbf{R}) T(\mathbf{K}, t)$. The (canonical) momentum operator $\hat{\mathbf{p}}$ simply gives $\hbar \mathbf{K}$ acting on this wave function and the plane wave cancels, leaving

$$\frac{1}{2m}(\hbar\mathbf{K} - e\mathbf{F}t)^2 T(\mathbf{K}, t) = \varepsilon_0(\mathbf{K} - e\mathbf{F}t/\hbar) T(\mathbf{K}, t) = i\hbar \frac{\partial}{\partial t} T(\mathbf{K}, t). \tag{6.17}$$

This is of first order and trivial to integrate, leading to

$$\Psi_{\mathbf{K}}(\mathbf{R}, t) = \exp\left\{i\left[\mathbf{K} \cdot \mathbf{R} - \frac{1}{\hbar} \int^t \varepsilon_0(\mathbf{K} - e\mathbf{F}t'/\hbar) dt'\right]\right\}. \tag{6.18}$$

Some features of this are easy to explain. The temporal part can be viewed as a generalization of the usual form $\exp(-i\varepsilon t/\hbar)$ to the case where the energy is a function of time. The momentum that appears within $\varepsilon(\mathbf{K})$ is not a constant but changes as $\hbar\mathbf{K} - e\mathbf{F}t'$. This shows uniform acceleration with a force $-e\mathbf{F}$, which is just what we expect for a charge $-e$ in a field $\mathbf{F}$. The density is also uniform over all space, reflecting the uniformity of the electric field. On the other hand, we might expect that the acceleration would be reflected in the spatial part of Ψ through a changing wave vector, but $\mathbf{K}$ here is constant. This is due to the two momenta mentioned in Section 6.1: the energy depends on the mechanical momentum, which changes under acceleration due to the electric field, but the spatial wave function depends on the canonical momentum, which is constant because the potential is constant throughout space.

When difficulties of interpretation arise like this, it is important to calculate physically observable quantities such as the current density. The modified expression (6.5), which contains the vector potential, gives $\mathbf{J}(t) = -e(\hbar\mathbf{K} - e\mathbf{F}t)/m$. This is constant over all space as we expect, and increases linearly with time to reflect the uniform acceleration.

Perhaps the picture with a vector potential is closer to the classical viewpoint, in showing that a particle is uniformly accelerated and that all points in space are equivalent. Unfortunately we have had to sacrifice our usual description in terms of stationary states to achieve this, which means that it is much harder to define quantities such as the density of states. It would be pretty to combine the two pictures and remove the dependence on the gauge used to represent the electric field, but this requires Green's functions.

6.2.3   NARROW BAND IN AN ELECTRIC FIELD

Before passing on to magnetic fields, we shall take a brief look at the effect of an electric field on an electron in a narrow band in a crystal. This will provide a new perspective on results such as Bloch oscillation in Section 2.2. Several important differences emerge when they are compared with those for free electrons. The importance of the 'narrow' band is that we shall neglect Zener tunnelling between bands, so the gaps must be wide.

Use the cosine approximation for a band of width W, $\varepsilon(k) = \frac{1}{2}W(1 - \cos ka)$. According to Section 3.10, k should be replaced by $-i\,\partial/\partial z$ to construct the effective Hamiltonian. Thus the Schrödinger equation for the envelope function of a one-dimensional electron in an electric field is

$$\left[\varepsilon\left(-i\frac{\partial}{\partial z} + \frac{eA}{\hbar}\right) - e\phi\right]\chi(z,t) = i\hbar\frac{\partial\chi}{\partial t}. \qquad (6.19)$$

Consider first a vector potential $A = -Ft$. The Schrödinger equation is the same as that for free electrons (equation 6.16) except for the form of $\varepsilon(k)$, and the form of the solution is identical to equation (6.18) with an integral over $\varepsilon(k - eFt'/\hbar)$. An important difference arises from the periodic nature of $\varepsilon(k)$, which makes the wave function periodic in time. The period is $2\pi/a$ in k, which becomes $(2\pi\hbar)/(eFa)$ in time, the same as what we found before for Bloch oscillations (equation 2.14). In fact this is a justification for the treatment of the dynamics of electrons in a band used in Section 2.2, where it was asserted that electric and magnetic fields drove the crystal momentum. It is clear from the Schrödinger equation (6.19) that a constant electric field causes the crystal momentum k to increase linearly, as long as the field is represented by a vector potential and the coupling to other bands is ignored.

The results appear quite different if a scalar potential is used, although the calculation is a little too complicated to give here. Both the top and bottom of the band are tilted like eFz, as in Figure 2.7. An electron of constant energy is therefore restricted to a finite region of space, although its wave function decays in the band gaps like the free electrons in Figure 6.1. Again the wave functions have the same functional form and slide along z and E. The difference is that free electrons can be slid continuously, but wave functions in a crystal can be moved only in multiples of the lattice constant a. Thus the energies form a discrete *Stark ladder* of separation eFa, the analogue of Bloch oscillations in a vector potential. The separation depends on the lattice constant but not on the original bandwidth. The local density of states can be calculated, and the $E^{-1/2}$ features at the edges are again blurred as in Figure 6.2(a).

The wave function becomes restricted to fewer atoms as the field increases and the bands are tilted more steeply. In very high fields the change in energy between adjacent sites exceeds the original width of the band, $eFa > W$, and the wave functions become *Stark localized* on single sites. The energies of adjacent atoms are now so different that tunnelling between them, which forms the band (Section 7.7), has almost disappeared. This has been detected optically in a superlattice, in the same way as the Franz–Keldysh effect.

The nature of conduction along the electric field also changes dramatically. We normally think that scattering impedes transport, as in low fields. In Stark localization, however, an electron can travel along the superlattice only by jumping between localized states and this requires the emission of energy. Transport is therefore *promoted* by inelastic scattering. This picture in real space is analogous to the argument in Section 2.2 that scattering is necessary to disrupt Bloch oscillations and permit transport.

A severe limitation is that this calculation has ignored the coupling to higher bands. Consider the superlattice shown in Figure 6.4. When there is no field, the levels in adjacent wells are aligned and tunnelling between them forms bands. These are separated by broad gaps if the barriers are wide to give weak tunnelling. An electric field pulls the levels out of alignment and reduces tunnelling, restricting the wave function to a finite number of wells as discussed earlier. However, it is possible to align the lowest energy level in a well with the *second* level in an adjacent well with an appropriate value of electric field. These two levels will then be strongly coupled to form a resonance, although they gave rise to separate bands in low fields. There is now a staircase of coupled levels, shown in Figure 6.4(b), which has been used as the basis of an intersubband *quantum cascade* laser.

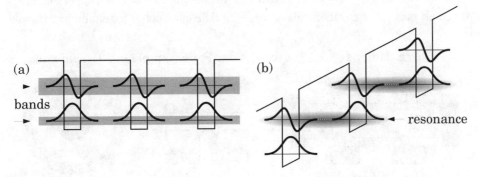

(a)

bands

(b)

resonance

FIGURE 6.4. Effect of a strong electric field on a superlattice. (a) Tunnelling between the same levels in adjacent wells forms widely separated bands at equilibrium. (b) A high electric field can cause resonant alignment of *different* levels in adjacent wells.

6.3 Conductivity and Resistivity Tensors

Before embarking on the quantum-mechanical description of electrons in a magnetic field, it is important to discuss briefly the distribution of electric field and current in the presence of a magnetic field. Consider a two-dimensional system in the xy-plane with a magnetic field along z.

The current density and electric field are related by $\mathbf{J} = \sigma\mathbf{F}$ in the absence of a magnetic field, where σ is the conductivity. It is a scalar if the system is homogeneous and isotropic, which we shall take to be the case. A magnetic field gives rise to the Hall effect, an electric field perpendicular to the current, so the current and electric field are no longer parallel. The scalar σ must be replaced by a conductivity tensor (a 2×2 matrix) giving

$$\mathbf{J} = \sigma\mathbf{F}, \quad \text{or} \quad \begin{pmatrix} J_x \\ J_y \end{pmatrix} = \begin{pmatrix} \sigma_{xx} & \sigma_{xy} \\ \sigma_{yx} & \sigma_{yy} \end{pmatrix} \begin{pmatrix} F_x \\ F_y \end{pmatrix}. \tag{6.20}$$

This gives the current density as a response to an electric field; the inverse relation is given by the resistivity tensor, $\mathbf{F} = \rho\mathbf{J}$.

The diagonal elements of σ are equal and are even functions of B, while the off-diagonal elements are equal and opposite, being odd functions of B. We also know that the resistivity tensor is the reciprocal of the conductivity, so they can be written as

$$\sigma = \begin{pmatrix} \sigma_L & -\sigma_T \\ \sigma_T & \sigma_L \end{pmatrix}, \quad \rho = \frac{1}{\sigma_L^2 + \sigma_T^2} \begin{pmatrix} \sigma_L & \sigma_T \\ -\sigma_T & \sigma_L \end{pmatrix}, \tag{6.21}$$

where the signs are appropriate for electrons. Both matrices are diagonal in the absence of a magnetic field and we find the familiar result $\rho_L = 1/\sigma_L$.

Now apply a weak magnetic field to a Hall bar carrying a current (Figure 6.5(a)). Use the Drude model (equation 2.15) where the conductivity in the absence of a magnetic field is $\sigma_0 = ne^2\tau/m$. Recall that n is the number of carriers per unit area, m is their mass, and τ is their relaxation time. The current has to remain in the same

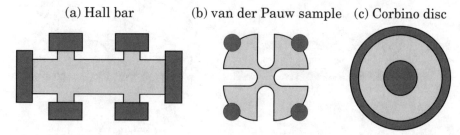

(a) Hall bar   (b) van der Pauw sample   (c) Corbino disc

FIGURE 6.5. Samples commonly used for measuring the conductivity of semiconductors: (a) Hall bar, (b) van der Pauw sample, and (c) Corbino disc. The dark areas are the contacts for measuring voltage or current, and the light areas are the active regions of the sample.

direction in the middle of the sample, so within the classical picture the Lorentz force due to the magnetic field must be balanced by a transverse electric (Hall) field. Thus $F_y = B_z v_x = -B_z J_x/ne$. The sign comes from that of the carriers. Note that we are imposing a current and calculating the electric field that results. This gives elements of the *resistivity* tensor

$$\rho = \begin{pmatrix} \rho_0 & B/ne \\ -B/ne & \rho_0 \end{pmatrix} = \rho_0 \begin{pmatrix} 1 & \omega_c \tau \\ -\omega_c \tau & 1 \end{pmatrix}, \tag{6.22}$$

where $\rho_0 = 1/\sigma_0$ and the cyclotron frequency $\omega_c = eB/m$ has been introduced. The longitudinal resistivity is unaffected in this approximation. The conventional Hall constant is $R_H = F_y/J_x B_z = \rho_{yx}/B_z = -1/ne$ and gives the density of carriers with the sign of their charge. This partly explains the technological importance of the Hall effect. Another feature is that the dimensions of the sample do not enter when the Hall constant is expressed in terms of voltage and current rather than electric field and current density, $R_H = V_y/I_x B_z$. For the Hall bar, $V_y = L_y F_y$ and $I_x = L_y J_x$, where L_y is the width of the sample, which cancels.

Inverting ρ gives the conductivity,

$$\sigma = \frac{\sigma_0}{1 + (\omega_c \tau)^2} \begin{pmatrix} 1 & -\omega_c \tau \\ \omega_c \tau & 1 \end{pmatrix}. \tag{6.23}$$

In this case the diagonal elements are affected by the magnetic field.

Note that $\omega_c \tau = eB\tau/m = \mu B$, where μ is the mobility. This relation looks useful but its validity is limited because the Drude model contains only a single relaxation time. We shall see in Section 8.2 that different relaxation times can be defined, whose values may be an order of magnitude apart in a two-dimensional electron gas.

Several channels of carriers often contribute to the current. There are both electrons and holes in an intrinsic bulk semiconductor, there may be more than one subband occupied in a quasi-two-dimensional electron gas, or there may be an unwanted parallel channel of electrons in a modulation-doped structure (Section 9.2). All channels respond to the same electric field so the total current density is found by adding the conductivity tensors, before inverting to get the overall resistivity tensor. The longitudinal resistivity gains a quadratic dependence on magnetic field, which is taken as a signature of parallel channels of conduction.

Consider two parallel channels of electrons with densities $n_{1,2}$ and mobilities $\mu_{1,2}$. Define an effective number density by $n_{\text{eff}} = 1/eR_H = B/e\rho_T$ and an effective mobility by $\mu_{\text{eff}} = 1/n_{\text{eff}} e\rho_L$. Then in low fields, where $\omega_c \tau \ll 1$ for both channels,

$$n_{\text{eff}} = \frac{(n_1 \mu_1 + n_2 \mu_2)^2}{n_1 \mu_1^2 + n_2 \mu_2^2}, \qquad \mu_{\text{eff}} = \frac{n_1 \mu_1^2 + n_2 \mu_2^2}{n_1 \mu_1 + n_2 \mu_2}. \tag{6.24}$$

In very high fields, where $\omega_c \tau \gg 1$ for both channels, the simple result $n_{\text{eff}} = n_1 + n_2$ is recovered although this may be masked by quantum-mechanical effects to be discussed shortly. These results are useful in practice to resolve parallel channels; see Figure 9.12 for an example.

Surprising behaviour occurs in strong magnetic fields where it is possible to achieve the condition $|\sigma_T| \gg \sigma_L$. This is particularly important in the quantum Hall effect. In this limit

$$\sigma = \begin{pmatrix} \sigma_L & \sigma_T \\ -\sigma_T & \sigma_L \end{pmatrix}, \qquad \rho \approx \begin{pmatrix} \sigma_L/\sigma_T^2 & -1/\sigma_T \\ 1/\sigma_T & \sigma_L/\sigma_T^2 \end{pmatrix}. \qquad (6.25)$$

Now the longitudinal components of σ and ρ are in *direct* rather than inverse proportion to one another. The longitudinal components of both the conductivity and resistivity can vanish together in this limit, quite unlike the behaviour in the absence of a magnetic field. The current and electric field are perpendicular to each other if $|\sigma_T| \gg \sigma_L$, so the current flows along equipotentials rather than along field lines.

This strongly modifies the distribution of the current and potential inside a rectangular bar, as shown in Figure 6.6. Far from the ends, the current must flow along the length of the sample. This is the same as the direction of the electric field when no magnetic field is applied and there is a uniform electric field along the sample. The uniform current distribution continues to the contacts at either end.

This changes in a high field where $|\sigma_T| \gg \sigma_L$ and current flows along equipotentials. Thus the electric field runs across the width of the sample, giving a Hall voltage between the long faces, and there is no longitudinal field. This cannot

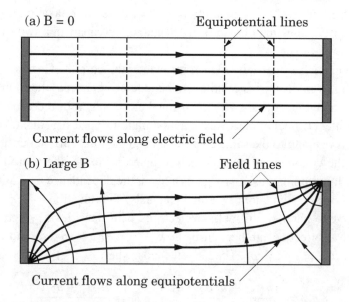

FIGURE 6.6. Electric field, current flow, and equipotentials inside a long rectangular sample with contacts across each end. (a) In the absence of a magnetic field the current is uniform throughout the sample and runs along the electric field. (b) In a strong magnetic field, where $|\sigma_T| \gg \sigma_L$, the current runs along equipotentials.

hold near the ends of the bar because the contacts are also equipotentials. The distributions of current and field distort, while remaining mutually orthogonal, such that all current enters and leaves at two opposite corners of the sample, which are the only places where a potential difference can occur along the edge (Figure 6.6(b)). The left-hand contact is at the same potential as the top of the sample, whereas the right-hand contact is tied to the bottom, so they measure the Hall voltage despite the 'two-probe' geometry. Although this is a mathematically correct solution, it is clear that it must break down near the vertices where the current enters and leaves because neither the current density nor the electric field can be infinite. Nevertheless, experiments have shown that all the power is dissipated near these two points. We shall return to the distribution of current when we study the quantum Hall effect in Section 6.6.

6.4 Uniform Magnetic Field

The quantum-mechanical behaviour of free electrons in an electric field is not too far from that expected classically, although it is obscured by the dependence on the gauge. The effect of a magnetic field is more dramatic, particularly on a two-dimensional system. Classically a magnetic field **B** has no effect on the motion parallel to the field, which we shall take to be along the z-axis as usual. The electrons execute circular trajectories in the plane normal to **B** at constant angular frequency

$$\omega_{\mathrm{c}} = \left| \frac{eB}{m} \right|, \qquad (6.26)$$

known as the *cyclotron frequency*. The radius of the orbit, the *cyclotron radius*, is given by

$$R_{\mathrm{c}} = \frac{v}{\omega_{\mathrm{c}}} = \frac{\sqrt{2mE}}{|eB|}, \qquad (6.27)$$

where v is the (constant) magnitude of the velocity and E is the kinetic energy. An important feature is that the period of the motion is independent of the energy, but the amplitude R_{c} is proportional to the square root of the energy. This sounds like a harmonic oscillator (Section 4.3), and the quantum-mechanical results reflect this.

6.4.1 SOLUTION IN LANDAU GAUGE

The simplest algebra is in Landau gauge where there is only one component of vector potential. Take $\mathbf{A} = (0, Bx, 0)$. The Schrödinger equation (6.3) becomes

$$\left\{ \frac{1}{2m} \left[-\hbar^2 \frac{\partial^2}{\partial x^2} + \left(-i\hbar \frac{\partial}{\partial y} + eBx \right)^2 - \hbar^2 \frac{\partial^2}{\partial z^2} \right] + V(z) \right\} \psi(\mathbf{R}) = E\psi(\mathbf{R}).$$

$$(6.28)$$

Expanding the inner bracket gives

$$\left[-\frac{\hbar^2}{2m}\nabla^2 - \frac{ie\hbar Bx}{m}\frac{\partial}{\partial y} + \frac{(eBx)^2}{2m} + V(z)\right]\psi(\mathbf{R}) = E\psi(\mathbf{R}). \tag{6.29}$$

The magnetic field produces two terms: a parabolic magnetic potential, which tends to confine the wave functions in x, and a first derivative, which couples x and y and is reminiscent of the Lorentz force.

This latter term is imaginary, which is important because it spoils time-reversal invariance (Section 5.3), which relies on the Hamiltonian being real. We must also reverse a magnetic field if the same properties are to hold in its presence. Thus if $\Psi(t; B)$ is a solution of Schrödinger's equation with a magnetic field B, time-reversal invariance states that $\Psi^*(-t; -B)$ must also be a solution.

The potential $V(z)$ is absent for free electrons, or it might be a potential well trapping electrons in this dimension to form a 2DEG. This potential is additive so the motion along z can be factored out of the equation and is not disturbed by the magnetic field. We shall therefore drop z and solve the equation in two dimensions; our final wave function should be multiplied by the z-function and the corresponding energy should be added.

The vector potential does not depend on y, which suggests that the wave function should be a product of a plane wave of y with some unknown function of x, $u(x)\exp(iky)$. On substituting this into the Schrödinger equation (6.28) we find that the the plane wave cancels from both sides, confirming that it is a correct guess. It leaves an equation in x only:

$$\left[-\frac{\hbar^2}{2m}\frac{d^2}{dx^2} + \tfrac{1}{2}m\omega_c^2\left(x + \frac{\hbar k}{eB}\right)^2\right]u(x) = \varepsilon\,u(x). \tag{6.30}$$

This is the Schrödinger equation for a one-dimensional harmonic oscillator (Section 4.3), as predicted. Note that the wave function in y has not left the usual contribution to the energy on the right-hand side because this term has been absorbed into the potential energy. The cyclotron frequency $\omega_c = |eB/m|$ appears as in the classical case, and the only difference compared with our previous treatment of the oscillator is that the vertex of the parabolic potential is displaced to $x_k = -\hbar k/eB$. The length scale (equation 4.27) is called the *magnetic length*, given by

$$l_B = \sqrt{\frac{\hbar}{m\omega_c}} = \sqrt{\frac{\hbar}{|eB|}}. \tag{6.31}$$

This depends only on the field, not on the mass of the particle; a typical value is $l_B \approx 26\,\text{nm}$ in $B = 1\,\text{T}$.

We can now write down the energies and wave functions for motion in the xy-plane:

$$\varepsilon_{nk} = (n - \tfrac{1}{2})\hbar\omega_c \qquad \text{(independently of } k), \qquad (6.32)$$

$$\phi_{nk}(x, y) \propto H_{n-1}\left(\frac{x - x_k}{l_B}\right) \exp\left[-\frac{(x - x_k)^2}{2l_B^2}\right] \exp(iky), \qquad (6.33)$$

where $n = 1, 2, 3, \ldots$, and the H_n are Hermite polynomials. The wave functions are not normalized but the factors were given in equation (4.35). The lowest wave function is simply a Gaussian function whose probability distribution has a standard deviation of $l_B/\sqrt{2}$.

This solution has several unusual properties. The first is that the energy depends only on n, not on k. States with the same n but different k are degenerate. The density of states collapses from the constant for the two-dimensional electron gas to a series of δ-functions called *Landau levels* at energies given by equation (6.32).

Another feature is that the wave functions form parallel strips in the y-direction, spaced equally along the x-axis, along the lines of $\mathbf{A}$. This is perplexing because there is nothing special about any direction in the xy-plane, and the distinction between x and y is entirely due to the gauge chosen for the vector potential. One could equally choose a gauge where the strips were parallel to the x-axis or, as in the next section, where circular symmetry is restored. The important point is that all states within a Landau level are degenerate (have the same energy), so any mixture of them is also a solution to the Schrödinger equation with the same energy. By mixing them up, we can transform the wave functions from one set to another. This is a crude way of looking at a gauge transformation, which we shall consider in a little more detail in Section 6.4.8.

There are usually no gaps between the wave function strips. Let the sample be a rectangle of dimensions $L_x \times L_y$. Periodic boundary conditions in the y-direction (not always safe in a magnetic field) give the usual condition $k = (2\pi/L_y)j$, where j is an integer. The separation between adjacent values of x_k is $\Delta x_k = (2\pi/L_y)(\hbar/eB) = 2\pi l_B^2/L_y$. The spread of each wave function is set by l_B, so the ratio of separation to spread is $\Delta x_k/l_B = 2\pi l_B/L_y$. Now $l_B \approx 26$ nm in a field of 1 T, so it is clear that $\Delta x_k \ll l_B$ in any field where the Landau levels are resolved and that a large number of wave functions overlap in space.

Although l_B sets the scale of the lowest state in the parabolic potential from the magnetic field, higher states spread farther. Fortunately we can calculate this without the wave functions by using the result that the expectation values of the kinetic and potential energies are equal for a harmonic oscillator, and each is equal to half the total energy. The potential energy function is $\tfrac{1}{2}m\omega_c^2(\Delta x)^2$, where Δx is measured from the minimum, so

$$\left\langle \tfrac{1}{2}m\omega_c^2(\Delta x)^2 \right\rangle = \tfrac{1}{2}\varepsilon_n, \qquad \left\langle (\Delta x)^2 \right\rangle = \frac{\varepsilon_n}{m\omega_c^2}. \qquad (6.34)$$

This agrees with the classical formula (6.27) for the cyclotron radius up to a factor of 2, which is not surprising given the different nature of the states.

The plane wave $\exp(iky)$ gives the impression that the states carry a current that depends on k and therefore on the position x_k. However, we must be careful because the expression (6.5) for the current contains the vector potential as well as the derivative of the wave function. Substitution shows that the two terms cancel and the state carries no net current. There is, however, a circulating current that produces a magnetic moment. This connects with the classical result, where the magnetic field caused the electrons to orbit but to carry no net current. A quicker way of seeing that the electrons carry no net current is to remember that the group velocity depends on $d\varepsilon/dk$. This vanishes because of the degeneracy, but the argument is not rigorous because it ignores the displacement along x when k is changed.

Before looking at the consequences of the Landau levels in the density of states, we'll look briefly at the solution using a different vector potential that is closer to the classical result.

6.4.2 SOLUTION IN SYMMETRIC GAUGE

A more symmetric solution can be found if we choose the symmetric gauge $\mathbf{A} = \frac{1}{2}\mathbf{B} \times \mathbf{R}$ to represent the magnetic field. The lines of $\mathbf{A}$ now form circles about the origin, and cylindrical polar coordinates (r, θ) take advantage of this symmetry, with $A_\theta = \frac{1}{2}Br$ and $A_r = A_z = 0$. Rewriting the Schrödinger equation in polar coordinates (Section 4.7) gives

$$\left[-\frac{\hbar^2}{2m}\left(\frac{\partial^2}{\partial r^2} + \frac{1}{r}\frac{\partial}{\partial r} + \frac{1}{r^2}\frac{\partial^2}{\partial \theta^2} \right) - \frac{i\hbar eB}{2m}\frac{\partial}{\partial \theta} + \frac{e^2 B^2 r^2}{8m} \right] \psi(r, \theta) = \varepsilon\psi(r, \theta).$$

(6.35)

In this case there is rotational invariance instead of translational invariance along y as in Landau gauge, and we can take the angular part of the wave function as $\exp(il\theta)$. Substitution leads to an equation for the radial function $v(r)$:

$$\left[-\frac{\hbar^2}{2m}\left(\frac{d^2}{dr^2} + \frac{1}{r}\frac{d}{dr} \right) + \frac{\hbar^2 l^2}{2mr^2} + \frac{1}{8}m\omega_c^2 r^2 \right] v(r) = (\varepsilon - \frac{1}{2}l\hbar\omega_c)v(r).$$

(6.36)

This is the equation for a two-dimensional harmonic oscillator (Section 4.7) with an additional term in the energy on the right. Solution of this equation leads to the following energies and wave functions (without normalization):

$$\varepsilon_{nl} = (n + \tfrac{1}{2}l + \tfrac{1}{2}|l| - \tfrac{1}{2})\hbar\omega_c,$$

(6.37)

$$\psi_{nl}(r, \theta) = \exp(il\theta) \exp\left(-\frac{r^2}{4l_B^2} \right) r^{|l|} L_{n-1}^{(|l|)}\left(\frac{r^2}{2l_B^2} \right).$$

(6.38)

The functions $L_n^{(l)}$ are associated Laguerre polynomials (Abramowitz and Stegun 1972, chapter 22).

The wave functions have the same general form as in Landau gauge with a Gaussian decay modulated by a polynomial, and again they run along the lines of **A**. The centrifugal potential pushes states with large $|l|$ away from the origin, so they occupy annular regions rather than disks. The form of the energy is rather strange, as it increases with l for $l > 0$ but is independent of l for $l < 0$. However, the allowed values of ε are the same and again the levels are highly degenerate.

The wave functions are simpler in Landau gauge and this is usually used in practice except for problems such as quantum dots where the rotational symmetry is important. In fact the 'symmetric' gauge is not as symmetric as one might like because the states all circulate about the same point in space, unlike the classical case where the orbits could be anywhere. This gauge preserves the rotational symmetry of space but at the expense of its translational symmetry.

6.4.3 SPIN OF THE ELECTRON

Up to this point, the spin of the electron has been ignored except for the fact that it doubles the number of states available. This is not acceptable in a magnetic field because there is a magnetic moment associated with the spin. The spin can be 'up' or 'down' so the moment is either parallel or antiparallel to the field. These configurations have different energies due to the moment of $\pm\frac{1}{2}g\mu_B B$, where $\mu_B = e\hbar/2m_0$ is the *Bohr magneton*. The factor of $\frac{1}{2}$ appears because the spin carries angular momentum of $\pm\frac{1}{2}\hbar$, unlike the integral multiples of $\hbar$ for orbital motion.

The g-factor is close to 2 for the spin of free electrons, in contrast to the value of 1 for orbital motion, and the prediction of its precise value is a triumph of quantum electrodynamics. However, g may take quite different values in a solid because of the band structure, particularly the spin–orbit coupling. It is negative for bulk GaAs, $g = -0.44$, but $g \approx +0.4$ for $Al_{0.3}Ga_{0.7}As$ and the value lies between these in a heterostructure. Further complications include the exchange enhancement, which can raise g substantially as a function of the magnetic field and the density of carriers, and arises from their mutual interaction (Section 9.3.2).

Typically the Landau levels contain both spin states at low fields, split partly into two as the field is raised, and the levels associated with the two spins become completely separate at the highest fields (Figure 6.10). In silicon there is an additional splitting of the Landau levels that arises because the electrons in the two-dimensional electron gas occupy two [001]-valleys, but this is not present in GaAs.

6.4.4 LANDAU LEVELS

Probably the most important aspect of the solution of the Schrödinger equation in a magnetic field is the energy spectrum. The continuous density of states for free electrons is replaced by a sequence of δ-functions. These Landau levels each contain a large number of degenerate states. We shall now explore the consequences of this spectrum.

First, we need to know the number of states in each level. Suppose that the system is a rectangle of dimensions $L_x \times L_y$ and use Landau gauge. Periodic boundary conditions along y give the usual condition $k = (2\pi/L_y)j$, where j is an integer. The restriction from L_x comes about because the wave functions are centred on $x_k = -\hbar k/eB = -2\pi\hbar j/eBL_y$. Requiring x_k to be within the sample means

$$-L_x < \frac{2\pi\hbar j}{eBL_y} < 0, \qquad \frac{eBL_xL_y}{h} < j < 0. \tag{6.39}$$

Thus the allowed number of states in each Landau level per unit area is $n_B = eB/h$. It is conventional *not* to include the usual factor of 2 for spin because up and down spins are not degenerate in a magnetic field. This can be written in a number of more illuminating ways. The quantum of magnetic flux is defined by $\Phi_0 = h/e$ (although $h/2e$ is often used in superconductivity). The number of states per unit area can then be written as B/Φ_0 or, in a sample of area A, as $(AB)/\Phi_0 = \Phi/\Phi_0$, where Φ is the total applied flux. Thus there is one state (of each spin) in each Landau level for each quantum of magnetic flux that passes through the sample. Hence each state in a Landau level occupies an area $h/eB = 2\pi l_B^2$. Another relation comes from introducing the cyclotron energy $\hbar\omega_c$, giving

$$2n_B = \frac{2eB}{h} = \frac{2m\omega_c}{2\pi\hbar} = \frac{m}{\pi\hbar^2}\,\hbar\omega_c. \tag{6.40}$$

Thus each Landau level (counting both spins) contains the same number of states as the original two-dimensional band over a range of $\hbar\omega_c$. As this is also the separation between the Landau levels, it shows the important result that the magnetic field does not change the density of states when averaged over many Landau levels. Each block of width $\hbar\omega_c$ has collapsed into a δ-function at its centre, as shown in Figure 6.7(a).

The density of states of the Landau levels consists of sharp δ-functions only in an ideal system where electrons are never scattered, either by other electrons or by impurities, phonons, and the like. It is more realistic to assume that an electron typically survives for a finite time τ_i between scattering events. Then the energy can be defined only within a precision of $\Gamma = \hbar/\tau_i$, as for the quasi-bound state in a resonant-tunnelling structure. The Landau levels acquire a width Γ, which might be defined precisely as the standard deviation or the full width at half-maximum. The precise shape of Landau levels in real systems remains controversial; common assumptions are a Gaussian or Lorentzian profile. In either case, we would not expect to see strong changes in the density of states unless the separation of the Landau levels exceeds their width, $\hbar\omega_c > \Gamma$ (Figure 6.7(b) and (c)). This can also be written as $\omega_c\tau_i > 1$, which means that an electron must survive for at least a complete orbit in the magnetic field before the density of states splits up. This boundary separates

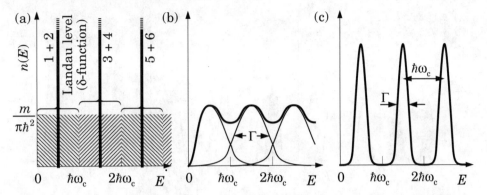

FIGURE 6.7. Density of states in a magnetic field, neglecting spin splitting. (a) The states in each range $\hbar\omega_c$ are squeezed into a δ-function Landau level. (b) Landau levels have a non-zero width Γ in a more realistic picture and overlap if $\hbar\omega_c < \Gamma$. (c) The levels become distinct when $\hbar\omega_c > \Gamma$.

semiclassical effects seen at low magnetic fields from the quantum-mechanical effects due to Landau levels at high fields.

The time τ_i that appears in these expressions, called the *single-particle lifetime* or *quantum lifetime*, is *not* the same as the *transport lifetime* that appears in the mobility $\mu = e\tau_{tr}/m$. The difference is that all collisions contribute equally to τ_i, but the contributions to τ_{tr} depend on the change of direction that occurs. This will be discussed in Section 8.2.

As the magnetic field is raised from zero, the separation between the Landau levels grows and so does the number of states that each level holds, as shown in Figure 6.8. Most experiments are performed with the density of electrons n_{2D} held constant, so the number of occupied Landau levels must change. The number of occupied Landau levels or filling factor ν can be written in many ways:

$$\nu = \frac{n_{2D}}{n_B} = \frac{hn_{2D}}{eB} = \frac{\Phi_0 n_{2D}}{B} = 2\pi l_B^2 n_{2D}. \qquad (6.41)$$

This definition counts the two spins as *separate* levels, which is common but not universal.

In general the filling factor ν is not an integer, and at zero temperature there will be n Landau full levels, where n is the largest integer below ν, and the top level will be partly filled. Raising B further causes the Landau levels to move up in energy and the number of states in each to grow, so fewer electrons occupy the top level. The top level becomes empty when $\nu = n$, at a field

$$B_n = \frac{hn_{2D}}{en} \qquad (6.42)$$

where there are exactly n full Landau levels. After this, n falls by 1 and the next level begins to empty. This is illustrated in Figure 6.8 for $\nu = 4$, $\frac{8}{3}$, and 2, where the spin splitting has been neglected so each Landau level contains $2n_B$ electrons.

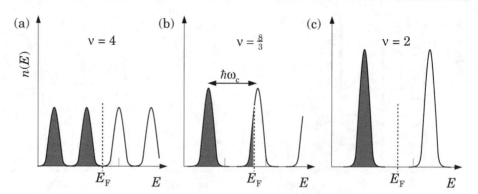

FIGURE 6.8. Occupation of Landau levels in a magnetic field neglecting the spin splitting, showing how the Fermi level moves to maintain a constant density of electrons. The fields are in the ratio $2:3:4$ and give $\nu = 4$, $\frac{8}{3}$, and 2.

Eventually all electrons lie in the lowest Landau level when $\nu < 1$ and this is called the *magnetic quantum limit*. This is a spin-split level so all electrons have their spins aligned in the same direction. This contrasts with an electron gas with no magnetic field, where there are equal numbers of up and down spins.

The Fermi level E_F moves with the density of states to keep the number of electrons constant. At the fields B_n, where ν is exactly an integer n (or an even integer in the absence of spin splitting), E_F lies in the empty region between Landau levels and is equal to its value before the field was applied E_F^0. It moves away from this value as soon as the field is changed. Suppose that the magnetic field lies in such a range that $n < \nu < n + 1$, that is, $B_{n+1} < B < B_n$. There are n completely filled Landau levels with level $n + 1$ partly filled to give the correct density of electrons. Thus the Fermi level lies inside level $n + 1$, and $E_F \approx (n + \frac{1}{2})\hbar\omega_c$ to within the width Γ. Raising B causes E_F to rise linearly until the field reaches B_n, at which point level $n + 1$ becomes empty and the Fermi level falls back to the Landau level below. The movement of E_F is illustrated in Figure 6.9 for a perfect system with δ-function Landau levels, neglecting spin splitting. This can be measured experimentally as a contact potential between the 2DEG and a metal gate on the surface of a heterostructure (Section 9.1.4). The sharp jumps are rounded if the levels are broadened or the temperature rises; clearly we need $k_B T \ll \hbar\omega_c$ to observe well-resolved Landau levels.

An important conclusion is that it is difficult to put the Fermi level into the gaps between Landau levels. The density of states (Figure 6.7(c)) was plotted with E as the independent variable, and it is easy to imagine that E_F can be positioned at will. Unfortunately the need to keep n_{2D} constant 'pins' E_F within Landau levels, foiling attempts to put E_F in a gap.

The position of E_F has a qualitative effect on the electronic behaviour of a 2DEG. If E_F lies within a Landau level, the density of states at the Fermi level is high and a small change in energy induces a large change in density. The system is said to be

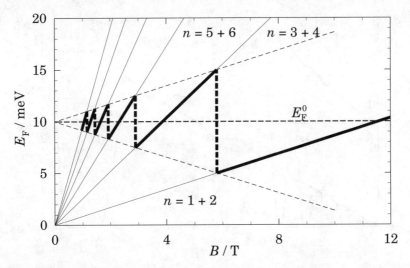

FIGURE 6.9. Variation of the Fermi level as a function of magnetic field for a two-dimensional electron gas in GaAs with $E_F^0 = 10$ meV before the field was applied. Spin splitting is neglected. The fan of thin lines shows the Landau levels, while the discontinuous thick line is E_F.

compressible. The opposite behaviour occurs if E_F lies within a gap. Now a small change in energy has no effect at all on the density of the 2DEG because the density of states at the Fermi level is zero, and the system is said to be *incompressible.* This distinction is important in theories of the quantum Hall effect.

6.4.5   SHUBNIKOV–DE HAAS EFFECT

Figure 6.8 shows that the density of states at the Fermi level changes as a function of B. It drops to zero when $\nu = n$ (assuming that there are clear gaps between Landau levels) and has maxima when $\nu \approx n + \frac{1}{2}$. This is reflected in many observable quantities. The simplest of these to measure is the longitudinal resistance R_{xx}, and a typical measurement is shown in Figure 6.10. The resistivity is constant at low magnetic fields but develops strong oscillations with zeros at high fields. This is the *Shubnikov–de Haas effect,* a striking signature of the quantum-mechanical behaviour of electrons in a magnetic field.

Minima occur at the fields B_n when $\nu = n$ (equation 6.42). This is because longitudinal conduction occurs at the Fermi level (Section 5.4) and therefore disappears when the density of states goes to zero at $B = B_n$. Under these conditions $\rho_L \propto \sigma_L$ and therefore vanishes too. Thus a plot of n against the measured values of $1/B_n$ should give a straight line of slope $(h/e)n_{2D}$ passing through the origin. This *fan diagram* is a standard way of measuring the density and also provides a clear demonstration that a system is two-dimensional. Alternatively, the conductivity should have periodic structure when considered as a function of $1/B$, and this can be brought out by calculating its power spectrum. A large peak arises due

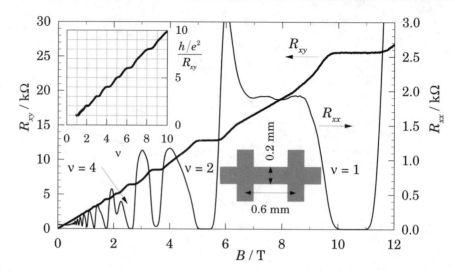

FIGURE 6.10. Longitudinal and transverse (Hall) resistivity, R_{xx} and R_{xy}, of a two-dimensional electron gas of density $n_{2D} = 2.6 \times 10^{15} \, \mathrm{m}^{-2}$ as a function of magnetic field. The measurements were made at $T = 1.13 \, \mathrm{K}$. The inset shows $1/R_{xy}$ divided by the quantum unit of conductance e^2/h as a function of the filling factor ν. [Data kindly supplied by Dr A. R. Long, University of Glasgow.]

to the Shubnikov–de Haas effect and its location gives n_{2D}. This can also reveal the occupation of higher electric subbands from the motion along z; the density of electrons will be different and give rise to a distinct period in $1/B$.

Further analysis of the Shubnikov–de Haas oscillations gives the temperature and lifetime τ_i of the 2DEG. The latter is sometimes expressed as a 'quantum mobility' $\mu_i = e\tau_i/m$. The sample shown had $\mu = e\tau_{tr}/m = 130 \, \mathrm{m}^2 \, \mathrm{V}^{-1} \, \mathrm{s}^{-1}$ and $\mu_i = e\tau_i/m = 7 \, \mathrm{m}^2 \, \mathrm{V}^{-1} \, \mathrm{s}^{-1}$. The factor of 20 between these emphasizes the importance of using the correct lifetime.

The evolution of spin splitting is clear in Figure 6.10. Landau levels are doubly degenerate, containing both spins, for $B < 2 \, \mathrm{T}$. There is a double peak around $2 \, \mathrm{T}$ as the spin-split minimum at $\nu = 5$ begins to emerge, and R_{xx} falls almost to zero when $\nu = 3$ around $3.5 \, \mathrm{T}$. There is a very strong minimum between 10 and $11 \, \mathrm{T}$ where $\nu = 1$. All electrons lie in the lowest Landau level, the magnetic quantum limit, for larger fields.

6.4.6 INTEGER QUANTUM HALL EFFECT

The Hall effect is the standard way of measuring the concentration of carriers in a semiconductor and we have studied it for low magnetic fields. Figure 6.10 shows its remarkable behaviour in high fields. The transverse resistance $R_{xy} = \rho_T = BR_H = B/en_{2D}$ at low fields as expected (remember that the geometry drops out). At integral filling factors, where $B = B_n$, this gives $R_{xy}^{-1} = (e^2/h)n$, the same form of quantized conductance as found in quasi-one-dimensional systems

(Section 5.7.1). This should hold whenever the Fermi level lies between Landau levels, but we have just seen that this condition is extremely difficult to achieve. Therefore this argument does not prepare us for the most significant feature of the experimental results: there are broad plateaus at the corresponding values of the Hall resistance that coincide with the minima in the longitudinal resistance. This is the *integer quantum Hall effect* and the quantization appears to be exact within experimental accuracy. A theory will be given in Section 6.6.

The inset to Figure 6.10 shows the dimensionless quantities $(h/e^2)/R_{xy}$ plotted against $v = hn_{2D}/eB$. Classically this should be a straight line of unit slope. Although this holds on average, the quantum Hall effect leads to plateaus when v is integral. Odd integers correspond to spin-split levels whose plateaus fade more rapidly as the field is reduced.

Other quantities that depend on the density of states at the Fermi level also show oscillatory behaviour in a magnetic field. An example is the magnetic susceptibility, where the oscillations are called the de Haas–van Alphen effect, an extremely important probe of the Fermi surface in metals. Another example is the electronic contribution to the specific heat. Unfortunately both are difficult to measure in low-dimensional systems because they are masked by the contributions from the substrate.

In three dimensions we must include the free motion along z parallel to **B**. The Landau levels become the bottoms of one-dimensional subbands, with $1/\sqrt{E}$ divergences in the density of states rather than δ-functions. Similar effects are seen as the magnetic field is varied, leading to changes in the density of states at the Fermi level, but they are much less dramatic. We shall see in Section 6.5 the changes that result when the electrons are further confined to become quasi-one-dimensional.

6.4.7 CROSSED ELECTRIC AND MAGNETIC FIELDS

The classical motion of a charged particle in perpendicular (crossed) electric and magnetic fields has two components. There is a steady drift motion with velocity $\mathbf{v}_d = \mathbf{F} \times \mathbf{B}/B^2$ at right angles to both fields, with superposed cyclotron orbits of frequency ω_c. An electron released from rest executes a cycloid,

$$x(t) = -\frac{mF}{eB^2}(1 - \cos \omega_c t), \qquad y(t) = -\frac{mF}{eB^2}(\omega_c t - \sin \omega_c t). \qquad (6.43)$$

This is for **B** along z and **F** along x, which fits best with our calculations in Landau gauge. The drift velocity $v_d = F/B$ along $-y$. The particle is initially accelerated along $-x$ by the electric field but this motion is bent by the magnetic field to give a net drift along $-y$ at right angles to both fields. This is shown in example (iii) of Figure 6.11, which has been rotated to a more convenient orientation. Other trajectories start from different initial conditions.

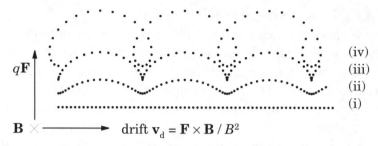

FIGURE 6.11. Classical motion of a charged particle in crossed electric and magnetic fields, with **B** normal to the page and equal intervals of time between the symbols. The curves correspond to different initial velocities and energies, with (iii) showing the cycloid for a particle initially at rest.

Fortunately the problem is also simple in quantum mechanics when built upon our previous results. The field adds a potential energy eFx to the Schrödinger equation (6.28). Motion along z can again be decoupled and removed, and a plane wave can be taken out for y. The resulting Schrödinger equation for $u(x)$ is the same as equation (6.30) but with the extra potential energy. This is linear, so it simply shifts the vertex of the parabolic potential to give

$$\left\{ -\frac{\hbar^2}{2m}\frac{d^2}{dx^2} + \tfrac{1}{2}m\omega_c^2 \left[x + \left(\frac{\hbar k}{eB} + \frac{eF}{m\omega_c^2} \right) \right]^2 - \frac{\hbar k F}{B} - \frac{m F^2}{2B^2} \right\} u(x) = \varepsilon\, u(x).$$

(6.44)

The vertex of the parabola is displaced to

$$x_k = -\left(\frac{\hbar k}{eB} + \frac{eF}{m\omega_c^2} \right) = \frac{m v_d - \hbar k}{eB}.$$

(6.45)

The set of wave functions (6.33) is shifted rigidly by the electric field, and the displacement of the parabola changes the energies to

$$\varepsilon_{nk} = (n - \tfrac{1}{2})\hbar\omega_c - \frac{\hbar k F}{B} - \frac{m F^2}{2B^2}$$

(6.46)

$$= (n - \tfrac{1}{2})\hbar\omega_c + eF x_k + \tfrac{1}{2}m v_d^2.$$

(6.47)

The degeneracy with respect to k has now been broken: there is a contribution from the electrostatic potential that depends on the shift x_k of the wave function, and an additional kinetic energy $\tfrac{1}{2}m v_d^2$ from the drift velocity. This is the same as the classical result except that the energy of cyclotron oscillations is quantized.

The shift of the parabola by the electric field causes the states to acquire a velocity v_d along y. This is the Hall effect again. The current density from n_{2D} electrons per unit area is $J_y = -e n_{2D} v_d = e n_{2D} F_x / B_z$. Thus $\sigma_L = \sigma_{xx} = 0$ since there is no current along x, and $\rho_T = 1/\sigma_T = B_z / e n_{2D}$, which is the classical Hall coefficient.

Note that *all* electrons contribute to the current, not just those near the Fermi level as in longitudinal conduction.

6.4.8 GAUGE INVARIANCE

For both electric and magnetic fields, we have seen that very different solutions can be found for the same physical situation by choosing different gauges – different scalar and vector potentials. We derived the general form for a gauge transformation of the potentials in equation (6.2); to this should be added a corresponding change in wave function for a particle of charge q:

$$\Psi \to \Psi \exp\left(\frac{iq\chi}{\hbar}\right). \tag{6.48}$$

The operator $\hat{\mathbf{p}}$ in the Schrödinger equation brings down an extra term $q\,\mathrm{grad}\,\chi$, which cancels the change in $\mathbf{A}$, while the time derivative produces $-q\,d\chi/dt$, which cancels the change in ϕ. Thus the new wave function is a solution to the Schrödinger equation with the transformed potentials. This cancellation is clearer if the scalar potential is taken over to the right-hand side to write the Schrödinger equation in the form

$$\frac{(\hat{\mathbf{p}} - q\mathbf{A})^2}{2m}\Psi = \left(i\hbar\frac{\partial}{\partial t} - q\phi\right)\Psi. \tag{6.49}$$

We saw before that the left-hand side can be interpreted as kinetic energy in a uniform field, and the right-hand side is the difference of total and potential energy, so this grouping appears natural.

Gauge invariance demands that physical results are not affected by such a gauge transformation. The density is unaffected because it depends on $|\Psi|^2$, from which the phase factor due to the gauge transformation vanishes. The current is also invariant because the changes cancel between the two parts of $\hat{\mathbf{p}} - q\mathbf{A}$. The energy levels in a magnetic field also remain unchanged, as we found in Section 6.4.2. On the other hand, we saw in Section 6.2 that it is not possible to define energy levels in an electric field with a vector potential, so the energy levels in a scalar potential depend on the specific choice of gauge and cannot themselves be observable.

For an example of a simple gauge transformation, consider a uniform static electric field. This can be described by $\phi = -\mathbf{F}\cdot\mathbf{R}, \mathbf{A} = 0$ or $\phi = 0, \mathbf{A} = -\mathbf{F}t$. The function $\chi = \mathbf{F}\cdot\mathbf{R}t$ switches a pure vector potential to a pure scalar potential. Thus we can multiply the wave functions (6.18) in a vector potential by the phase factor $\exp(-ie\mathbf{F}\cdot\mathbf{R}t/\hbar)$ and they become solutions of the Schrödinger equation with a scalar potential. They are not eigenstates because they still describe accelerating functions rather than standing waves, and are known as Houston functions.

Gauge invariance may appear to be something of a nuisance. This is often the case in practice, where it is easy to make approximations that hold only in a particular gauge and lead to disastrous results. Fortunately this is a misleading picture and obscures the deep significance associated with gauge symmetry, the freedom to make transformations in the fields and wave functions that leave the physics intact. For example, the presence of the charge q in the exponent when the wave function is transformed leads to the conservation of charge. In particle physics, the form of theories such as quantum chromodynamics (QCD) is set by demanding that they be invariant with respect to transformations of the wave function that are generalizations of equation (6.48); this in turn means that fields analogous to the scalar and vector potentials must be introduced to couple to the particles. Returning to semiconductors, the quantized Hall effect was first explained using gauge invariance. It is not appropriate to delve further into gauge invariance here, but simply to note that it should be regarded as an important symmetry of nature and not simply as an annoyance.

6.4.9 THE AHARONOV–BOHM EFFECT

It often seems a nuisance that the Hamiltonian contains potentials rather than the fields themselves. However, it has the startling consequence that it is possible for electrons to be affected by a vector potential *even though they never feel the magnetic field itself*. This gives rise to the *Aharonov–Bohm effect*. Consider the interferometer shown in Figure 6.12(a). Electrons enter from a wave guide at the left, divide into two equal amplitudes each going around an arm of the loop, recombine and interfere on the right, and leave through another wave guide.

Now suppose that a magnetic field is generated in a small solenoid positioned *entirely inside the loop* so that no field passes through the wave guides. Although

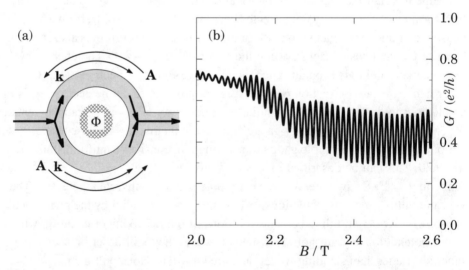

FIGURE 6.12. (a) Schematic layout of an interferometer to demonstrate the Aharonov–Bohm effect, with a small solenoid carrying magnetic flux Φ enclosed between the two arms. (b) Transmission around an 'antidot' as a function of magnetic field showing oscillations due to the Aharonov–Bohm effect. [Redrawn from Ford et al. (1994).]

$\mathbf{B} = \mathbf{0}$ outside the solenoid, there is a vector potential A_θ running in loops around it. Its magnitude at radius r is given by Stokes's theorem:

$$2\pi r A_\theta(r) = \oint_{\text{loop}} \mathbf{A} \cdot d\mathbf{l} = \int_{\text{area}} (\text{curl}\,\mathbf{A}) \cdot d\mathbf{S} = \int_{\text{area}} \mathbf{B} \cdot d\mathbf{S} = \Phi, \qquad (6.50)$$

where Φ is the flux within the solenoid. Thus $A_\theta = \Phi/2\pi r$.

Next consider the effect of $\mathbf{A}$ on an electron travelling on either side of the solenoid. Its energy, and therefore its mechanical momentum $\hbar\mathbf{k}$, is not affected, but this momentum is given by $\mathbf{p} + e\mathbf{A}$ and it is $\mathbf{p}$ that appears in the phase of the wave function. In the top branch, $\mathbf{k}$ and $\mathbf{A}$ are in opposite directions so $\mathbf{p}$ increases and the phase of the electron changes more rapidly; $\mathbf{k}$ and $\mathbf{A}$ are in the same direction on the bottom so $\mathbf{p}$ is reduced. The vector potential therefore induces a change in phase between the two paths even though the electrons do not pass through the magnetic field. The change in phase affects the interference at the right-hand junction and therefore the transmission coefficient and conductance of the device. The difference in phase $\Delta\phi$ is given by

$$\hbar\Delta\phi = \int_{\text{top}} \mathbf{p} \cdot d\mathbf{l} - \int_{\text{bottom}} \mathbf{p} \cdot d\mathbf{l} = -e\left(\int_{\text{top}} \mathbf{A} \cdot d\mathbf{l} - \int_{\text{bottom}} \mathbf{A} \cdot d\mathbf{l} \right)$$

$$= e\oint_{\text{loop}} \mathbf{A} \cdot d\mathbf{l} = e\Phi. \qquad (6.51)$$

Thus $\Delta\phi = (e/\hbar)\Phi = 2\pi(\Phi/\Phi_0)$, where $\Phi_0 = h/e$ is the quantum of flux.

Interference is therefore periodic in the number of flux quanta passing through the loop; it is constructive when Φ is a multiple of Φ_0 and destructive halfway between. This is the Aharonov–Bohm effect. In perfectly one-dimensional wave guides this would reduce the transmission coefficient and conductance to zero, giving $G \propto \cos^2(\pi\Phi/\Phi_0)$. Real devices are far from perfect but nevertheless show clear modulation of the conductance, as shown in Figure 6.12(b). To be pedantic such experiments are not a strict test of the Aharonov–Bohm effect because the magnetic field penetrates the arms of the interferometer, not just the area enclosed by them. This leads to additional structure at higher magnetic fields, but the enclosed flux dominates at low fields. The Aharonov–Bohm effect has been confirmed by experiments performed in electron microscopes under more rigorous conditions.

6.5 Magnetic Field in a Narrow Channel

Further effects appear when a magnetic field is applied to electrons in a narrow channel or quantum wire. Start with a 2DEG in the xy-plane, and then confine the electrons in x leaving them free along y. The Schrödinger equation (6.30) in Landau

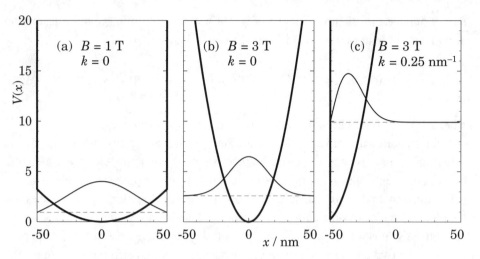

FIGURE 6.13. Potential energy and lowest eigenstate in a magnetic field for an electron with wave number k in a hard-walled wire of width $0.1\ \mu$m in GaAs.

gauge gains the additional confining potential $V(x)$ to become

$$\left[-\frac{\hbar^2}{2m}\frac{d^2}{dx^2} + \tfrac{1}{2}m\omega_c^2(x - x_k)^2 + V(x)\right]u(x) = \varepsilon\,u(x),\qquad(6.52)$$

with $x_k = -\hbar k/eB$. The magnetic potential can be rewritten as $(\hbar k + eBx)^2/(2m)$, which reduces to the usual kinetic energy for motion along y when $B = 0$.

This problem can be solved analytically for a parabolic confining potential, but an infinitely deep square well shows the physics more clearly. Let $V(x) = 0$ for $|x| < \tfrac{1}{2}a$ with an infinite potential outside. For zero field the energies are $\varepsilon_{nk}(0) = \hbar^2 n^2\pi^2/2ma^2 + \hbar^2 k^2/2m$. Consider $k = 0$ first, in which case the magnetic potential is a parabola centred on the origin. A weak magnetic field (Figure 6.13(a)) raises the energy of the state and distorts the wave function a little but the walls still provide the main confinement. Increasing the field strengthens the magnetic potential until this confines the electrons more tightly. The wave function now decays almost to zero in the parabola before it meets the wall (Figure 6.13(b)). In this limit the energy is close to its value for an unconfined two-dimensional system in a magnetic field, $\varepsilon_{n0}(B) \approx (n - \tfrac{1}{2})\hbar\omega_c$.

Now raise k, which displaces the vertex of the parabola to x_k. This adds $\hbar^2 k^2/2m$ to the energy when $B = 0$. At large fields, where electrons are confined at $k = 0$ by the magnetic potential rather than the hard walls, a shift of the parabola along x simply displaces the wave function and leaves its energy unaffected. This is the same as in the Landau level of an unconfined system, and holds until the wave function hits the wall of the wire. For large k the vertex of the magnetic parabola lies outside the wire altogether, as in Figure 6.13(c). The electron now sits in a roughly triangular potential well, its energy rises, and the degeneracy is lost.

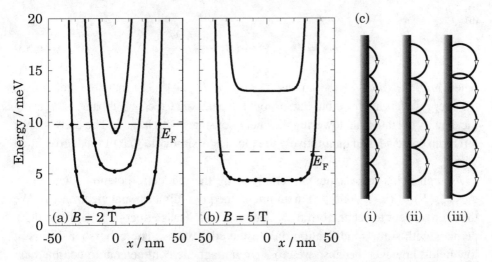

FIGURE 6.14. Energies $\varepsilon_{nk}(B)$ of electrons in a hard-walled wire in a magnetic field of (a) 2 T and (b) 5 T, plotted against the guiding centre $\langle x_{nk} \rangle$. For clarity, only the lowest three bands are shown for $B = 2$ T. The dots represent some of the occupied states and the dashed line is the Fermi level. (c) Classical skipping orbits along the edge of a wire.

The states squeezed against the side of the wire, with higher energy than those in the middle, are called *edge states*. They are of enormous importance for transport in a magnetic field, as we shall see shortly. An illuminating way of displaying them is to plot their energy against the average position of a wave function $\langle x_{nk} \rangle$. This is shown in Figure 6.14. At high magnetic fields the energy is nearly constant when $\langle x_{nk} \rangle$ lies well within the wire but rises rapidly as it approaches the edge. Under most conditions the only states at the Fermi level are edge states.

The classical behaviour of electrons near the edge of a hard-walled wire is shown in Figure 6.14(c). Electrons deep inside the wire execute circular cyclotron orbits with no net drift. Near the edge, however, the orbits are interrupted when the electron hits the boundary. The result is a *skipping orbit* that bounces along the boundary. The electron thus acquires a net drift velocity, which is higher for orbits closer to the edge.

The states on opposite edges travel in opposite directions. This can be interpreted in terms of the Lorentz force from a classical point of view, because the confining potential acts like an electric field in the opposite direction on the two edges. It is also a consequence of the broken time-reversal invariance in a magnetic field. When $B = 0$, this symmetry requires that forward- and backward-going states occupy the same region of space, as in the quantum point contact (Section 5.7.1). The states can separate when $B \neq 0$; this suppresses scattering between them and is another ingredient of the quantum Hall effect.

A parabolic confining potential in a wire, $V(x) = \frac{1}{2}m\omega_0^2 x^2$, can be solved exactly for the obvious reason that the magnetic potential is also parabolic. The energies are

$$\varepsilon_{nk} = (n - \tfrac{1}{2})\hbar\omega(B) + \frac{\hbar^2 k^2}{2m(B)}, \tag{6.53}$$

where $\omega(B) = (\omega_0^2 + \omega_c^2)^{1/2}$ and $m(B) = m[\omega(B)/\omega_0]^2$. The magnetic field raises the energy levels at $k = 0$ but flattens out the parabola for each subband, increasing the density of states and lowering the energy of states with large k. The derivation is left as an exercise, and qualitatively the physics is similar to that of the hard-walled wire.

At high fields most states are confined by the magnetic potential, with energies $\varepsilon_{nk}(B) \approx (n - \tfrac{1}{2})\hbar\omega_c$, as in an unconfined two-dimensional electron gas. We therefore expect to see the ordinary Shubnikov–de Haas effect in a magnetic field, perhaps with some modification due to the edge states. The behaviour differs at low fields, however, because we cross from magnetic confinement to confinement by the wire along x. In particular, the number of magnetic oscillations is given by the number of subbands occupied when $B = 0$, rather than the infinite series for a perfect two-dimensional electron gas without scattering. This can be shown by a fan plot of n against B_n^{-1}, the reciprocal of the magnetic fields at the minima of σ_{xx}. An example is shown in Figure 6.15. The sample is a wire confined by the negative bias V_g on split gates, whose width can be controlled by V_g. The wire is broad at smaller negative bias, giving a straight line as expected for an unconfined two-dimensional system (equation 6.42). Curvature develops at low fields as the wire is made narrower and the electric subbands become important. This process is termed *magnetic depopulation* of the subbands, which change from electric to magnetic in nature as the field is raised.

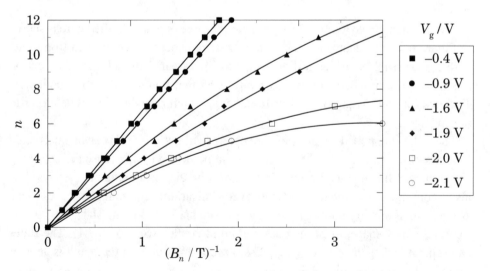

FIGURE 6.15. Fan diagram showing magnetic depopulation in a gated wire at different values of the bias V_g, which controls the width of the wire. The lines are given by the theory of Berggren et al. (1986). [Redrawn from Ford et al. (1988).]

6.5.1 QUANTUM DOTS AND SUPERLATTICES

Similar arguments can be used to find the effect of a magnetic field on electrons in a quantum dot, where they are confined in all three dimensions. Often the electrons are tightly confined in the direction z of growth but spread farther in the xy-plane. Most quantum dots are more or less circular so it is convenient to use the symmetric gauge and the results of Section 6.4.2. Again we can add the confining potential to the Schrödinger equation (6.36). In the case of a hard wall, the energies of the states with $l = 0$ are raised by the parabolic magnetic potential and eventually become Landau levels. The angular momentum l is analogous to the wave number k in a wire, and the centrifugal potential squeezes the electrons against the outer edge of the dot, assisted by the $\frac{1}{2}\hbar\omega_c l$ term, which lowers the energy for negative l. Thus edge states arise with a preferred direction.

The case of a parabolic confining potential, $V(r) = \frac{1}{2}m\omega_0^2 r^2$, can again be solved exactly. This is a famous result, dating from Darwin and Fock in 1928, and the energy levels are

$$\varepsilon_{nl} = (2n + |l| - 1)[(\hbar\omega_0)^2 + (\tfrac{1}{2}\hbar\omega_c)^2]^{1/2} + (\tfrac{1}{2}\hbar\omega_c)l \qquad (6.54)$$

for $n = 1, 2, 3, \ldots$ and $l = 0, \pm 1, \pm 2, \ldots$. The magnetic field has two effects on the spectrum. First, all states rise by the same factor because the magnetic potential adds to the confinement. The second term is the Zeeman energy of a dipole in a field. For $l \neq 0$ the electrons circulate in the dot and act like a circulating current to generate a magnetic dipole $\mu_B l$, where μ_B is the Bohr magneton. The energy of this dipole in a magnetic field is $\mu_B l B = \frac{1}{2}\hbar\omega_c l$, explaining the second term in equation (6.54). It breaks the degeneracy at $B = 0$ where the nth level of a harmonic oscillator holds n states (Section 4.7.2). The spectrum is illustrated in Figure 6.16 for $\hbar\omega_0 = 2$ meV. Experiments have verified this structure.

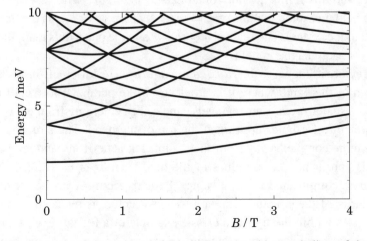

FIGURE 6.16. Energy levels in a magnetic field of a GaAs dot with a parabolic confining potential giving $\hbar\omega_0 = 2$ meV.

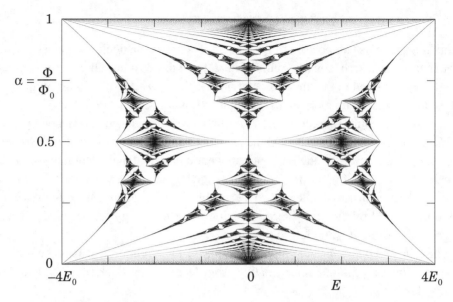

FIGURE 6.17. The Hofstadter butterfly: energy levels of a two-dimensional crystal in a magnetic field. Points are plotted for the allowed energies as a function of α, the fractional part of the number of flux quanta that penetrate each unit cell. The band occupies $\pm 4E_0$ in the absence of a magnetic field. [Figure kindly supplied by Prof. A. MacKinnon, Imperial College, London, based on the work of Hofstadter (1976).]

Finally, in case it should appear that motion in a magnetic field is simple, Figure 6.17 shows the remarkable spectrum of a periodic system, a two-dimensional crystal, in a strong magnetic field. The energies in the absence of a field are given by the tight-binding approximation as $\varepsilon(\mathbf{k}) = 2E_0(\cos k_x a + \cos k_y a)$, the two-dimensional version of the simple model (equation 2.9) that we used to explore the dynamics of electrons in one dimension. All energies in the range $\pm 4E_0$ are permitted, and this forms the horizontal axis of the figure. The vertical axis is the magnetic field measured in units of $\alpha = \Phi/\Phi_0$, where $\Phi = Ba^2$ is the flux through a unit cell of the crystal. The scale is $0 \le \alpha \le 1$ and the black dots mark the allowed energies for each magnetic field.

The structure is known as the *Hofstadter butterfly*. The lower left-hand part shows the bottom of the band, where it is an electron-like parabola, for small magnetic fields. The usual array of Landau levels can be seen emerging from the corner, and similar structure for holes appears at the top of the band. The structure becomes infinitely more complex away from such points but is far from random. Large gaps and bands appear at special values of the field such as $\alpha = 1/N$, with smaller gaps at more complicated rational values. Perhaps the most striking feature is the self-similarity of the structure: the large-scale structure is repeated at smaller and smaller scales within the figure, a classic example of a fractal. Each Landau level evolves into a treelike structure that ends on a band at a special field, from which new sets of Landau levels radiate.

A disappointing feature of this beautiful result is the magnitude of the magnetic field required. A typical crystal has $a < 1$ nm, which needs $B > 1000$ T to put a significant fraction of a flux quantum through each cell. An artificial superlattice offers a better chance, with $a \approx 50$ nm and $B \approx 1$ T. Unfortunately we then have to worry about the broadening of the energy levels as in the Shubnikov–de Haas effect. This requires a relatively strong periodic potential, which seems to be rather beyond what can be achieved in current structures without ruining the mean free path. The situation is similar to that for the observation of Bloch oscillations in an electric field. Fortunately the shortfall is not large, and we may hope to see experimental confirmation of this remarkable spectrum in the not too distant future.

6.6 The Quantum Hall Effect

Experimental measurements of the (integer) quantum Hall effect were shown in Figure 6.10. The key features are the plateaus in the Hall resistance at $\rho_T = (1/n)(h/e^2)$, while at the same time the longitudinal resistance almost vanishes. The fine-structure constant $\alpha = \mu_0 c e^2 / 2h \approx 1/137$ can be deduced from the quantum Hall effect, as both μ_0 and c are defined quantities. The value agrees with measurements using quite different techniques, as well as calculations in quantum electrodynamics, to an accuracy of 3×10^{-7}. The quantum Hall effect has now been adopted as a standard of resistance with the definition $R_K = h/e^2 = 25\,812.807\,\Omega$. The longitudinal resistivity in the Hall plateaus is lower than in any material other than superconductors, with values as low as $10^{-10}\,\Omega/\square$.

These results do not depend on the material and have been verified in devices made from Si, GaAs, and other semiconductors. They are also far more accurate than many of the traditional approximations made in the theory of electronic structure and transport, such as the effective-mass approximation, so a satisfactory theory must transcend these simplifications. Several arguments have been presented to explain the quantum Hall effect, varying from those based on gauge invariance to the picture based on edge states that we shall now discuss. This builds on the earlier theory of coherent transport in samples with many leads (Section 5.7.2).

Consider the Hall bar in a strong magnetic field shown in Figure 6.18(a). Current passes between probes 1 and 2, while the others are voltage probes and draw no current. Suppose that the magnetic field is such that the Fermi level lies between Landau levels in the middle of the wire, as in Figure 6.14(b). In this case the only states at the Fermi level are the edge states. There are N of these, given by the number of occupied Landau levels in the centre. The sets of edge states on opposite sides of the wire are well separated and travel in opposite directions, as shown in Figure 6.18(a). These edge states play the part of the 'modes' considered in Section 5.7. This identification of the current-carrying states is crucial.

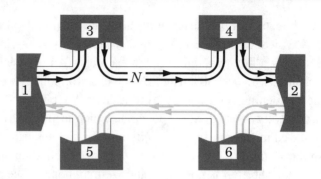

FIGURE 6.18. A Hall bar in a strong magnetic field, showing the propagation of edge states. A negative bias on contact 1 injects extra electrons into the N edge states that leave it (only two of which are drawn); the electrons depart through the other current probe (2).

Now apply a negative bias V_1 to contact 1 so that a conventional current flows from probe 2 to 1. This raises the Fermi level of the electrons leaving contact 1 by $-eV_1$ and therefore injects some extra electrons into the edge states that leave this contact, shown by the dark lines in Figure 6.18(a). Assume that these edge states run along the upper edge of the sample without scattering and enter the voltage probe 3. This is not allowed to draw a net current, so its Fermi level must rise to inject an equal current into the edge states that leave it. This requires $V_3 = V_1$. In the same way, $V_4 = V_1$ too: all the contacts along the top edge come to the same potential. Likewise, all the contacts on the bottom are at the same potential as the other current probe, $V_2 = 0$, because their edge states (shown by the light grey lines) carry no extra current.

A current $-(e^2/h)V_1$ is injected into each of the edge states on the top; this differs by a factor of 2 from the value used earlier because it is more appropriate to treat up and down spins separately as the Landau levels may be split. Thus the total current flowing $I = -N(e^2/h)V_1$, and the Hall resistance $(V_5 - V_3)/I = -V_1/I = (1/N)(e^2/h)$, the quantized value. Also, the longitudinal resistance $(V_4 - V_3)/I = 0$. Thus we have 'proved' the quantum Hall effect. The argument can be made more formal using the Landauer–Büttiker equations, but we shall save our effort for a more interesting example shortly.

Now consider more carefully the large number of assumptions in this picture. The most crucial is that the edge states suffer no scattering. Forward scattering occurs if an electron is scattered from one edge state to another state propagating along the same edge. This has no effect on the total current because the electron is still going in the same direction, and the transmission coefficient (summed over all edge states) is therefore unaffected. It has been found experimentally that such scattering can be very weak, with a mean free path of tens of micrometers. Scattering into edge states travelling in the opposite direction, however, will destroy the quantization and must be avoided. Fortunately, such edge states are on the other side of the sample and scattering will be extremely weak provided that the Fermi level is between Landau

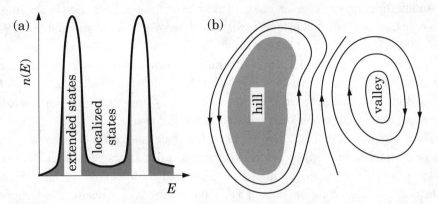

FIGURE 6.19. (a) Density of states of a Landau level in a disordered system showing a band of extended states in the centre of each level with localized states in between. (b) Edge states localized in a slowly varying potential, with a hill on the left and a hollow on the right.

levels in the centre of the sample. Thus the quantization is robust unlike that of a point contact in the absence of a magnetic field (Section 5.7.1).

Unfortunately this attractive argument has a severe problem. We saw in Section 6.4.4 that the Fermi level adjusts itself to lie within a Landau level for almost all values of the magnetic field in the bulk, and the situation for a broad wire is similar. Thus the previous argument holds only for very narrow ranges of B and is unable to explain the width of the plateaus in the Hall resistance. The other essential ingredient is, perhaps surprisingly, disorder due to a random potential from impurities or defects at the interface. This causes many of the states in the Landau levels to become *localized*, meaning that they are restricted to a small area of the sample.

There are several mechanisms that lead to localization depending on the nature of the random potential. A short-ranged random potential, one that varies rapidly in space on the scale of l_B, leads to *Anderson localization*, illustrated in Figure 6.19(a) for the density of states. The centre of each Landau level contains extended states, which propagate throughout the sample as we have assumed before, but the states in the tails are localized like bound states and play no part in conduction. Now we no longer require the stringent condition that the Fermi level lie between Landau levels for the quantum Hall effect, only that it lie within the localized states. This explains the existence of plateaus but makes the origin of the quantization more mysterious, because the Hall conductivity takes the value expected if *all* electrons contribute, whereas we have now argued that the localized electrons take no part! The resolution is that the extended states compensate for the localized states by carrying more current than they would in a clean system. Detailed calculations confirm this surprising behaviour, and show that electrons are accelerated around scattering potentials to increase their average velocity.

Another picture of disorder holds if the random potential is slowly varying in space. In this case the Landau levels drift up and down in energy to remain $(n - \frac{1}{2})\hbar\omega_c$ above the random potential at each point. In this case most of the edge states lie in loops around hills or hollows in the potential as shown in Figure 6.19(b). Such states are effectively localized and cannot conduct. Very few states percolate through the whole sample to connect the probes, and the transitions between Hall plateaus, where $\rho_L > 0$, occur when such states lie at the Fermi level.

These arguments indicate that the best Hall plateaus should be seen not in the cleanest samples, but in those with moderate disorder. It has been found that $\mu \approx 10\,\mathrm{m^2\,V^{-1}\,s^{-1}}$ is about the optimum for GaAs and $\mu \approx 1\,\mathrm{m^2\,V^{-1}\,s^{-1}}$ for Si. The transitions may then have only about 5% of the width of the plateaus, an appropriate characteristic for metrology.

Closer scrutiny reveals that many more problems have been 'swept under the carpet'. For example, how is the Hall voltage distributed across the device? The profile of the potential that we have derived for the Hall bar in Figure 6.18 looks superficially like that which we derived classically in Section 6.3 from the conductivity tensor. However, current flows uniformly through the sample in the classical picture (Figure 6.6(b)), whereas we have just argued that it is all carried by edge states. Moreover, the simple picture of edge states does not take proper account of compressible and incompressible regions (Section 6.4.4).

Another question is whether the Hall current is driven by a difference in *chemical* or *electrostatic* potential. We have assumed that edge states are controlled by the chemical potentials of the contacts, whereas the current was driven by an electric field in the classical case. We calculated the Hall current due to crossed electric and magnetic fields in Section 6.4.7 and there it was again the electric field that drove the current. Fortunately the Hall conductivity at small currents, where the response is linear, does not depend on the division of the electrochemical potential between its electrostatic and chemical components. Many features of the quantum Hall effect await a full understanding.

6.6.1 EDGE STATES AND BARRIERS

We have seen that the integer quantum Hall effect can be described in terms of edge states, which act as well-defined 'modes' in the parlance of tunnelling. Numerous pretty experiments have been carried out on these states and analyzed in terms of the Landauer–Büttiker equations (Section 5.7). An example is provided by the slightly more complicated sample shown in Figure 6.20. This has a barrier in the middle that raises the energy of all states such that only M edge states can propagate rather than N in the rest of the sample. Assume that there is no scattering between edge states. Of the N states that emerge from probe 3, M pass through the barrier to probe 4 while $N - M$ are reflected and enter probe 5 instead. Thus $N - M$ of the states entering probe 5 carry extra current due to the bias while the other M, which

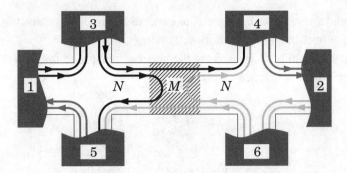

FIGURE 6.20. A Hall bar with a barrier across the middle that transmits only M of the N edge states.

originate from probe 6, do not. Probe 5 is a voltage probe so this current must be balanced by the states that leave 5 and enter 1. A voltage V_5 therefore develops, but its magnitude is less than V_1, as shown by the medium grey colour of the lines leaving the contact. The contact shares out electrons among the edge states so it is not entirely passive, even though the total current is unaffected. The same happens at contact 4.

We can write down the transmission coefficients and equations for the currents using equation (5.108). As in the simple Hall bar, we expect that $V_3 = V_1$ and $V_6 = V_2$ and can check this. Fortunately, very few of the transmission coefficients are non-zero:

$$T_{15} = T_{24} = T_{31} = T_{62} = N, \quad T_{43} = T_{56} = M, \quad T_{46} = T_{53} = N - M. \quad (6.55)$$

These obey the 'row-and-column' sum rule (5.107). We would like to know the Hall resistance $(V_5 - V_3)/I \equiv R_{21,53}$, the four-terminal resistance $(V_4 - V_3)/I \equiv R_{21,43}$, and the two-terminal resistance $(V_2 - V_1)/I \equiv R_{21,21}$.

First, the current into probe 3 is given by equation (5.108). This becomes $0 = (h/e^2)I_3 = NV_3 - NV_1$, so $V_3 = V_1$. Thus we confirm our previous argument that the probes along the top of the Hall bar come to the same voltage as the left current probe (in the absence of a barrier).

Next, calculate the four-terminal resistance between probes 4 and 3, denoted $R_{21,43}$, where the first pair of indices gives the current probes and the second pair the voltage probes. The equations for I_2 and I_4 are

$$(h/e^2)I = (h/e^2)I_2 = NV_2 - NV_4, \qquad (6.56)$$

$$0 = (h/e^2)I_4 = NV_4 - MV_3 - (N - M)V_6. \qquad (6.57)$$

Eliminating V_2, and remembering that $V_3 = V_1$ and $V_6 = V_2$, gives

$$R_{21,43} = \frac{V_4 - V_3}{I} = \frac{h}{e^2}\left(\frac{1}{M} - \frac{1}{N}\right). \qquad (6.58)$$

The longitudinal resistance is nonzero, as expected from the insertion of the barrier. It returns to zero if the barrier is removed and $M = N$.

Further results can be derived from the general four-probe formula (5.119). For example, the mixed resistance

$$R_{21,54} = \frac{h}{e^2}\left(\frac{2}{N} - \frac{1}{M}\right).$$ (6.59)

The sum $V_{53} + V_{34} = V_{54}$ so this agrees with the previous results.

6.6.2 THE FRACTIONAL QUANTUM HALL EFFECT

Figure 6.21 shows a measurement of the fractional quantum Hall effect. Unlike the integer effect, which requires disorder to give width to the plateaus, the fractional effect is seen only at low temperature in samples with very high mobility. The features are again plateaus in ρ_T associated with minima in ρ_L. The fractional effect appears at filling factors $\nu = p/q$, where p and q are integers with q odd. Particularly strong features occur when $\nu = \frac{2}{3}, \frac{3}{5}$, and $\frac{2}{5}$.

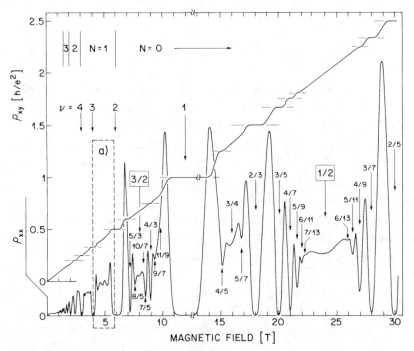

FIGURE 6.21. Longitudinal resistivity $\rho_{xx} = \rho_L$ and transverse resistivity $\rho_{xy} = \rho_T$ of a high-mobility two-dimensional electron gas at 150 mK, showing the fractional quantum Hall effect. The filling factor ν is indicated, and ρ_{xx} is reduced by a factor of 2.5 at high fields. [From Willett et al. (1987).]

The physical origin of the fractional quantum Hall effect is quite different from that of the integer effect. All electrons are in the lowest Landau level when $\nu \leq 1$ and, according to our previous theory, all have the same kinetic energy $\frac{1}{2}\hbar\omega_c$ and their spins are aligned. Physics that we have ignored now comes into play, and we have as usual neglected the Coulomb repulsion between electrons. It has been shown that the electrons can arrange themselves in a particularly favourable configuration to minimize this energy at values of ν, which explains some of the observed fractions. These highly correlated states have remarkable properties; their excitations may carry a fractional charge such as $e^* = \frac{1}{3}e$, for example.

Much recent work has focussed on the physics of half-filled Landau levels. Returning to Figure 6.21, we see that both ρ_T and ρ_L measured in either direction from $\nu = \frac{1}{2}$ resemble the same quantities measured from $B = 0$. This has been explained in terms of *composite fermions*, electrons with two flux tubes attached. Each flux tube is like an infinitesimal solenoid carrying flux Φ_0. The composite fermions move in an effective field B^* that vanishes when $\nu = \frac{1}{2}$. Theory and experiment are developing rapidly.

Finally, another phenomenon appears at the lowest filling factors, roughly $\nu < \frac{1}{7}$, where the longitudinal resistance rises rapidly. This is believed to be due to the localization of electrons, not by disorder but due to their Coulomb repulsion. It leads to the formation of a *Wigner crystal* in which electrons sit on a lattice rather than roam through the 2DEG. The density at which a 2DEG should crystallize when $B = 0$ is extremely low, and in practice the properties are dominated by disorder. A strong magnetic field makes crystallization more favourable because, as mentioned before, the kinetic energy is quenched when all electrons occupy the lowest Landau level and electron–electron interaction rules.

The discussion has focussed, as usual, on electrons but many experiments have been performed with holes. These have a higher effective mass, which reduces their kinetic energy and therefore increases the importance of electron–electron interaction. Other systems of current interest are closely spaced layers of electrons and/or holes in which new effects due to interaction between the layers have been seen. There will, no doubt, be many further surprises to come.

Further Reading

Weisbuch and Vinter (1991) show how the effect of an electric field can be harnessed as an electro-optic modulator. Stradling and Klipstein (1990) cover the use of the Hall effect as a diagnostic tool for semiconducting structures.

Bastard (1988) gives more details of the motion of electrons in electric and magnetic fields. A useful survey of the integer and fractional quantum Hall effects, from both the experimental and theoretical standpoints, is given by Prange and Girvin (1990).

EXERCISES

6.1   Show that expressions (6.4) and (6.5) for the current are consistent with the continuity equation, by repeating the derivation of section 1.4 with the Schrödinger equation (6.3).

6.2   Calculate the Franz–Keldysh effect on absorption between the valence and conduction bands in GaAs. The mass that appears in the density of states in this case is the optical effective mass (Section 1.3.1) given by $1/m_{eh} = 1/m_e + 1/m_h$, where m_h could be that for light or heavy holes. How large an electric field is needed to get a significant tail in the absorption edge (0.1 eV, say)?

6.3   What scale of electric field is needed to induce Stark localization in a typical superlattice? Use the example from Section 5.6. Is it practicable?

6.4   Do the usual ways of measuring the conductivity and Hall effect in semiconductors measure σ or ρ? Typical geometries are shown in Figure 6.5.

6.5   Consider a Hall bar of length L_x and width L_y where we pass a current I_x and measure voltages V_x and V_y. Calculate V_x and V_y using the result (6.22) from the Drude model in a magnetic field.

A two-dimensional electron gas is fabricated into a Hall bar with $L_x = 0.5$ mm and $L_y = 0.1$ mm. It gives $V_x = 0.13$ mV and $V_y = 0.31$ mV with $I_x = 1\,\mu$A and $B = 0.15$ T. Find the concentration and mobility of the 2DEG. Is the use of the low-field formulas justified?

6.6   Derive equation (6.24) for the Hall effect with two channels of electrons in low fields, and show that $n_{eff} = n_1 + n_2$ in high fields. Show also that ρ_L gains a quadratic term in B at low fields. How do these results change for a channel of electrons in parallel with one of holes?

Suppose that channel 2 has ten times the density, but only one-hundredth the mobility, of channel 1. How easy will it be to detect channel 2 using the Hall effect? (This situation often occurs in modulation-doped layers.)

6.7   Explain, using the relation between $\mathbf{J}$ and $\mathbf{F}$, how it is possible for the longitudinal components of both the conductivity and resistivity to vanish together in high magnetic fields.

6.8   What happens to the distribution of current and electric field in a Corbino ring (Figure 6.5(c)) as the magnetic field is raised to give $\sigma_T > \sigma_L$?

6.9   Show that the current density of an electron in a magnetic field, using Landau gauge, is

$$J_y = -\frac{e^2 B}{m L_y}(x - x_k)|u_n(x - x_k)|^2, \qquad \text{(E6.1)}$$

where $u_n(x - x_k)$ is the nth harmonic-oscillator wave function, centred on $x = x_k$. Demonstrate that there is no net current, despite the travelling wave $\exp(iky)$ in the wave function (6.33), but that there is a circulating current.

6.10 Show that the wave functions (without normalization) of the lowest Landau level in symmetric gauge can be written in the compact form $z^m \exp(-\frac{1}{4}|z|^2)$, where $m \geq 0$ and $z = (x - iy)/l_B$ (not quite the usual definition in complex numbers!). Note that $L_0^{(l)}(t) = 1$ for all l. This basis has proved particularly useful in the study of the fractional quantum Hall effect.

6.11 What would happen to the energies of Landau levels, including spin splitting, if electrons in a 2DEG had $g = 2$ as if they were free?

6.12 Confirm the density of electrons for the sample in Figure 6.10, using both the Hall effect at low fields and the minima of the Shubnikov–de Haas effect (fan diagram) at large fields. Deduce also the current driven through the sample from the plateaus in the Hall voltage.

6.13 Use the data in Figure 6.10 to estimate the width Γ of the Landau levels and the single-particle lifetime. How does this compare with the transport lifetime from the mobility? (The theory of these scattering rates will be given in Chapter 9.)

6.14 Verify that there are no physical consequences of constant vector and scalar potentials, since these produce no fields. What do they do to the energy, momentum, and wave functions of free electrons?

6.15 What is the effective diameter of the antidot used in the Aharonov–Bohm experiment of Figure 6.12?

6.16 Calculate the energy levels of a wire with parabolic confinement, equation (6.53), and the position of the guiding centre. This should be a good model for narrow wires with few electrons. A rough value is $\hbar\omega_0 = 3$ meV, so what range of fields is involved?

There are now two frequencies, ω_0 and $\omega(B)$. Which enters the wave function, and which enters the energy as a function of guiding centre?

6.17 Estimate the width of the narrowest wire shown in the fan diagram, Figure 6.15.

6.18 If you are feeling brave, derive the energies of the parabolic dot in a magnetic field, equation (6.54).

6.19 Suppose that there are 12 electrons in the parabolic dot whose energy levels are plotted in Figure 6.16. How would the energy of the highest filled state change as a function of magnetic field, assuming that the lowest states were always occupied? Spin splitting may be neglected, so each level is doubly degenerate.

6.20 Show that the two-probe resistance of the Hall bar with a barrier (Section 6.6.1) is $R_{21,21} = (h/e^2)(1/M)$, depending only on the number of states transmitted through the barrier, and that the Hall resistance $R_{21,53} = R_{21,64} = (h/e^2)(1/N)$, giving the total number of edge states.

6.21 Derive equation (6.59) for the Hall bar with a barrier using the Landauer–Büttiker four-probe formulas. This need be applied to leads 1, 2, 4, and 5 only, because we know that leads 3 and 6 are uninteresting. The transmission coefficients are given by equation (6.55) upon putting $3 \to 1$ and $6 \to 2$. Show that the determinant $S = MN^2$ and hence that $R_{21,54} = (h/e^2)(2/N - 1/M)$.

APPROXIMATE METHODS

7

Few problems in physics and engineering can be solved exactly, and one has to resort to approximate or numerical methods. Consider, for example, an electron in a square well whose potential is tilted by applying an electric field. The energy and wave function of its states change only slightly if the field is small. The lowest state becomes polarized to the deeper part of the well, causing a quadratic reduction in its energy. Perturbation theory provides a framework for calculating such changes, and this example is discussed in Section 7.2.

This approach works well if the potential can be divided into a 'large' part that can be solved exactly and a 'small' perturbation. Other methods must be used if this is not the case. The WKB method described in Section 7.4 is applicable to potentials that vary slowly in space, and is closely related to classical mechanics. The variational method (Section 7.5) gives only the energy of the ground state but has unrivalled accuracy and can include electron–electron interaction and other complications.

There are many applications to band structure. The $\mathbf{k} \cdot \mathbf{p}$ method in Section 7.3 gives the form of energy bands near a gap, the most important region in a semiconductor. Two general methods take opposite points of view. The tight-binding method (Section 7.7) is based on a picture of isolated atoms brought together to form the solid, where the bands originate from atomic levels. In contrast, the nearly free electron method in Section 7.8 assumes that the solid perturbs the motion of the electrons only weakly. It shows that any periodic potential, however weak, creates band gaps.

Before getting into the physics, it is convenient to get some mathematics out of the way. Our previous work has been based on the Schrödinger differential equation and eigenfunctions, but much of perturbation theory is more convenient within a formulation based on matrices rather than differential operators.

7.1 The Matrix Formulation of
Quantum Mechanics

We noted in Section 1.5.1 that the terminology of eigenstates and eigenvalues, used to describe solutions to the Schrödinger equation, is very similar to that used for matrix equations. It is simple to make this connection closer and rewrite the Schrödinger equation with matrices rather than differential operators. This results in quantities called (not surprisingly) matrix elements, which occur throughout perturbation theory.

The crucial property on which this transformation rests is that of *orthogonality*, which was described in Section 1.6. This means that it is possible to define a set of functions such that

$$\int \phi_m^*(x)\, \phi_n(x)\, dx = \delta_{mn}\,. \tag{7.1}$$

The integral vanishes if the two states are different and gives unity if they are the same (this means that the functions are orthonormal, both orthogonal and normalized to unity). The integral is over the range of interest, which may be finite or infinite and in one, two, or three dimensions. The choice of functions depends on the region of interest; sine and cosine waves or complex exponentials are an obvious choice for a one-dimensional system extending from $-\infty$ to ∞, or sine waves in a well from 0 to a. In these two examples the results are familiar from Fourier theory. However, the choice of functions is much wider than this, and the solutions to *any* Schrödinger equation can be made orthonormal.

The Schrödinger equation can be written symbolically as $\hat{H}\psi = E\psi$, where $\hat{H}$ is a differential operator for which we have encountered many forms already. Expand ψ in terms of some complete set of functions ϕ_n:

$$\psi = \sum_n a_n \phi_n\,. \tag{7.2}$$

This is written with a sum over n, which would be appropriate for a system restricted to a finite region of space, such as a particle in a box with $\phi_n(x) = (2/a)^{1/2} \sin(n\pi x/a)$. The sum may become an integral for an infinite region as in the case of a Fourier transform using $\phi_k(x) = \exp(ikx)$.

Now substitute this expansion into the Schrödinger equation. The result is

$$\hat{H}\sum_n a_n \phi_n = \sum_n a_n \hat{H}\phi_n = E\sum_n a_n \phi_n\,. \tag{7.3}$$

Next multiply from the left by ϕ_m^* and integrate. This gives

$$\sum_n a_n \int \phi_m^* \hat{H}\phi_n = E\sum_n a_n \int \phi_m^* \phi_n\,. \tag{7.4}$$

It is assumed that the summation and integral or differential operators can be freely reordered. The integral on the right is precisely the definition of orthonormality and immediately reduces to δ_{mn}. All the terms in the summation then vanish except for $n = m$ so we are left with Ea_m on the right. The left-hand side does not simplify this way, but we can define a quantity

$$H_{mn} = \langle m|\hat{H}|n\rangle = \int \phi_m^* \hat{H}\phi_n\,, \tag{7.5}$$

which is called the *matrix element* of $\hat{H}$ between the states m and n. In terms of these, our Schrödinger equation reduces to

$$\sum_n H_{mn}a_n = Ea_m . \tag{7.6}$$

This is now an eigenvalue equation for the matrix H,

$$\mathbf{Ha} = E\mathbf{a}, \tag{7.7}$$

where E is the eigenvalue and $\mathbf{a}$ is the eigenvector. This is the matrix equivalence of our original differential Schrödinger equation. It can also be written as $(E\mathsf{I} - \mathsf{H})\mathbf{a} = \mathbf{0}$, where I is the unit matrix. The condition for this equation to have nontrivial solutions is that the determinant should vanish,

$$\det |E\mathsf{I} - \mathsf{H}| = 0. \tag{7.8}$$

This is often known for historical reasons as the *secular equation*. Its roots determine the allowed energies E_n. The unit matrix I is often dropped.

This shows that there is a strong relation between the description of states as functions in space such as $\psi(x)$, which we have used up to now, and a description based on matrices where we expand the wave function in terms of some orthonormal set and use the amplitudes to describe the state. Unfortunately the matrix equation is not quite as simple as 'ordinary' eigenvalue problems because the dimensions are infinite. However, almost all approximate methods truncate the basis of states used to expand the wave function and the matrix becomes finite.

The notation $\langle m|\hat{H}|n\rangle$ for matrix elements is due to Dirac. Here $|n\rangle$, called a *ket*, denotes a state and the *bra* $\langle n|$ is its complex conjugate. Note the economy of notation compared with the more straightforward forms such as $\phi_n(x)$: the Dirac form retains only the n that labels the states. The fact that state n can be expressed as $\phi_n(x)$ is not important because we could instead choose to expand the state in terms of some complete set of functions, as in equation (7.2), and the expansion parameters would be just as valid a description of the wave function as $\phi_n(x)$. Dirac notation discards this unimportant detail. In fact the notation has a deeper mathematical significance that bridges the gap between matrices and differential equations.

Although any set of states can be used to reduce the differential operator $\hat{H}$ to a matrix H_{mn}, a particularly convenient choice is the eigenfunctions of $\hat{H}$ (if we know them!). These are the stationary states of the Schrödinger equation, $\hat{H}\phi_n = \varepsilon_n\phi_n$. Then

$$H_{mn} = \int \phi_m^* \hat{H}\phi_n = \int \phi_m^* \varepsilon_n \phi_n = \varepsilon_n \delta_{mn} . \tag{7.9}$$

The last form follows from the orthonormality of the states ϕ_n, and shows that the Hamiltonian is diagonal in this basis. Thus another way of describing the eigenstates of any operator, not just the Hamiltonian, is to say that these states 'diagonalize the operator'.

It may happen that another operator is also diagonal in the set of states that diagonalize $\hat{H}$. The same states are therefore eigenfunctions of both operators. This can happen only if the two operators commute (Section 1.6). In this case the physical quantity corresponding to the second operator is said to be a constant of the motion. Take free electrons in one dimension as an example. Plane waves $\exp(ikx)$ simultaneously diagonalize both the Hamiltonian $(-\hbar^2/2m)\partial^2/\partial x^2$ and the momentum operator $-i\hbar\,\partial/\partial x$ (Section 1.5). This shows that they have both a definite energy and momentum.

We have now turned the Hamiltonian (differential) operator into a matrix and the Schrödinger equation into a matrix equation. The same process can be repeated on any operator. As we shall see in the following sections, perturbation theory consists of splitting the Hamiltonian into a large part $\hat{H}_0$, which we know how to diagonalize, and a small perturbation $\hat{V}$, which enters expressions for the energy or wave function in terms of matrix elements between the eigenstates of $\hat{H}_0$.

Differential operators that correspond to observable quantities must be Hermitian to ensure that their eigenvalues, which represent measurable values, are real (Section 1.5.2). The corresponding matrices must also be Hermitian, which means that $\mathbf{M}^\dagger = \mathbf{M}$, or $M_{mn} = M_{nm}^*$.

Armed with these preliminaries, we can now approach perturbation theory.

7.2 Time-Independent Perturbation Theory

Start with a familiar example, an electron in a quantum well. Figure 7.1(a) shows the usual finite rectangular potential well, which we solved in Chapter 4. It is trivial if the depth is infinite, mildly tedious if it is finite. Now suppose that an electric field is applied to the sample, tilting the potential as shown in Figure 7.1(b). The ground state changes as shown in the figure; its wave function becomes asymmetric and the mean position $\langle x \rangle$ moves away from the centre of the well (taken as $x = 0$) towards the side whose energy has been lowered by the electric field. The electric field polarizes the electron and generates a dipole moment $p = -e\langle x \rangle$. We expect the dipole to be proportional to the field F for small fields and define a polarizability α (with units of volume) by $p = \epsilon_0 \alpha F$. The shift of the wave function lowers its energy by $-\frac{1}{2}\epsilon_0 \alpha F^2$, the energy of an induced dipole.

It happens in this case that the problem can be solved exactly. The potential in Figure 7.1(b) is linear everywhere and the wave function is therefore composed of Airy functions matched at the edges of the well. This is tedious and requires extensive tables or a computer. Such precision may be unnecessary, and we shall see in Section 7.2.3 that an exact solution may even be undesirable. To find the polarizability α we can assume that the applied change in potential energy is small

(a)   (b)

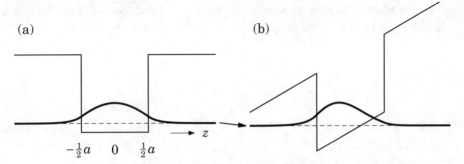

$-\frac{1}{2}a$ 0 $\frac{1}{2}a$

FIGURE 7.1. A quantum well (a) with flat potentials and (b) in an electric field.

and find an expansion of the wave function and energy in powers of the applied electric field. The foregoing discussion shows that we need either

(i) the change in wave function to *first* order in the field, or
(ii) the change in energy to *second* order in the field.

These limited requirements turn out to be very common, and we shall now develop a scheme to calculate these quantities. Unfortunately this sort of perturbation theory can fail, usually for good physical reasons, and we shall see that it does not give the full story even for the simple quantum well in an electric field.

We shall now derive the general results for this perturbation theory, then apply them to the example of the quantum well in an electric field.

7.2.1 GENERAL THEORY

The general idea is to split the Hamiltonian $\hat{H}$ of the system that we wish to solve into two parts, $\hat{H} = \hat{H}_0 + \hat{V}$. These are

(i) $\hat{H}_0$, the unperturbed system, which is 'large' and can be solved exactly (like the well before the field was applied); and
(ii) $\hat{V}$, the *perturbation*, which must be 'small'; the precise meaning of 'small' will become clear later.

Thus the starting point is that we know the solutions to the Schrödinger equation

$$\hat{H}_0\phi_n = \varepsilon_n\phi_n \tag{7.10}$$

and wish to find the solutions to

$$\hat{H}\psi_n = (\hat{H}_0 + \hat{V})\psi_n = E_n\psi_n. \tag{7.11}$$

The idea of perturbation theory is to expand the energy and wave function in powers of the small potential $\hat{V}$. To aid the bookkeeping, write $\hat{H} = \hat{H}_0 + \lambda\hat{V}$ so that powers of λ identify degrees of smallness. We set $\lambda = 1$ at the end. The energy and wave functions have the expansions

$$E_n = E_n^{(0)} + \lambda E_n^{(1)} + \lambda^2 E_n^{(2)} + \cdots, \tag{7.12}$$

$$\psi_n = \psi_n^{(0)} + \lambda \psi_n^{(1)} + \lambda^2 \psi_n^{(2)} + \cdots. \tag{7.13}$$

The subscript labels the state and the superscript labels the power of λ, which is the degree of smallness.

Substituting these expansions into the Schrödinger equation gives

$$(\hat{H}_0 + \lambda \hat{V})(\psi_n^{(0)} + \lambda \psi_n^{(1)} + \cdots) = (E_n^{(0)} + \lambda E_n^{(1)} + \cdots)(\psi_n^{(0)} + \lambda \psi_n^{(1)} + \cdots). \tag{7.14}$$

This must hold for all values of λ, which means that the coefficients of powers of λ on both sides must be the same. For powers 0, 1, and 2 this gives

$$\hat{H}_0 \psi_n^{(0)} = E_n^{(0)} \psi_n^{(0)}, \tag{7.15}$$

$$\hat{V} \psi_n^{(0)} + \hat{H}_0 \psi_n^{(1)} = E_n^{(1)} \psi_n^{(0)} + E_n^{(0)} \psi_n^{(1)}, \tag{7.16}$$

$$\hat{V} \psi_n^{(1)} + \hat{H}_0 \psi_n^{(2)} = E_n^{(2)} \psi_n^{(0)} + E_n^{(1)} \psi_n^{(1)} + E_n^{(0)} \psi_n^{(2)}. \tag{7.17}$$

The first of these, equation (7.15), is just the unperturbed Schrödinger equation so we see immediately that

$$\psi_n^{(0)} = \phi_n, \qquad E_n^{(0)} = \varepsilon_n. \tag{7.18}$$

This is fairly obvious: the zeroth-order estimates of the wave function and energy are just the unperturbed ones.

Inserting these results into the first-order equation (7.16) and gathering terms gives

$$(\hat{H}_0 - \varepsilon_n)\psi_n^{(1)} = (E_n^{(1)} - \hat{V})\phi_n. \tag{7.19}$$

To make further progress we need to use the results of the previous section. Expand $\psi_n^{(1)}$ in terms of the complete set ϕ_n, which are the eigenfunctions of $\hat{H}_0$. Write

$$\psi_n^{(1)} = \sum_k a_{nk}^{(1)} \phi_k. \tag{7.20}$$

The indices are proliferating, but subscripts consistently label the states involved while the superscript is again the order of smallness. With this expansion, the first-order equation (7.19) becomes

$$\sum_k (\hat{H}_0 - \varepsilon_n) a_{nk}^{(1)} \phi_k = (E_n^{(1)} - \hat{V})\phi_n. \tag{7.21}$$

The operator $\hat{H}_0$ has been taken inside the sum on the left because we know that its action on ϕ_k is simply $\hat{H}_0 \phi_k = \varepsilon_k \phi_k$. Thus $\hat{H}_0$ disappears, a result of the expansion of $\psi_n^{(1)}$. We now have

$$\sum_k a_{nk}^{(1)} (\varepsilon_k - \varepsilon_n) \phi_k = E_n^{(1)} \phi_n - \hat{V} \phi_n . \qquad (7.22)$$

The next step is to use the orthonormality of the states ϕ_n and reduce $\hat{V}$ to a matrix element. Multiply equation (7.22) on the left by ϕ_n^* and integrate. On the left-hand side this picks out the term with $k = n$ because the others all vanish by orthogonality. The difference $(\varepsilon_k - \varepsilon_n)$ causes this remaining term to vanish too. The ϕ_n with $E_n^{(1)}$ on the right-hand side integrates to unity, and the term with the potential becomes a matrix element V_{nn} to give

$$E_n^{(1)} = \int \phi_n^* \hat{V} \phi_n \equiv V_{nn} . \qquad (7.23)$$

Equation (7.23) shows that the first-order change to the energy is simply the expectation value of the perturbing potential $\hat{V}$ in the unperturbed state ϕ_n, another predictable result.

To get the coefficients of the wave function, multiply equation (7.22) on the left by ϕ_m^*, where $m \neq n$, and integrate. Only $k = m$ survives on the left, and does not vanish this time. The term with $E_n^{(1)}$ disappears from the right by orthogonality, leaving the matrix element V_{mn}. Thus

$$a_{nm}^{(1)} (\varepsilon_m - \varepsilon_n) = - \int \phi_m^* \hat{V} \phi_n \equiv -V_{mn} \qquad (7.24)$$

so $a_{nm}^{(1)} = V_{mn} / (\varepsilon_n - \varepsilon_m)$. The change in wave function to first order is therefore

$$\psi_n^{(1)} = a_{nn}^{(1)} \phi_n + \sum_{k, k \neq n} \frac{V_{kn}}{\varepsilon_n - \varepsilon_k} \phi_k . \qquad (7.25)$$

Note that we can't determine the coefficient $a_{nn}^{(1)}$ by this route. Fortunately it turns out that it will not be needed. It affects only the normalization, so we shall take $a_{nn}^{(1)} = 0$.

Now continue to second order to find the next correction to the energy. With the results that we have already found, equation (7.17) becomes

$$(\hat{H}_0 - \varepsilon_n) \psi_n^{(2)} = (V_{nn} - \hat{V}) \sum_k a_{nk}^{(1)} \phi_k + E_n^{(2)} \phi_n . \qquad (7.26)$$

Expand the wave function as

$$\psi_n^{(2)} = \sum_k a_{nk}^{(2)} \phi_k, \qquad (7.27)$$

which gives

$$(\hat{H}_0 - \varepsilon_n) \sum_k a_{nk}^{(2)} \phi_k = (V_{nn} - \hat{V}) \sum_k a_{nk}^{(1)} \phi_k + E_n^{(2)} \phi_n. \qquad (7.28)$$

We can take $\hat{H}_0$ inside the summation and replace it by ε_k as before. Next, multiply on the left by ϕ_n^* and integrate. The left-hand side vanishes to leave

$$E_n^{(2)} = \sum_k a_{nk}^{(1)} \int \phi_n^* \hat{V} \phi_k - V_{nn} a_{nn}^{(1)} = \sum_k a_{nk}^{(1)} V_{nk} - V_{nn} a_{nn}^{(1)}. \qquad (7.29)$$

The unknown coefficient $a_{nn}^{(1)}$ cancels. Inserting the other coefficients $a_{nk}^{(1)}$ gives

$$E_n^{(2)} = \sum_{k, k \neq n} a_{nk}^{(1)} V_{nk} = \sum_{k, k \neq n} \frac{V_{nk} V_{kn}}{\varepsilon_n - \varepsilon_k}. \qquad (7.30)$$

This can be simplified a little if we remember that V is a Hermitian matrix, so $V_{nk} = V_{kn}^*$ and the numerator can be written as $|V_{nk}|^2$ or $|V_{kn}|^2$.

We can now collect our results for the expansion of the wave function to first order and the energy to second order:

$$E_n = \varepsilon_n + V_{nn} + \sum_{k, k \neq n} \frac{|V_{kn}|^2}{\varepsilon_n - \varepsilon_k} + \cdots, \qquad (7.31)$$

$$\psi_n = \phi_n + \sum_{k, k \neq n} \frac{V_{kn}}{\varepsilon_n - \varepsilon_k} \phi_k + \cdots. \qquad (7.32)$$

The process can be carried to higher order but it is rarely necessary, and we shall instead look at the implications of these formulas.

(i) We can now see what is meant by 'small'. If the series is to converge rapidly at first, the matrix elements must be much smaller than the energy denominators. In other words, the coupling between states induced by the perturbation must be smaller than the separation between the energy levels. This is hardly surprising, because the states of the original Hamiltonian will be nothing like those of the perturbed system if this condition is not met.

(ii) Disaster strikes if an energy denominator vanishes. This happens if the state in which we are interested is degenerate with another one, and a different approach must then be used (Section 7.6).

(iii) Many matrix elements V_{kn} often vanish because of symmetry. The same matrix elements appear in the calculation of scattering rates by time-dependent perturbation theory, where the restrictions due to symmetry are known as *selection rules*. Our quantum well in an electric field provides a simple example. Its potential is symmetric in x about its midpoint, so the states are alternately symmetric and antisymmetric (even or odd in x). The electric field gives a perturbation $V(x) = eFx$, so the matrix elements take the form

$$V_{mn} = eF \int \phi_m^*(x)\, x\, \phi_n(x)\, dx. \qquad (7.33)$$

The complex conjugate can be dropped because the functions are real. Now x is an odd function so the product $\phi_m \phi_n$ must also be odd if the overall integrand is to be even and give a non-zero result. Thus one of the states must be even and the other odd, a simple example of a selection rule. We shall encounter these again in optical spectra (Section 8.7). In particular, the diagonal matrix elements ($n = m$) vanish.

(iv) The first-order change in energy V_{nn} may be of either sign. However, it often vanishes by symmetry and we must use the second term. In the case of the lowest state, which is often of interest, all the energy denominators $\varepsilon_n - \varepsilon_k$ are negative. The numerators are of course positive, so the second-order change to the energy of the lowest state is always negative. The perturbation mixes other states into the wave function, and the system can always use this freedom to lower its energy.

7.2.2 QUANTUM WELL IN AN ELECTRIC FIELD

We can now solve the problem posed at the beginning of this section, the quantum well in an electric field. Take for simplicity an infinitely deep well of width a centred on the origin, so $\varepsilon_n = (\hbar^2/2m)(n\pi/a)^2$ and the ϕ_n are alternately cosine and sine waves. We will need the matrix elements V_{k1} for the change in energy of the lowest state, where the perturbation due to the electric field is $\hat{V} = eFx$. As we saw before, the symmetry of the wave functions and perturbation can be used to show that the diagonal elements V_{kk} all vanish and that V_{k1} exists only for even k. Thus the electric field couples the even ground state only to those higher states that have odd symmetry (even k), and there is no change in energy to first order. This is obvious physically because the change in energy should not depend on the sign of F, so a quadratic term is the lowest to appear. To second order we must evaluate

$$\Delta E_1 = -\sum_{k=1}^{\infty} \frac{|V_{2k,1}|^2}{\varepsilon_{2k} - \varepsilon_1}, \qquad (7.34)$$

where only the non-zero terms are retained, and it is shown explicitly that the energy is lowered.

For a lower bound on the shift, calculate the first term alone ($k = 1$). This needs the dipole matrix element (equation E1.4)

$$V_{21} = \frac{2}{a} \int_{-a/2}^{a/2} \sin\left(\frac{2\pi x}{a}\right)(eFx)\cos\left(\frac{\pi x}{a}\right) dx = \frac{16}{9\pi^2}(eFa). \qquad (7.35)$$

The energy denominator is $\varepsilon_2 - \varepsilon_1 = 3\varepsilon_1$, so our bound on the shift in energy is

$$-\Delta E_1 > \frac{256}{243\pi^4} \frac{(eFa)^2}{\varepsilon_1}. \tag{7.36}$$

This can be put in words as

$$\left(\frac{-\text{change in energy}}{\text{difference in energy}}\right) \approx \frac{1}{10}\left(\frac{\text{energy drop across well}}{\text{difference in energy}}\right)^2. \tag{7.37}$$

This shows the energies that appear naturally in the problem. The perturbation is the electric field, so its matrix element must be proportional to the voltage drop across the active region. The electron starts in the middle and stays within the well, so we might guess $V_{21} \approx \frac{1}{4}(eFa)$. The denominator is the difference in energy between the state of interest and the nearest to which it couples. An estimate using these values differs from equation (7.36) only by a factor of 2. In most cases it is straightforward to estimate the energies that enter by perturbation theory in this way, although one still has to evaluate the integrals to find the absolute magnitude.

The change in energy can also be written in terms of the polarizability α. The quantum well is immersed in a semiconductor of dielectric constant ϵ_b, so it is consistent to define α by $\Delta E_1 = -\frac{1}{2}\epsilon_0\epsilon_b\alpha F^2$. Thus

$$\alpha = \frac{512\,e^2}{243\,\pi^4\epsilon_0\epsilon_b} \frac{a^2}{\varepsilon_1} = \frac{4096}{243\,\pi^5} \frac{a^4}{a_B}. \tag{7.38}$$

The Bohr radius a_B (equation 4.67) has been introduced into the second expression for α because it is the natural unit of length in a semiconductor and conveniently absorbs all the parameters such as the effective mass.

We have considered only the effect of the nearest state on the shift in energy and should worry about the contribution from the others. It happens in this simple case that the sum over all states can be evaluated analytically. It changes the prefactor in equation (7.36) from $256/243\,\pi^4 \approx 0.010\,815$ to $(15 - \pi^2)/48\pi^2 \approx 0.010\,829$. The contribution from higher states is small and this is common. The difference in energies suppresses the effect of higher states through the denominator, as $1/k^2$ here, and it also happens in this case that the matrix elements fall off as $1/k^3$. Thus the series converges very rapidly and the first term alone provides an excellent estimate. Unfortunately this may not be the case in a finite quantum well because the higher states are not bound.

The effect of an electric field on optical absorption in a quantum well is shown in Figure 7.2. The energy of both electrons and holes is reduced (in the appropriate sense) by the electric field. This shift of the absorption line is known as the *quantum-confined Stark effect*. It is clearly related to the Franz–Keldysh effect (Section 6.2.1), which was the change in the absorption edge for free rather than confined electrons. The quantum-confined Stark effect blends into the Franz–Keldysh effect as the

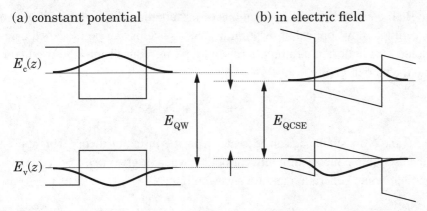

FIGURE 7.2. (a) A quantum well with flat bands, showing the energy E_{QW} for absorption between the bound states in the well. (b) Bands tilted by an applied electric field, which lowers the energy of both bound states and reduces the absorption energy to E_{QCSE}. The band gap has been reduced for clarity.

width of the well is increased and the separation between the bound states falls. This also happens as the electric length (equation 4.39) becomes smaller than the width of the well and provides the more important confinement. Figure 7.3 shows a measurement of the quantum-confined Stark effect, compared with a theory similar to that just developed.

The quantum-confined Stark effect can be used to construct electro-optic devices, using either the change in absorption directly or the associated change in refractive index demanded by the Kramers–Kronig relations (Section 10.1.1). A full calculation requires another important ingredient, the effect of the electric field on excitons, to be discussed in Section 10.7.4.

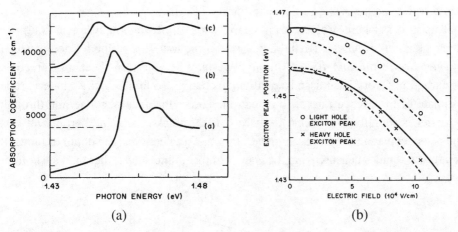

FIGURE 7.3. (a) Absorption spectra of a multiquantum well as a function of normal electric field. The GaAs wells were 9.5 nm wide, separated by 9.8 nm barriers of $Al_{0.32}Ga_{0.68}As$. The fields were (a) 1.0, (b) 4.7, and (c) 7.3 MV m^{-1}. The two peaks on each curve are due to the light and heavy holes. (b) Position of the peaks in energy as a function of the electric field; the lines are theoretical estimates. [From Miller et al. (1985).]

We shall encounter many applications of perturbation theory later. To give one more example now, optical absorption in many systems can be modelled using a harmonic oscillator. Anharmonic terms may be introduced into the oscillator to extend the model, giving

$$V(x) = \tfrac{1}{2}Kx^2 + Bx^3 + Cx^4 + \cdots, \tag{7.39}$$

where B and C are small and can be treated using perturbation theory. This approach is often used in the theory of nonlinear optics, where it gives rise to effects such as second-harmonic generation and four-wave mixing.

7.2.3 A CAUTIONARY TALE

Perturbation theory usually works well and gives sensible answers to many problems. However, it is well to be aware that it can go wrong. Look at a quantum well in an electric field again, this time considering a larger region of the semiconductor (Figure 7.4). The electric field tilts the barriers such that one gets higher but the other now has only a finite thickness above the energy of the bound state in the well. Thus an electron in the 'bound' state can tunnel out, like the resonant state in a double-barrier structure (Section 5.5). The shift in energy that we have calculated appears to be nonsense because the electron is no longer bound! Fortunately the lifetime of an electron in the well may be very long, because the barrier will remain thick provided that the field is not too large. Optical experiments will not be affected provided that electrons and holes recombine before tunnelling out of the wells. Only in the highest electric fields does the resonant nature of the state become important and broaden the absorption line, as in Figure 7.3(a).

This process where an electron is able to escape from a previously bound state after the application of an electric field, because of the lowering of the barrier, is called *Fowler–Nordheim tunnelling*. Although we may be able to neglect it in practice, perturbation theory has missed a qualitative change. To find out why, make a crude estimate of the escape rate of the electron. The probability of tunnelling through a barrier is roughly $\exp(-2\kappa L)$, where L is the thickness of the barrier and κ is the decay constant. Here we can put $L \approx V_0/eF$, where V_0 is the depth of the well (really we should measure from the energy of the bound state rather than the bottom

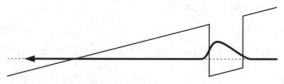

FIGURE 7.4. Quantum well after applying an electric field, showing how the previously bound state can now tunnel out.

of the well). The decay rate κ is not constant, but we can use its value halfway through the barrier as a rough estimate, so $\hbar^2\kappa^2/2m \approx \frac{1}{2}V_0$. These values give

$$T \approx \exp\left[-\frac{(4mV_0^3)^{1/2}}{eF\hbar}\right]. \tag{7.40}$$

This vanishes as the field goes to zero, and the important mathematical feature is that it does so in a non-analytic way ($T(F)$ has an essential singularity at $F = 0$). Thus it is impossible to expand T as a power series in the perturbation F, which was the assumption made to set up our perturbation theory (equation 7.12). Such a theory will therefore never reveal tunnelling out of the well. Zener tunnelling (Section 2.2) is a similar problem.

The moral of this is that perturbation theory should be used only to calculate numbers for an effect that is already well understood. Although this is almost always the case in practice, it is well to bear in mind that non-perturbative effects like Fowler–Nordheim tunnelling exist.

7.3 k · p Theory

We know that most processes in a semiconductor occur near the top of the valence band and the bottom of the conduction band. Although we have happily used parabolas for the bands in this region, full calculations (Section 2.6) showed that this is only a rough approximation for the conduction band and worse for the valence band, where the light and heavy holes are degenerate at Γ. It would be desirable to have a more accurate description of the bands in these important regions without having to resort to numerical methods. The **k · p** method and its extensions address this aim.

Recall that Bloch's theorem states that the wave function in a crystal can be written as the product $\phi_{n\mathbf{K}}(\mathbf{R}) = u_{n\mathbf{K}}(\mathbf{R})\exp(i\mathbf{K} \cdot \mathbf{R})$, where $u_{n\mathbf{K}}(\mathbf{R})$ is a periodic function (Section 2.1). We also argued, in deriving the effective-mass approximation, that it might be a good approximation to assume that $u_{n\mathbf{K}}(\mathbf{R})$ was constant over a small region of **K**-space (equation 3.5). Without going so far, we can take this as a hint that it might be easier to find approximate solutions for the slowly varying function $u_{n\mathbf{K}}(\mathbf{R})$ than for $\phi_{n\mathbf{K}}(\mathbf{R})$.

Let the Schrödinger equation for the Bloch functions of the crystal be

$$\left[\frac{\hat{\mathbf{p}}^2}{2m_0} + V_{\text{per}}(\mathbf{R})\right]\phi_{n\mathbf{K}}(\mathbf{R}) = \varepsilon_n(\mathbf{K})\phi_{n\mathbf{K}}(\mathbf{R}), \tag{7.41}$$

where the momentum operator $\hat{\mathbf{p}} = -i\hbar\nabla$ as usual and m_0 is the mass of free electrons. Substitute $\phi_{n\mathbf{K}}(\mathbf{R}) = u_{n\mathbf{K}}(\mathbf{R})\exp(i\mathbf{K} \cdot \mathbf{R})$ into this. The derivatives of $\hat{\mathbf{p}}$

acting on the plane wave simply give $\hbar \mathbf{K}$ and the wave then cancels out. This leaves an equation for the periodic part $u_{n\mathbf{K}}(\mathbf{R})$ alone:

$$\left\{ \left[\frac{\hat{\mathbf{p}}^2}{2m_0} + V_{\text{per}}(\mathbf{R}) \right] + \left[\frac{\hbar}{m_0} \mathbf{K} \cdot \hat{\mathbf{p}} + \frac{\hbar^2 K^2}{2m_0} \right] \right\} u_{n\mathbf{K}}(\mathbf{R}) = \varepsilon_n(\mathbf{K}) u_{n\mathbf{K}}(\mathbf{R}). \quad (7.42)$$

Suppose that we have solved this at $\mathbf{K} = \mathbf{0}$, and know the set of wave functions $u_{n0}(\mathbf{R})$ and energies $\varepsilon_n(\mathbf{0})$. General theory tells us that these form a complete set of functions. We can therefore use them as a basis in which to expand the solutions at some other value of $\mathbf{K}$, giving a matrix equation. Alternatively, we can use perturbation theory, as we are mainly concerned with small values of $|\mathbf{K}|$. The terms that depend on $\mathbf{K}$ in equation (7.42) are viewed as a perturbation away from the solution at $\mathbf{K} = \mathbf{0}$. One is simply the change in energy of a free electron, while the other contains the operator $\mathbf{K} \cdot \hat{\mathbf{p}}$, which gives the method its name. Remember that $\mathbf{K}$ is the Bloch wave vector, which we are treating as the perturbation, while $\hat{\mathbf{p}}$ is the momentum operator. Thus $\mathbf{K} \cdot \hat{\mathbf{p}} = k_x(-i\hbar \, \partial/\partial x) +$ terms in y and z.

Focus on a particular band n, which we shall assume not to be degenerate with any other band at $\mathbf{K} = \mathbf{0}$. It is easy to show that diagonal matrix elements such as $\langle n\mathbf{0}|\mathbf{K} \cdot \hat{\mathbf{p}}|n\mathbf{0}\rangle$ vanish. Second-order perturbation theory gives

$$\varepsilon_n(\mathbf{K}) \approx \varepsilon_n(\mathbf{0}) + \frac{\hbar^2 K^2}{2m_0} + \frac{\hbar^2}{m_0^2} \sum_{m, m \neq n} \frac{|\langle m\mathbf{0}|\mathbf{K} \cdot \hat{\mathbf{p}}|n\mathbf{0}\rangle|^2}{\varepsilon_n(\mathbf{0}) - \varepsilon_m(\mathbf{0})}. \quad (7.43)$$

Although this appears rather cumbersome, it is quadratic in $\mathbf{K}$ and can always be diagonalized to give the forms of conduction band that we saw in Section 2.6.4. It reduces to a scalar for the Γ-valley of GaAs with $\varepsilon_c(\mathbf{K}) \approx E_c + \hbar^2 K^2/2m_0 m_e$.

We can get some idea of the shapes of the bands from the form of equation (7.43), even without knowing the exact wave functions. Consider the Γ-valley in the conduction band of GaAs. The largest contributions should come from bands close in energy. The top of the valence band provides the nearest states at Γ (although the next-highest conduction bands are not far away). The matrix elements do not vanish because the conduction band is symmetric (s-like), the valence band is antisymmetric (p-like), and the operator $\hat{\mathbf{p}}$ is also antisymmetric. The energy denominator is positive because the states that we are mixing in are lower in energy than the original state (unlike the case of the ground state that we studied in Section 7.2.2). Thus the correction increases the energy of the electron as a function of $\mathbf{K}$, which decreases its effective mass.

Take $\mathbf{K}$ along x to make a rough estimate of its magnitude. The sum is over the states at the top of the valence band, which arise from the three p orbitals (Section 2.6.3). The matrix element is conventionally written as $\langle S|\hat{p}_x|X \rangle = (im_0/\hbar)P$ (note that Bastard (1988) omits the $\hbar$). It contains a conduction-band state S built out of s orbitals and the valence-band state X, which is built out of p_x orbitals. This matrix element appears in other properties such as optical absorption. The other two matrix elements such as $\langle S|\hat{p}_x|Y\rangle$ vanish by symmetry. Equation (7.43) shows the the energy of the conduction band is

$$\varepsilon_{\mathrm{c}}(\mathbf{K}) \approx E_{\mathrm{c}} + \frac{\hbar^2 K^2}{2m_0} + \frac{\hbar^2}{m_0^2} \frac{|K(im_0/\hbar)P|^2}{E_{\mathrm{c}} - E_{\mathrm{v}}} = E_{\mathrm{c}} + \frac{\hbar^2 K^2}{2m_0}\left(1 + \frac{2m_0 P^2}{\hbar^2 E_{\mathrm{g}}}\right).$$
(7.44)

The effective mass is given by the expression in parentheses and can be written as $1/m_{\mathrm{e}} \approx 1 + E_P/E_{\mathrm{g}}$ where E_{g} is the band gap at Γ and $E_P = 2m_0 P^2/\hbar^2$.

Finally we need an estimate of E_P. The operator $\hat{\mathbf{p}}$ is a derivative, so roughly it picks out the wave vector of the states around the band gap. The edge of the Brillouin zone (X) is at $k_x = \pi/a$ and Figure 2.16 shows that the top of the valence band has been folded in once from the boundary, so the wave number in P is roughly $2\pi/a$. The lattice constant $a \approx 0.5$ nm giving $E_P \approx 22$ eV. In fact this value holds remarkably well for the common semiconductors (Appendix 2). The band gap $E_{\mathrm{g}} \approx 1.4$ eV for GaAs, so we predict $m_{\mathrm{e}} \approx 0.061$, an excellent result for so little effort.

The trend that light masses go with narrow gaps is also correct. Unfortunately this also limits the validity of the method, because a small effective mass means that the kinetic energy increases rapidly with k and soon becomes comparable with a narrow band gap. The change in energy due to the perturbation $\mathbf{K}$ is then larger than the separation of the unperturbed states and perturbation theory becomes inaccurate. A related approach due to Kane is better under these circumstances and will be discussed in Section 10.2.

The same matrix elements appear in the expression for the energies of the valence band, but the reversals of the energy denominators mean that they have the opposite effect. This coupling to the conduction band overcomes $\hbar^2 K^2/2m_0$, which tends to push the band upwards, and the valence band bends downwards to give the expected behaviour for holes. Unfortunately we can make no more than this general statement because the valence bands are degenerate at Γ and the theory cannot be applied directly. We shall return to this with the Kane model.

The $\mathbf{k} \cdot \mathbf{p}$ method can also be used at points away from Γ. In this case there will generally be a linear term in $\mathbf{K}$, which gives the group velocity of an electron in the band, as well as the quadratic terms that we have considered.

7.4 WKB Theory

Often one needs to solve the Schrödinger equation for a system where the potential $V(x)$ varies 'slowly' in space. An example is the potential that confines electrons in a wire defined by split gates on the surface of a heterostructure (Figure 3.17(c)). Detailed calculations show that this potential is roughly parabolic at the edges with a flat section in the middle as shown in Figure 7.5, rather like the cross-section through

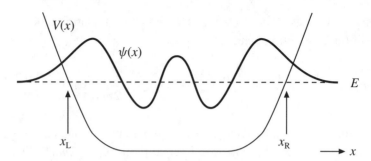

FIGURE 7.5. Simplified version of the potential that confines electrons in a quantum wire ('bathtub potential'), which can be analyzed using the WKB method. The classical turning points x_L and x_R occur where $V(x) = E$.

a bathtub. The width of the potential well at the Fermi energy might be $0.2\,\mu m$ or more, large compared with the Fermi wavelength of $0.05\,\mu m$. Classically a particle of energy E would bounce backwards and forwards between the two *classical turning points* x_L and x_R where $V(x) = E$. In quantum mechanics the particle can tunnel into the barrier and the restriction of its motion leads to quantization of its energy.

Classically the kinetic energy at a point x is $E - V(x)$ and varies slowly in space. If it did not vary at all, the quantum-mechanical solution would be just a plane wave $\exp(ikx)$ with $k = \sqrt{2m(E - V)}/\hbar$. An obvious guess for a system where $V(x)$ varies slowly would be to let the wave number vary with the local kinetic energy, $k(x) = \sqrt{2m[E - V(x)]}/\hbar$. The phase of the plane wave would also have to change from the simple product kx to an integral $\int k(x)\,dx$ to allow for the variation in k. This is the basis of the WKB method, named after Wentzel, Kramers, and Brillouin. It is also associated with Jeffreys and Rayleigh, and goes by the name of the quasi-classical approximation in the Russian literature. We shall now put these ideas on a firmer footing and test the method on a triangular well.

7.4.1 GENERAL THEORY

We wish to solve the usual one-dimensional Schrödinger equation,

$$\left[-\frac{\hbar^2}{2m}\frac{d^2}{dx^2} + V(x) \right] \psi(x) = E\psi(x), \tag{7.45}$$

where $V(x)$ is slowly varying in a sense that we shall need to make precise. Our preliminary argument shows that the phase of the wave function is likely to be the crucial part to get right, so rewrite $\psi(x) = \exp[i\chi(x)]$. Substituting this into the Schrödinger equation yields an equation for the phase $\chi(x)$:

$$[\chi'(x)]^2 - i\chi''(x) = \frac{2m}{\hbar^2}[E - V(x)] \equiv k^2(x), \tag{7.46}$$

where the local wave number $k(x)$ has been introduced in favour of $V(x)$. This is exact, and approximations must be introduced to make further progress. The second derivative $\chi''(x)$ should be small because the potential and wave number are supposed to vary slowly. Neglecting it gives

$$\chi(x) = \pm \int^x k(x')\,dx', \tag{7.47}$$

which is exactly the form of the phase that we postulated. It will be accurate if $|\chi''(x)| \ll [\chi'(x)]^2$. Using the approximation $\chi' = k$, this becomes

$$\left|\frac{dk}{dx}\right| \ll k^2, \qquad \left|\frac{1}{k}\frac{dk}{dx}\right| \ll |k|. \tag{7.48}$$

Now the wavelength is $2\pi/k$, so this inequality requires that the change in k per wavelength be much less than k itself. This is a reasonable definition of a slowly varying system. Unfortunately disaster strikes near the classical turning points because $V(x)$ approaches E there, k drops to zero, and the wavelength goes to infinity, so the inequality cannot possibly be satisfied. We shall therefore have to patch the WKB solution around the turning points.

Before doing this, it is useful to go one step further in the WKB solution. Our equation for $\chi(x)$ was

$$[\chi'(x)]^2 = k^2(x) + i\chi''(x) \approx k^2(x) \pm ik'(x), \tag{7.49}$$

where the first approximation for χ has been used in the second derivative. Taking the square root and making a binomial expansion gives

$$\chi'(x) \approx \pm k(x)\sqrt{1 + \frac{ik'(x)}{k^2(x)}} \approx \pm k(x) + \frac{ik'(x)}{2k(x)}, \tag{7.50}$$

which can be integrated to find

$$\chi(x) = \pm \int k(x)\,dx + \frac{i}{2}\ln k(x). \tag{7.51}$$

Thus

$$\psi(x) \approx \frac{1}{\sqrt{k(x)}} \exp\left[\pm i \int^x k(x')\,dx'\right], \tag{7.52}$$

which is the form in which the WKB approximation is usually quoted. It has been derived for positive kinetic energies but can be extended to negative energies (tunnelling rather than propagating waves) by replacing $k(x)$ with $\kappa(x) = \sqrt{[V(x) - E]}/\hbar$ and removing the i in the exponential. For bound states the complex exponential is replaced by a sine or cosine.

The prefactor of $1/\sqrt{k(x)}$ helps to conserve current for propagating states. For a genuine plane wave $A \exp(ikx)$ we know that $J = (\hbar k/m)|A|^2$; in WKB the

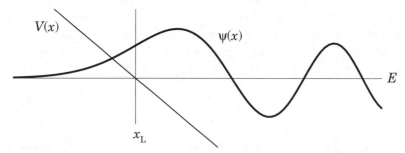

FIGURE 7.6. Matching the WKB wave functions on either side of a classical turning point x_L.

prefactor cancels the change in current that would be caused by a change in k. It ensures that a particle has a lower density (spends less time) in regions where it is moving faster. The way in which it conserves current also draws attention to a weakness of the WKB approximation: it ignores reflections. It is assumed that the particle can follow a change in potential purely by changing its wave number, whereas a (small) part of the wave is always reflected.

One important use of the WKB method is to estimate the rate of tunnelling through barriers. We want the probability rather than the amplitude so the wave function must be squared, and the WKB estimate is therefore

$$T \approx \exp\left[-2\int_{x_L}^{x_R} \kappa(x)\,dx\right]. \tag{7.53}$$

The points x_L and x_R are the edges of the barrier, defined by $V(x) = E$. This is an obvious generalization of $\exp(-2\kappa d)$. An example of this will be treated in Section 7.4.3.

We must address the problem of turning points if we wish to apply WKB to bound states. Consider the situation at x_L shown in Figure 7.6 where a classical particle cannot penetrate the barrier to the left. The WKB method is not valid in this region because the wavelength goes to infinity and the changes in $k(x)$ cannot be considered small. A rigorous approach, which it is not appropriate to repeat here, avoids the turning points by making an excursion into the complex plane. Another way of looking at this is to note that the potential is linear for a small region around the turning point, and we know that Airy functions are solutions to the wave equation in such a potential. We can therefore imagine patching our WKB solutions for $x \ll x_L$ and $x \gg x_L$ onto an Airy function that fills the region between. The problem sounds complicated but fortunately the results are simple:

$$\psi(x) \sim \frac{2}{\sqrt{k(x)}} \cos\left[\int_{x_L}^{x} k(x')\,dx' - \frac{\pi}{4}\right] \quad (x \gg x_L), \tag{7.54}$$

$$\psi(x) \sim \frac{1}{\sqrt{\kappa(x)}} \exp\left[-\int_{x}^{x_L} \kappa(x')\,dx'\right] \quad (x \ll x_L). \tag{7.55}$$

The important features are the factor of 2 and the phase of $-\frac{1}{4}\pi$. It is easy to remember these from the sketch of the wave function in Figure 7.6. The oscillating wave clearly has a larger amplitude than the exponential extrapolated back to $x = x_L$; we just have to remember that it is a factor of 2. Similarly, the cosine has to start with an upward slope in order to match to the decaying exponential. This means that the phase of the cosine must be between $-\frac{1}{2}\pi$ and 0; it is right in the middle.

The constant in the phase is important in determining bound states, which is the second common application of WKB. The boundary condition in a well with infinitely steep walls is that the wave function goes to zero at the boundaries, so an exact number of half-wavelengths must fit between them. Thus the change in phase between x_L and x_R must be a multiple of π, and the condition for quantized states is

$$\int_{x_L}^{x_R} k(x)\,dx = n\pi, \qquad n = 1, 2, 3, \dots. \tag{7.56}$$

Note that x_L, x_R, and $k(x)$ are all functions of energy.

On the other hand, consider a well with soft walls such as the bathtub in Figure 7.5. An allowed state must now obey the matching condition (7.54) at each end, which requires a phase of $\pm\frac{1}{4}\pi$. Now the condition for quantization is

$$\int_{x_L}^{x_R} k(x)\,dx = (n - \tfrac{1}{2})\pi, \qquad n = 1, 2, 3, \dots. \tag{7.57}$$

Obviously one gets $(n - \frac{1}{4})\pi$ when one wall is soft and the other is hard. In general the precision increases with n as the influence of the turning points, where WKB is least accurate, decreases.

7.4.2 BOUND STATES IN A TRIANGULAR WELL

We shall first use the WKB method to estimate the energies of the bound states in a triangular well (Figure 4.6). This has $V(x) = eFx$ for $x > 0$ and a hard wall at $x = 0$. We solved this exactly in Section 4.4 but the energies depend on the zeros of $\mathrm{Ai}(x)$, which must be found numerically, so a simple analytic approximation would be desirable. The left-hand turning point is $x_L = 0$, a hard boundary, while the right-hand boundary is soft and given by $V(x_R) = E$ or $x_R = E/eF$. The condition for quantization is

$$(n - \tfrac{1}{4})\pi = \int_{x_L}^{x_R} k(x)\,dx = \int_0^{E/eF} \left[\frac{2m}{\hbar^2}(E - eFx)\right]^{1/2} dx$$

$$= \left[\frac{2mE}{\hbar^2}\right]^{1/2} \frac{E}{eF} \int_0^1 \sqrt{1-s}\,ds, \tag{7.58}$$

TABLE 7.1 A comparison of various approximate methods for energy levels in a triangular potential, in units of $\varepsilon_0 = [(eF\hbar)^2/(2m)]^{1/3}$, and the exact results from the Airy function.

n	Airy function (exact)	WKB	Variational (Fang–Howard)	Variational (Gaussian)
1	2.3381	2.3203	2.4764	2.3448
2	4.0879	4.0818		
3	5.5206	5.5172		
$\vdots$	$\vdots$	$\vdots$		
10	12.8288	12.8281		

where $s = x/x_R$. The integral is trivial and we get

$$\varepsilon_n \approx [\tfrac{3}{2}\pi(n - \tfrac{1}{4})]^{2/3}\left[\frac{(eF\hbar)^2}{2m}\right]^{1/3}. \tag{7.59}$$

A few values are shown in Table 7.1. The accuracy increases with n as expected but the error is less than 1% even for $n = 1$ so this is a remarkably good approximation.

Although the numerical results attest to the accuracy of the WKB approximation for this potential, we should check it using the inequality (7.48). This required $|k'(x)| \ll k^2(x)$. Now

$$\frac{dk}{dx} = \frac{1}{2k}\frac{dk^2}{dx} = \frac{1}{2k}\frac{2meF}{\hbar^2} = \frac{1}{2kx_0^3}, \tag{7.60}$$

where x_0 is the length scale associated with a triangular potential (Section 4.4). Thus the condition for WKB to be valid is $k \gg (1/x_0)$. If we evaluate k at the midpoint of the motion, $x = \tfrac{1}{2}E/(eF)$, the inequality becomes $E \gg \varepsilon_0$, where $\varepsilon_0 = [(eF\hbar)^2/(2m)]^{1/3}$ is the energy scale of the linear potential. This is satisfied (at least with '>' instead of '$\gg$') even for the lowest state, so we confirm the accuracy of the method.

7.4.3 TUNNELLING THROUGH A SCHOTTKY BARRIER

As an application of the WKB method to tunnelling, consider the potential shown in Figure 7.7. This is the Schottky barrier that forms between n-GaAs and a metal. States at the metal–semiconductor interface pin the Fermi level E_F at an energy V_b below the conduction band E_c of the semiconductor at $x = 0$. The barrier $V_b \approx 0.7\,\text{eV}$ for GaAs and varies only slightly for different metals. The influence of the surface is lost for large x and the Fermi level sits close to

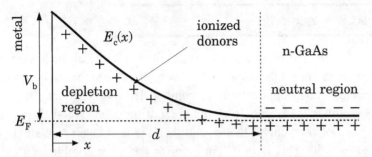

FIGURE 7.7. Schottky barrier in the conduction band $E_c(x)$ between a metal and n-GaAs. The potential is parabolic with height V_b and thickness d.

the edge of the conduction band in neutral, (heavily) n-doped material. Between lies a depletion layer with space charge density $+eN_D$ due to ionized donors. This provides the potential difference needed to restore $E_c(z)$ from $E_F + V_b$ to roughly E_F. Solution of Poisson's equation with this constant charge density shows that the band is parabolic, $V(x) = V_b[1 - (x/d)]^2$, where the thickness d is given by $V_b = e^2 N_D d^2 / 2\epsilon_0 \epsilon_b$. The Schottky barrier will be discussed further in Section 9.1.

We shall now estimate the transmission coefficient of this barrier. This is an important practical problem because tunnelling is not desirable if the metal is to act as a gate, which should be well insulated from the active region of the device. On the other hand, the metal might be an ohmic contact and in this case the resistance of the barrier should be as low as possible.

Consider electrons of low energy, $E \approx E_F$. The WKB estimate of the probability of tunnelling (equation 7.53) gives

$$T \approx \exp\left\{-2 \int_0^d \left[\frac{2mV_b}{\hbar^2}\left(1 - \frac{x}{d}\right)^2\right]^{1/2} dx\right\} = \exp\left[-\left(\frac{2mV_b}{\hbar^2}\right)^{1/2} d\right].$$

$$(7.61)$$

The decay length $(\hbar^2/2mV_b)^{1/2} \approx 1$ nm in GaAs. Thus d must be very small, demanding very high doping, if the barrier is to be reasonably transparent. This partly explains the painful variability of ohmic contacts on GaAs, only too well known!

To calculate the current (Section 5.4) we must take into account the distribution in energy of the incoming electrons, a few of which have energies well above E_F. Some can pass right over the barrier and give rise to a thermionic current, whose magnitude is dominated by the Boltzmann factor $\exp(-V_b/k_B T)$. Even those with less high energies can tunnel through the barrier more readily than those near the bottom, so it takes considerably more work to calculate the current except at very low or high temperatures.

7.5 Variational Method

The variational method provides an estimate of the energy of the lowest state of a system. Although this might seem a rather specialized task, the method is important for its accuracy, for its applicability to complicated problems, and because many numerical methods are available to minimize a function. We shall go through a simple version of the general theory and then apply it again to a single electron in a triangular well. This is in preparation for the more complicated calculation in Section 9.3.3 of the energy levels and density of a 2DEG at a heterojunction containing many electrons, where a self-consistent method is needed to encompass both the Schrödinger and Poisson equations.

7.5.1 GENERAL THEORY

We want to find the lowest energy level of a system, which obeys

$$\hat{H}\phi_1 = \varepsilon_1 \phi_1 . \tag{7.62}$$

Multiply both sides from the left by ϕ_1^* and integrate, which gives

$$\int \phi_1^* \hat{H}\phi_1 = \int \phi_1^* \varepsilon_1 \phi_1 = \varepsilon_1 \int \phi_1^* \phi_1 . \tag{7.63}$$

Thus the energy of this state can be found from the quotient

$$\varepsilon_1 = \frac{\int \phi_1^* \hat{H}\phi_1}{\int \phi_1^* \phi_1}. \tag{7.64}$$

Of course, the denominator is often unity by normalization. The variational principle asserts that

$$\varepsilon_1 \leq \frac{\int \psi^* \hat{H}\psi}{\int \psi^* \psi}, \tag{7.65}$$

where ψ is *any* wave function that obeys the appropriate boundary conditions. Thus we can keep trying different functions ψ to make the energy as low as possible, knowing that we can never go below the true energy ε_1.

The proof is simple. Let ϕ_n and ε_n be the true eigenfunctions and eigenvalues of $\hat{H}$. The arbitrary wave function ψ can be expanded in terms of these as

$$\psi = \sum_n a_n \phi_n , \tag{7.66}$$

where a_n are unknown coefficients. Now substitute this into the variational expression (7.65). First, the denominator gives

$$\int \psi^* \psi = \int \left(\sum_m a_m \phi_m \right)^* \left(\sum_n a_n \phi_n \right)$$

$$= \sum_{m,n} a_m^* a_n \int \phi_m^* \phi_n = \sum_{m,n} a_m^* a_n \delta_{mn} = \sum_n |a_n|^2. \qquad (7.67)$$

The integral is simple because the states ϕ_n are orthonormal. The numerator is similar:

$$\int \psi^* \hat{H} \psi = \sum_{m,n} a_m^* a_n \int \phi_m^* \hat{H} \phi_n = \sum_{m,n} a_m^* a_n \int \phi_m^* \varepsilon_n \phi_n$$

$$= \sum_{m,n} a_m^* a_n \varepsilon_n \delta_{mn} = \sum_n \varepsilon_n |a_n|^2. \qquad (7.68)$$

Thus the variational expression becomes

$$\frac{\int \psi^* \hat{H} \psi}{\int \psi^* \psi} = \frac{\sum_n \varepsilon_n |a_n|^2}{\sum_n |a_n|^2}. \qquad (7.69)$$

The energy ε_1 is the lowest, so $\varepsilon_n \geq \varepsilon_1$ and the quotient satisfies the inequality

$$\frac{\int \psi^* \hat{H} \psi}{\int \psi^* \psi} = \frac{\sum_n \varepsilon_n |a_n|^2}{\sum_n |a_n|^2} \geq \frac{\sum_n \varepsilon_1 |a_n|^2}{\sum_n |a_n|^2} = \frac{\varepsilon_1 \sum_n |a_n|^2}{\sum_n |a_n|^2} = \varepsilon_1. \qquad (7.70)$$

This completes the proof of the variational principle (7.65).

The method is slightly more general than this because it can be used to find the lowest state with a given symmetry, not just the overall lowest state. For example, it can be used to find the lowest even and lowest odd states in a symmetric potential well by ensuring that the trial functions ψ have the correct symmetry.

The numerator of expression (7.69) shows one reason why the variational method is accurate. The coefficients are all squared, so first-order errors in the wave function, mixing in functions other than ϕ_1, lead only to second-order errors in the energy. Thus a reasonable estimate for the energy may emerge from a rather poor stab at the wave function, but the method works much better if the trial wave function is chosen well. In practice, two approaches are often used. One is to guess the form of the wave function but with a small number of adjustable parameters, which are then chosen to minimize the energy. We shall use this method in the next section for the triangular well. A more 'brute force' approach is to write ψ as a sum $\sum_n^N b_n \xi_n$ over a finite number N of functions ξ_n. The variational principle becomes

$$\varepsilon_1 \leq \frac{\sum_{m,n}^N b_m^* H_{mn} b_n}{\sum_{m,n}^N b_m^* S_{mn} b_n}, \qquad H_{mn} = \int \xi_m^* \hat{H} \xi_n, \qquad S_{mn} = \int \xi_m^* \xi_n. \qquad (7.71)$$

Note that the functions ξ_n need not be orthonormal; it is often better to choose functions whose matrix elements are easy to calculate. Efficient numerical methods are available to minimize such expressions and the number of functions can be raised until sufficient accuracy is reached.

7.5.2 BOUND STATE IN A TRIANGULAR WELL

As an example that can be done analytically, we shall again estimate the energy of the lowest state in a triangular well. The important step is the choice of a suitable wave function. We know that it should vanish at $x = 0$ and decay roughly like an exponential as $x \to \infty$. A simple choice that satisfies both criteria is

$$\psi(x) = x \exp(-\tfrac{1}{2}bx). \tag{7.72}$$

This is known as the *Fang–Howard wave function* in the theory of the 2DEG, as these authors first applied it to the inversion layer in a silicon MOSFET. The parameter b is unknown; rather than guess a value at the start we shall leave it free and obtain an estimate for the energy that will be a function of b. We can then find the minimum of this function and get the best estimate possible given the form of this wave function.

The denominator of the variational principle (7.65) requires the integral

$$\int |\psi|^2 dx = \int_0^\infty x^2 e^{-bx} dx = \frac{2}{b^3}. \tag{7.73}$$

The numerator is more complicated but all the integrals reduce to factorials:

$$\int \psi^* \hat{H} \psi = \int_0^\infty x\, e^{-bx/2} \left[-\frac{\hbar^2}{2m}\frac{d^2}{dx^2} + eFx \right] x\, e^{-bx/2}\, dx$$

$$= \int_0^\infty \left[\frac{\hbar^2}{2m} bx(1 - \tfrac{1}{4}bx) + eFx^3 \right] e^{-bx} dx$$

$$= \frac{\hbar^2}{4mb} + \frac{6eF}{b^4}. \tag{7.74}$$

Dividing the two expressions shows that

$$\varepsilon_1 \le \frac{\hbar^2 b^2}{8m} + \frac{3eF}{b}. \tag{7.75}$$

The best estimate of ε_1 is the minimum of the expression on the right (for positive b!), and elementary calculus gives

$$b^3 = 6\left(\frac{2meF}{\hbar^2}\right), \qquad \varepsilon_1 \le \left(\frac{243}{16}\right)^{1/3}\left[\frac{(eF\hbar)^2}{2m}\right]^{1/3}. \tag{7.76}$$

The prefactor is 2.4764, which is not far from the exact result 2.3381 although WKB happens to work better here. Results using the different methods are compared in Table 7.1. A more accurate estimate can be obtained with more parameters or by choosing a better functional form (see the exercises).

7.6 Degenerate Perturbation Theory

The perturbation theory in Section 7.2 fails if the state of interest happens to be degenerate with another state, because the coefficients that mix the unperturbed wave functions (equation 7.32) contain the difference in energy in their denominator and become infinite. A different strategy is needed. This prediction that mixing diverges suggests that the strongest effect of a perturbation will be to mix up the degenerate states. Thus we should start by focussing attention on these and calculating the smaller effect of non-degenerate states later if needed. By restricting attention to the degenerate states and ignoring the others we reduce the infinite-matrix form of the Schrödinger equation to a very small matrix, whose dimensions are given by the number of degenerate states. Often this can be solved exactly.

Another point of view leads to the same conclusion. As an example of a system with degenerate states, consider a two-dimensional particle in a box or square quantum dot, with side length a in the xy-plane. It is centred on the origin with an infinite potential outside. The energies of the unperturbed dot are

$$\varepsilon_{p,q} = \frac{\hbar^2 \pi^2}{2ma^2}(p^2 + q^2) \tag{7.77}$$

with $\varepsilon_{q,p} = \varepsilon_{p,q}$. The lowest state is non-degenerate but the next level is doubly degenerate with wave functions

$$\phi_{1,2} = \frac{2}{a}\cos\frac{\pi x}{a}\sin\frac{2\pi y}{a}, \qquad \phi_{2,1} = \frac{2}{a}\sin\frac{2\pi x}{a}\cos\frac{\pi y}{a}. \tag{7.78}$$

The density $|\phi|^2$ of these two wave functions is sketched in Figure 7.8(a) and (b). Although this is an obvious choice for the eigenstates of this energy, it is not the only one. Because the states are degenerate, any linear combination of them will also be a solution to the Schrödinger equation with the same energy. Another simple pair is given by $(\phi_{1,2} \pm \phi_{2,1})/\sqrt{2}$, whose density is plotted in Figure 7.8(c) and (d). There is no reason to prefer any specific choice in the unperturbed quantum dot.

This freedom to choose the states vanishes if we add a perturbation that breaks the square symmetry in a suitable way. Consider $V(x, y) = -Kxy$ with $K > 0$. This lowers the energy in the upper right and lower left quarters of the square and raises it in the other two. Such a potential could be produced by a positive bias on gates near the top right and bottom left corners or by elastic stress; errors in fabrication that led to a trapezoidal dot would have a similar effect. Now it *does*

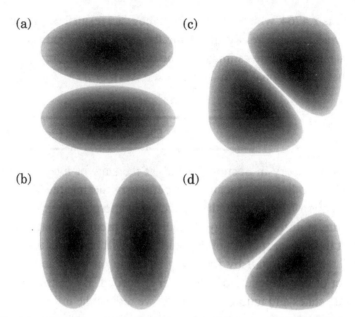

FIGURE 7.8. Two choices for the density of the lowest degenerate pair of wave functions in a square quantum dot. Figures (a) and (b) are the initial choices $\phi_{1,2}$ and $\phi_{2,1}$ while (c) and (d) are the combinations $(\phi_{1,2} \pm \phi_{2,1})/\sqrt{2}$.

matter what states we choose. The energies of the original choice, $\phi_{1,2}$ and $\phi_{2,1}$ are not changed by the perturbation because they are distributed equally over top and bottom, and left and right. The linear combinations are affected, however; $(\phi_{1,2} + \phi_{2,1})/\sqrt{2}$ is concentrated in the upper right and lower left so its energy falls due to the perturbation, while the energy of $(\phi_{1,2} - \phi_{2,1})/\sqrt{2}$ rises. The degeneracy has been broken because of the reduced symmetry, and it is important to choose states that respect the symmetry of the perturbation.

Although the answer is obvious in this case, we can confirm it by working through the algebra. The aim is to solve the Schrödinger equation exactly, including the original Hamiltonian $\hat{H}_0$ and the perturbation $\hat{V}$, but restricting attention to the degenerate states of which there are only two here. Put $\phi_A \equiv \phi_{1,2}$, $\phi_B \equiv \phi_{2,1}$, and $\varepsilon \equiv \varepsilon_{1,2}$ to reduce the number of subscripts. We seek a wave function of the form $a_A \phi_A + a_B \phi_B$ whose coefficients can be written as a vector $\mathbf{a}$ that obeys a matrix Schrödinger equation $\mathsf{H}\mathbf{a} = E\mathbf{a}$. The matrix H contains the matrix elements of the full Hamiltonian $\hat{H} = \hat{H}_0 + \hat{V}$ between the two states of interest. Its first element is

$$H_{AA} = \langle A|\hat{H}_0 + \hat{V}|A\rangle = \langle A|\hat{H}_0|A\rangle + \langle A|\hat{V}|A\rangle = \varepsilon + 0. \tag{7.79}$$

The ε appears because ϕ_A is an eigenstate of $\hat{H}_0$, and the matrix element of $\hat{V}$ vanishes by symmetry. Next we need

$$\langle A|\hat{H}_0 + \hat{V}|B\rangle = \varepsilon\langle A|B\rangle + \langle A|(-Kxy)|B\rangle = 0 - \left(\frac{16}{9\pi^2}\right)^2 Ka^2 \equiv -\Delta. \tag{7.80}$$

The term from $\hat{H}_0$ vanishes because A and B are eigenstates and orthogonal, while the matrix element of $\hat{V}$ involves integrals such as (7.35). The other two elements follow similarly. The matrix Schrödinger equation (7.6) becomes

$$\begin{pmatrix} \varepsilon & -\Delta \\ -\Delta & \varepsilon \end{pmatrix} \mathbf{a} = E\mathbf{a}, \tag{7.81}$$

and the condition (7.8) for solutions is

$$0 = \det |E - \mathsf{H}| = \det \begin{vmatrix} E - \varepsilon & \Delta \\ \Delta & E - \varepsilon \end{vmatrix} = (E - \varepsilon)^2 - \Delta^2. \tag{7.82}$$

Thus $E = \varepsilon \pm \Delta$. The eigenvector corresponding to $E = \varepsilon - \Delta$ is $(1, 1)$ as we expected, and similarly $(1, -1)$ corresponds to $E = \varepsilon + \Delta$. These are the states sketched in Figure 7.8(c) and (d). The change in energy is linear in the perturbation K or Δ, although the expectation value vanished in both the original states and we would have found a quadratic change if the states had not been degenerate.

There are not many direct applications of degenerate perturbation theory to low-dimensional systems. However, the main principle is widely used, that one can restrict the Schrödinger equation to a small number of states that lie close in energy and solve the restricted equation exactly. Often the states lying further away in energy are ignored completely, as we have just done. Occasionally it is necessary to include them by merging the techniques of degenerate and non-degenerate perturbation theory, and a systematic approach is due to Löwdin.

Two important applications of these ideas are to band structure in solids, the tight-binding and nearly free electron pictures.

7.7 Band Structure: Tight Binding

We looked exhaustively at the solution of the Schrödinger equation for a single finite potential well in Section 4.2, and used T-matrices to solve the electronic structure of a superlattice in Section 5.6. Although the solution of the superlattice was exact, it cannot be generalized simply to other potentials, nor to more than one dimension. The tight-binding model is an approximate picture of band structure based on the idea of starting with the energy levels of atoms and bringing them closer and closer together to form a crystal. It complements the nearly free electron model, which starts from the opposite viewpoint and will be described in the next section. We shall first solve the problem of two 'atoms' being brought close together, and then that of the crystal. The 'atoms' might really be atoms, but we shall treat potential wells in a heterostructure.

Start with a single potential well (or atom) centred on the origin and choose the zero of energy at the plateau, so the potential well has negative energy. The Hamiltonian can be written as $\hat{H} = \hat{T} + \hat{V}$, where $\hat{T} = -(\hbar^2/2m)(d^2/dx^2)$ is the kinetic energy operator and $\hat{V}$ is the potential energy. Let ε and ϕ be the energy and wave function of the lowest state in the well. We know how to find these for a square well, but might have to calculate them numerically for a more complicated profile.

7.7.1 TWO WELLS: DIATOMIC MOLECULE

Next consider the problem of two wells, shown in Figure 7.9. One is centred on x_L and the other on x_R, and the Hamiltonian can be written as $\hat{H} = \hat{T} + \hat{V}_L + \hat{V}_R$, where V_L and V_R are the left and right potential wells. Of course this can still be solved analytically for square wells, by explicitly matching the wave function across all the potential jumps or by using T-matrices. However, it seems a reasonable guess that the wave function of the lowest two states of the double well will consist almost entirely of a mixture of the lowest states in each of the two individual wells, ϕ_L and ϕ_R. In the spirit of degenerate perturbation theory we shall restrict attention to these two states and ignore the rest. The details are not quite the same as straightforward degenerate perturbation theory because ϕ_L and ϕ_R are solutions to *different* Schrödinger equations,

$$(\hat{T} + \hat{V}_L)\phi_L = \varepsilon \phi_L , \qquad (\hat{T} + \hat{V}_R)\phi_R = \varepsilon \phi_R . \tag{7.83}$$

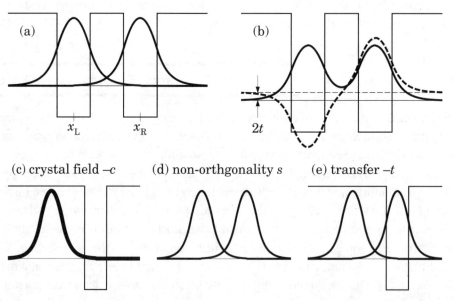

FIGURE 7.9. Two potential wells, analogous to a simple diatomic molecule. (a) Wave functions and energy levels of individual wells; (b) wave functions and energies of even and odd states of coupled wells, showing splitting of approximately $\pm t$; matrix elements that give rise to (c) crystal field $-c$ (the heavy line is a reminder that the wave function is squared), (d) non-orthogonality s, and (e) transfer $-t$.

The wave functions ϕ_L and ϕ_R are therefore not orthogonal, as is clear from Figure 7.9. However, the idea is exactly the same, to write the desired wave function ψ as a sum over the ϕs with coefficients to be found, $\psi = \sum_n a_n \phi_n$, where n runs over L and R.

The restricted Schrödinger equation is

$$\hat{H} \sum_n a_n \phi_n = E \sum_n a_n \phi_n . \tag{7.84}$$

As usual, multiply both sides on the left by ϕ_m^* and integrate to get matrices,

$$\sum_n H_{mn} a_n = E \sum_n S_{mn} a_n , \tag{7.85}$$

whose elements are

$$H_{mn} = \int \phi_m^* \hat{H} \phi_n , \quad S_{mn} = \int \phi_m^* \phi_n . \tag{7.86}$$

A new feature is the appearance of the matrix S. This is not just the unit matrix because the states are not orthogonal. The matrix equation to be solved is now $\mathsf{H}a = E\mathsf{S}a$, a generalized eigenvalue problem.

We need the matrix elements to continue, whose components are shown pictorially in Figure 7.9(c)–(e). Start with H_{LL}:

$$H_{LL} = \int \phi_L^* (\hat{T} + \hat{V}_L + \hat{V}_R) \phi_L \, dx = \varepsilon + \int \phi_L^* \hat{V}_R \phi_L \, dx \equiv \varepsilon - c. \tag{7.87}$$

The term with $\hat{T} + \hat{V}_L$ simply gives ε using the Schrödinger equation (7.83) for ϕ_L. The remaining term gives the expectation value of the added potential $\hat{V}_R$ for the wave function ϕ_L and is called the *crystal field*. Denote it as $-c$ to remind us that it is negative because the potential wells are attractive. The other diagonal element H_{RR} is the same. The two off-diagonal terms are also equal:

$$H_{RL} = \int \phi_R^* (\hat{T} + \hat{V}_L + \hat{V}_R) \phi_L \, dx = \varepsilon \int \phi_R^* \phi_L \, dx + \int \phi_R^* \hat{V}_R \phi_L \, dx = \varepsilon s - t. \tag{7.88}$$

The lack of orthogonality gives rise to the term s, which would otherwise have vanished. The other term $-t$ is the most important and is called the *transfer*, *tunnelling*, or *overlap integral*. It contains the product of the two wave functions and one of the potentials and 'transfers' an electron from one well to the other. This interpretation will become more clear when we derive the golden rule in the next chapter. It is also negative here although this depends on the wave functions. On the right-hand side of the equation, the diagonal elements of S are unity by normalization of the wave functions and the off-diagonal elements are both s. Thus

$$\mathsf{H} = \begin{pmatrix} \varepsilon - c & \varepsilon s - t \\ \varepsilon s - t & \varepsilon - c \end{pmatrix}, \quad \mathsf{S} = \begin{pmatrix} 1 & s \\ s & 1 \end{pmatrix}, \tag{7.89}$$

and the energies are given by the secular equation

$$\det |ES - H| = \det \begin{vmatrix} (E - \varepsilon) + c & (E - \varepsilon)s + t \\ (E - \varepsilon)s + t & (E - \varepsilon) + c \end{vmatrix} = 0. \qquad (7.90)$$

This has roots

$$E_- = \varepsilon - \frac{c}{1+s} - \frac{t}{1+s}, \qquad E_+ = \varepsilon - \frac{c}{1-s} + \frac{t}{1-s}. \qquad (7.91)$$

The approximation of neglecting all higher states will be good only if the overlap between the wells is weak. In this case the non-orthogonality factor $s \ll 1$ and the denominators can be expanded with the binomial theorem, giving $E \approx \varepsilon + (st - c) \pm t$. The tunnelling integral t splits the energy levels. Their average energy is shifted both by the crystal field and by a product involving non-orthogonality; both terms are usually of higher order in the tunnelling and are therefore neglected. It is easy to show that the eigenvector corresponding to the lower energy E_- (assuming $t > 0$) is $(1, 1)/\sqrt{2}$, which is the even combination $(\phi_L + \phi_R)/\sqrt{2}$; E_+ corresponds to the odd combination $(\phi_L - \phi_R)/\sqrt{2}$. In molecular terms, these would be *bonding* and *antibonding* molecular orbitals, shown in Figure 7.9(b). The state with higher energy has a node in the wave function between the atoms that increases its kinetic energy.

Splitting *always* occurs, however weak the tunnelling: a quantum state can always lower its energy by spreading out over a larger region of space, say, by using two wells rather than one. If an electron is initially in one of the wells, it will oscillate between them with an angular frequency of $|E_+ - E_-|/\hbar = 2|t|\hbar$ (Section 1.5). The period of oscillation will be very long if t is small and it would be difficult to tell that the wells were coupled.

7.7.2   ROW OF WELLS: TIGHT-BINDING SOLID

Having solved two wells, it is only a small step to solve the problem of an infinite number of wells arranged to form a superlattice or one-dimensional crystal. Bringing a pair of wells together caused the energy level to split into two levels separated by $2t$, and the splitting increases as the separation decreases and the overlap t grows. Bringing N wells together causes their common energy level to split into N values and these merge into a continuous band as $N \to \infty$. This is the tight-binding model of a solid.

The Hamiltonian is

$$\hat{H} = \hat{T} + \sum_n \hat{V}_n, \qquad (7.92)$$

where $\hat{V}_n$ is the potential of well (or ion) n. The orbital associated with well n obeys $(\hat{T} + \hat{V}_n)\phi_n = \varepsilon \phi_n$. The orbitals are the same except for their location, so $\phi_n = \phi(x - X_n)$, where X_n is the position of well n. Again we write the wave function of the crystal as a sum $\psi = \sum_n a_n \phi_n$, neglecting contributions from other orbitals. We will clearly end up with the same equation as for two wells, $\mathbf{Ha} = E\mathbf{Sa}$, except that the rank of the matrix is infinite rather than 2. Fortunately this is offset by the fact that we know the wave functions from Bloch's theorem (equation 2.2), which tells us that the coefficients a_n are simply phase factors. Thus

$$\psi_k = \sum_n a_n^k \phi_n, \qquad a_n^k = e^{ikx} = e^{ikna}, \tag{7.93}$$

and row m of $\mathbf{Ha} = E\mathbf{Sa}$ becomes

$$\sum_n H_{mn} e^{ikna} = E(k) \sum_n S_{mn} e^{ikna}. \tag{7.94}$$

This must be the same for every site so m should eventually drop out.

The matrix elements of the Hamiltonian are

$$H_{mn} = \int \phi_m^* \left[\hat{T} + V_n + \sum_{l, l \neq n} V_l \right] \phi_n = \varepsilon S_{mn} + \sum_{l, l \neq n} V_{mn}^l. \tag{7.95}$$

The term $l = n$ in the sum over potentials has been separated out so that we can use $(\hat{T} + V_n)\phi_n = \varepsilon \phi_n$. The remaining matrix elements V_{mn}^l are more complicated than those for the pair of wells because they can involve three sites, m and n for the two orbitals and l for the well. Fortunately most of these are very small and can be neglected. We shall make the simplest approximation and assume that the three indices must be restricted to two adjacent sites only, which leads to the same matrix elements as for the pair of wells. The diagonal element becomes

$$H_{mm} = \varepsilon S_{mm} + V_{mm}^{m-1} + V_{mm}^{m+1} = \varepsilon - 2c. \tag{7.96}$$

Here c is again the crystal field, the expectation value of the potential from well $m \pm 1$ in the orbital on site m. The only off-diagonal elements to survive are $H_{m, m \pm 1}$. Consider $H_{m, m+1}$. The well l can be only m or $m + 1$, but $l = m + 1$ has already been taken out of the sum to give ε. This leaves only $l = m$, which gives

$$H_{m, m+1} = \varepsilon S_{m, m+1} + V_{m, m+1}^m = \varepsilon s - t. \tag{7.97}$$

Again s is the non-orthogonality integral for wave functions on adjacent sites and t is the transfer integral.

We can now complete the sums required in equation (7.94). The left-hand side becomes

$$\sum_n H_{mn} e^{ikna} \approx (H_{mm} + H_{m, m+1} e^{ika} + H_{m, m-1} e^{-ika}) e^{ikma} \tag{7.98}$$

$$= [\varepsilon - 2c + 2(\varepsilon s - t) \cos ka] e^{ikma}. \tag{7.99}$$

Likewise the sum on the right-hand side gives

$$\sum_n S_{mn} e^{ikna} \approx (1 + 2s \cos ka) e^{ikma}. \tag{7.100}$$

The site m can now be cancelled from both sides, as asserted before. Then the energy is given by

$$E(k) = \frac{\sum_n H_{mn} e^{ikna}}{\sum_n S_{mn} e^{ikna}} = \varepsilon - 2 \frac{c + t \cos ka}{1 + 2s \cos ka} \approx \varepsilon - 2t \cos ka. \tag{7.101}$$

The non-orthogonality and crystal-field terms are usually dropped as in the final expression. We have now derived the cosine approximation for a narrow band (small t) that we have used several times already, in Section 2.2, for instance. The transfer integral gives the band its width of $4t$. The band is 'upside down' if $t < 0$, which would be the case for states in the well with $n = 2$ rather than $n = 1$.

The method can readily be extended to two or three dimensions and the phase factors in equation (7.98) become $\exp(i\mathbf{K} \cdot \mathbf{R}_j)$ summed over nearest neighbours $\mathbf{R}_j$. For example, a two-dimensional square lattice yields

$$E(\mathbf{k}) \approx \varepsilon - 2t(\cos k_x a + \cos k_y a). \tag{7.102}$$

The bandwidth is $4dt$ in d dimensions. We can also go beyond nearest neighbours to introduce higher Fourier components into $E(\mathbf{k})$.

The tight-binding model shows that the width of bands depends on the strength of tunnelling from one well (or atom) to another. Electrons that are tightly bound to their atoms give rise to narrow bands, whereas loosely bound electrons cause wide bands. The quantitative picture that we have used breaks down if the atoms come too close, because the bands broaden so much that they overlap and the bands cannot be treated separately. In practice the tight-binding picture is poor even for the upper valence bands of semiconductors unless next-nearest neighbours and beyond are included. However, it is often used as a way of parameterizing the bands. One can write down the tight-binding bands as a function of the tunnelling integrals between nearest neighbours, next-nearest neighbours, and perhaps beyond, and use these integrals as adjustable parameters to fit the bands to the results of a more complete calculation or to experiment. This gives a simple functional form for $E(\mathbf{K})$, which can then be used to calculate other quantities such as the optical response.

7.8 Band Structure: Nearly Free Electrons

The tight-binding model shows how band structure develops as atoms are brought together from far apart to form a closely spaced crystal. The nearly free electron model, as its name implies, takes the opposite point of view: we start with free electrons, add a weak periodic potential, and see how this leads to the formation of energy bands and gaps. This is closer in spirit to the description of band structure given in Chapter 2, and the nearly free electron approach is a better quantitative basis for describing the band structure of the common semiconductors.

Consider a one-dimensional crystal of period a and length L, which we shall send to infinity at the end. The unperturbed system of free electrons has energy $\varepsilon_0(k) = \hbar^2 k^2 / 2m_0$ and wave functions $\phi_k(x) = \exp(ikx)/\sqrt{L}$. Now add the periodic potential of the crystal as a perturbation. The periodicity of the potential in a crystal means that it can be expanded as a Fourier series,

$$V(x) = \sum_{n=-\infty}^{\infty} V_n \exp\left(\frac{2\pi i n x}{a}\right) \equiv \sum_n V_n \exp(i G_n x), \qquad (7.103)$$

where the $G_n = (2\pi/a)n$ are the reciprocal lattice 'vectors'. The potential $V(x)$ is real, which implies $V_{-n} = V_n^*$.

As the potential is supposed to be weak we can estimate its effect on the energies using the usual perturbation theory of Section 7.2. Thus

$$E(k) \approx \varepsilon_0(k) + V_{kk} + \sum_{k',k'\neq k} \frac{|V_{k'k}|^2}{\varepsilon_0(k) - \varepsilon_0(k')}. \qquad (7.104)$$

The matrix elements are

$$V_{k'k} = \int \phi_{k'}^*(x) V(x) \phi_k(x)\, dx = \sum_n V_n \frac{1}{L} \int_0^L e^{-ik'x} e^{i G_n x} e^{ikx} dx. \qquad (7.105)$$

The integral vanishes unless the total wave number is zero, which requires $k' = k + G_n$, and in this case the integral cancels the factor of $1/L$. The states involved are illustrated in Figure 7.10(a). A specific case is $G_n = 0$, which shows that $V_{kk} = V_0$ for all k. Now V_0 is just the average potential and shifts all states equally, so we shall drop it. Our perturbation expansion becomes

$$E(k) \approx \varepsilon_0(k) + \sum_{n,n\neq 0} \frac{|V_{k+G_n,k}|^2}{\varepsilon_0(k) - \varepsilon_0(k + G_n)}, \qquad (7.106)$$

and the corresponding wave function is

$$\psi_k = \phi_k + \sum_{n,n\neq 0} \frac{V_{k+G_n,k}\, \phi_{k+G_n}}{\varepsilon_0(k) - \varepsilon_0(k + G_n)} = e^{ikx}\left[1 + \sum_{n,n\neq 0} \frac{V_n e^{i G_n x}}{\varepsilon_0(k) - \varepsilon_0(k + G_n)}\right]. \qquad (7.107)$$

The function inside the square brackets is a Fourier series like in equation (7.103) and is consequently periodic. We have therefore 'proved' Bloch's theorem in the

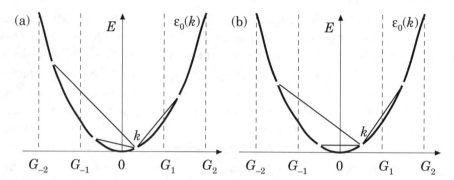

FIGURE 7.10. States mixed into k by the periodic potential of a one-dimensional crystal. Non-degenerate perturbation theory is good in (a) where all the states have different energies, but fails in (b) where one of the states has become degenerate and leads to the formation of a band gap.

form (2.2), which stated that the wave function in a crystal can be written in the form of a plane wave multiplied by a function with the period of the lattice. The wave function also satisfies the obvious check that it reverts to a simple plane wave in an empty lattice where $V(x) = 0$.

We must always check the energy denominators when using perturbation theory. Disaster strikes if the two energies become equal, that is, $\varepsilon_0(k + G_n) = \varepsilon_0(k)$. The only degeneracy in one dimension is $\varepsilon_0(-k) = \varepsilon_0(k)$, so the denominator vanishes when $k + G_n = -k$, or

$$k = -\tfrac{1}{2}G_n, \qquad k = \frac{n\pi}{a}, \qquad n = \pm 1, \pm 2, \dots. \tag{7.108}$$

Perturbation theory breaks down at these points in k-space *however weak the periodic potential* (Figure 7.10(b)). The failure arises from the symmetry of the lattice, as discussed in Section 2.1.1.

These values of k are clearly interesting so it is distressing that the theory fails at them. Consider the effect of a particular Fourier component G_n. Since the failure is the result of degeneracy between the states with wave numbers k and $k + G_n$, an obvious approach is to solve the Schrödinger equation exactly for these two states alone (treating the other states with perturbation theory if desired). This should be valid for k close to $-\tfrac{1}{2}G_n$. Thus $\psi \approx a_k \phi_k(x) + a_{k+G_n} \phi_{k+G_n}(x)$ and the only relevant Fourier components of the potential are V_n and $V_{-n} = V_n^*$. The Schrödinger equation becomes the 2×2 matrix

$$\begin{pmatrix} \varepsilon_0(k) & V_n^* \\ V_n & \varepsilon_0(k + G_n) \end{pmatrix} \begin{pmatrix} a_k \\ a_{k+G_n} \end{pmatrix} = E(k) \begin{pmatrix} a_k \\ a_{k+G_n} \end{pmatrix}. \tag{7.109}$$

Its eigenvalues satisfy

$$\det \begin{vmatrix} E(k) - \varepsilon_0(k) & -V_n^* \\ -V_n & E(k) - \varepsilon_0(k + G_n) \end{vmatrix} = 0, \tag{7.110}$$

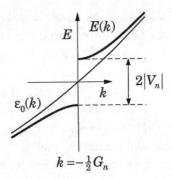

FIGURE 7.11. Expanded view of $E(k)$ near an energy gap (thick line) at $k = -\frac{1}{2}G_n$ according to nearly free electron theory. The thin curve is the energy of free electrons.

whence

$$E(k) = \frac{\varepsilon_0(k) + \varepsilon_0(k + G_n)}{2} \pm \sqrt{\left[\frac{\varepsilon_0(k) - \varepsilon_0(k + G_n)}{2}\right]^2 + |V_n|^2}. \qquad (7.111)$$

This is plotted in Figure 7.11. For k away from $-\frac{1}{2}G_n$, where $|\varepsilon_0(k) - \varepsilon_0(k + G_n)| \gg |V_n|$, this becomes

$$E(k) \approx \varepsilon_0(k) + \frac{|V_n|^2}{\varepsilon_0(k) - \varepsilon_0(k + G_n)} \qquad (7.112)$$

or the same with k and $k + G_n$ interchanged. This is the dominant correction from the ordinary perturbation expansion (7.106). Close to $k = -\frac{1}{2}G_n$ we have $|\varepsilon_0(k) - \varepsilon_0(k + G_n)| \ll |V_n|$ and

$$E(k) \approx \frac{\varepsilon_0(k) + \varepsilon_0(k + G_n)}{2} \pm \left\{ |V_n| + \frac{[\varepsilon_0(k) - \varepsilon_0(k + G_n)]^2}{8|V_n|} \right\}. \qquad (7.113)$$

In particular, $E(-\frac{1}{2}G_n) = \varepsilon_0(-\frac{1}{2}G_n) \pm |V_n|$, so we see that a gap of width $2|V_n|$ has opened up at $k = -\frac{1}{2}G_n$.

This is the most important result of nearly free electron theory. It demonstrates that band gaps occur at half the reciprocal lattice vectors, where states k and $k + G_n$ become degenerate, and shows that the width of the gap is twice the corresponding Fourier component of the lattice potential. The method can be extended to two or three dimensions with the results described in Section 2.4. Again the band gaps due to the Fourier component $\mathbf{G}$ occur at the degeneracy $\varepsilon_0(\mathbf{K}) = \varepsilon_0(\mathbf{K} + \mathbf{G})$, which now defines a plane normal to the vector joining $-\mathbf{G}$ to the origin and bisecting it. More than two waves may become degenerate and should be retained when calculating the gap in the corners of the Brillouin zone.

Suppose that we wish to apply nearly free electron theory to a semiconductor such as silicon. There are four valence electrons so they see an ionic core of charge $+4$. The Coulomb potential from this core is around 10 eV within the unit cell, and is not screened by the usual factor of ϵ_b because it is the potential seen by the valence electrons themselves. Clearly the periodic potential is large and it might therefore appear that nearly free electron theory is rather useless. Fortunately, it turns out that the real potential can be replaced by a much weaker *pseudopotential*, which is chosen so that it scatters the valence electrons in exactly the same way. Band structure can then be calculated using these weaker pseudopotentials. The theory is sophisticated and pseudopotentials can be calculated ab initio. In the simpler empirical pseudopotential method, a small number of Fourier components are adjusted so that a few critical features of the resulting band structure agree with experiment. Only three Fourier components are needed to get good agreement for silicon, showing that the nearly free electron approximation with pseudopotentials is remarkably good for the common semiconductors.

Further Reading

The methods described in this chapter are all standard and further details can be found in any book on quantum mechanics such as Merzbacher (1970), Gasiorowicz (1974), or Bransden and Joachain (1989). Similarly, the theory of band structure is covered in books on solid state physics, including Ashcroft and Mermin (1976), Kittel (1995), and Myers (1990). Yu and Cardona (1996) give a detailed discussion of the band structure of semiconductors.

Bastard (1988) and Weisbuch and Vinter (1991) both contain numerous examples of perturbation theory applied to low-dimensional systems.

Mathews and Walker (1970) give an illuminating account of the WKB method and the treatment of turning points. There are also some impressive (and difficult) examples of its use in the Russian literature; see, for example, Landau, Lifshitz, and Pitaevskii (1977).

Löwdin perturbation theory, a systematic way of dividing states into those nearby and far away in energy, is described well by Chuang (1995).

EXERCISES

7.1   An electron is in the lowest state of the GaAs quantum well shown in Figure 7.12(a). The total width is 15 nm, and the middle 5 nm is 100 meV deeper than the rest. Estimate the energy of the lowest state, assuming that the well is infinitely deep. How should the potential of the well be partitioned between $\hat{H}_0$ and $\hat{V}$, and does it have a significant effect? How would you treat a similar well with a barrier rather than a deeper well in the middle (Figure 7.12(b)), and when would you expect perturbation theory to be appropriate?

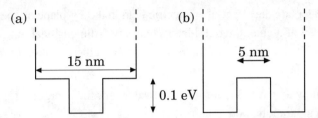

FIGURE 7.12. Stepped quantum wells, with (a) an inner deeper well and (b) a barrier.

7.2   Show that the polarizability of the lowest state in a quantum well tilted by an electric field, equation (7.38), can also be found by considering the change in wave functions and the resulting dipole moment $-e\langle x \rangle$.

7.3   Calculate the expected shift in the absorption lines seen in Figure 7.3. Use the approximation of infinitely deep wells and assume that the holes are either purely heavy or purely light. (The real wells have finite depth, which permits the electrons and holes to tunnel into the barriers, making the wave functions wider; this increases their polarizability, so the simplification should give too small a shift.)

7.4   Although most quantum wells have a flat bottom, other profiles can be grown. Three examples are shown in Figure 7.13. How would you expect the quantum-confined Stark effect to be changed in these systems? In particular, would the change in energy still be quadratic in the strength of the field?

7.5   What happens to the energy of the second state in an infinitely deep well when an electric field is applied?

7.6   A problem in Chapter 4 concerned the energy levels in a parabolic potential grown into $Al_xGa_{1-x}As$. Estimate the error of assuming that the parabolic potential continues upwards for ever rather than turning into a plateau in the barriers outside the well.

7.7   Calculate the change in energy of the lowest state in an infinitely deep quantum well due to a small magnetic field (Figure 6.13(a)). Take the transverse wave vector to be zero.

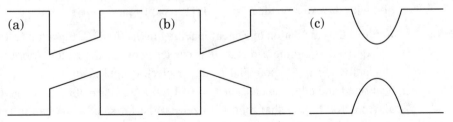

FIGURE 7.13. Modified profiles for observing the quantum-confined Stark effect: (a) built-in field; (b) tapered band gap; (c) parabolic wells.

7.8   Calculate the effect of an electric field in the xy-plane on the lowest energy
level of a quantum dot, idealized as a two-dimensional infinitely deep well
($|x|, |y| < \frac{1}{2}a$). Does the result depend on the direction of the field within
the plane?

7.9   Show that WKB gives the exact values for the energies of the bound states
in a parabolic potential.

7.10   Use WKB to estimate the energy of the lowest state in a symmetric triangular
potential well, with $V(x) = |eFx|$. The exact result can be found using Airy
functions.

7.11   A split gate on a GaAs heterostructure produces a parabolic potential in
the 2DEG whose energy levels are separated by $\varepsilon_p = 2\,\text{meV}$ when few
electrons are present. As more electrons enter, the parabola turns into a
'bathtub' with a flat region of width $w = 50\,\text{nm}$ (Figure 7.5). Assume that
the parabolic parts retain the same curvature. Use the WKB method to show
that the new energy levels are given by the solutions to

$$\frac{E}{\varepsilon_p} + \sqrt{\frac{E}{\varepsilon_w}} = (n - \tfrac{1}{2}), \tag{E7.1}$$

where $\varepsilon_w = \hbar^2 \pi^2 / 2mw^2$.

7.12   What value of N_D is needed for the tunnelling current to dominate the
thermionic current in a Schottky barrier on GaAs at room temperature?
Consider only the exponential terms. The distinction is important because
a barrier dominated by tunnelling has a roughly ohmic $I(V)$ characteristic,
whereas the thermionic current gives a diode.

7.13   Use the WKB method to estimate the escape rate from a quantum well in an
electric field due to Fowler–Nordheim tunnelling (Section 7.2.3). How large
a field can be applied before the escape of electrons or holes by tunnelling
limits the lifetime of an exciton (around 1 ps from phonon scattering)? Is
your estimate consistent with the experiment shown in Figure 7.3?

7.14   Estimate the transmission coefficient through a parabolic barrier $V(x) =
-\frac{1}{2}m\omega_0^2 x^2$ for $E < 0$ using the WKB method. Compare your approxima-
tion with the exact result $T(E) = 1/[1 + \exp(-2\pi E/\hbar\omega_0)]$.

7.15   Improve the calculation of Zener tunnelling in Section 2.2. Equation (2.16)
was derived assuming a rectangular barrier between the bands. More re-
alistically, the linear potential from the electric field gives $V(x) = eFx$
near one band edge, taken as at $x = 0$. Then $V(x) = eFx(1 - x/d)$ gives
a symmetric barrier that returns to zero at the far side of the gap in real
space $d = E_g/eF$. Use the WKB method to estimate the rate of tunnelling
through this barrier.

7.16   Repeat the variational calculation for the triangular well but with a Gaussian decay, $\psi(x) = x \exp[-\frac{1}{2}(bx)^2]$. Show that this gives a better estimate of the energy (because it is lower) with a prefactor of 2.3448.

In fact neither of these decays is correct; we know from the theory of Airy functions (Appendix 5) that the true decay contains $x^{3/2}$. You might like to try the variational calculation with $\psi(x) = x \exp[-\frac{1}{2}(bx)^{3/2}]$. The integrals give fractional factorials and you will need $\Gamma(\frac{5}{3}) \equiv \frac{2}{3}! = 0.902\,745$.

The result is disappointing, because the prefactor of 2.3472 is higher than that from the Gaussian decay. The reason is that the integral that gives the energy is dominated by the regions where the wave function is highest, which is around $x = 1/b$. Clearly the Gaussian wave function is more accurate in this region, and its incorrect decay is not significant because the wave function and the contribution to the integral for the energy are small there.

7.17   Estimate the energy of the lowest state in a symmetric triangular potential well, with $V(x) = |eFx|$, using the variational method. A Gaussian function $\psi(x) = \exp[-\frac{1}{2}(bx)^2]$ is an obvious guess for the wave function but you might like to try others such as sech (bx).

7.18   Use the variational method to estimate the polarizability of an electron in an infinitely deep one-dimensional well, calculated by perturbation theory in Section 7.2.2. An electric field breaks the symmetry of the even function $\phi_1(x)$ by mixing in an odd part, so a suitable variational function is $(1 + \lambda x)\phi_1(x)$ with an adjustable parameter λ. The energy of this can be minimized for the square well (many terms vanish by symmetry) and the quadratic term in the dependence on electric field gives the polarizability.

Alternatively, the calculation can be carried through for a general symmetric well, giving $\alpha = 16\pi \langle x^2 \rangle^2 / a_B$. This requires some manipulation of operators and commutators similar to that which we shall use for the f-sum rule in Section 10.1.3. For a particle in a box of width a, $\langle x^2 \rangle = a^2(1 - 6/\pi^2)/12$ and the result can be compared with equation (7.38).

7.19   The perturbation $-Kxy$ applied to the square quantum dot in Section 7.6 describes a saddle point with a particular orientation. How do the results depend on the orientation of the saddle point? Qualitatively, what would be the effect of the same perturbation on similar states in a circular quantum dot?

7.20   Extend the calculation of Section 7.7.1 to two wells of *unequal* depth, with eigenvalues $\varepsilon \pm \Delta$ when isolated. Ignore non-orthogonality and crystal-field terms. The levels of the isolated wells cross as Δ changes sign but show that the coupled wells *anticross*, as sketched in Figure 7.14, however weak the coupling.

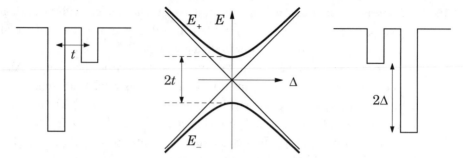

FIGURE 7.14. Anticrossing behaviour of two wells coupled by t (thick lines) as their relative depth $\pm\Delta$ is changed. The thin lines show the eigenvalues of the uncoupled wells.

7.21   Show that the results of the previous calculation agree with non-degenerate perturbation theory when $|\Delta| \gg |t|$.

7.22   In a superlattice of InAs and GaSb, the top of the valence band in GaSb lies *above* the bottom of the conduction band in InAs by $\Delta \approx 150\,\mathrm{meV}$ (Figure 3.5). Consider the effect of this on the band structure for k in the plane of the wells (normal to growth). If the wells were uncoupled we would have $\varepsilon_c(k) = E_c + \hbar^2 k^2/2m_0 m_e$ and $\varepsilon_v(k) = E_v - \hbar^2 k^2/2m_0 m_h$, where the apparent 'gap' $E_c - E_v$ is negative so the bands would overlap. Estimate the effect of coupling between the bands, supposing that this can be modelled by a constant term, by writing down and solving a simple Hamiltonian matrix. Sketch the bands and show that the anticrossing restores a positive energy gap. (The zero-point energy in the quantum wells should really be added to E_v and E_c).

7.23   Consider a one-dimensional 'molecule' whose wells are δ-functions of strength S separated by a distance a, giving $V(x) = -S[\delta(x + \frac{1}{2}a) + \delta(x - \frac{1}{2}a)]$. The pair of wells has an even and an odd bound state. We know from Section 4.2 that the wave function of a single such well at the origin is proportional to $\exp(-\kappa|x|)$ with $\kappa = mS/\hbar^2$ and energy $E = -\hbar^2\kappa^2/2m$. Normalize this wave function and show that the parameters of the tight-binding model of the two wells are

$$t = S\kappa \exp(-\kappa a), \quad s = (1 + \kappa a)\exp(-\kappa a), \quad c = S\kappa \exp(-2\kappa a).$$

$$(\mathrm{E7.2})$$

Show that the even and odd states are split by $2S\kappa \exp(-\kappa a)$ in energy and that their mean is raised by $S\kappa^2 a \exp(-2\kappa a)$. The crystal-field term lowers the average energy, but the other term involving non-orthogonality dominates here. In any case the shift of the mean is clearly of second order in the tunnelling and can usually be neglected.

Now solve the model exactly. The wave function is cosh or sinh between the wells and a decaying exponential outside them. All have the same decay constant, which is determined by the balance between the discontinuity in derivative at the potential wells and the strength of the δ-function. Show that

$$\kappa_{e}(1 + \tanh \tfrac{1}{2}\kappa_{e}a) = 2\kappa = \kappa_{o}(1 + \coth \tfrac{1}{2}\kappa_{o}a), \qquad (E7.3)$$

where κ_{e}, κ_{o}, and κ are the decay constants for the even solution, odd solution, and single well. Show that the splitting of the energy levels and their average agree to leading order with the tight-binding model.

7.24 Consider a well 5 nm wide and 0.3 eV deep in GaAs. The lowest state in this well has a binding energy of 0.210 eV so $\kappa = 0.61$ nm^{-1} outside the well. Estimate the splitting when two such wells are separated by a 5 nm barrier (taking the effective mass as 0.067 everywhere). Do not aim for great accuracy but make drastic approximations to get a simple expression for t that will tell you the order of magnitude of the splitting.

7.25 Extend the previous exercise to estimate the position and width of the lowest band in a superlattice of alternating 5 nm wells and 5 nm barriers of height 0.3 eV in GaAs. A numerical calculation puts the edges of the bands at -0.215 eV and -0.205 eV, measured from the tops of the barriers.

7.26 Calculate the effective mass near a band gap from equation (7.113). How are the mass and gap related, and do these results agree with the predictions of the $\mathbf{k} \cdot \mathbf{p}$ method?

7.27 How well does the nearly free electron model predict the band structure of the Kronig–Penney model plotted in Figure 5.18?

7.28 Estimate the width of the gap near the corner of the first Brillouin zone for a square two-dimensional crystal. How does this compare with the gap at the middle of a face?

8 SCATTERING RATES: THE GOLDEN RULE

Fermi's golden rule is one of the most important tools of quantum mechanics. It gives the general formula for transition rates, the rates at which particles are 'scattered' from one state to another by a perturbation. 'Scattered' is in quotation marks because it is a much more general concept than one might guess. An obvious example is provided by impurities in a crystal, which scatter an electron from one Bloch state to another. They change its momentum but not its energy. Similarly, phonons (vibrations of the lattice) also scatter electrons, but in this case they change the energy of the electrons as well as their momentum. A less obvious example is the absorption of light, which can be viewed as a scattering process in which an electron collides with a photon. The converse process also occurs, where an electron loses energy to a photon, and gives rise to spontaneous and stimulated emission. Thus scattering is a remarkably general concept.

The examples suggest that there are two broad classes of scattering processes that we should treat:

(i) potentials that are constant in time, such as impurities in a crystal, which do not change the energy of the particle being scattered;

(ii) potentials that vary harmonically in time as $\cos \omega_q t$, such as phonons and photons, which change the energy of the particle by $\pm \hbar \omega_q$.

The change in energy should of course be predicted, not put in by hand. We shall now develop the theory for these two cases, with elastic scattering of electrons from impurities as the example of a constant potential, and scattering by phonons as the first example of a harmonic potential. Finally we shall calculate the optical conductivity, although the full glory of optical phenomena in low-dimensional systems is reserved for Chapter 10.

8.1 Golden Rule for Static Potentials

We shall first treat a constant perturbation such as the potential from impurities in a crystal. An obvious question is why one doesn't just use the time-independent perturbation theory that we have already developed in Section 7.2, which would give us the eigenstates of the system with impurities. The answer is that both approaches are valid but have different applications.

Suppose that we found the exact eigenstates of a system containing a random distribution of impurities. These eigenstates would be extremely complicated mixtures of the states of the unperturbed system, making calculations cumbersome. Moreover the states would be different from sample to sample, because the impurities would be in different positions. This is an ideal way of approaching very small systems where we expect to see results specific to each sample. Measurements such as magnetoresistance provide a 'fingerprint' that characterizes the distribution of impurities. This defines the *mesoscopic* regime. It is dominated by interference between electron waves and therefore requires samples that are smaller than the distance over which the phase of an electron is destroyed by collisions with phonons or other electrons. Typically this means submicrometre structures at helium temperatures (below 4 K).

In the macroscopic regime, however, we expect quantities such as the conductivity to be the same for different samples. The information contained in individual eigenstates is averaged away (and is also unmanageable). Fermi's golden rule offers a different perspective in this limit. We continue to use the eigenstates of the clean system (pure material), Bloch waves for a crystal or plane waves for free electrons. These are no longer true eigenstates of the system with impurities, so an electron that starts in one eigenstate will not remain purely in that state for ever. Instead, other states will mix in as the electron propagates forwards in time. The probability of being found in one of these other states increases linearly with time, and the rate is the desired scattering or transition rate.

More formally, we again assume that the Hamiltonian $\hat{H}$ can be divided into a large unperturbed part $\hat{H}_0$, which is constant in time and can be solved exactly, and a small perturbation $\hat{V}(t)$ that is turned on at $t = 0$. The eigenstates of $\hat{H}_0$ are ϕ_j with energies ε_j. Before the perturbation is turned on, the electron is in an initial state i whose time-dependent wave function is

$$\Psi(t) = \Phi_i(t) = \phi_i \exp\left(-\frac{i\varepsilon_i t}{\hbar}\right). \tag{8.1}$$

For $t > 0$ we need the solution to

$$\hat{H}\Psi(t) = \left[\hat{H}_0 + \hat{V}(t)\right]\Psi(t) = i\hbar\frac{\partial}{\partial t}\Psi(t), \tag{8.2}$$

subject to the boundary condition that $\Psi = \Phi_i$ at $t = 0$. As usual, the method is to expand the exact solution in terms of the solutions of the unperturbed problem; the difference is that the expansion parameters now depend on time. Thus

$$\Psi(t) = \sum_j a_j(t)\Phi_j(t), \tag{8.3}$$

where $a_j(t)$ is the probability amplitude for being in state j at time t, with initial value $a_j(t = 0) = \delta_{ij}$. Substituting the expansion (8.3) into the Schrödinger equation (8.2) gives

$$\left[\hat{H}_0 + \hat{V}(t)\right] \sum_j a_j(t)\Phi_j(t) = i\hbar \frac{\partial}{\partial t} \sum_j a_j(t)\Phi_j(t), \tag{8.4}$$

or

$$\sum_j a_j(t)\hat{H}_0\Phi_j(t) + \sum_j a_j(t)\hat{V}(t)\Phi_j(t)$$

$$= i\hbar \sum_j a_j(t)\frac{\partial \Phi_j(t)}{\partial t} + i\hbar \sum_j \frac{da_j(t)}{dt}\Phi_j(t). \tag{8.5}$$

The first term on each side cancels because $\Phi_j(t)$ is a solution to the time-dependent Schrödinger equation with $\hat{H}_0$. Turning the equation around and writing explicitly the dependence of $\Phi_j(t)$ on time leaves

$$i\hbar \sum_j \frac{da_j(t)}{dt}\phi_j \exp\left(-\frac{i\varepsilon_j t}{\hbar}\right) = \sum_j a_j(t)\hat{V}(t)\phi_j \exp\left(-\frac{i\varepsilon_j t}{\hbar}\right). \tag{8.6}$$

Many of these terms can be eliminated in the usual way by taking matrix elements. Multiply throughout by ϕ_f^* (for 'final' state) and integrate over space. All the terms on the left vanish except for $j = f$ to give

$$i\hbar \frac{da_f(t)}{dt}\exp\left(-\frac{i\varepsilon_f t}{\hbar}\right) = \sum_j a_j(t)\exp\left(-\frac{i\varepsilon_j t}{\hbar}\right) \int \phi_f^*\hat{V}(t)\phi_j$$

$$= \sum_j a_j(t)\exp\left(-\frac{i\varepsilon_j t}{\hbar}\right) V_{fj}(t). \tag{8.7}$$

Thus

$$\frac{da_f(t)}{dt} = \frac{1}{i\hbar} \sum_j a_j(t) V_{fj}(t)\exp\left(\frac{i\varepsilon_{fj} t}{\hbar}\right), \tag{8.8}$$

where $\varepsilon_{fj} = (\varepsilon_f - \varepsilon_j)$ is the difference in energy between states f and j.

Equation (8.8) is equivalent to the original Schrödinger equation and remains exact. It is now time to make approximations. The zeroth-order approximation is to neglect $\hat{V}$ altogether, so $a_j(t) = \delta_{ij}$. Use this on the right-hand side of equation (8.8) to get a first-order result,

$$\frac{da_f(t)}{dt} = \frac{1}{i\hbar} V_{fi}(t)e^{i\varepsilon_{fi} t/\hbar}. \tag{8.9}$$

This can be integrated for $f \neq i$, remembering that $a_f = 0$ at $t = 0$, to give

$$a_f(t) = \frac{1}{i\hbar} \int_0^t V_{fi}(t')e^{i\varepsilon_{fi} t'/\hbar}dt'. \tag{8.10}$$

This is the general expression within Fermi's golden rule for the probability amplitude of state j at time t. The approximation assumes that the transition rate is small, so that the initial state can always be taken as being nearly full and the final states are nearly empty.

We have not yet made any assumptions about the dependence of the perturbation $\hat{V}(t)$ on time. Let it be constant. In this case V_{fi} can be pulled out of the integral in equation (8.10) to give

$$a_f(t) = -V_{fi}\frac{\exp(i\varepsilon_{fi}t/\hbar) - 1}{\varepsilon_{fi}} = -i\exp(i\varepsilon_{fi}t/2\hbar)V_{fi}\frac{\sin(\varepsilon_{fi}t/2\hbar)}{\varepsilon_{fi}/2}. \qquad (8.11)$$

The probability of finding the electron in the final state is

$$|a_f(t)|^2 = |V_{fi}|^2\left[\frac{\sin(\varepsilon_{fi}t/2\hbar)}{\varepsilon_{fi}/2}\right]^2 = \frac{|V_{fi}|^2 t^2}{\hbar^2}\mathrm{sinc}^2\left(\frac{\varepsilon_{fi}t}{2\hbar}\right), \qquad (8.12)$$

with the notation $\mathrm{sinc}\,\theta = (\sin\theta)/\theta$.

This is an interesting result, plotted in Figure 8.1 as a function of the difference in energy ε_{fi}. The probability of being found in any particular final state oscillates as a function of time with constant amplitude. The sinc function becomes narrower in energy like $1/t$, while the prefactor causes its height to grow like t^2. Thus it becomes infinitely high and narrow as $t \to \infty$, which is reminiscent of the behaviour of a δ-function. The standard integral

$$\int_{-\infty}^{\infty} \mathrm{sinc}^2 x\, dx = \pi \qquad (8.13)$$

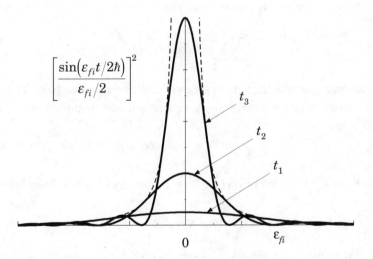

FIGURE 8.1. Probability of being found in a final state as a function of the difference in energy ε_{fi} at three times t_1, t_2, and t_3 in the ratio $1 : 2 : 4$. The broken line is the envelope of the sinc^2 functions at different times.

shows that

$$\int_{-\infty}^{\infty} t^2 \text{sinc}^2(\varepsilon_{fi} t/2\hbar) d\varepsilon_{fi} = 2\pi \hbar t, \qquad (8.14)$$

which in turn means that

$$t^2 \text{sinc}^2(\varepsilon_{fi} t/2\hbar) \rightarrow 2\pi \hbar t\, \delta(\varepsilon_{fi}) \qquad (8.15)$$

as $t \rightarrow \infty$. Thus the probability of being found in a final state with energy ε_f at large times goes like

$$|a_f(t)|^2 \sim \frac{2\pi}{\hbar} |V_{fi}|^2 \delta(\varepsilon_{fi}) t. \qquad (8.16)$$

This increases linearly with time, so there is a constant transition rate from state i to f given by

$$W_{fi} = \frac{2\pi}{\hbar} |V_{fi}|^2 \delta(\varepsilon_f - \varepsilon_i). \qquad (8.17)$$

This is *Fermi's golden rule*. It shows that the energy of the final state must be the same as the energy of the initial state, as expected from the conservation of energy, although this holds exactly only in the limit of large times.

At short times the 'δ-function' has a width in energy of $\hbar/t$ so energy need not be conserved exactly. This width may be important in systems with strong scattering because the electron may not live for long in its 'final' state before it is scattered again and the limit $t \rightarrow \infty$ cannot be taken. In such cases $\delta(E)$ may be replaced by a function $A(E)$ of unit area and width $\hbar/\tau$, where τ is a measure of the lifetime between scattering events. This effect is known as *collisional broadening*.

An alternative form of the golden rule without the δ-function is often used. Instead of considering the rate of scattering from a *particular* initial state, we can sum equation (8.17) to obtain the rate from *any* initial state to the same final state. This gives

$$\frac{2\pi}{\hbar} \sum_i |V_{fi}|^2 \delta(\varepsilon_f - \varepsilon_i). \qquad (8.18)$$

The δ-function constrains the sum to states of energy very close to ε_f. Provided that the matrix element is similar for all these states it can be pulled out of the sum, which becomes

$$\sum_i \delta(\varepsilon_f - \varepsilon_i) = N(\varepsilon_f), \qquad (8.19)$$

where we have used the definition (1.95) of the density of states. Thus the transition rate is

$$W_{fi} = \frac{2\pi}{\hbar} |V_{fi}|^2 N(\varepsilon_f), \qquad \varepsilon_f = \varepsilon_i. \qquad (8.20)$$

This form of Fermi's golden rule is entirely equivalent to the earlier one because the δ-function is meaningful only inside an integral. Its derivation emphasizes an important point: there must be a *continuum* of initial or final states for Fermi's golden rule to be used, or the density of states cannot be defined and the δ-function makes no sense.

We shall now apply this formula to the scattering of electrons by impurities.

8.2 Impurity Scattering

Impurity scattering limits the mobility of electrons at low temperature when there are few phonons present. The nature of the potential varies widely. Charged impurities such as ionized donors and acceptors have a long-range Coulomb potential, whereas neutral impurities have complicated short-range potentials. These two cases have different effects on the total scattering rates. Alloy scattering due to the random arrangement of Al and Ga in (Al,Ga)As and interface-roughness scattering can be treated in a similar way. We shall examine the details of electrons in a 2DEG in Chapter 9; here we develop the general theory.

Consider free electrons in two dimensions. Lower-case vectors such as $\mathbf{r}$ are used for position according to our standard notation. Put the system in a box of finite area A; physical results should not depend on the value of A and this will take some care. Define the initial and final states to be the plane waves

$$\phi_i = A^{-1/2}\exp(i\mathbf{k}\cdot\mathbf{r}), \qquad \phi_f = A^{-1/2}\exp[i(\mathbf{k}+\mathbf{q})\cdot\mathbf{r}], \qquad (8.21)$$

with extra momentum $\hbar\mathbf{q}$ after the scattering event. The perturbation is simply the extra potential energy from the impurity $V(\mathbf{r})$, so the matrix element is

$$V_{fi} = \int \phi_f^* \hat{V}\phi_i = \frac{1}{A}\int e^{-i(\mathbf{k}+\mathbf{q})\cdot\mathbf{r}}V(\mathbf{r})e^{i\mathbf{k}\cdot\mathbf{r}}d^2\mathbf{r} = \frac{1}{A}\int V(\mathbf{r})e^{-i\mathbf{q}\cdot\mathbf{r}}d^2\mathbf{r}$$
$$= A^{-1}\tilde{V}(\mathbf{q}), \qquad (8.22)$$

where $\tilde{V}(\mathbf{q})$ is the two-dimensional Fourier transform of the scattering potential. Inserting this matrix element into the golden rule, equation (8.17), shows that the scattering rate from $\mathbf{k}$ to $\mathbf{k}+\mathbf{q}$ is

$$W_{\mathbf{k}+\mathbf{q},\mathbf{k}} = \frac{1}{A^2}\frac{2\pi}{\hbar}|\tilde{V}(\mathbf{q})|^2\delta[\varepsilon(\mathbf{k}+\mathbf{q}) - \varepsilon(\mathbf{k})]. \qquad (8.23)$$

The simple result that the scattering rate is proportional to the squared modulus of the Fourier transform of the scattering potential is known as the *Born approximation*. It is very widely used, and detailed conditions for its validity are given in books on scattering theory.

An odd feature is the factor of $1/A^2$ in equation (8.23), which implies that the effect of the impurity diminishes as the system becomes larger. This is not surprising, because a single impurity becomes less prominent in a bigger system, but physical consequences of scattering such as the mobility should not depend on the size of the system.

Consider the total scattering rate for an electron, the rate for scattering from $\mathbf{k}$ to *any* final state. This is denoted by $1/\tau_i$ and is understood to be a function of $\mathbf{k}$. There are many different lifetimes that can be defined for an electron and this one has many names. We shall call τ_i the *single-particle lifetime* (against impurity scattering, to be precise); *quantum lifetime* is also used. Assume that the final state is guaranteed to be empty, so that we do not have to worry about occupation (Fermi) factors. Then the total scattering rate due to a single impurity is given by summing the rate (8.23) from the Born approximation over all wave vectors,

$$\left(\frac{1}{\tau_i}\right)_{1 \text{ impurity}} = \sum_{\mathbf{q}} W_{\mathbf{k}+\mathbf{q},\mathbf{k}}. \tag{8.24}$$

The sum over $\mathbf{q}$ can be converted into an integral, as in Section 1.7, using

$$\sum_{\mathbf{q}} \rightarrow \frac{A}{(2\pi)^2} \int d^2\mathbf{q}. \tag{8.25}$$

Here is one factor of A, because the density of final states is proportional to the area of the system. There is no factor of 2 for spin because the potential is assumed not to flip the spin of the electron, so there is no choice of spin in the final state.

Next, there will not be a single impurity in any sample of real material, but a very large number. If their average density is $n_{\text{imp}}^{(2D)}$ per unit area, the total number of impurities in our sample of area A is $N_{\text{imp}}^{(2D)} = A n_{\text{imp}}^{(2D)}$. Unfortunately it is not straightforward to combine the scattering due to many impurities, because electrons are waves and there is interference between waves scattered by nearby impurities. Such interference is particularly important in the mesoscopic regime and depends on the precise configuration of impurities. It is particularly strong in one dimension, where resonant tunnelling provides a dramatic example of interference (Section 5.5).

Fortunately it is usually permissible to ignore interference in large samples at high temperature and to assume that the scattering due to each impurity is independent of the others. The total rate is then given by multiplying that of a single impurity by $N_{\text{imp}}^{(2D)}$,

$$\frac{1}{\tau_i} = \left[A n_{\text{imp}}^{(2D)}\right] \frac{A}{(2\pi)^2} \int \frac{1}{A^2} \frac{2\pi}{\hbar} |\tilde{V}(\mathbf{q})|^2 \delta[\varepsilon(\mathbf{k}+\mathbf{q}) - \varepsilon(\mathbf{k})] d^2\mathbf{q}$$

$$= n_{\text{imp}}^{(2D)} \frac{2\pi}{\hbar} \int |\tilde{V}(\mathbf{q})|^2 \delta[\varepsilon(\mathbf{k}+\mathbf{q}) - \varepsilon(\mathbf{k})] \frac{d^2\mathbf{q}}{(2\pi)^2}. \tag{8.26}$$

The factors of A have vanished to leave the standard formula for the single-particle lifetime of an electron.

An important point is that the single-particle lifetime τ_i is *not* the time that appears in the conductivity or mobility. These contain the *transport lifetime* τ_{tr}, so $\mu = e\tau_{tr}/m$, for example. The difference between τ_i and τ_{tr} lies in the weighting of

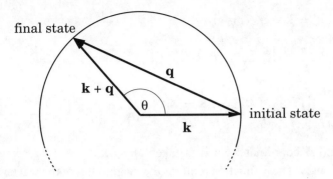

FIGURE 8.2. Relation in **k**-space between the wave vector **q** and angle θ through which an electron is scattered by an impurity.

different collisions. Figure 8.2 shows the geometry of the scattering process and how the wave vector **q** is related to the angle θ through which the electron is scattered. Conservation of energy requires that the wave vectors before and after the collision have the same magnitude, $|\mathbf{k} + \mathbf{q}| = |\mathbf{k}|$, so the vectors lie on a circle in **k**-space. Elementary trigonometry shows that

$$q = 2k \sin \frac{\theta}{2}. \tag{8.27}$$

The single-particle lifetime (equation 8.26) contains a sum over all scattering processes, equally weighted. This means that small-angle scattering in which θ is tiny counts as much as backscattering events where $\theta = \pi$ and the electron's direction is reversed. However, backscattering has a much larger effect on current than small-angle scattering. Given that the component of the electron's motion parallel to its original direction is proportional to $\cos \theta$, one might guess that the efficacy of a scattering event depends on the change in this cosine, $1 - \cos \theta$. A complete calculation confirms this. Equation (8.27) shows that $1 - \cos \theta = q^2/2k^2$, so the transport scattering rate is given by

$$\frac{1}{\tau_{\text{tr}}} = n_{\text{imp}}^{\text{(2D)}} \frac{2\pi}{\hbar} \int \frac{q^2}{2k^2} \left| \tilde{V}(\mathbf{q}) \right|^2 \delta[\varepsilon(\mathbf{k} + \mathbf{q}) - \varepsilon(\mathbf{k})] \frac{d^2 \mathbf{q}}{(2\pi)^2}. \tag{8.28}$$

The additional factor favours scattering through large angles.

Scattering is said to be isotropic if $\tilde{V}(\mathbf{q})$ is independent of **q**. The angular weighting then has no effect and $\tau_{\text{tr}} = \tau_{\text{i}}$. This is characteristic of short-range potentials. The opposite is often the case for charged impurities in low-dimensional systems. Here $\tilde{V}(\mathbf{q})$ falls rapidly as **q** increases and there may be an order of magnitude difference between τ_{i} and τ_{tr}.

These expressions for the total scattering rates are general. They can usually be simplified because most potentials have circular symmetry; exceptions include scattering from charged dipoles. Start from equation (8.28) for the transport lifetime. It is convenient to rewrite the integral temporarily in terms of the wave vector of the final state $\mathbf{k}' = \mathbf{k} + \mathbf{q}$. This gives

$$\frac{1}{\tau_{\text{tr}}} = n_{\text{imp}}^{(2D)} \frac{2\pi}{\hbar} \int \frac{|\mathbf{k}' - \mathbf{k}|^2}{2k^2} \left|\tilde{V}(|\mathbf{k}' - \mathbf{k}|)\right|^2 \delta[\varepsilon(\mathbf{k}') - \varepsilon(\mathbf{k})] \frac{d^2\mathbf{k}'}{(2\pi)^2} \tag{8.29}$$

$$= \frac{n_{\text{imp}}^{(2D)}}{2\pi\hbar} \int_{-\pi}^{\pi} d\theta \int_0^{\infty} k' \, dk' \frac{|\mathbf{k}' - \mathbf{k}|^2}{2k^2} \left|\tilde{V}\left(|\mathbf{k}' - \mathbf{k}|\right)\right|^2 \delta\left(\frac{\hbar^2 k'^2}{2m} - \frac{\hbar^2 k^2}{2m}\right).$$

The axis of polar coordinates for $\mathbf{k}'$ is taken along $\mathbf{k}$, so θ is the scattering angle defined previously. The δ-function requires the magnitudes of the wave vectors to be equal to conserve energy, $k' = k$, but we must treat the argument of the δ-function properly. The general rule is that we should divide by the derivative of the function inside the δ-function, $\hbar^2 k'/m$ here. (Really we are just changing the variable of integration to the energy of the final state $\varepsilon' = \hbar^2 k'^2/2m$.) Finally, we need the magnitude of the wave vector inside the potential and the weighting factor. The cosine formula gives

$$|\mathbf{k}' - \mathbf{k}|^2 = k'^2 + k^2 - 2kk' \cos\theta. \tag{8.30}$$

The δ-function forces $k' = k$ so this reduces to the usual expression $|\mathbf{k}' - \mathbf{k}| = 2k \sin(\frac{1}{2}\theta) = q$. Thus the integral over k' in equation (8.29) yields

$$\frac{1}{\tau_{\text{tr}}} = n_{\text{imp}}^{(2D)} \frac{m}{\pi\hbar^3} \int_0^{\pi} \left|\tilde{V}[2k \sin(\frac{1}{2}\theta)]\right|^2 (1 - \cos\theta) d\theta. \tag{8.31}$$

The integral over θ has been reduced to the range $(0, \pi)$ and doubled to compensate. The rate can alternatively be written in terms of q as

$$\frac{1}{\tau_{\text{tr}}} = n_{\text{imp}}^{(2D)} \frac{m}{2\pi\hbar^3 k^3} \int_0^{2k} \left|\tilde{V}(q)\right|^2 \frac{q^2 \, dq}{\sqrt{1 - (q/2k)^2}}. \tag{8.32}$$

Note that q does not run to infinity. Figure 8.2 shows that the largest value of q that satisfies the conservation of energy is $2k$, which corresponds to backscattering. The angular factor $1 - \cos\theta = q^2/2k^2$ can be removed from these formulas to give the single-particle rate.

Conduction takes place at the Fermi energy in a metal, so we can simply put $k = k_F$ in these formulas. The same can be done with semiconductors if they are degenerate, such as a 2DEG at low temperature, but an average over the active range of k must be made if this is not true.

Scattering can also be described in terms of *cross sections* (areas in three dimensions or lengths in two dimensions). Associated with the single-particle lifetime is a mean free path $l_i = v\tau_i$, where $v = \hbar k/m$. Suppose that each impurity can be represented as a line of length σ_i perpendicular to the velocity of the electron. An electron will hit any impurity within an area $l_i\sigma_i$ while travelling a distance l_i. By the definition of the mean free path there should be exactly one impurity within this area on average, so $n_{imp}^{(2D)} l_i \sigma_i = 1$ and $\sigma_i = m/\hbar k n_{imp}^{(2D)} \tau_i$. Comparison with equation (8.31), after removing the weighting factor for the transport rate, shows that

$$\sigma_i = \frac{m^2}{\pi \hbar^4 k} \int_0^{\pi} \left| \tilde{V}[2k \sin(\tfrac{1}{2}\theta)] \right|^2 d\theta. \tag{8.33}$$

The total cross-section σ_i can instead be written as an integral over a differential cross-section $\sigma(\theta)$,

$$\sigma_i = \int_{-\pi}^{\pi} \sigma(\theta)\, d\theta, \qquad \sigma(\theta) = \frac{m^2}{2\pi \hbar^4 k} \left| \tilde{V}[2k \sin(\tfrac{1}{2}\theta)] \right|^2. \tag{8.34}$$

A transport cross-section σ_{tr} can also be defined by introducing the usual factor of $1 - \cos\theta$ into the integral in equation (8.34). Cross-sections are physically appealing because their magnitude gives the apparent size of a scattering object (which may be different from its physical size).

The next step is to calculate the Fourier transform required for the scattering rate, $|\tilde{V}(q)|^2$.

8.2.1 SCATTERING BY A SHORT-RANGE IMPURITY

A simple example of two-dimensional impurity scattering is provided by a circular barrier of radius a. This is a short-range potential and might be used as a simple model of a *neutral* impurity such as an Al atom that has diffused from a barrier into a GaAs well. The contrasting case of scattering by the long-range potential of a remote ionized impurity will be treated in Chapter 9. Both have rotational symmetry.

The potential is defined by

$$V(r) = \begin{cases} V_0 & \text{if } r < a, \\ 0 & \text{if } r > a. \end{cases} \tag{8.35}$$

Its Fourier transform is

$$\tilde{V}(q) = \int V(r) e^{-i\mathbf{q}\cdot\mathbf{r}} d^2\mathbf{r} = \int_0^{\infty} dr\, r\, V(r) \int_0^{2\pi} d\theta\, e^{-iqr\cos\theta}, \tag{8.36}$$

where θ is the angle between $\mathbf{q}$ and $\mathbf{r}$. Unfortunately the integral over θ gives a Bessel function, a pervasive feature of two-dimensional Fourier transforms, leaving

$$\tilde{V}(q) = 2\pi \int_0^{\infty} V(r)\, J_0(qr)\, r\, dr. \tag{8.37}$$

In the specific case of the circular barrier this becomes

$$\tilde{V}(q) = 2\pi V_0 \int_0^{a} J_0(qr) r\, dr = \pi a^2 V_0 \left(\frac{2 J_1(qa)}{qa} \right), \tag{8.38}$$

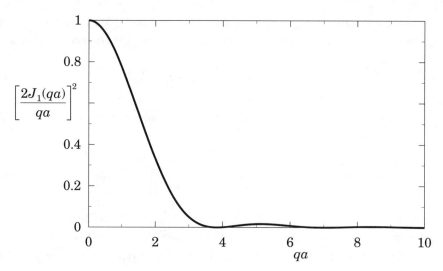

FIGURE 8.3. Scattering rate of a circular barrier of radius a as a function of the scattering wave number q.

where equation (9.1.30) from Abramowitz and Stegun (1972) has been used for the integral, which yields another Bessel function.

The scattering rate is proportional to the square of this and is plotted in Figure 8.3. It shows decaying oscillations that arise from the sharp edge of the potential and are absent from a more smoothly varying form. The function $2J_1(x)/x$ is rather like $\operatorname{sinc} x$ and has a limiting value of 1 as $x \to 0$. Thus $\tilde{V}(q) \approx \pi a^2 V_0$ for small q and scattering becomes isotropic. This widely applicable result can be seen from equation (8.36), which gives

$$\lim_{\mathbf{q} \to 0} \tilde{V}(\mathbf{q}) = \int V(\mathbf{r}) \, d^2\mathbf{r}, \qquad (8.39)$$

an integral over the potential. Unfortunately some potentials are too long-range for the integral to exist and do not show this limit; an important example is Rutherford scattering from an unscreened Coulomb potential. The total cross-section at small k reduces to $\sigma_i = \sigma_{tr} = (\pi m a^2 V_0 / \hbar^2)^2 / k$. This is very different from the size $2a$ of the obstacle, and diverges as $k \to 0$.

Figure 8.3 shows that scattering falls off for large values of q, roughly for $q > \pi/a$, a general relation between the size of the obstacle and the maximum wave number for scattering. The scattering angle θ is given by equation (8.27) and becomes $\theta = q/k$ for small angles. The maximum scattering angle is roughly π/ka, which decreases as the energy of the incident electron increases, a reassuring result.

A neutral impurity in a semiconductor is of atomic dimensions so scattering is nearly independent of q for $q < 10^9 \, \mathrm{m}^{-1}$. Scattering from neutral impurities can therefore be treated as isotropic in a 2DEG, where $k_F \approx 10^8 \, \mathrm{m}^{-1}$.

A feature of the scattering rate within the Born approximation is that it does not depend on the sign of the potential: a well of depth V_0 scatters in the same way as a barrier of height V_0. This result fails for large V_0 because the Born approximation becomes inaccurate. For example, the circular barrier is impenetrable when its height greatly exceeds the energy of the incident electron and scattering becomes independent of V_0. More accurate methods, such as phase shifts, must be used in this case and scattering from an impenetrable circular obstacle can be solved exactly. Rigorous theory also shows that there are upper bounds to the cross-section, so it is not possible to design objects that scatter with arbitrary strength. For example, isotropic scattering cannot give a cross-section greater than $4/k$. Further problems arise with an attractive potential, particularly when there is a bound state near the top of the well. This is worrisome, since all two-dimensional potential wells have bound states. Fortunately exact calculations show that the Born approximation remains acceptable under a wide range of conditions.

If the radius a of the circular barrier is reduced to zero we obtain a potential $V(r) = S\delta(\mathbf{r})$ with a two-dimensional δ-function. The Fourier transform is a constant, $\tilde{V}(q) = S$, and scattering is isotropic. Calculations are often greatly simplified in this limit – the single-particle and transport cross-sections are identical, for example – so it is widely used in theory. Unfortunately scattering in a 2DEG is dominated by remote ionized impurities whose cross-section is far from isotropic, but this does not seem to curtail the use of δ-function potentials as a model.

8.3 Golden Rule for Oscillating Potentials

The second kind of perturbation varies harmonically in time,

$$\hat{V}(t) = 2\hat{V}\cos\omega_0 t = \hat{V}(e^{-i\omega_0 t} + e^{i\omega_0 t}). \tag{8.40}$$

The factor of 2 is included to simplify later expressions but is easily lost – beware! The perturbation $\hat{V}$ contains operators that act on the wave function but is not dependent on time. The matrix elements in equation (8.8) also have the form $V_{fi}(t) = 2V_{fi}\cos\omega_0 t$, and the first-order integral (8.10) for the coefficients becomes

$$a_f(t) = \frac{1}{i\hbar}V_{fi}\int_0^t (e^{-i\omega_0 t'} + e^{i\omega_0 t'})e^{i\varepsilon_{fi}t'/\hbar}dt'$$

$$= -V_{fi}\left[\frac{e^{i(\varepsilon_{fi}-\hbar\omega_0)t/\hbar} - 1}{(\varepsilon_{fi} - \hbar\omega_0)} + \frac{e^{i(\varepsilon_{fi}+\hbar\omega_0)t/\hbar} - 1}{(\varepsilon_{fi} + \hbar\omega_0)}\right]. \tag{8.41}$$

The probability of being found in the final state is therefore

$$|a_f(t)|^2 = \frac{|V_{fi}|^2 t^2}{\hbar^2}\left\{\text{sinc}^2\frac{(\varepsilon_{fi} - \hbar\omega_0)t}{2\hbar} + \text{sinc}^2\frac{(\varepsilon_{fi} + \hbar\omega_0)t}{2\hbar}\right. \tag{8.42}$$

$$\left. + 2\cos\omega_0 t \,\,\text{sinc}\frac{(\varepsilon_{fi} - \hbar\omega_0)t}{2\hbar}\,\text{sinc}\frac{(\varepsilon_{fi} + \hbar\omega_0)t}{2\hbar}\right\}.$$

The first term in braces looks the same as for the static perturbation, equation (8.12), except that it is centred on $\varepsilon_{fi} = \hbar\omega_0$ rather than zero, or $\varepsilon_f = \varepsilon_i + \hbar\omega_0$. Thus the final state has absorbed energy $\hbar\omega_0$ from the perturbation. Similarly the second term gives $\varepsilon_f = \varepsilon_i - \hbar\omega_0$, a loss of energy to the perturbation. The third term contains the interference of these two events and can be ignored for $\omega_0 t \gg 1$, which is usually the case. In this limit the separation of the two sinc^2 functions, $2\hbar\omega_0$, will be much greater than their width $\hbar/t$, so they may be treated independently. This means that the two exponential components $e^{\pm i\omega_0 t}$ of $2\cos\omega_0 t$ separate. The derivation proceeds exactly as for the static case, and the final result for the transition rate induced by $\hat{V}e^{-i\omega_0 t}$ is

$$W_{fi} = \frac{2\pi}{\hbar}|V_{fi}|^2\delta(\varepsilon_f - \varepsilon_i - \hbar\omega_0). \tag{8.43}$$

This is Fermi's golden rule for a harmonic perturbation, and shows that the $e^{-i\omega_0 t}$ component leads to the absorption of energy by the electron. It can instead be written with the density of states as

$$W_{fi} = \frac{2\pi}{\hbar}|V_{fi}|^2 N(\varepsilon_f), \qquad \varepsilon_f = \varepsilon_i + \hbar\omega_0. \tag{8.44}$$

The results for the other part of the cosine, $\hat{V}e^{+i\omega_0 t}$, are identical except that $\varepsilon_f = \varepsilon_i - \hbar\omega_0$, corresponding to the emission of energy from the electron.

Two important applications of the golden rule are to scattering by phonons (lattice vibrations) and photons (light), which will be covered in the next two sections. Unfortunately we have to cheat a little because both photons and phonons should themselves be quantized, whereas we have treated the harmonic potential classically. In the case of phonons we shall calculate the scattering rate due to a single phonon and multiply the result by the Bose–Einstein distribution. For optical absorption we shall ignore the problem completely and treat light as a classical wave.

8.4 Phonon Scattering

Phonons were described in Section 2.8 and are the quanta of lattice vibrations in a solid. They are the dominant scattering mechanism for electrons at room temperature. There are many ways in which electrons and phonons interact although all arise from the same basic effect: the motion of the ions due to the phonon induces electric and magnetic fields that affect the motion of the electrons and scatter them. We shall consider two important and contrasting cases, the deformation coupling to LA (longitudinal acoustic) phonons and the polar coupling to LO (longitudinal optic) phonons.

8.4.1 LONGITUDINAL ACOUSTIC PHONONS AND THE DEFORMATION POTENTIAL

Longitudinal acoustic phonons are like sound waves for long wavelengths (small q). The simplest coupling to electrons for such phonons is through the deformation potential. The phonon compresses and dilates alternating regions of the solid. A uniform compression or dilation of the crystal causes the edge of each electronic energy band to move up or down proportionally to the strain. The constant of proportionality is called the *deformation potential* Ξ. (A more complicated expression is needed if the bands are degenerate or lack spherical symmetry.) Longitudinal strain is defined as the fractional increase in length of the object, so for one unit cell of the chain (Figure 2.21(a)) it is given by $(u_j - u_{j-1})/a$. For long wavelengths we can treat the one-dimensional chain of atoms as a continuum and the strain becomes a derivative. The displacement of an atom at z is given by equation (2.30), $u(z) = U_0 \cos(qz - \omega_q t)$, so the longitudinal strain is

$$e(z) = \frac{\partial u}{\partial z} = -U_0 q \sin(qz - \omega_q t). \tag{8.45}$$

The potential energy can be calculated as though the strain were uniform in each region provided that the wavelength is long. This gives

$$V(z) = \Xi e(z) = -U_0 q\, \Xi \sin(qz - \omega_q t) = -\sqrt{\frac{2\hbar}{\Omega \rho \omega_q}}\, q\, \Xi \sin(qz - \omega_q t), \tag{8.46}$$

where the amplitude U_0 for a single phonon has been inserted from equation (2.30). Since $\omega_q = v_s q$ for long-wavelength acoustic phonons, this finally becomes

$$\hat{V}(z, t) = i\sqrt{\frac{\hbar q}{2\Omega \rho v_s}}\, \Xi (e^{iqz} e^{-i\omega_q t} - e^{-iqz} e^{i\omega_q t}). \tag{8.47}$$

This is the form of the perturbing potential caused by the phonon to be used in Fermi's golden rule.

One more ingredient, the number of phonons, is needed to get the total scattering rates. The number of phonons that occupy a mode of wave vector $\mathbf{Q}$ at equilibrium is given by the Bose–Einstein distribution (1.127):

$$N_{\mathbf{Q}} = \left[\exp\left(\frac{\hbar \omega_{\mathbf{Q}}}{k_B T} \right) - 1 \right]^{-1}. \tag{8.48}$$

The rate for a single phonon must be multiplied by the number of phonons in a mode $N_{\mathbf{Q}}$ to calculate the total rate of scattering due to absorption. The factor for emission of phonons is $N_{\mathbf{Q}} + 1$; the 1 describes *spontaneous* emission while the $N_{\mathbf{Q}}$ describes *stimulated* emission. (Spontaneous emission can instead be viewed as emission stimulated by vacuum fluctuations.) This asymmetry between the rates for absorption and emission is vital to ensure that electrons and phonons remain in equilibrium with one another.

Scattering in low-dimensional systems is complicated because the phonons usually retain their three-dimensional nature, even if the electrons are confined. We shall therefore look first at the interaction between three-dimensional electrons and phonons and defer the two-dimensional electron gas to Section 9.6.3. Put the electrons in a finite box of volume Ω where the wave functions are plane waves, scattering from $\mathbf{K}$ to $\mathbf{K'}$. Start with the first term in equation (8.47). This depends on time like $\exp(-i\omega_\mathbf{Q} t)$ and causes the electron to absorb energy from the phonon. The condition for energy conservation in the δ-function is therefore $\varepsilon(\mathbf{K'}) = \varepsilon(\mathbf{K}) + \hbar\omega_\mathbf{Q}$. The matrix element is

$$V_{\mathbf{K'K}}^+ = i\sqrt{\frac{\hbar Q}{2\Omega\rho v_s}}\, \Xi\, \frac{1}{\Omega}\int_\Omega e^{-i\mathbf{K'}\cdot\mathbf{R}} e^{i\mathbf{Q}\cdot\mathbf{R}} e^{i\mathbf{K}\cdot\mathbf{R}} d^3\mathbf{R}. \tag{8.49}$$

The integral gives Ω, which cancels the $1/\Omega$ from the normalization of the wave functions, if the wave vectors sum to $\mathbf{0}$, and zero otherwise. This requires $\mathbf{K'} = \mathbf{K} + \mathbf{Q}$, the expected result for the conservation of momentum. We must also include a factor $N_\mathbf{Q}$ to account for the number of phonons for absorption. Thus the scattering rate from $\mathbf{K}$ to $\mathbf{K'}$ caused by the absorption of longitudinal acoustic phonons is

$$W_{\mathbf{K'K}}^+ = \frac{2\pi}{\hbar} N_\mathbf{Q} \frac{\hbar Q}{2\Omega\rho v_s} \Xi^2 \delta_{\mathbf{K'},\mathbf{K}+\mathbf{Q}}\, \delta[\varepsilon(\mathbf{K}+\mathbf{Q}) - \varepsilon(\mathbf{K}) - \hbar\omega_\mathbf{Q}]. \tag{8.50}$$

The relation $\mathbf{K'} = \mathbf{K} + \mathbf{Q}$ enforced by the δ-function for the wave vectors has been used to replace $\mathbf{K'}$ inside the δ-function for energy. The electron has gained wave vector $\mathbf{Q}$ and energy $\hbar\omega_\mathbf{Q} = \hbar v_s q$ from the phonon.

The second half of the perturbation (equation 8.47) causes the electron to lose both momentum and energy to the phonon. The only differences are the signs in the δ-functions and the $N_\mathbf{Q} + 1$ in the phonon occupation factor to include spontaneous as well as stimulated emission. Thus

$$W_{\mathbf{K'K}}^- = \frac{2\pi}{\hbar} (N_\mathbf{Q} + 1) \frac{\hbar Q}{2\Omega\rho v_s} \Xi^2 \delta_{\mathbf{K'},\mathbf{K}-\mathbf{Q}}\, \delta[\varepsilon(\mathbf{K}-\mathbf{Q}) - \varepsilon(\mathbf{K}) + \hbar\omega_\mathbf{Q}]. \tag{8.51}$$

Unfortunately the change in energy prevents us from writing a simple formula for the contribution of phonons to the transport scattering rate, as we did for impurities. However, we can find out whether acoustic phonons carry away significant energy. Consider a 2DEG of density $n_{2D} = 3 \times 10^{15}\,\mathrm{m}^{-2}$, for which $E_F \approx 10\,\mathrm{meV}$ and $k_F \approx 0.14\,\mathrm{nm}^{-1}$. Backscattering provides the phonon with the largest wave vector, $q = 2k_F$, whose energy is $\hbar\omega_{2k_F}$. The velocity of sound is around $5\,\mathrm{km\,s}^{-1}$, so the energy is $2\hbar v_s k_F \approx 0.9\,\mathrm{meV}$. The conclusion is that acoustic phonons do not carry away much energy, and they are often treated in the *quasi-elastic approximation* where this energy is neglected. The same conclusion is reached for non-degenerate electrons at room temperature.

As a simple example we can calculate the total scattering rate for an electron in a three-dimensional gas using the quasi-elastic approximation. The condition for energy conservation becomes $\varepsilon(\mathbf{K} + \mathbf{Q}) = \varepsilon(\mathbf{K})$, as for impurities. Also, the number of phonons in the relevant modes is large, so we can use the non-degenerate limit of the Bose distribution, $N_\mathbf{Q} \approx (N_\mathbf{Q} + 1) \approx k_B T / \hbar \omega_\mathbf{Q} \gg 1$. These approximations mean that the rates for emission and absorption are nearly equal and together give the total rate for scattering from $\mathbf{K}$ to $\mathbf{K} + \mathbf{Q}$,

$$W_{\mathbf{K}+\mathbf{Q},\mathbf{K}} = 2 \frac{2\pi}{\hbar} \frac{k_B T}{\hbar v_s Q} \frac{\hbar Q}{2\Omega \rho v_s} \Xi^2 \, \delta[\varepsilon(\mathbf{K} + \mathbf{Q}) - \varepsilon(\mathbf{K})]$$

$$= \frac{2\pi}{\hbar} \frac{1}{\Omega} \frac{\Xi^2 k_B T}{\rho v_s^2} \delta[\varepsilon(\mathbf{K} + \mathbf{Q}) - \varepsilon(\mathbf{K})]. \qquad (8.52)$$

After all the simplifications this resembles the scattering rate (8.23) for impurities. In fact this is the particularly simple case of isotropic scattering because $\mathbf{Q}$ has vanished from the matrix element. Summing over all $\mathbf{Q}$ gives the total scattering rate for an electron with wave vector $\mathbf{K}$,

$$\frac{1}{\tau} = \frac{2\pi}{\hbar} \frac{\Xi^2 k_B T}{\rho v_s^2} \left\{ \frac{1}{\Omega} \sum_\mathbf{Q} \delta[\varepsilon(\mathbf{K} + \mathbf{Q}) - \varepsilon(\mathbf{K})] \right\}. \qquad (8.53)$$

The sum gives the density of states as in equation (1.97) except that there is no summation over spin. We could have reached this result more directly by using the 'density-of-states' form of Fermi's golden rule. Thus the scattering rate reduces to

$$\frac{1}{\tau} = \frac{\pi \Xi^2 k_B T n[\varepsilon(\mathbf{K})]}{\hbar \rho v_s^2}. \qquad (8.54)$$

The single-particle and transport lifetimes are identical for isotropic scattering so no subscript is needed on τ. A typical energy in a non-degenerate three-dimensional electron gas is $\varepsilon(\mathbf{k}) = \frac{3}{2} k_B T$. The density of states $n(E) \propto \sqrt{E} \propto \sqrt{T}$, so we reach the well-known result $\mu \propto T^{-3/2}$.

There are other mechanisms by which electrons and LA phonons interact. For example, the motion of ionic cores generates a magnetic field. More important, strain may generate a piezoelectric potential in compound semiconductors, and there remains considerable dispute over the relative importance of deformation-potential and piezoelectric coupling in GaAs.

8.4.2   LONGITUDINAL OPTIC PHONONS AND POLAR COUPLING

Optic phonons occur in crystals with more than one atom per unit cell (Section 2.8.3). They provide a contrast to acoustic phonons because their energy is large (36 meV at $Q = 0$ in GaAs). They also illustrate a quite different way of coupling to the electrons. The name 'optic' arises because the two atoms in a unit cell move in opposite directions (Figure 2.24(b)). This sets up an electric field if the atoms carry a charge, and we shall consider only this polar case, which includes the III–V semiconductors.

Since the *relative* displacement of the two atoms in a unit cell is the relevant coordinate for an optic phonon, denote this by $u_j(t) = U_0 \cos(\mathbf{Q} \cdot \mathbf{R} - \omega_\mathbf{Q} t)$. The dispersion relation (Figure 2.23) is nearly flat around $\mathbf{Q} = \mathbf{0}$ so we can put $\omega_\mathbf{Q} = \omega_{\mathrm{LO}}$, a constant. The prefactor U_0 for a single phonon can be found in the same way as for the acoustic phonon in Section 2.8.1, with the result

$$U_0 = \sqrt{\frac{2\hbar}{N_{\mathrm{cells}}\mu\omega_{\mathrm{LO}}}} = \sqrt{\frac{2\hbar}{\Omega n_{\mathrm{cells}}\mu\omega_{\mathrm{LO}}}}, \tag{8.55}$$

where n_{cells} is the number of unit cells (pairs of ions) per unit volume. This is similar to the result (2.29) for acoustic phonons, the main difference being the appearance of the *reduced mass* defined by

$$\frac{1}{\mu} = \frac{1}{m} + \frac{1}{M}, \tag{8.56}$$

where m and M are the two individual masses. The two ionic masses are very similar in GaAs and $\mu \approx \frac{1}{2}m_{\mathrm{ion}}$.

Let the two ions carry opposite charges $\pm Q_{\mathrm{eff}}e$; materials such as GaAs are only weakly ionic so the effective charge $Q_{\mathrm{eff}} \ll 1$. As each pair of ions vibrates, its sets up an electric dipole $p(t) = Q_{\mathrm{eff}}eu(t)$. This in turn generates a polarization field $P(t) = n_{\mathrm{cells}}p(t)$. No free charge has been introduced so the displacement $\mathbf{D} = \mathbf{0}$ and the general relation $\mathbf{D} = \epsilon_0\mathbf{E} + \mathbf{P}$ reduces to $E = -P/\epsilon_0$. The electric potential is the integral of E, which introduces a factor of $1/Q$. The potential energy for an electron is finally given by

$$V(\mathbf{R}, t) = \frac{e^2 Q_{\mathrm{eff}}}{\epsilon_0 Q}\sqrt{\frac{2\hbar n_{\mathrm{cells}}}{\Omega\mu\omega_{\mathrm{LO}}}}\sin(\mathbf{Q} \cdot \mathbf{R} - \omega_{\mathrm{LO}}t). \tag{8.57}$$

The scattering rate can then be calculated as for acoustic phonons, and the rate for absorption of optic phonons is

$$W_{\mathbf{K}'\mathbf{K}}^+ = \frac{2\pi}{\hbar}N_{\mathrm{LO}}\left(\frac{e^2}{\epsilon_0}\right)^2\frac{\hbar n_{\mathrm{cells}}Q_{\mathrm{eff}}^2}{2\Omega\mu\omega_{\mathrm{LO}}Q^2}\,\delta_{\mathbf{K}',\mathbf{K}+\mathbf{Q}}\,\delta[\varepsilon(\mathbf{K}+\mathbf{Q}) - \varepsilon(\mathbf{K}) - \hbar\omega_{\mathrm{LO}}], \tag{8.58}$$

where N_{LO} is the occupation number of LO phonons given by the Bose–Einstein distribution.

There are two important differences from the result for acoustic phonons. First is the factor of $1/Q^2$. This causes the scattering rate of polar optic phonons to fall off rapidly as a function of Q, in contrast with acoustic phonons, where the coupling grows like Q. Second, the energy of $\hbar\omega_{LO}$ exchanged with the phonon is large. This means that absorption of optic phonons is significant only at high temperatures (typically above 77 K), as they are frozen out below this. However, spontaneous emission is possible at lower temperatures if the electron has sufficient energy, and optic phonons cool hot electrons effectively. The scattering rate as a function of the energy of an electron increases rapidly at the threshold $\varepsilon(K) = \hbar\omega_{LO}$ where spontaneous emission becomes possible, but then declines slowly as the factor of $1/Q^2$ causes the coupling to become weaker. Another important effect of LO phonons is to ionize excitons at room temperature, which we shall consider in Section 10.7.

Although the effective charge Q_{eff} seems an obvious factor to include in the scattering rate, it is usually expressed in terms of the dielectric constant at low and high frequencies:

$$Q_{\text{eff}}^2 = \frac{\epsilon_0}{e^2} \frac{\omega_{LO}^2 \mu}{n_{\text{cells}}} \left[\frac{1}{\epsilon(\infty)} - \frac{1}{\epsilon(0)} \right]. \tag{8.59}$$

Only the electrons can respond to an electric field at frequencies much higher than ω_{LO} because the ions are too heavy, giving a dielectric constant $\epsilon(\infty)$. At frequencies below ω_{LO} the ions can also respond to an applied field and, being charged, make an additional contribution to the screening, so the dielectric constant $\epsilon(0)$ becomes larger. For GaAs $\epsilon(0) = 12.90$ and $\epsilon(\infty) = 10.92$. The factor in front of the square brackets in equation (8.59) is close to 3 for the III–V semiconductors, so we find $Q_{\text{eff}} \approx 0.2$ in GaAs. The same dielectric constants also appear in the *Lyddane–Sachs–Teller* relation between the frequencies of transverse and longitudinal optic phonons,

$$\frac{\omega_{TO}^2}{\omega_{LO}^2} = \frac{\epsilon(\infty)}{\epsilon(0)}. \tag{8.60}$$

The frequencies are different because the electric field set up by longitudinal vibrations gives an additional restoring force that raises their frequency. Purely transverse vibrations do not generate dipoles so this effect is absent.

As in the case of acoustic phonons, there are other mechanisms by which optic phonons can interact with electrons. A deformation potential can again be defined and is important in elemental semiconductors where there is no polar coupling.

Before leaving phonon scattering we should take a closer look at the matrix elements because our treatment of the wave functions has been rather sloppy. We included only a plane wave, in the spirit of effective-mass theory, but omitted the Bloch function $u_c(\mathbf{R})$ that should multiply it. The integral in equation (8.49) should be replaced by

$$\int_{\Omega} u_{\mathrm{c}}^{*}(\mathbf{R}) e^{-i\mathbf{K}'\cdot\mathbf{R}} e^{i\mathbf{Q}\cdot\mathbf{R}} u_{\mathrm{c}}(\mathbf{R}) e^{i\mathbf{K}\cdot\mathbf{R}} d^{3}\mathbf{R}$$

$$= \sum_{mn} u_{cm}^{*} u_{cn} \int_{\Omega} e^{-i(\mathbf{K}'+\mathbf{G}_m)\cdot\mathbf{R}} e^{i\mathbf{Q}\cdot\mathbf{R}} e^{i(\mathbf{K}+\mathbf{G}_n)\cdot\mathbf{R}} d^{3}\mathbf{R}. \qquad (8.61)$$

The Bloch functions have the periodicity of the lattice and have therefore been expanded on the right-hand side as Fourier series with the wave vectors $\mathbf{G}_n$ of the reciprocal lattice. The general term survives the integration if $\mathbf{K}' = \mathbf{K} + \mathbf{Q} + \mathbf{G}_n - \mathbf{G}_m$. Thus $\mathbf{K}'$ and $\mathbf{K} + \mathbf{Q}$ need not be identical but may differ by a reciprocal lattice vector. This is called *umklapp scattering*. It is particularly important for intervalley scattering in Si, where the shortest route between two X-valleys may involve a reciprocal lattice vector. Umklapp scattering is shown naturally in the repeated zone scheme of band structure.

A full treatment of electron–phonon scattering has to go far beyond this brief discussion. The interaction is screened by electrons or holes if there is a dense gas of them, and this can have a marked effect on the coupling. The scattering of holes is difficult to calculate because of the complicated nature of the valence bands. Intervalley scattering must also be included in a material with degenerate valleys in the conduction band such as Si, and this may be umklapp scattering as just discussed. Scattering from Γ to higher valleys (X and L) also becomes important for hot electrons in GaAs, where it gives rise to the Gunn effect (Section 2.6.4.3).

8.5 Optical Absorption

The optical properties of low-dimensional systems provide their most widespread applications and fill Chapter 10. Here we shall use the golden rule to derive a general formula for optical absorption and apply it to two contrasting examples: transitions from the valence band to the conduction band where there is a continuous range of absorption, and absorption between the levels of a quantum well where only discrete frequencies are permitted. We face the same problem as in electron–phonon scattering, which is that the photons should be quantized as well as the electrons. Here we shall ignore quantization of the photon field and treat it simply as a classical electromagnetic wave. This is adequate for traditional measurements of optical absorption but would need to be improved for lasers where a correct treatment of spontaneous and stimulated emission is needed. Also, the calculations are limited to linear effects and do not describe phenomena such as frequency doubling seen with light of very high intensity.

For clarity, the formulas will be written down as though they describe an infinite homogeneous medium. This is obviously not true for a heterostructure and one should calculate the electromagnetic fields appropriately, applying matching conditions at interfaces. We shall assume that this has been done, using the laws of classical electromagnetism. In practice this step may be far from trivial, even for common structures such as a rib wave guide.

8.5.1 MACROSCOPIC EQUATIONS

In classical electromagnetism, the response to an electric field is described by both a current $\mathbf{J}$ and a polarization $\mathbf{P}$, which is usually absorbed into the displacement $\mathbf{D} = \epsilon_0\mathbf{E} + \mathbf{P}$. Let the electric field be a single Fourier component with angular frequency ω and wave vector $\mathbf{Q}$, given by the real part of $\mathbf{E}(\mathbf{R}, t) = \mathbf{E}(\mathbf{Q}, \omega)\exp[i(\mathbf{Q} \cdot \mathbf{R} - \omega t)]$. A reminder about signs is in order here. This book employs the 'quantum-mechanics convention' for the dependence on time, $e^{-i\omega t}$, but standard notation in electrical engineering is $e^{j\omega t}$. Unfortunately this difference in sign is pervasive. Users of $e^{j\omega t}$ often define $\tilde{\epsilon}_r = \epsilon_1 - j\epsilon_2$ to ensure that the physically meaningful sign of ϵ_2, which represents dissipation, is unchanged. A general rule is to change i to $-j$ *everywhere*. Take care!

The fourth Maxwell equation is

$$\text{curl }\mathbf{H} = \frac{\partial \mathbf{D}}{\partial t} + \mathbf{J} = -i\omega\mathbf{D} + \mathbf{J}, \tag{8.62}$$

where the second form follows from the harmonic dependence on time. The fields $\mathbf{D}$ and $\mathbf{J}$ are given by the constitutive relations $\mathbf{D} = \epsilon_0\epsilon_r\mathbf{E}$ and $\mathbf{J} = \sigma\mathbf{E}$, where ϵ_r is the dielectric function (relative permittivity) and σ is the conductivity. Both are functions of frequency and wave vector; they should be second-rank tensors rather than scalars but we shall assume that the material is isotropic so that this complication can be neglected for now. Although this simplification holds for the cubic semiconductors, it is not true for heterostructures such as quantum wells where we shall see that the response to an electric field normal to the well may be quite different from the response to a parallel field.

It is inconvenient to have two quantities so they are usually combined into a complex dielectric function $\tilde{\epsilon}_r$ or a complex conductivity $\tilde{\sigma}$. Then the fourth Maxwell equation can be written in the two forms

$$\text{curl }\mathbf{H} = (-i\omega\epsilon_0\epsilon_r + \sigma)\mathbf{E} = -i\omega\epsilon_0\tilde{\epsilon}_r\mathbf{E} = (\tilde{\sigma} - i\omega\epsilon_0)\mathbf{E}. \tag{8.63}$$

The form with conductivity has two terms because free space has $\tilde{\epsilon}_r = 1$ but $\tilde{\sigma} = 0$. The complex dielectric function and conductivity are related by

$$\tilde{\sigma} = -i\omega\epsilon_0(\tilde{\epsilon}_r - 1), \qquad \tilde{\epsilon}_r = 1 + \frac{i}{\omega\epsilon_0}\tilde{\sigma}. \tag{8.64}$$

They are entirely equivalent, and one uses whichever seems the more appropriate. Their real and imaginary parts are denoted $\tilde{\epsilon}_r = \epsilon_1 + i\epsilon_2$ and $\tilde{\sigma} = \sigma_1 + i\sigma_2$. The dielectric function and conductivity that we introduced first are just the real parts of the corresponding complex quantities.

There is an important physical distinction between the real and imaginary parts: σ_1 and ϵ_2 describe conduction, which is a dissipative process, whereas the other part describes polarization, which does not absorb energy. This means that we usually expect $\sigma_1 > 0$, as a negative sign would imply that energy was released when the system was stimulated. Another way of looking at this is to note that σ_1 gives a current in phase with the electric field, and hence dissipation, whereas σ_2 gives a current in quadrature, which is associated with an oscillating transfer of energy but no flow on average.

The complex refractive index $\tilde{n}_r = n_r + i\kappa_r$ is often used for optical properties, defined by $\tilde{\epsilon}_r = \tilde{n}_r^2$. It relates the wave number and frequency of a light wave through $k = \tilde{n}_r \omega / c = (n_r + i\kappa_r)\omega/c$. Thus n_r affects the wavelength and velocity while κ_r gives the rate at which the wave decays with distance. The real and imaginary parts are related by $\epsilon_1 = n_r^2 - \kappa_r^2$ and $\epsilon_2 = 2n_r\kappa_r$. Rapid attenuation (large κ_r) may result from a negative ϵ_1, as in a plasma below its plasma frequency, and not necessarily from absorption (large ϵ_2). In other words, energy may be reflected rather than absorbed.

One might guess that $\epsilon_1(\omega)$ and $\epsilon_2(\omega)$ are independent functions, since one describes polarization and the other dissipation, which are apparently distinct physical processes. This turns out to be quite wrong: $\epsilon_2(\omega)$ is entirely determined if $\epsilon_1(\omega)$ is known for all frequencies and vice versa. This remarkable result is known as the Kramers–Kronig relation and is described in Appendix 6. We shall use it in Section 10.1 to derive the imaginary part of $\tilde{\sigma}$ from its real part, which is easier to calculate.

8.5.2 ELECTRON–PHOTON INTERACTION

The strategy of the calculation is to calculate the energy absorbed from a plane wave of light in two ways, one using Fermi's golden rule and the other from classical electromagnetism; equating the two gives the real part of the conductivity σ_1. Apply a light wave with an electric field $\mathbf{E}(\mathbf{R}, t) = 2\mathbf{e}E_0 \cos(\mathbf{Q} \cdot \mathbf{R} - \omega t)$. Assume that the wave does not decay as it passes through the sample. There is a magnetic field too but its effect is usually much smaller and we shall ignore it. The unit vector $\mathbf{e}$ gives the polarization of the electric field, which must lie in the plane normal to the direction of propagation, so $\mathbf{e} \perp \mathbf{Q}$. This wave is plane polarized but a complex $\mathbf{e}$ can be used to describe more general waves. Classically, the electric field induces a current whose in-phase component is $\mathbf{J} = \sigma_1 \mathbf{E}$, which dissipates energy at a rate $\mathbf{J} \cdot \mathbf{E}$ per unit volume. Averaging over time, the total power dissipated in volume Ω is

$$P = 2\sigma_1 \Omega E_0^2 . \qquad (8.65)$$

We shall now derive a quantum-mechanical expression using Fermi's golden rule.

The Hamiltonian contains scalar and vector potentials rather than the fields directly (Section 6.1). Here we use $\mathbf{E} = -\partial\mathbf{A}/\partial t$, which requires a vector potential $\mathbf{A}(\mathbf{R}, t) = (2e E_0/\omega)\sin(\mathbf{Q}\cdot\mathbf{R} - \omega t)$. A scalar potential would give the wrong kind of field. Consider a single Fourier component, $\phi(\mathbf{R}, t) \propto \exp[i(\mathbf{Q}\cdot\mathbf{R} - \omega t)]$. This gives $\mathbf{E} = -\operatorname{grad}\phi \propto i\mathbf{Q}\exp[i(\mathbf{Q}\cdot\mathbf{R} - \omega t)]$. The important feature is that the electric field is in the same direction as the direction of travel $\mathbf{Q}$. This is called a longitudinal field in contrast to the transverse nature of an electromagnetic wave.

The Schrödinger equation for an electron (charge $-e$) is

$$\left[\frac{(\hat{\mathbf{p}} + e\mathbf{A})^2}{2m_0} + V_{\text{crystal}}\right]\Psi = i\hbar\frac{\partial\Psi}{\partial t}, \tag{8.66}$$

where V_{crystal} is the potential of the crystal or heterostructure under study. The perturbation due to the light wave is the difference between this Hamiltonian and that without $\mathbf{A}$, which gives

$$\hat{V} = \frac{e}{2m_0}(\mathbf{A}\cdot\hat{\mathbf{p}} + \hat{\mathbf{p}}\cdot\mathbf{A} + e\mathbf{A}^2). \tag{8.67}$$

Note that the order of $\mathbf{A}$ and $\hat{\mathbf{p}}$ has been respected, as $\hat{\mathbf{p}}$ is an operator and cannot be moved around like a simple number. The term with $\mathbf{A}^2$ is of second order in the electric field and will be neglected; it can be shown rigorously to vanish within the electric-dipole approximation that we shall shortly make. The $\hat{\mathbf{p}}\cdot\mathbf{A}$ term also needs a little thought, remembering that it is a differential operator that acts on the wave function Ψ. It means

$$\hat{\mathbf{p}}\cdot\mathbf{A}\Psi = -i\hbar\nabla\cdot(\mathbf{A}\Psi) = -i\hbar[(\nabla\cdot\mathbf{A})\Psi + \mathbf{A}\cdot(\nabla\Psi)], \tag{8.68}$$

where the rule for the divergence of the product of a vector and a scalar has been used. Now $\nabla\cdot\mathbf{A} = 0$ for our transverse wave so only $\mathbf{A}\cdot\hat{\mathbf{p}}\Psi$ remains and the total perturbation is $\hat{V} = (e/m_0)\mathbf{A}\cdot\hat{\mathbf{p}}$. This form of interaction may look rather odd, but remember that the current operator (1.42) is something like $(e/m_0)\hat{\mathbf{p}}$. Thus our perturbation can be written roughly in the form $\mathbf{A}\cdot\mathbf{J}$, which is used in classical electrodynamics and is closely related to the simplest expression $\mathbf{E}\cdot\mathbf{J}$.

Expanding the perturbation gives

$$\hat{V}(\mathbf{R}, t) = \frac{e E_0}{i m_0\omega}[e^{i(\mathbf{Q}\cdot\mathbf{R} - \omega t)} - e^{-i(\mathbf{Q}\cdot\mathbf{R} - \omega t)}](\mathbf{e}\cdot\hat{\mathbf{p}}). \tag{8.69}$$

This is very close to the form (8.47) of the electron–phonon perturbation. The first term represents the absorption of a photon by the electron, increasing the electron's energy by $\hbar\omega$ and its momentum by $\hbar\mathbf{Q}$. Similarly, the second term describes the emission of a photon so the electron loses both energy and momentum.

In optical phenomena the formulas can usually be simplified because the photon carries negligible momentum and induces vertical transitions in bands (Section 2.7).

Typical optical experiments are done with light of visible or nearby wavelengths, which have wavelengths in the hundreds of nanometres. This is much longer than a unit cell in a semiconductor crystal or the width of a quantum well, which set the scale of the electronic wave functions. Thus we can usually neglect the momentum of the photon and set $\mathbf{Q} = \mathbf{0}$ in the perturbation, treating the electric field as constant across the electronic states. This is the *electric-dipole approximation* (and is the opposite of the quasi-elastic approximation that we made for acoustic phonons).

We can now put the perturbation into Fermi's golden rule. The transition rate from a state i to another state j due to the absorption of a photon is

$$W_{ji} = \frac{2\pi}{\hbar} \left(\frac{eE_0}{m_0\omega}\right)^2 |\langle j|\mathbf{e}\cdot\hat{\mathbf{p}}|i\rangle|^2 \delta(E_j - E_i - \hbar\omega). \qquad (8.70)$$

The matrix element looks a little complicated. The middle expands to

$$\mathbf{e}\cdot\hat{\mathbf{p}} = -i\hbar \left(e_x\frac{\partial}{\partial x} + e_y\frac{\partial}{\partial y} + e_z\frac{\partial}{\partial z}\right). \qquad (8.71)$$

For example, if the electric field were polarized purely along z, $\mathbf{e} = (0, 0, 1)$ and $\mathbf{e}\cdot\hat{\mathbf{p}} = -i\hbar\,\partial/\partial z$.

The power absorbed by this transition is the product of the transition rate and the energy of each photon $\hbar\omega$. The next step is to sum over all possible initial and final states to get the total power. We must include Fermi factors $f(E_i)$ to ensure that the initial state is filled and $[1 - f(E_j)]$ to ensure that the final state is empty, otherwise the transition cannot take place. Thus the system absorbs power at a rate

$$P_+ = \frac{2\pi}{\hbar}\hbar\omega\left(\frac{eE_0}{m_0\omega}\right)^2 2\sum_{i,j}|\langle j|\mathbf{e}\cdot\hat{\mathbf{p}}|i\rangle|^2 f(E_i)[1 - f(E_j)]\delta(E_j - E_i - \hbar\omega). \qquad (8.72)$$

The factor of 2 in front of the summation is for spin, which must be counted only once as the spin is not changed by the optical transition (this factor is always a nuisance and it is difficult to know the best place for it; conventions vary).

The system also emits energy at a rate P_- due to the $e^{+i\omega t}$ term in the perturbation. This is identical to P_+ except for the change in sign of ω:

$$P_- = -\frac{2\pi}{\hbar}\hbar\omega\left(\frac{eE_0}{m_0\omega}\right)^2 2\sum_{i,j}|\langle j|\mathbf{e}\cdot\hat{\mathbf{p}}|i\rangle|^2 f(E_i)[1 - f(E_j)]\delta(E_j - E_i + \hbar\omega). \qquad (8.73)$$

The argument of the δ-function can be restored to the form in P_+ by interchanging the labels i and j, which are only dummy summation indices. The matrix elements are not affected but the Fermi functions change to $f(E_j)[1 - f(E_i)]$. These factors and the overall sign are now the only differences between P_+ and P_-, and together they give

$$f(E_i)[1 - f(E_j)] - f(E_j)[1 - f(E_i)] = f(E_i) - f(E_j). \qquad (8.74)$$

Thus the overall rate of absorption of energy is

$$P = \frac{2\pi}{\hbar} \hbar\omega \left(\frac{eE_0}{m_0\omega}\right)^2 2 \sum_{i,j} |\langle j|\mathbf{e} \cdot \hat{\mathbf{p}}|i\rangle|^2 [f(E_i) - f(E_j)]\delta(E_j - E_i - \hbar\omega).$$

(8.75)

Comparison with the classical result (8.65) shows that

$$\sigma_1(\omega) = \frac{\pi e^2}{m_0^2\omega} \frac{2}{\Omega} \sum_{i,j} |\langle j|\mathbf{e} \cdot \hat{\mathbf{p}}|i\rangle|^2 [f(E_i) - f(E_j)]\delta(E_j - E_i - \hbar\omega). \qquad (8.76)$$

This is the general formula for the real part of the conductivity, which can be converted to the imaginary part of the dielectric function using equation (8.64). The factor of 2 is written in front of the summation as a reminder that it arises from the sum over spins. Although the volume Ω of the system appears in the denominator, it must cancel after the summation as it did for impurity scattering. The mass in equation (8.76) is m_0, that for *free* electrons, rather than an effective mass. This is because it comes from the kinetic energy term of the original Schrödinger equation rather than an effective mass approximation. We shall derive the corresponding formula for ϵ_1 or σ_2 in Section 10.1.

We argued earlier that we normally expect $\sigma_1 > 0$. The only factors that affect the sign are ω and the occupation functions. The δ-function forces $E_j > E_i$ (if $\omega > 0$) and a system in equilibrium has $f(E_i) > f(E_j)$. This guarantees that $\sigma_1 > 0$ and the system will absorb energy from light. It is easy to show that σ_1 is an even function of ω and is therefore always positive in equilibrium. In a laser, on the other hand, one arranges for the population of some states to be *inverted* so that $f(E_j) > f(E_i)$ and in this case the system can release energy to amplify an incident beam of light (Section 10.6).

8.6 Interband Absorption

We saw in Section 2.7 that optical absorption is a convenient way of measuring direct band gaps and can now calculate this for the simple conduction and valence bands shown in Figure 8.4. Assume that the material is three-dimensional, undoped, and at zero temperature so the valence band is full and the conduction band is empty. The sums over states i and j in equation (8.76) now each become a sum over the bands with another over the Bloch wave vector $\mathbf{K}$ within each band. In our example we can assume that the initial state is in the valence band, $i = \text{v}\mathbf{K}$, and the final state is in the conduction band, $j = \text{c}\mathbf{K}'$, as the occupation factors give 1 in this case but vanish otherwise. Thus

$$\sigma_1(\omega) = \frac{\pi e^2}{m^2\omega} \frac{2}{\Omega} \sum_{\mathbf{K},\mathbf{K}'} \left|\langle c\mathbf{K}'|\mathbf{e} \cdot \hat{\mathbf{p}}|0\text{v}\mathbf{K}\rangle\right|^2 \delta[\varepsilon_c(\mathbf{K}') - \varepsilon_v(\mathbf{K}) - \hbar\omega]. \qquad (8.77)$$

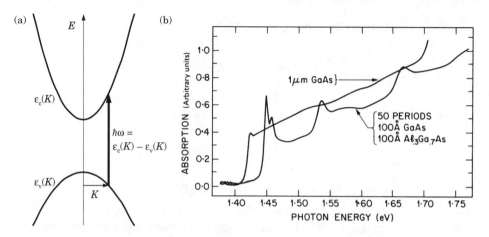

FIGURE 8.4. (a) Absorption from the valence band to the conduction band in a semiconductor with a direct band gap. (b) Absorption of two- and three-dimensional samples of GaAs at room temperature. The active volume of the two-dimensional (multiquantum well) sample is only half that of the three-dimensional one. [Reprinted from Schmitt-Rink, Chemla, and Miller (1989) by kind permission of Taylor and Francis.]

Look at the matrix element first. The Bloch states can be written in the form

$$\psi_{vK}(\mathbf{R}) = \Omega^{-1/2} e^{i\mathbf{K}\cdot\mathbf{R}} u_{vK}(\mathbf{R}), \tag{8.78}$$

where $u_{vK}(\mathbf{R})$ is periodic from one unit cell of the crystal to the next. The factor of $\Omega^{-1/2}$ normalizes the states in such a way that $u_{vK}(\mathbf{R}) = 1$ in the limit of free electrons. The operator $\hat{\mathbf{p}} = -i\hbar\nabla$ takes the first derivative, so the product rule gives

$$\hat{\mathbf{p}}\psi_{vK}(\mathbf{R}) = \Omega^{-1/2} e^{i\mathbf{K}\cdot\mathbf{R}} [\hbar\mathbf{K} u_{vK}(\mathbf{R}) + \hat{\mathbf{p}} u_{vK}(\mathbf{R})]. \tag{8.79}$$

Thus the matrix element is

$$\langle c\mathbf{K}'|\mathbf{e}\cdot\hat{\mathbf{p}}|v\mathbf{K}\rangle = \frac{1}{\Omega}\mathbf{e}\cdot\int u_{c\mathbf{K}'}^*(\mathbf{R}) \, e^{i(\mathbf{K}-\mathbf{K}')\cdot\mathbf{R}} \, [\hbar\mathbf{K} u_{vK}(\mathbf{R}) + \hat{\mathbf{p}} u_{vK}(\mathbf{R})] d^3\mathbf{R}. \tag{8.80}$$

The first term in square brackets vanishes after integration because it reduces to the product of two states that must be orthogonal. The remaining term also vanishes unless $\mathbf{K}' = \mathbf{K}$ because of the plane waves in the wave functions; the proof is along the same lines as the discussion of umklapp scattering earlier. We now have $\langle c\mathbf{K}'|\mathbf{e}\cdot\hat{\mathbf{p}}|v\mathbf{K}\rangle = \delta_{\mathbf{K}',\mathbf{K}}\, p_{cv}(\mathbf{K})$, where

$$p_{cv}(\mathbf{K}) = \frac{1}{\Omega}\int u_{c\mathbf{K}}^*(\mathbf{R}) \, (\mathbf{e}\cdot\hat{\mathbf{p}}) u_{vK}(\mathbf{R}) d^3\mathbf{R}. \tag{8.81}$$

We can again use the fact that $u(\mathbf{R})$ is periodic between unit cells to reduce the integral to the volume of a unit cell rather than the crystal,

$$p_{cv}(\mathbf{K}) = \frac{1}{\Omega_{cell}}\int_{cell} u_{c\mathbf{K}}^*(\mathbf{R}) \, (\mathbf{e}\cdot\hat{\mathbf{p}}) u_{vK}(\mathbf{R}) d^3\mathbf{R}. \tag{8.82}$$

The expression for the conductivity now becomes

$$\sigma_1(\omega) = \frac{\pi e^2}{m_0^2 \omega} \frac{2}{\Omega} \sum_{\mathbf{K}} |p_{\mathrm{cv}}(\mathbf{K})|^2 \delta[\varepsilon_{\mathrm{c}}(\mathbf{K}) - \varepsilon_{\mathrm{v}}(\mathbf{K}) - \hbar\omega]. \qquad (8.83)$$

This is the general result for σ_1 in a crystal except that we have removed the occupation functions and sums over bands, which can easily be restored.

We have several times assumed that the periodic part of the Bloch functions are only weakly dependent on $\mathbf{K}$ near the edge of a band (equation 3.5), and put $u_{n\mathbf{K}} \approx u_{n0}$. The matrix element reduces to $p_{\mathrm{cv}}(\mathbf{0})$ if we repeat this assumption here, and can be pulled out of the summation. It leaves

$$\sigma_1(\omega) \approx \frac{\pi e^2}{m_0^2 \omega} |p_{\mathrm{cv}}(\mathbf{0})|^2 \left\{ \frac{2}{\Omega} \sum_{\mathbf{K}} \delta[\varepsilon_{\mathrm{c}}(\mathbf{K}) - \varepsilon_{\mathrm{v}}(\mathbf{K}) - \hbar\omega] \right\}. \qquad (8.84)$$

The summation in braces defines a density of states per unit volume (Section 1.7.3) known as the *optical joint density of states* $n_{\mathrm{opt}}(E)$. It involves the energies of two bands but is almost identical in form to the usual result and contains a single factor of 2 for spin. The conductivity can now be written in the simple form

$$\sigma_1(\omega) \approx \frac{\pi e^2}{m_0^2 \omega} |p_{\mathrm{cv}}(\mathbf{0})|^2 n_{\mathrm{opt}}(\hbar\omega). \qquad (8.85)$$

The matrix element is the same as that in $\mathbf{k} \cdot \mathbf{p}$ theory (Section 7.3) and we can write $p_{\mathrm{cv}}(\mathbf{0}) = (im_0/\hbar)P$. The strength of absorption is closely related to other quantities that depend on P such as the effective masses, and σ_1 can be rewritten in terms of E_P, which varies little among the common semiconductors. The differences arise from the density of states.

The optical joint density of states looks much like an ordinary density of states, and its behaviour dominates $\sigma_1(\omega)$. Suppose that the two bands are parabolic and spherically symmetric in the region of interest, an acceptable approximation near Γ in GaAs, with

$$\varepsilon_{\mathrm{v}}(\mathbf{K}) = E_{\mathrm{v}} - \frac{\hbar^2 K^2}{2m_0 m_{\mathrm{h}}}, \qquad \varepsilon_{\mathrm{c}}(\mathbf{K}) = E_{\mathrm{c}} + \frac{\hbar^2 K^2}{2m_0 m_{\mathrm{e}}}. \qquad (8.86)$$

Then the difference of energies that appears in the definition of $n_{\mathrm{opt}}(E)$ is

$$\varepsilon_{\mathrm{c}}(\mathbf{K}) - \varepsilon_{\mathrm{v}}(\mathbf{K}) = (E_{\mathrm{c}} - E_{\mathrm{v}}) + \frac{\hbar^2 K^2}{2m_0} \left(\frac{1}{m_{\mathrm{e}}} + \frac{1}{m_{\mathrm{h}}} \right). \qquad (8.87)$$

Therefore $n_{\mathrm{opt}}(E)$ will be the familiar density of states for free electrons but with the bottom of the band offset to the band gap $E_{\mathrm{g}} = E_{\mathrm{c}} - E_{\mathrm{v}}$ and with the optical joint effective mass given by the reduced mass $1/m_{\mathrm{eh}} = 1/m_{\mathrm{e}} + 1/m_{\mathrm{h}}$. In three dimensions the density of states rises from zero like $(E - E_{\mathrm{g}})^{1/2}$, and equation (8.85) shows that the optical conductivity just above the absorption edge at $\hbar\omega = E_{\mathrm{g}}$ should do the same.

Data for GaAs are shown in Figure 8.4(b). Unfortunately, the shape of the absorption curve does not agree well with our predictions. The problem is caused by excitons again and will be addressed in Section 10.7.

This calculation can be extended to low-dimensional systems if we don't worry too much about the matrix element and polarization of the light (Chapter 10). The major change is in $n_{\mathrm{opt}}(E)$, which behaves like the density of states for free electrons (Figure 1.9). Thus the absorption edge in a quantum well should be a steplike feature, and a wire should show a divergence like $(\hbar\omega - E_{\mathrm{g}})^{-1/2}$. Again these results are not quantitatively correct but the trend is extremely important: the optical absorption edge becomes a stronger feature as the number of dimensions is reduced. The absorption edge is also shifted to higher energies to include the energy of the electron and hole above the bottom of the quantum well. The energy levels are discrete in a quantum dot and the absorption reflects this, being a series of lines rather than a band.

8.7   Absorption in a Quantum Well

As a contrasting example, consider absorption between bound states in the quantum well aligned along z, shown in Figure 8.5. This system is immersed in the conduction band of a heterostructure and we ought to use effective-mass wave functions. We shall defer this complication to Section 10.5 as it has no qualitative effect on the behaviour.

We know from Section 4.5 that the states factorize into a product of a bound state in z and a transverse plane wave,

$$\psi_{i\mathbf{k}}(\mathbf{R}) = A^{-1/2}\phi_i(z)\exp(i\mathbf{k}\cdot\mathbf{r}), \qquad E_i(\mathbf{k}) = \varepsilon_i + \frac{\hbar^2 k^2}{2m}, \qquad (8.88)$$

where $\mathbf{r} = (x, y)$ and A is the area of the quantum well. Let L be the thickness of the whole sample (not just the well) along z so the volume $\Omega = AL$. Each state i for motion along z gives rise to a subband of energies.

Consider the matrix element between two such states $\langle j\mathbf{k}'|\mathbf{e}\cdot\hat{\mathbf{p}}|i\mathbf{k}\rangle$. The form of this depends strongly on $\mathbf{e}$, the polarization of the light. Suppose first that $\mathbf{e} = (1, 0, 0)$ so that the electric field is parallel to x, in the plane of the quantum well. The direction of propagation might be along y or z, in the plane of the quantum well or normal to it (see Figure 10.6 for examples of polarization and propagation). In this case, $\mathbf{e}\cdot\hat{\mathbf{p}} = -i\hbar\,\partial/\partial x$ and

$$(\mathbf{e}\cdot\hat{\mathbf{p}})\psi_{i\mathbf{k}}(\mathbf{R}) = \hbar k_x \psi_{i\mathbf{k}}(\mathbf{R}) \qquad (8.89)$$

because the momentum operator affects only the plane wave in $\psi_{i\mathbf{k}}(\mathbf{R})$. Then

$$\langle j\mathbf{k}'|\mathbf{e}\cdot\hat{\mathbf{p}}|i\mathbf{k}\rangle = \hbar k_x \langle j\mathbf{k}'|i\mathbf{k}\rangle = 0; \qquad (8.90)$$

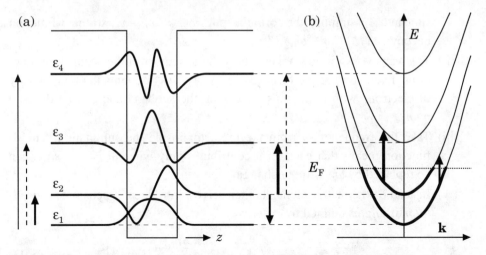

FIGURE 8.5. Absorption by transitions between states in a quantum well. (a) Wave functions along z with energy levels. The thicknesses of the arrows are rough indications of the oscillator strengths of the transitions, with broken lines signifying forbidden transitions. (b) Band structure in the transverse **k**-plane showing the vertical nature of the allowed transitions, which must go from filled to empty states.

the integral is over the product of two states and vanishes by orthogonality. No light is absorbed with this polarization and the same obviously holds for **e** along y. Thus light that propagates normal to the layers, a convenient orientation for experiments (although not for wave guides), cannot be absorbed in these transitions.

The results are different for $\mathbf{e} = (0, 0, 1)$, the electric field normal to the quantum well, which requires light to propagate in the plane of the well. Now $\mathbf{e} \cdot \hat{\mathbf{p}} = -i\hbar\, \partial/\partial z$, which affects only the wave function of the bound state. Thus

$$\langle j\mathbf{k}'|\mathbf{e} \cdot \hat{\mathbf{p}}|i\mathbf{k}\rangle = \frac{1}{A} \int dz \int d^2 r\, \phi_j^*(z)\, e^{i(\mathbf{k}-\mathbf{k}')\cdot\mathbf{r}}\, \hat{p}_z \phi_i(z). \tag{8.91}$$

The integral over **r** gives A if $\mathbf{k}' = \mathbf{k}$ and zero otherwise, so the two-dimensional wave vector is conserved. Thus optical transitions are again vertical in **k**, as shown in Figure 8.5. The remaining matrix element can be abbreviated to $\langle j|\hat{p}_z|i\rangle$. Substituting this matrix element and the energies from equation (8.88) into the general formula (8.76) gives

$$\sigma_1(\omega) = \frac{\pi e^2}{m^2\omega} \frac{2}{\Omega} \sum_{i,j,\mathbf{k}} |\langle j|\hat{p}_z|i\rangle|^2 \{f[E_i(\mathbf{k})] - f[E_j(\mathbf{k})]\}$$

$$\times \delta[E_j(\mathbf{k}) - E_i(\mathbf{k}) - \hbar\omega]. \tag{8.92}$$

This can be reduced to a simple formula in a few steps.

(i) The wave vector **k** cancels from the δ-function because $E_j(\mathbf{k}) - E_i(\mathbf{k}) = \varepsilon_j - \varepsilon_i$, assuming an ideal system where the mass is the same for each subband. To simplify the formulas put $\hbar\omega_{ji} = \varepsilon_j - \varepsilon_i$, so the δ-function becomes $(1/\hbar)\delta(\omega - \omega_{ji})$.

(ii) The wave vector $\mathbf{k}$ now appears only in the sums over the occupation functions. Each sum, of the form $(2/A)\sum_{\mathbf{k}} f[E_j(\mathbf{k})]$, counts the total density of electrons with both spins occupying the subband and can be denoted by n_j.

(iii) Take the frequency ω from the prefactor inside the summation. The δ-function means that each term contributes only when $\omega = \omega_{ji}$, so ω can be replaced by ω_{ji} without changing the value.

(iv) Instead of the matrix elements themselves, introduce dimensionless *oscillator strengths* defined by

$$f_{ji} = \frac{2}{m\hbar\omega_{ji}}|\langle j|\hat{p}_z|i\rangle|^2 = \frac{2m\omega_{ji}}{\hbar}|\langle j|z|i\rangle|^2. \qquad (8.93)$$

The second form of the oscillator strength, with a matrix element of z rather than $\hat{p}_z$, follows from a little trickery with operators that is left as an exercise. The diagonal oscillator strength is best excluded from the summation to ensure that we omit transitions from a state to itself.

After these manipulations we are left with the simple formula

$$\sigma_1(\omega) = \frac{\pi e^2}{2mL} \sum_{\substack{i,j \\ j\neq i}} f_{ji}(n_i - n_j)\delta(\omega - \omega_{ji}). \qquad (8.94)$$

Absorption is seen only at frequencies corresponding to the separation of bound states in the well. Thus there are discrete lines, despite the continuous spectrum of states available, because of the restriction to vertical transitions. The lines will be broadened by any difference in effective mass between the subbands (Section 4.9), and transitions into the continuum above the quantum well occur at high energies.

There is an overall factor of $1/L$ because the sample contains only one quantum well, whose influence diminishes as the sample becomes thicker. Multiquantum wells are used to enhance absorption in practice, and the length L should be replaced by the period of the structure. An experimental example of intersubband absorption in a multiquantum well is shown in Figure 8.6.

The strength of each transition depends on the wave functions, through the oscillator strength, and on the filling of the subbands. The strengths can be manipulated through f_{ji} by changing the shape of the quantum well, or by modifying the occupations through doping, injection of carriers, pumping, or simply a change in temperature. A population inversion ($n_j > n_i$) can make $\sigma_1(\omega) < 0$ in some range of frequencies. However, there is a remarkable result that the total absorbing power of a sample, defined by the integral of $\sigma_1(\omega)$ over all frequencies, depends only on

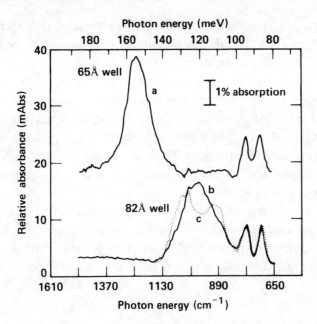

FIGURE 8.6. Measurement of absorption between subbands within a quantum well at room temperature, using layers with 50 quantum wells of width 6.5 nm for (a) and 8.2 nm for (b) and (c). Each well contains 4×10^{15} m^{-2} electrons. The smaller peaks on the right are due to phonons. [Reprinted with permission from West and Eglash (1985). Copyright 1985 American Institute of Physics.]

its total density of electrons:

$$\int_0^\infty \sigma_1(\omega)\, d\omega = \frac{\pi e^2 n_{2D}}{2mL}, \qquad n_{2D} = \sum_j n_j. \qquad (8.95)$$

This is an example of a *sum rule* and follows from the famous Thomas–Reiche–Kuhn f-sum rule on the oscillator strengths,

$$\sum_{j,\, j \neq i} f_{ji} = 1. \qquad (8.96)$$

Thus a change in σ_1 at one frequency must be compensated by opposite changes elsewhere. There are many such constraints on $\tilde{\epsilon}_r(\omega)$, which will be considered in more detail in Section 10.1.3.

Oscillator strengths are commonly used in expressions for optical properties, and the general formula (8.76) can be rewritten as

$$\sigma_1(\omega) = \frac{\pi e^2}{2m} \frac{2}{\Omega} \sum_{i,j} f_{ji}[f(E_i) - f(E_j)]\delta(\omega - \omega_{ji}). \qquad (8.97)$$

It can alternatively be expressed as a sum over transitions rather than initial and final states (Section 10.1.2). The formula (8.83) for interband absorption can also be written in this way, although the notation is perhaps less appropriate for a continuous range of energies. Other results, such as the $\mathbf{k} \cdot \mathbf{p}$ expansion of the energy bands (equation 7.43), can also be expressed in terms of oscillator strengths.

One might worry about the applicability of the golden rule in the case of discrete energies, because it is valid only if a continuum of energies is involved. Fortunately the radiation field usually provides such a continuum. An exception arises if the electromagnetic field is trapped within a microcavity or a periodic structure, both of which induce gaps in its spectrum. There may then be no continuum of energies, the golden rule fails, and the electrodynamics show interesting new physics with potential applications to optoelectronic devices.

Returning to the quantum well, we have found that optical absorption occurs only between different subbands and therefore at energies given by the differences between the bound states in the well. Also, the light must have a component of electric field along z. The remaining task is to evaluate the matrix element

$$\langle j|\hat{p}_z|i\rangle = -i\hbar \int \phi_j^*(z) \frac{d}{dz} \phi_i(z)\, dz. \tag{8.98}$$

An important result follows from the symmetry of the quantum well. We know that the wave functions in a symmetric well such as that shown in Figure 8.5, where $V(-z) = V(z)$, are either even or odd in z. The derivative changes the parity, and the matrix element will be non-zero only if one state is even and the other odd. This is a *selection rule* that governs which transitions can be seen in optical absorption, and is exactly the same result that we found for the polarizability of a quantum well in an electric field (Section 7.2.2). Thus absorption is permitted from the lowest state ($n = 1$) to $n = 2, 4, \ldots$ but not to odd values of n. This result applies to any symmetric well and can be defeated by deliberately growing an asymmetric well.

To conclude, we shall calculate the oscillator strength from the transition $n = 1$ to $n = 2$ in an infinitely deep well at $0 < z < a$. We need the matrix element

$$\langle 2|\hat{p}_z|1\rangle = -i\hbar \frac{2}{a} \int_0^a \sin \frac{2\pi z}{a} \left(\frac{d}{dz} \sin \frac{\pi z}{a} \right) dz = \frac{-8i\hbar}{3a}. \tag{8.99}$$

Alternatively, from equation (E1.4) or (8.93),

$$\langle 2|z|1\rangle = -\frac{16a}{9\pi^2}. \tag{8.100}$$

The latter form, the *dipole matrix element*, is more commonly quoted. It depends on the width of the well and becomes large in a wide well. An obvious hope is that this 'giant' dipole means in turn that absorption will be strong. The oscillator strength is

$$f_{21} = \frac{2m}{\hbar^2}(\varepsilon_2 - \varepsilon_1)\left|\langle 2|z|1\rangle\right|^2 = \frac{256}{27\pi^2} \approx 0.96, \tag{8.101}$$

which can be inserted into equation (8.94) to get σ_1. An amusing point is that the oscillator strength is independent of a and does not increase for a wide well.

However, we can also see that this is a strong transition because its oscillator strength of 0.96 exhausts most of the f-sum rule and the other transitions from state 1 must be much weaker. Thus the lowest transition in a quantum well may be an efficient way of absorbing light. In practice there are many obstacles to be overcome, such as scattering from phonons and escape from the wells, and a complete theory cannot ignore excitons.

We have now seen two contrasting examples of optical absorption, one giving rise to a continuous band of absorption and the other giving discrete energies reflecting bound states in a well. These features are combined in transitions in a quantum well from the valence band to the conduction band, probably the most important case in practice, and this will be covered in Chapter 10.

Before leaving Fermi's golden rule we shall take a brief excursion into diagrams. This is emphatically *not* a treatment of formal perturbation theory and Green's functions, but merely an illustration of the link between Fermi's golden rule and simple Feynman diagrams. It also shows how the two types of perturbation theory that we have developed in these two chapters are related. In particular we shall see that our results from perturbation theory can be combined to give (nearly) the 'Born approximation to the self-energy', which is possibly the quantity calculated most often with Green's functions. This section is not essential for the remaining chapters.

We have derived two apparently different forms of perturbation theory for static potentials in the last two chapters, the time-independent theory in Section 7.2 and Fermi's golden rule in Section 8.1. The first gives the change in energy and wave function of an eigenstate, whereas the second gives the lifetime (scattering rate) with respect to the original states. Although these processes appear rather different they are in fact closely related. Consider the effect of an impurity potential whose Fourier transform is $\tilde{V}(\mathbf{q})$ on plane waves of energy $\varepsilon(\mathbf{k})$. In this case equations (7.31) for the change in energy and (8.24) for the total scattering rate (or inverse single-particle lifetime) become

$$\bar{\Sigma}(\mathbf{k}) \equiv \Delta\varepsilon(\mathbf{k}) = \sum_{\mathbf{q},\mathbf{q}\neq 0} \frac{|\tilde{V}(\mathbf{q})|^2}{\varepsilon(\mathbf{k}) - \varepsilon(\mathbf{k}+\mathbf{q})},$$

$$\Gamma(\mathbf{k}) \equiv \frac{\hbar}{\tau_i} = 2\pi \sum_{\mathbf{q}} |\tilde{V}(\mathbf{q})|^2 \delta[\varepsilon(\mathbf{k}) - \varepsilon(\mathbf{k}+\mathbf{q})]. \quad (8.102)$$

The scattering rate has been converted to an energy Γ by multiplying by $\hbar$. The two formulas are similar and can be combined to give a single complex energy $\Sigma(\mathbf{k})$, defined by

$$\Sigma = \bar{\Sigma} - \tfrac{1}{2}i\Gamma = \sum_{\mathbf{q}} |\tilde{V}(\mathbf{q})|^2 \left\{ \mathcal{P}\frac{1}{\varepsilon(\mathbf{k}) - \varepsilon(\mathbf{k}+\mathbf{q})} - i\pi\,\delta[\varepsilon(\mathbf{k}) - \varepsilon(\mathbf{k}+\mathbf{q})] \right\}$$

$$= \sum_{\mathbf{q}} \frac{|\tilde{V}(\mathbf{q})|^2}{\varepsilon(\mathbf{k}) - \varepsilon(\mathbf{k}+\mathbf{q}) + i0_+}. \tag{8.103}$$

The notation '0_+' means an infinitesimally small positive quantity. The second form, where the principal part of the reciprocal and the δ-function have been combined, is entirely equivalent in meaning to the first; the same trick will be used for the complex dielectric function and explained in Section 10.1.2. The '$\mathcal{P}$' is essentially a reminder to omit the term with $\mathbf{q} = \mathbf{0}$ that would lead to a vanishing denominator (it really means the Cauchy principal part of an integral over energy). We are now very close to the (retarded) *self-energy*, a central quantity in field theory, which we have calculated within the Born approximation. More precisely, the self-energy has $\mathbf{k}$ and E as independent variables:

$$\Sigma(\mathbf{k}, E) = \sum_{\mathbf{q}} \frac{|\tilde{V}(\mathbf{q})|^2}{E - \varepsilon(\mathbf{k}+\mathbf{q}) + i0_+}. \tag{8.104}$$

We have evaluated it at the unperturbed energy of the state k by setting $E = \varepsilon(\mathbf{k})$.

We already know that the real part of the self-energy shifts the energy of the state; what about the imaginary part? Putting the complex energy into the standard time dependence $\exp(-iEt/\hbar)$ gives

$$\Psi(t) \propto \exp\left(-\frac{i[\varepsilon(\mathbf{k}) + \Sigma(\mathbf{k})]t}{\hbar} \right)$$

$$= \exp\left(-\frac{i[\varepsilon(\mathbf{k}) + \bar{\Sigma}(\mathbf{k})]t}{\hbar} \right) \exp\left(-\frac{\Gamma(\mathbf{k})t}{2\hbar} \right). \tag{8.105}$$

The imaginary part causes the wave function to decay as a function of time, and the density $|\Psi(t)|^2$ goes like $\exp(-\Gamma t/\hbar)$ with a lifetime $\tau_i = \hbar/\Gamma$. Thus the fraction of the wave function in the original state decays as the probability of scattering into another state increases, which is exactly what we calculated in Fermi's golden rule.

The imaginary part of the self-energy therefore gives the rate of real transitions out of the original state, which had wave vector $\mathbf{k}$ in our example of impurity scattering. It is natural to interpret the real part of Σ, which shifts the energy, in the same way and say that it arises from *virtual* transitions. The picture is that the electron makes a transition from state $\mathbf{k}$ to $\mathbf{k} + \mathbf{q}$ but, because it has the wrong energy for this state, it returns after a finite time $\hbar/[\varepsilon(\mathbf{k}) - \varepsilon(\mathbf{k}+\mathbf{q})]$. The more nearly energy is conserved, the longer the electron can remain in the other state, and the bigger is the contribution to the change in energy and wave function.

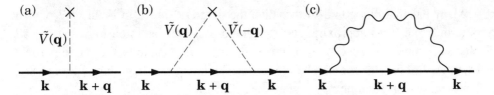

FIGURE 8.7. Diagrams to illustrate an electron scattering from an impurity or a phonon. (a) Single vertex where an electron enters with wave vector $\mathbf{k}$ and leaves with $\mathbf{k} + \mathbf{q}$. The matrix element for the interaction is $\tilde{V}(\mathbf{q})$. (b) Combination of two such vertices to give a self-energy. (c) Self-energy for electron–phonon scattering; the phonon is represented by the wavy line.

We now have a unified description of the self-energy in terms of transitions to other states, both real and virtual processes, and this picture translates directly into diagrams. Figure 8.7(a) is an obvious representation of the scattering of an electron by an impurity. The electron is shown as a straight line. It comes in with wave vector $\mathbf{k}$ and energy $\varepsilon(\mathbf{k})$, interacts at the vertex with the impurity of strength $\tilde{V}(\mathbf{q})$, and leaves with wave vector $\mathbf{k} + \mathbf{q}$. Conservation of momentum fixes the wave vector of the outgoing electron but its energy differs from the initial value in virtual transitions. The impurity is shown as a cross and is separated from the vertex by a dashed line for clarity.

Now, both second-order perturbation theory and the golden rule, and therefore the self-energy, contain two matrix elements like $|\tilde{V}(\mathbf{q})|^2$, and one might wonder what has happened to the second one. The answer is that this diagram contains only half the action. The matrix elements are Hermitian, so the complex conjugate can be rewritten as $V_{fi}^* = V_{if}$, corresponding to the process with the initial and final states reversed. For the impurity this means $\tilde{V}^*(\mathbf{q}) = \tilde{V}(-\mathbf{q})$. Putting the two processes together gives the diagram in Figure 8.7(b). The two vertices are joined by dashed lines to the same impurity to remind us that it is the same potential acting twice. This would be exactly the form of a self-energy diagram in many-body theory if the lines on the ends, corresponding to the initial state, were omitted. We can also rewrite the expression for the self-energy to reflect these two individual events as

$$\Sigma(\mathbf{k}, E) = \sum_{\mathbf{q}} \tilde{V}(-\mathbf{q}) \frac{1}{E - \varepsilon(\mathbf{k} + \mathbf{q}) + i0_+} \tilde{V}(\mathbf{q}). \qquad (8.106)$$

The electron scatters by gaining wave vector $\mathbf{q}$, propagates for some time with wave vector $\mathbf{k} + \mathbf{q}$, and finally loses $\mathbf{q}$ again to return to its initial state. Indeed the quotient in the middle of equation (8.106) is called the (retarded) single-particle propagator or Green's function. This is a good place to stop and refer interested readers to a book on many-body theory!

Phonons can be pictured in a similar way (Figure 8.7(c)). The phonon is shown as a wavy line joining the two vertices. In this case both energy and momentum are transferred at the vertices, and the phonon should itself be described by a Green's function, which leads to the phonon numbers for emission and absorption.

Many calculations go no further than the diagram shown in Figure 8.7, which is termed the Born approximation for phonons as well as impurities, but not all physical properties reduce to something so simple. For example, the single-particle lifetime τ_i of an electron can be found by calculating the diagrams of Figure 8.7, but the transport lifetime τ_{tr} is much more difficult.

Further Reading

Fermi's golden rule is derived in textbooks on quantum mechanics such as Merzbacher (1970), Gasiorowicz (1974), Landau, Lifshitz, and Pitaevskii (1977), and Bransden and Joachain (1989). Datta (1989) derives the golden rule using scattering rates in semiconductors as an example.

Bastard (1988) and Weisbuch and Vinter (1991) both describe scattering rates in a variety of low-dimensional systems. An exhaustive description of scattering rates in semiconductors, with an interesting treatment of impurities, is given by Ridley (1993).

A standard reference on the use of diagrams and Green's functions is by Mahan (1990). The breadth of coverage is impressive. Rickayzen (1980) gives a more concise, but also less comprehensive, account.

EXERCISES

8.1   An electron is in the lowest state of a GaAs quantum well of width 15 nm, which can be treated as an infinitely deep one-dimensional system. Suddenly the middle 5 nm becomes 100 meV deeper. Use the golden rule to derive an expression for the rate at which the electron scatters into other states, and evaluate the rate for the lowest allowed transition. How would the result differ if the middle rose instead of fell, and does perturbation theory look appropriate?

Repeat the calculation of scattering rates, supposing that the initially flat-bottomed well was suddenly tilted by an electric field $F = 1\,\mathrm{MV\,m^{-1}}$. How is this calculation related to that in Section 7.2.2 for the energy levels of a quantum well in an electric field?

8.2   Derive the scattering rate for transport in a three-dimensional system with spherically symmetric potentials,

$$\frac{1}{\tau_{tr}} = n_{\mathrm{imp}}^{(3D)} \frac{m}{4\pi\hbar^3 K^3} \int_0^{2K} |\tilde{V}(Q)|^2 Q^3 \, dQ \tag{E8.1}$$

$$= n_{\mathrm{imp}}^{(3D)} \frac{mK}{2\pi\hbar^3} \int_0^{\pi} \left|\tilde{V}[2K\sin(\tfrac{1}{2}\theta)]\right|^2 (1-\cos\theta)\sin\theta \, d\theta. \tag{E8.2}$$

The method is very similar to that used for equations (8.31) and (8.32).

8.3   The circular barrier provides a crude model to estimate alloy scattering for
a 2DEG in AlGaAs. Divide the plane into unit cells; those that contain
GaAs have zero potential, whereas those with AlAs have a potential of
1 eV, which is roughly the difference in energy of the Γ-valley in the two
materials. Approximate the unit cell by a circle and calculate the scattering
rate and mobility in $Al_{0.3}Ga_{0.7}As$. (Only a small part of the wave function
penetrates into the AlGaAs in a typical GaAs–AlGaAs heterostructure so
the effect is much smaller than that in this calculation.)

A superior approximation might be to measure energies from the average
potential energy, rather than that of GaAs. In this case both the GaAs cells
and the AlAs cells contribute to scattering. How much difference does this
make? A good theory should not depend on such arbitrary choices, and
established methods such as the coherent potential approximation (CPA)
are much more satisfactory.

8.4   Consider two-dimensional Rutherford scattering from an unscreened Coulomb potential $V(r) = -e^2/4\pi\epsilon_0\epsilon_b r$ in the plane of a 2DEG. Show that
the differential cross-section within the Born approximation is

$$\sigma(\theta) = \frac{\pi}{2k(a_B k)^2 \sin^2(\frac{1}{2}\theta)}, \tag{E8.3}$$

where a_B is the Bohr radius, which takes account of the effective mass and dielectric constant (equation 4.67). You will need the integral $\int_0^\infty J_0(x)\,dx =
1$. Do the single-particle or transport cross-sections exist?

This problem can be solved exactly in both classical and quantum
mechanics [F. Stern and W. E. Howard, *Physical Review* **163** (1967):
816–35]. This shows that the Born approximation omits a factor of
$(a_B k/\pi)\tanh(\pi/a_B k)$. The Born approximation is therefore reliable provided that $k \gg 1/a_B$; does this hold for a typical 2DEG with 3×10^{15} m^{-2}
electrons in GaAs? (In three dimensions the Born approximation is identical
to the exact result in both quantum and classical mechanics.)

8.5   Suppose that the long range of the Coulomb potential is cut off by an exponential factor, giving $V(r) = -(e^2/4\pi\epsilon_0\epsilon_b r)e^{-\lambda r}$. Calculate the scattering
rate in two dimensions; you will need the integral $\int_0^\infty e^{-\alpha x} J_0(\beta x)dx =
(\alpha^2 + \beta^2)^{-1/2}$. Show that the result has the same general features as the
circular barrier but without the oscillations. (Note that this is *not* the correct
form of a screened Coulomb potential in a two-dimensional system, which
we shall calculate in Section 9.4.)

8.6   Some time-dependent problems are simple enough to solve without recourse
to perturbation theory. A famous example is due to Tien and Gordon [*Physical Review* **129** (1963): 647–51]. Here is an adaptation to a low-dimensional

system. Suppose that we have a static quantum dot with eigenstates ϕ_n of energy ε_n. The dot is then driven by a radio-frequency field that adds a potential energy $V_0 \cos \omega_0 t$, where V_0 is constant in space. Show that the time-dependent eigenstates become

$$\Phi_n(t) = \phi_n \exp\left(-\frac{i\varepsilon_n t}{\hbar}\right) \exp\left[-i\left(\frac{V_0}{\hbar\omega_0}\right) \sin \omega_0 t\right]. \tag{E8.4}$$

The identity $\exp(iz \sin \theta) = \sum_{k=-\infty}^{\infty} J_k(z) \exp(ik\theta)$ allows this to be rewritten as

$$\Phi_n(t) = \phi_n \sum_{k=-\infty}^{\infty} J_k\left(\frac{V_0}{\hbar\omega_0}\right) \exp\left[-\frac{i(\varepsilon_n + k\hbar\omega_0)t}{\hbar}\right]. \tag{E8.5}$$

Thus each state develops 'sidebands' at frequencies shifted by $k\hbar\omega_0$, with intensities $J_k^2(V_0/\hbar\omega_0)$. This set of energies may allow additional transitions if the dot is coupled to some other probe of its electronic structure.

8.7 Consider a non-degenerate electron gas, and compare the thermal energy to that exchanged with an acoustic phonon in a typical scattering event. When are the quasi-elastic approximation and the associated scattering rate valid?

8.8 Compare the predictions of equation (8.54) for the mobility limited by acoustic phonons with data for semiconductors at room temperature (Appendix 2). Scattering at room temperature in GaAs is dominated by polar LO phonons so this result is inapplicable, but it should work for Si. Take $m_e = 0.3$, $\Xi = 10$ eV, and $\rho v_s^2 = 1.4 \times 10^{11}$ Jm^{-3}.

8.9 Estimate the scattering rate due to absorption of polar LO phonons in GaAs. Assume that the initial state of the electron has very low energy so that its final state has energy only just above $\hbar\omega_{LO}$. In this case the change in wave vector $Q \approx K_{LO}$, where $\hbar^2 K_{LO}^2/2m = \hbar\omega_{LO}$. Show that the scattering rate is roughly

$$W_{K+Q,K}^{+} = \frac{2\pi}{\hbar} N_{LO} \frac{e^2 \hbar\omega_{LO}}{2\Omega\epsilon_0 K_{LO}^2} \left[\frac{1}{\epsilon(\infty)} - \frac{1}{\epsilon(0)}\right]$$
$$\times \delta[\varepsilon(K+Q) - \varepsilon(K) - \hbar\omega_{LO}]. \tag{E8.6}$$

Again $\mathbf{Q}$ has vanished from all but the δ-function so the scattering is isotropic. This must be summed over all phonons, which gives a density of states and reduces to

$$\frac{1}{\tau_{LO}} = \frac{e^2 K_{LO}}{4\pi\epsilon_0\hbar} \left[\frac{1}{\epsilon(\infty)} - \frac{1}{\epsilon(0)}\right] N_{LO}. \tag{E8.7}$$

How well does the mobility compare with experiment for GaAs?

8.10 Our derivation of optical absorption treated the electromagnetic wave classically, which is not always appropriate. The conductivity can instead be written with a photon number rather than E_0^2, in analogy with the phonon case. A similar trick may be used to calculate the conductivity due to a single photon, and then multiply by N_ω for absorption and $N_\omega + 1$ for emission. We need to find E_0 for a single photon, which can be done as for the phonon in Section 2.8.1. Remember that there is also a magnetic field, which carries equal energy. Show that $E_0 = (\hbar\omega/2\epsilon_1\epsilon_0\Omega)^{1/2}$ and write down the formula for $\sigma_1(\omega)$ in terms of photon numbers. Some care is needed with this result, which may need to be summed over all electromagnetic modes with the same frequency but different directions of propagation. Both polarizations must also be included.

8.11 Calculate the *Burstein shift* in the absorption edge of a direct semiconductor with parabolic bands due to heavy doping (p-type, say), at such a low temperature that the holes are degenerate. Neglect excitons. Note that the shift is not simply the Fermi energy of the holes and involves the mass of both conduction and valence bands. What should happen to the shape of the absorption edge?

8.12 Prove equation (8.95) from the f-sum rule. It is easier if you notice that $\sigma_1(\omega)$ is an even function of ω and extend the integral to the range $-\infty$ to ∞. The relation $f_{ij} = -f_{ji}$ also helps.

8.13 Prove that the two forms of the oscillator strength in equation (8.93) are equivalent. First, show that the commutator $[z, \hat{H}] = (i\hbar/m)\hat{p}_z$ provided that the momentum operator in $\hat{H}$ appears only in $\hat{p}^2/2m$. You will need to evaluate $[z, \hat{p}_z^2]$, which can be done most simply by adding and subtracting $\hat{p}_z z \hat{p}_z$. This operator relation can be applied to matrix elements, so $\langle j|\hat{p}_z|i\rangle = (m/i\hbar)\langle j|[z, \hat{H}]|i\rangle$. Expand the commutator and use the property that $|i\rangle$ and $|j\rangle$ are eigenstates of the Hamiltonian to show that

$$\langle j|\hat{p}_z|i\rangle = \frac{im(\varepsilon_j - \varepsilon_i)}{\hbar}\langle j|z|i\rangle. \qquad \text{(E8.8)}$$

8.14 If you have solved the previous problem, you will have no difficulty in proving the f-sum rule. The method is to evaluate in two ways the expectation value of a double commutator in state i,

$$\langle i \,|\, [[z, \hat{H}], z] \,|\, i\rangle. \qquad \text{(E8.9)}$$

First, use the previous result for $[z, \hat{H}]$ and commute this with z to obtain $\hbar^2/m$. This is just a constant and the remaining integral $\langle i|i\rangle$ is unity by normalization. In the second evaluation, multiply out the commutator to get four terms. Insert complete sets of energy eigenstates between the operators, using the closure relation (1.79),

$$\delta(z - z') = \sum_{j=1}^{\infty} \phi_j(z)\, \phi_j^*(z') \equiv \sum_{j=1}^{\infty} |j\rangle\langle j|, \qquad \text{(E8.10)}$$

and integrate over the extra variable z' introduced. The matrix elements of the Hamiltonian operator are trivial between energy eigenstates and the only non-trivial matrix elements are those of z. Show that the expectation value reduces to

$$2 \sum_j (\varepsilon_j - \varepsilon_i)\big|\langle j|z|i\rangle\big|^2. \qquad \text{(E8.11)}$$

Equating the two expressions gives the f-sum rule. (The proof can be made much more general than this.)

8.15   The infinitely deep well is so simple a system that it is possible to verify the f-sum rule directly. Try this for $i = 1$; you will need the rapidly converging series

$$\sum_{n=1}^{\infty} \frac{n^2}{(4n^2 - 1)^3} = \frac{\pi^2}{256}. \qquad \text{(E8.12)}$$

You might also like to look at transitions starting from the second level $i = 2$. The lowest oscillator strength is negative, $f_{12} = -f_{21} \approx -0.96$, but the sum over f_{j2} must still give unity.

8.16   What are the selection rules for transitions from the lowest state in a parabolic well? Calculate the oscillator strength of the first allowed transition, and deduce from the f-sum rule that all others must vanish.

8.17   Light with $\hbar\omega > E_g$ is incident on undoped GaAs. Its intensity (proportional to the square of the amplitude) decays like $\exp(-\alpha z)$. Derive the relation $\alpha = 2\omega\kappa_r/c = \sigma_1/c\epsilon_0 n_r$ and show that the decay length $\alpha^{-1} \approx 1\,\mu\text{m}$ for $\hbar\omega = E_g + 0.05\,\text{eV}$. Assume that the holes are heavy, and use data from Appendix 2 with $n_r \approx 3.5$.

8.18   How does absorption between subbands in a quantum well (equation 8.94) change as a function of the density of electrons at low temperature, as more than one subband becomes occupied? How do the results satisfy the sum rule (8.95) for the absorption integrated over all frequencies?

THE TWO-DIMENSIONAL ELECTRON GAS

The two-dimensional electron gas (2DEG) trapped at a doped heterojunction is the most important low-dimensional system for electronic transport. It forms the core of a field-effect transistor, which goes by many acronyms including modulation-doped field-effect transistor (MODFET) and high electron mobility transistor (HEMT). I shall use the first, which emphasizes the close relation to the silicon MOSFET but with MODulation doping. The silicon MOSFET is perhaps the most common electronic device, with electrons or holes trapped in an inversion layer at an interface between Si and SiO_2. Many of the ideas in this chapter were originally derived for the MOSFET but it has been almost completely superseded in physics experiments by the MODFET because of the enormous improvement in the mobility of electrons and holes. The highest mobility of electrons in a MOSFET is around $4\,\mathrm{m^2\,V^{-1}\,s^{-1}}$, whereas values over $1000\,\mathrm{m^2\,V^{-1}\,s^{-1}}$ have been achieved in a MODFET. These mobilities are measured at low temperature, where they are limited by scattering from impurities, defects, and interfaces rather than phonons. The almost perfect crystalline quality of III–V heterostructures and the ability to separate carriers from the impurities that provide them by modulation doping mainly account for this huge difference.

First we shall study the electrostatics of modulation-doped layers to get estimates of important quantities such as the density of electrons, and then we shall develop models for the energy level and wave function of a two-dimensional electron gas. After this we shall look at screening, the way in which the 2DEG responds to a perturbation such as an ionized impurity, and use the results to calculate the mobility of a 2DEG.

The distinction between two- and three-dimensional vectors is important throughout this chapter. Recall our convention that the three-dimensional position is $\mathbf{R} = (\mathbf{r}, z)$, where $\mathbf{r} = (x, y)$ is a vector in the plane of the 2DEG.

9.1 Band Diagram of Modulation-Doped Layers

We shall first calculate the energy bands through the layers of a simple MODFET such as that shown in Figure 3.11, including a bias v_G on a large gate covering

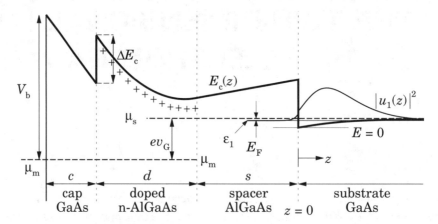

FIGURE 9.1. Self-consistent solution of the conduction band $E_c(z)$ through modulation-doped layers with a positive gate bias $v_G = 0.2$ V and $n_{2D} = 3 \times 10^{15}$ m^{-2} electrons in the 2DEG. [Modelling program courtesy of Prof. G. L. Snider, University of Notre Dame.]

the heterostructure. The basic structure, like that of all field-effect transistors, is that of a capacitor with the charge density on one 'plate' (the channel) controlled by the voltage on the other 'plate' (the metal gate). The equations will reflect this. Although the specific calculation is for GaAs and AlGaAs, the principles apply to other pairs of semiconductors.

A typical structure has the following layers, starting from the surface and going down, as shown in Figure 9.1:

(i) GaAs cap of thickness c and doping N_C (undoped or n-type) – a metal gate may be on top;

(ii) AlGaAs doped layer of thickness d and doping N_D (heavily n-type) – in δ-doped material this layer is very thin, ideally a monolayer;

(iii) AlGaAs spacer of thickness s, undoped;

(iv) GaAs substrate, which we shall initially take to be undoped.

The example in Figure 9.1 has the values $c = 10$ nm, $d = s = 20$ nm with barriers of Al$_{0.3}$Ga$_{0.7}$As and $V_b = 0.7$ eV at the surface. The doping $N_D = 1.35 \times 10^{24}$ m^{-3} was chosen to give $n_{2D} = 3 \times 10^{15}$ m^{-2} at $v_G = +0.2$ V.

The general principles used in sketching the band were described in Section 3.5. Briefly, we need to solve Poisson's equation and add the discontinuity ΔE_c at the heterojunctions. We can sketch the form of $E_c(z)$ by starting deep in the substrate, $z \rightarrow +\infty$, and working towards the surface. We have assumed that the substrate is undoped so the band has no curvature and must be flat. Working upwards, the first charge encountered is in the 2DEG and is negative. This causes the band to curve downwards and it reaches its maximum slope at the heterojunction, at which point we have passed through all the negative charge. As we enter the AlGaAs, $E_c(z)$ jumps up by ΔE_c but its slope remains constant through the spacer as no charge density is present. The positive charge of the donor ions causes the band to curve upwards in the doped layer, reversing its slope. There is a downward jump in $E_c(z)$ as we pass from the doped layer to the cap, and the band either continues to curve upwards or has constant slope, depending on whether the cap is n-doped or not. Finally, the band hits the metal gate on the surface with a large slope.

The behaviour at the surface is important. We could simply set $\mathbf{D}_\perp$ to be continuous if the surface behaved like a classical dielectric, and this is often appropriate for Si. GaAs, however, has a high density of *surface states* both on a free surface and at interfaces with a metal. They occupy a narrow band of energies near the middle of the band gap, at $V_b \approx 0.7\,\text{eV}$ below the conduction band. The Fermi level of an ungated layer always lies within this narrow band because it seems impossible to put enough charge into the surface states to fill or empty the band and allow the Fermi level to leave it. Thus the Fermi level is said to be *pinned* by the surface states at V_b below the conduction band. This in turn leads to the formation of a depletion layer under the surface. Other III–V compounds also exhibit surface states; they again lie in the gap for AlGaAs but in other cases, such as InAs, they lie within the conduction band and cause an accumulation layer to form at the surface.

There is little change in V_b when a metal is deposited on the surface to form a gate, typically only 0.1–0.2 eV. Thus, as in Figure 9.1, there is always a Schottky barrier that separates carriers in the semiconductor from those in the metal. The barrier must be made thin if the contact is intended to be ohmic, and tunnelling through such a barrier was considered in Section 7.4.3. In an FET, on the other hand, the depletion layer must be kept thick to prevent the gate from leaking.

The surface states therefore force $E_c(z)$ to meet the surface of GaAs at an energy V_b above the local Fermi level. The steep slope of the conduction band seen in Figure 9.1 arises from the high charge density in the surface states. This charge must be balanced by that from the donors, and we shall see that most of the donors in a MODFET are needed to neutralize the Schottky barrier; only a small fraction contribute electrons to the 2DEG.

A practical difficulty is that we need to know the charge density to solve Poisson's equation and deduce the energy bands, but we can't calculate the density of free carriers until we know the energy bands. Thus the calculation has to be performed *self-consistently*. Numerical methods are needed for an accurate calculation, such as that shown in Figure 9.1, but reasonably accurate estimates can be found analytically. We shall start with some sweeping simplifications and relax them later.

(i) Ignore the differences in dielectric constant and effective mass between GaAs and AlGaAs.

(ii) Assume that the donors are fully ionized. This may seem obvious, but it is often highly inaccurate in AlGaAs and will be revisited in Section 9.2.1.

(iii) Assume that the only free carriers are the electrons in the 2DEG. Again we shall see later what happens when this breaks down.

(iv) Assume that the electrons in the two-dimensional electron gas are degenerate and occupy only one subband. This requires low temperatures and low densities of electrons (roughly $n_{2D} < 6 \times 10^{15}\,\text{m}^{-2}$). Assume also that tunnelling into the spacer is negligible and that electrons are perfectly confined within the substrate.

The notation is as follows. The coordinate z measures the depth with $z = 0$ at the heterojunction where the electrons reside. The zero of energy and potential is taken at the GaAs side of this heterojunction, $z = 0_+$. The Fermi level in the metal gate is μ_m and that in the semiconductor is μ_s; the difference is set by the gate bias, $\mu_s - \mu_m = ev_G$ (note the sign, because the charge of the electron is $-e$). The density of the 2DEG is n_{2D} and its Fermi energy, measured from the bound state in the potential well at the heterojunction, is E_F; these are related by $n_{2D} = (m/\pi\hbar^2)E_F$ for our simple situation. Typically $E_F \approx 10\,\text{meV}$ in a structure designed for experiments in physics, although it is much larger in a practical transistor. This is very small compared with other energies in the problem, and is barely visible in Figure 9.1.

9.1.1 ELECTROSTATIC POTENTIAL

In a calculation it is easier to assume that the density is given and deduce the gate voltage needed to achieve this. We shall calculate the electrostatic potential by starting at the heterojunction that confines the electrons ($z = 0$), which we have chosen to be the zero of potential, and integrating up through the layers to the surface.

The spacer layer contains no charge so the electric field is constant and $\phi(z) = -Fz$, which vanishes as required at $z = 0$. According to Gauss's theorem the field F is generated by σ, the total charge density per unit area to the right on Figure 9.1 (larger values of z). This follows from the boundary condition that the electric field vanishes as $z \to \infty$. The only charge is in the 2DEG, giving $\sigma = -en_{2D}$. The outward electric field is $\sigma/\epsilon_0\epsilon_b$ so

$$F = \frac{en_{2D}}{\epsilon_0\epsilon_b}. \tag{9.1}$$

The doped layer contains a positive charge density eN_D from the donors, so the potential obeys Poisson's equation $\partial^2\phi/\partial z^2 = -eN_D/\epsilon_0\epsilon_b$. The solution to this is a quadratic function of z whose value and slope must be matched onto the solution for the spacer at $z = -s$. This gives

$$\phi(z) = -\frac{eN_D}{2\epsilon_0\epsilon_b}(z+s)^2 - \frac{en_{2D}}{\epsilon_0\epsilon_b}z. \tag{9.2}$$

The cap may again contain a positive charge density if donors are present. The calculation is exactly the same as that in the preceding section, and the result must be matched to the previous one at $z = -(s+d)$. The result is

$$\phi(z) = -\frac{eN_C}{2\epsilon_0\epsilon_b}[z+(d+s)]^2 + \frac{eN_D d}{\epsilon_0\epsilon_b}[z+(s+\tfrac{1}{2}d)] - \frac{en_{2D}}{\epsilon_0\epsilon_b}z. \tag{9.3}$$

The potential at the gate is

$$\phi[-(c+d+s)] = -\frac{eN_C c^2}{2\epsilon_0\epsilon_b} - \frac{eN_D d(c+\frac{1}{2}d)}{\epsilon_0\epsilon_b} + \frac{en_{2D}(c+d+s)}{\epsilon_0\epsilon_b}. \quad (9.4)$$

This could instead have been constructed using superposition.

9.1.2 CONDUCTION BAND AND GATE BIAS

The conduction band in GaAs is given by $E_c(z) = -e\phi(z)$, while the discontinuity gives $E_c(z) = \Delta E_c - e\phi(z)$ in AlGaAs. This gives the form of the energy bands in Figure 9.1 except for the channel $z > 0$, which we have not yet calculated.

We need to find the Fermi levels in the gate and channel, both measured with respect to the zero of energy at $z = 0_+$, to deduce the gate bias. In the semiconductor just under the gate at $z = -(c+d+s)_+$, the conduction band is $E_c = -e\phi$, where the potential is given by equation (9.4). At the gate, the Schottky barrier due to the surface states pins E_c at an energy V_b above the Fermi level μ_m, so

$$\mu_m = E_c(-(c+d+s)_+) - V_b$$

$$= \frac{e^2}{\epsilon_0\epsilon_b}\left[\tfrac{1}{2}N_C c^2 + N_D d(c+\tfrac{1}{2}d) - n_{2D}(c+d+s)\right] - V_b. \quad (9.5)$$

Next we need the Fermi level of the 2DEG. Figure 9.1 shows that this is given by the sum of the energy of the bound state in the potential well and the Fermi energy of the 2DEG, $\mu_s = \varepsilon_1(n_{2D}) + E_F(n_{2D})$. We know $E_F = (\pi\hbar^2/m)n_{2D}$ but have not yet calculated ε_1, the energy of the lowest state in the potential well at the heterojunction. Both energies are functions of the density of electrons. We shall study models of the 2DEG in Section 9.3 but for the moment we shall leave ε_1 as an unknown function. Thus

$$v_G = \frac{\mu_s - \mu_m}{e} = \frac{\varepsilon_1(n_{2D}) + E_F(n_{2D})}{e} + \frac{V_b}{e} \quad (9.6)$$

$$+ \frac{e}{\epsilon_0\epsilon_b}\left[n_{2D}(c+d+s) - \tfrac{1}{2}N_C c^2 - N_D d(c+\tfrac{1}{2}d)\right].$$

The discontinuity ΔE_c does not appear in this because both regions are in GaAs. Since E_F is a linear function of n_{2D}, provided that only one subband is occupied, it can be incorporated into the electrostatic term that contains n_{2D}. Pulling out the prefactor of $e/\epsilon_0\epsilon_b$ gives

$$\frac{E_F(n_{2D})}{e} = \frac{e}{\epsilon_0\epsilon_b}n_{2D}\frac{\pi\epsilon_0\epsilon_b\hbar^2}{me^2} = \frac{e}{\epsilon_0\epsilon_b}n_{2D}\frac{a_B}{4}. \quad (9.7)$$

Thus the density of states gives an apparent thickness of $\frac{1}{4}a_B$, where a_B is the scaled Bohr radius allowing for the dielectric constant and effective mass of the semiconductor (equation 4.67). In GaAs $a_B \approx 10$ nm so the density of states of the 2DEG makes the 2DEG appear electrically to be some 2.5 nm deeper than the physical layers. Usually this is a very small effect. The function $v_G(n_{2D})$ now becomes

$$v_G = \frac{\varepsilon_1(n_{2D})}{e} + \frac{V_b}{e} + \frac{e}{\epsilon_0 \epsilon_b} \left[n_{2D}(c + d + s + \tfrac{1}{4}a_B) - \tfrac{1}{2}N_C c^2 - N_D d(c + \tfrac{1}{2}d) \right].$$

(9.8)

Ungated layers can be treated by setting $v_G = 0$.

9.1.3 THRESHOLD VOLTAGE

Electrons just begin to populate the channel when the threshold voltage $v_G = v_T$ is applied to the gate. The band is flat both in the substrate and in the spacer because n_{2D} is vanishingly small. This follows from Gauss's theorem and the absence of charge in the channel, spacer, and substrate. Thus the band at threshold is as shown in Figure 9.2, for the same wafer as in Figure 9.1, which has the rather small value $v_T = -0.073$ V. This corresponds roughly to strong inversion in a conventional FET; we are considering low temperatures, whereas the usual definition of inversion in a MOSFET accounts for thermal population of the bands. The potential well has disappeared so both the energy level ε_1 and the Fermi energy E_F are zero, and Figure 9.2 shows that

$$\frac{V_b}{e} - v_T = \text{voltage drop across doped layers.}$$

(9.9)

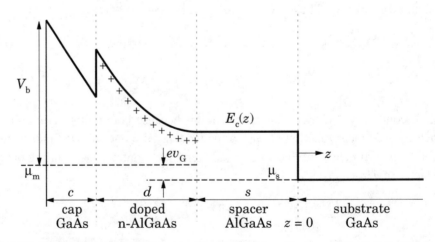

FIGURE 9.2. Band diagram though the same modulation-doped layers as in Figure 9.1 at threshold; $n_{2D} = 0$, the Fermi level brushes E_c in the channel, and $v_G = v_T < 0$. [Modelling program courtesy of Prof. G. L. Snider, University of Notre Dame.]

Equation (9.6) gives

$$v_T = \frac{V_b}{e} - \frac{e}{\epsilon_0 \epsilon_b} \left[\tfrac{1}{2} N_C c^2 + N_D d(c + \tfrac{1}{2}d) \right].$$
(9.10)

This can be derived more directly by solving Poisson's equation for Figure 9.2; n_{2D} never appears, and both the potential and its derivative vanish at $z = -s$.

The threshold voltage does not depend on the thickness of the spacer in this approximation, nor on the discontinuities in the conduction band. However, it depends strongly on the concentration and thickness of the doped layers. This can be a problem because the doping is often rather poorly controlled and can vary sufficiently over a single wafer to cause significant fluctuations in the threshold voltage. On the other hand, this dependence can be used to vary the threshold voltages, giving either depletion mode ($v_T < 0$) or enhancement mode ($v_T > 0$). In practice almost all devices are in depletion mode and, as we shall see in Section 9.2.1, non-ideal behaviour of the material can make this simple estimate of v_T highly inaccurate.

9.1.4 GATE–CHANNEL CAPACITANCE

An important term in the transconductance of an FET is the derivative $\partial n_{2D}/\partial v_G$ of the density of electrons in the channel with respect to the gate bias. This is the differential capacitance between the gate and channel, apart from a factor of e. From equation (9.8), the reciprocal is

$$\frac{1}{C_G} = \frac{1}{e} \frac{\partial v_G}{\partial n_{2D}} = \frac{1}{\epsilon_0 \epsilon_b} \left[(c + d + s) + \tfrac{1}{4} a_B + \frac{\epsilon_0 \epsilon_b}{e^2} \frac{\partial \varepsilon_1}{\partial n_{2D}} \right].$$
(9.11)

This resembles a parallel-plate capacitor, with several contributions to the apparent thickness of the dielectric in square brackets. Most of the spacing is contributed by the obvious insulator: the cap, doped, and spacer layers. The other two terms depend on the electronic structure of the 2DEG, a general result that is exploited in capacitance spectroscopy.

We have already seen that the density of states of the 2DEG adds the small extra thickness $\tfrac{1}{4} a_B$ to the 'electrical depth'. This contribution is halved if the Fermi level enters the second subband. In general we should return to equation (9.6), which shows that the Fermi energy contributes an apparent thickness $(\epsilon_0 \epsilon_b/e^2) \, dE_F/dn_{2D}$ to the differential capacitance. This can have rich structure. An important example is a 2DEG in a strong magnetic field where the density of states breaks up into Landau levels (Section 6.4.4). Capacitance spectroscopy has been used to determine the profile of the Landau levels, which are broadened from ideal δ-functions in a real sample (Figure 6.19).

The remaining term is the derivative of the energy of the lowest bound state in the well with respect to its occupation. We usually assume that energy levels are fixed as we pour electrons into a system, but the density of the 2DEG affects the shape of the well within which it is trapped. We shall calculate this properly in Section 9.3 but can see roughly what goes on from equation (9.1). This shows that the field in the spacer is directly proportional to the density of electrons in the 2DEG. The electric field is identical in the channel next to the heterojunction, so an increase in n_{2D} causes the potential well to become steeper and the energy of the bound state rises. Thus the derivative in equation (9.11) is positive and it is natural to associate this length with the thickness of the 2DEG,

$$h(n_{2D}) = \frac{\epsilon_0 \epsilon_b}{e^2} \frac{\partial \varepsilon_1}{\partial n_{2D}}. \tag{9.12}$$

The wave function is spread out along the z-direction and one might guess that the apparent depth of the 2DEG should be measured to the centre of its charge distribution, not just the interface. This is the main effect of h. Typically $h \approx$ 5–10 nm in GaAs; it varies only slowly and can often be treated as constant. The capacitance C_G is then also constant and the gate bias simplifies to

$$v_G = v_T + \frac{e}{\epsilon_0 \epsilon_b}(c + d + s + \tfrac{1}{4}a_B + h)n_{2D} = v_T + \frac{e n_{2D}}{C_G}. \tag{9.13}$$

Effectively we are putting $\varepsilon_1 \approx eFh$, which has an obvious electrostatic interpretation as a shifted sheet of charge. Setting $v_G = 0$ in equation (9.13) gives a rough value for the density of electrons in layers without a gate,

$$n_{2D} = \frac{\epsilon_0 \epsilon_b}{e} \frac{-v_T}{(c + d + s + \tfrac{1}{4}a_B + h)} = \frac{-v_T C_G}{e}. \tag{9.14}$$

A useful point comes from the magnitudes of the lengths that enter C_G in GaAs. The contribution from the density of states, which reflects the kinetic energy of the electrons, is very small, only about 2.5 nm. That from the layers, $c + d + s$, is rarely less than 25 nm and may be ten times this. Thus the dominant energies in this system are from classical electrostatics, with electronic energies playing only a minor role. This can provide a useful guide for estimating the behaviour of a 2DEG.

An important feature of doped heterostructures in GaAs–AlGaAs is that there are far more donors than mobile electrons because of the pinned surface. The band in Figure 9.1 has a minimum in the doped layer that is very close to the spacer, and only donors deeper than this supply electrons to the 2DEG. Electrons from the majority of donors, which lie above the minimum, populate the surface states and play no active role. This problem plagues GaAs devices but does not apply to Si, nor to some other III–V materials such as InAs.

9.2 Beyond the Simplest Model

The previous section has given a simple model of modulation-doped layers from which the density of electrons can easily be calculated as a function of the gate bias. Unfortunately it is often inadequate and we shall now address some of its deficiencies.

9.2.1 DONORS IN AlGaAs: DX CENTRES

We assumed before that the donors are fully ionized, as in classical semiconductors, and can be treated with a hydrogenic model (Section 4.7.5). This gives a wave function with a binding energy $E_D \approx 5\,\text{meV}$ and a radius $a_B \approx 10\,\text{nm}$ in GaAs, the values being slightly changed in AlGaAs. Regrettably this is far from true for the standard donor Si in AlGaAs, which can also exist in a second state, the *DX centre*. The lattice relaxes around the donor when an electron arrives, releasing extra energy so the electron is more deeply bound. Rough sketches of the crystal around the donor when empty and occupied are shown in Figure 9.3(b) and (c). The binding energy $E_{DD} \approx 0.12\,\text{eV}$ for $x = 0.3$, much larger than that for the hydrogenic state, and the radius of the wave function is reduced to atomic scale. The energy of the DX centre varies with x in $\text{Al}_x\text{Ga}_{1-x}\text{As}$; it lies within the conduction band for pure GaAs, emerges below it for $x > 0.2$, and may return to the X-minima of the conduction band as $x \to 1$.

Not only does it require a large amount of energy to remove an electron from a DX centre, but there is also a barrier to trapping the electron. This behaviour can be thoroughly explained with a configuration-coordinate diagram, but a crude picture is obtained by imagining that the potential around the donor looks like Figure 9.3(a). Clearly this cannot be characterized by a single energy; there is a barrier to occupation of the donor, another for release of the electron from the donor, and E_{DD}, which sets the occupation if the DX centres are in thermal equilibrium with the rest of the crystal. The high barriers to entry and exit mean that thermal equilibrium is difficult to achieve at low temperature, and it is found experimentally that the occupation of DX centres 'freezes' below $150\,\text{K}$.

Start with an ungated layer at low temperature to see the effect of DX centres on the band diagram. The example in Figure 9.4 has the same layers as in Figure 9.1 but with N_D raised to $3 \times 10^{24}\,\text{m}^{-3}$ and $E_{DD} = 0.12\,\text{eV}$. The DX centres are shown as a dashed line at an energy E_{DD} below $E_c(z)$ in Figure 9.4(a). If this energy falls

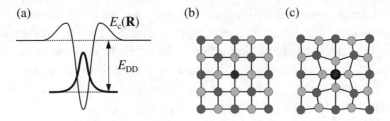

FIGURE 9.3. (a) Schematic picture of the conduction band $E_c(\mathbf{R})$ around a DX centre, with a deeply bound state and a barrier to trapping as well as emission. (b) Undistorted lattice around empty donor. (c) Distortion around occupied donor.

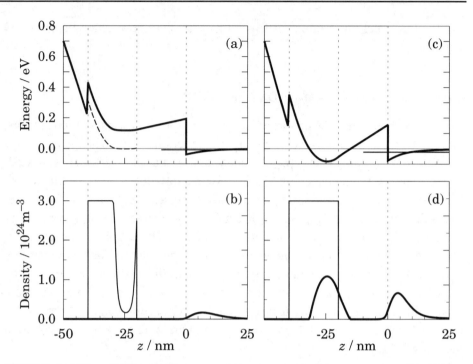

FIGURE 9.4. Effect of DX centres on an ungated heterostructure at low temperature. (a), (c) Conduction band (thick line) with energy of DX centres (dashed line) and energy level of 2DEG (thin line). (b), (d) Density of electrons (thick line) and ionized donors (thin line). The material is in the dark in (a) and (b), and illumination has ionized all donors in (c) and (d). [Modelling program courtesy of Prof. G. L. Snider, University of Notre Dame.]

below μ they trap electrons, the material becomes neutral, and the bands become flat. There is a very narrow band of ionized donors next to the spacer, to cancel the charge of the 2DEG, and then a roughly neutral layer where $E_c(z) \approx \mu + E_{DD}$ and most of the donors trap electrons in DX centres. Figure 9.4(b) shows that the density of ionized donors is reduced nearly to zero around $z \approx -25$ nm; the density of the 2DEG is also fairly low at 2.4×10^{15} m^{-2}. Finally, there is a thick band of ionized donors next to the cap, which produces the charge needed to cancel the Schottky barrier at the surface.

The neutral layer means that the 2DEG and the surface are decoupled, so n_{2D} does not depend on d as it did for the ideal layers discussed in Section 9.1. Consider the region around the 2DEG, shown in Figure 9.5, and assume that there is an abrupt change from neutrality to full ionization in the doped layer. There must be enough positive charge in the thin ionized strip to balance the negative charge of the 2DEG, so its thickness L obeys $N_D L = n_{2D}$. The figure shows that the discontinuity in the conduction band is given by a sum of energies,

$$\Delta E_c = \varepsilon_1 + E_F + E_{DD} + \frac{e^2 N_D L^2}{2\epsilon_0\epsilon_b} + \frac{e^2 n_{2D} s}{\epsilon_0\epsilon_b}, \qquad (9.15)$$

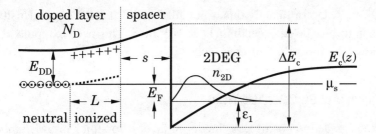

FIGURE 9.5. Conduction band around a doped heterojunction with DX centres.

where the final two terms are the electrostatic voltages across the thin ionized strip and spacer. The example in Figure 9.4(a) has $L = (2.4 \times 10^{15}\,\text{m}^{-2})/(3 \times 10^{24}\,\text{m}^{-3}) < 1\,\text{nm}$, so very few donors are needed to balance the 2DEG and the drop in voltage across the ionized layer is very small. This is the sole term that contains N_D so we reach the surprising conclusion that the density of electrons in the 2DEG is almost independent of doping under these conditions. Dropping this term, and writing the electronic energies as thicknesses, gives

$$n_{2D} \approx \frac{\epsilon_0 \epsilon_b}{e^2} \frac{\Delta E_c - E_{DD}}{s + h + \frac{1}{4} a_B}. \tag{9.16}$$

This shows that a thick spacer causes the density to decrease, falling as $1/s$ for large s. Thick spacers lead to high mobility because scattering by ionized donors in the n-AlGaAs is reduced, but they also give a low density in the 2DEG. Typical layers for physics experiments have thick spacers, sometimes over 100 nm, to promote the mobility; layers for practical transistors have spacers of only a few monolayers to raise the density of electrons.

There are two possible behaviours when a gate voltage is applied. The occupation of the DX centres is frozen at low temperatures so only the 2DEG responds to v_G. This is exactly as in the simple case and the differential capacitance is unchanged from the previous result (9.11). At higher temperatures, electrons can enter and leave the DX levels. A change in gate voltage now affects only the electrons in the doped layer, which provide the nearest mobile charge to the gate, and the capacitance gives the thickness of the depletion layer at the surface. The 2DEG is unaffected.

9.2.2 PERSISTENT PHOTOCONDUCTIVITY

The DX centres give rise to another important practical effect, persistent photo-conductivity. Light of energy below the band gap can excite electrons out of DX centres. This increases the density of both mobile electrons and ionized donors, and the barrier to retrapping means that the electrons remain free at low temperatures. This effect is widely used to control the density of a 2DEG at low temperatures. The form of E_c and the densities after illumination are shown in Figure 9.4(c) and (d). All the donors are now ionized and the density of the 2DEG has risen to

$6.8 \times 10^{15} \, \text{m}^{-2}$. Unfortunately this is not all: there is an even larger density of free electrons in the AlGaAs, exceeding $11 \times 10^{15} \, \text{m}^{-2}$. This is an example of parallel conduction, which we shall discuss next.

9.2.3 PARALLEL CONDUCTION

If the gate is made too positive, or the doped layer is too thick or too highly doped, the minimum in the conduction band within the doped AlGaAs layer falls below the Fermi level and a second population of electrons appears in the doped AlGaAs. This is called *parallel conduction* and is clear in Figure 9.4(c). With the structure considered as a transistor, the gate now controls a parasitic FET whose electrons travel through the n-AlGaAs rather than a MODFET with electrons in the 2DEG. The mobility is much lower in n-AlGaAs, mainly because it is a highly doped region rather than being carefully isolated from ionized donors, so the performance is poor. The 2DEG itself is almost perfectly screened from the gate and cannot be controlled until the gate is made sufficiently negative to empty the donors.

The maximum density of electrons that can occupy the 2DEG before the onset of parallel conduction can be calculated using the same arguments that led to equation (9.15). In this case the conduction band in the doped layer just brushes the Fermi level at the onset of parallel conduction, rather than being pinned E_{DD} above it, so we simply set $E_{\text{DD}} = 0$ in equation (9.16).

Layers are usually tested for the presence of parallel conduction using magnetotransport. One signature is a term proportional to B^2 in the Hall effect at small fields; another is that the minima in the Shubnikov–de Haas effect (Figure 6.10) do not fall to zero at large fields. However, the mobility of electrons is so much lower in n-AlGaAs than in a 2DEG that parallel conduction may not be easy to detect.

9.2.4 NEGATIVE HUBBARD *U*

There is another twist to the story of DX centres. We typically regard a donor as having two states, a positive ion D^+ (with its electron free in the conduction band) or a neutral state D^0 with an electron trapped. It is possible to add a second electron to D^0 to form a negative ion D^-, but the Coulomb repulsion between the two electrons (the Hubbard U, introduced in Section 1.8.5) means that the second electron is precariously bound. For a hydrogenic donor the second electron has a binding energy of only around 5% of that of the first.

This picture changes drastically if the addition of the second electron causes relaxation of the lattice as in Figure 9.3(c). The energy of D^- may then fall *below* that of D^0, implying that the second electron is more tightly bound than the first. It is as if the two electrons attract one another, giving a negative (effective) Hubbard U. It is widely believed that this applies to the DX centre. From a chemical point of view, it means that the disproportionation reaction $2D^0 \rightarrow D^- + D^+$ is favourable.

A neutral region of n-doped AlGaAs therefore contains equal numbers of D^+ and D^- rather than only D^0.

9.2.5 DOPED OR BIASSED SUBSTRATE

We have assumed for simplicity that the substrate is undoped, but it is usually lightly p-type. The effect of this on the band diagram is shown in Figure 9.6, including $E_v(z)$. Deep in the substrate the material is neutral, which at low temperature requires $\mu = E_v + E_A$, where E_A is the (small) binding energy of the acceptors. Since $\mu \approx E_c$ around the 2DEG, there must be a depletion region of ionized acceptors to produce a potential difference of roughly E_g/e (neglecting E_A, ε_1, and E_F). This requires a thickness L_{dep} given through Poisson's equation by

$$E_g \approx \frac{e^2 N_A L_{dep}^2}{2\epsilon_0 \epsilon_b}. \tag{9.17}$$

This can also be expressed as a total charge per unit area of $N_{dep} = N_A L_{dep} = \sqrt{2\epsilon_0\epsilon_b E_g N_A}/e$. If we take as an example $N_A = 10^{21}$ m^{-3}, then $L \approx 1.5\,\mu$m and $N_{dep} \approx 1.5 \times 10^{15}$ m^{-2}. Thus N_{dep} is not negligible compared with n_{2D}. If, as is often the case, the band diagram is dominated by electrostatic energies, n_{2D} will simply be reduced by the value of N_{dep}. This is the case for the layers in Figure 9.6, where substrate doping of $N_A = 10^{21}$ m^{-3} lowered n_{2D} from 3.0×10^{15} m^{-2} to 1.5×10^{15} m^{-2}. The ionized acceptors also degrade the mobility; see Section 9.6.1.

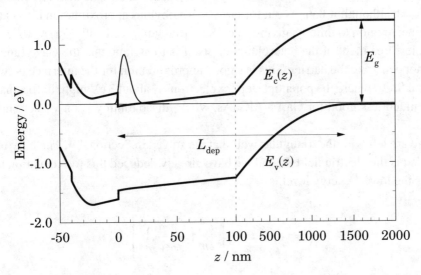

FIGURE 9.6. Bands of a doped heterostructure with a p-type substrate, showing a depletion region of thickness $L_{dep} \approx 1.5\,\mu$m and the density of electrons at the heterojunction. Note the change in scale at $z = 100$ nm. [Modelling program courtesy of Prof. G. L. Snider, University of Notre Dame.]

A second effect of a doped substrate is that the potential well at the heterojunction no longer disappears at threshold. The well is roughly triangular with an electric field $F_{\text{dep}} = eN_{\text{dep}}/\epsilon_0\epsilon_b$. The energy of the bound state within this well can be calculated as in Section 4.4. An electric field can also be generated by applying a bias to a contact or highly doped layer deep in the substrate. This substrate bias provides a convenient way to control the density of a 2DEG while retaining the freedom to confine the electrons with a patterned gate on the surface.

9.3   Electronic Structure of a 2DEG

We now need to return to the difficult problem that we skipped before, to calculate the electronic states of the 2DEG. Figure 9.7 shows an enlargement of Figure 9.1 with a numerical solution of the wave function and energy level at the heterojunction, compared with two approximate solutions. We can treat the triangular potential well quickly, since we have done the hard work already. Next comes a brief aside into the quantum mechanics of many-electron systems, which leads into the celebrated Fang–Howard variational treatment of the 2DEG.

9.3.1   TRIANGULAR-WELL MODEL

We know, both from qualitative arguments and numerical calculations, that the potential well in which the 2DEG is trapped is triangular close to the heterojunction at $z = 0$, although it flattens out for large z. An obvious approximation is to extend the linear region to infinity, giving a perfectly triangular well with $V(z) = eFz$. The simplest treatment of the heterojunction itself is to assume that the wave function cannot penetrate the barrier. This is a good approximation for the interface between Si and SiO_2, where the potential step for electrons is about 3 eV high; unfortunately it is much less good for GaAs–AlGaAs, whose discontinuity is only one-tenth of this.

We are left with the triangular well that we solved in Section 4.4. The only input needed is the electric field F, and we have already deduced this too (equation 9.1). Thus the lowest energy level is

$$\varepsilon_1 = c_1\varepsilon_0 = c_1\left[\frac{\hbar^2}{2m}\left(\frac{e^2 n_{2D}}{\epsilon_0\epsilon_b}\right)^2\right]^{1/3}, \qquad (9.18)$$

where $c_1 \approx 2.338$. This gives $\varepsilon_1 = 48\,\text{meV}$ for the example in Figure 9.7 with $n_{2D} = 3 \times 10^{15}\,\text{m}^{-2}$. It would not be sensible to use this approximation for higher energy levels because their density is mainly in the regions where the triangular

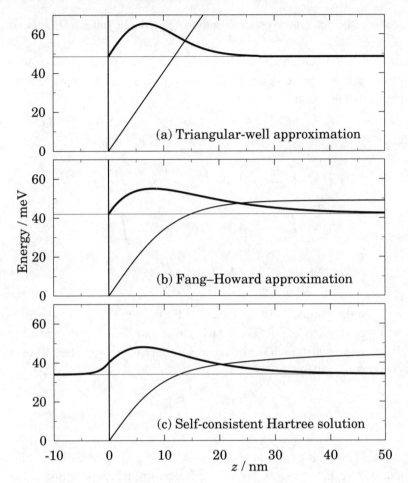

FIGURE 9.7. Three models for the potential well, energy level, and wave function at a hetero-junction, all with an undoped substrate and $n_{2D} = 3 \times 10^{15}\,\text{m}^{-2}$ electrons per unit area in the 2DEG. (a) Triangular approximation to the conduction band profile at a heterojunction. (b) Variational solution for a 2DEG using the Hartree approximation with a Fang–Howard wave function. (c) Numerical Hartree solution, showing penetration into the barrier of $Al_{0.3}Ga_{0.7}As$. [Modelling program courtesy of Prof. G. L. Snider, University of Notre Dame.]

approximation is poor. The corresponding wave function is

$$\phi_1(z) \propto \text{Ai}\left(\frac{eFz - \varepsilon_1}{\varepsilon_0}\right) \tag{9.19}$$

for $z > 0$, the length scale of which is $z_0 = \varepsilon_0/eF$.

The thickness defined by equation (9.12) is

$$h = \frac{\epsilon_0\epsilon_b}{e^2}\frac{\partial\varepsilon_1}{\partial n_{2D}} = \frac{2}{3}\frac{\varepsilon_1}{eF}. \tag{9.20}$$

For comparison we can also calculate the mean position of the electrons using the usual definition $\langle z \rangle = \int z|\phi_i(z)|^2 dz$. It happens in this case that the integrals can be done analytically, but a more general method using the Feynman–Hellman theorem is instructive.

Suppose that the Hamiltonian is a function of some parameter λ, and take the derivative of an eigenvalue $\varepsilon_n = \langle n|\hat{H}|n \rangle$ with respect to λ. If the wave function ϕ_n is normalized, this becomes

$$\frac{\partial \varepsilon_n}{\partial \lambda} = \frac{\partial}{\partial \lambda} \langle n|\hat{H}|n \rangle \equiv \frac{\partial}{\partial \lambda} \int \phi_n^* \hat{H} \phi_n$$

$$= \int \frac{\partial \phi_n^*}{\partial \lambda} \hat{H} \phi_n + \int \phi_n^* \frac{\partial \hat{H}}{\partial \lambda} \phi_n + \int \phi_n^* \hat{H} \frac{\partial \phi_n}{\partial \lambda}$$

$$= \int \frac{\partial \phi_n^*}{\partial \lambda} \varepsilon_n \phi_n + \int \phi_n^* \frac{\partial \hat{H}}{\partial \lambda} \phi_n + \int \phi_n^* \varepsilon_n \frac{\partial \phi_n}{\partial \lambda}$$

$$= \varepsilon_n \frac{\partial}{\partial \lambda} \int \phi_n^* \phi_n + \int \phi_n^* \frac{\partial \hat{H}}{\partial \lambda} \phi_n \, .$$

We have used the product rule, replacing $\hat{H}$ by ε_n in the outer two terms because ϕ_n is an eigenfunction of $\hat{H}$. These two terms can then be merged to give the normalization integral, which is a constant so its derivative vanishes. Thus only the last term remains, giving

$$\frac{\partial \varepsilon_n}{\partial \lambda} = \left\langle \frac{\partial \hat{H}}{\partial \lambda} \right\rangle_n \equiv \left\langle n \left| \frac{\partial \hat{H}}{\partial \lambda} \right| n \right\rangle, \tag{9.21}$$

which is the *Feynman–Hellman theorem*. In words, the derivative of the energy of a state with respect to some parameter is equal to the expectation value of the derivative of the Hamiltonian in that state. It is particularly useful in numerical work because $\partial \hat{H}/\partial \lambda$ can usually be found analytically and it is straightforward to evaluate its expectation value. This is a far better approach than numerical evaluation of the derivative $\partial \varepsilon_n/\partial \lambda$.

For the triangular well we choose $\lambda = F$, giving $\partial \hat{H}/\partial F = ez$, and obtain

$$\frac{\partial \varepsilon_n}{\partial F} = e \langle n|z|n \rangle = e \langle z \rangle_n \, . \tag{9.22}$$

Rewriting the left-hand side in terms of n_{2D} shows that the two definitions of thickness are identical, $h = \langle z \rangle$. Our example has $h = 8 \, \text{nm}$.

9.3.2 THE QUANTUM MECHANICS OF MANY ELECTRONS

We have seen that the energy of the bound state in a 2DEG depends on the density of electrons. To make further progress we should investigate systematically how the motion of one electron is affected by others. This requires a brief digression on the quantum mechanics of systems with many electrons and a review of the effects of the Coulomb repulsion between the electrons. We have generally neglected this repulsion, but it is disturbingly large! The distance between electrons is about 20 nm in a 2DEG of density 3×10^{15} m^{-2}, giving a Coulomb potential of 6 meV, which can be compared with the Fermi energy of 11 meV. It seems startling that we can obtain any useful results at all within the independent-electron approximation where this interaction is neglected. A full justification requires many-body theory and we shall only sketch some of the basic concepts.

In general, the wave function ψ of a system of N particles must be written as

$$\psi(\mathbf{R}_1, s_1; \mathbf{R}_2, s_2; \ldots; \mathbf{R}_N, s_N). \tag{9.23}$$

It is a function of the position and spin coordinates, $\mathbf{R}_j$ and s_j, of each particle j and cannot be separated into individual wave functions. There is an important symmetry that applies when the coordinates of two particles are interchanged. For fermions such as electrons, the wave function changes sign,

$$\psi(\ldots; \mathbf{R}_j, s_j; \ldots; \mathbf{R}_k, s_k; \ldots) = -\psi(\ldots; \mathbf{R}_k, s_k; \ldots; \mathbf{R}_j, s_j; \ldots). \tag{9.24}$$

This is a general statement of the Pauli exclusion principle. An important special case is that the wave function vanishes if the two coordinates are equal, $\mathbf{R}_j = \mathbf{R}_k$ and $s_j = s_k$. Thus two electrons of the same spin cannot be at the same point. Bosons obey equation (9.24) but with a $+$ sign rather than a $-$ sign.

The many-particle wave function obeys a complicated Schrödinger equation. There is a kinetic energy term for each particle plus two kinds of potential energy. Each electron feels the same external potential V_{ext}, which would arise from the ionized donors in a 2DEG. There is also a Coulomb repulsion between each pair of electrons. Thus the Hamiltonian is

$$\hat{H} = \sum_j \left[-\frac{\hbar^2}{2m} \nabla_j^2 + V_{\text{ext}}(\mathbf{R}_j) \right] + \frac{1}{2} \sum_{\substack{j,k \\ k \neq j}} \frac{e^2}{4\pi\epsilon |\mathbf{R}_j - \mathbf{R}_k|}, \tag{9.25}$$

where ∇_j means the gradient with respect to coordinate $\mathbf{R}_j$. The factor of $\frac{1}{2}$ in front of the second summation is to avoid double-counting the interaction, so that each pair of electrons is included only once. Numerical solutions to such Schrödinger equations can be obtained for a few particles, but it is an impossible problem for a large number and systematic approximations have been developed to treat it.

The simplest approach, with a clear classical interpretation, is the *Hartree approximation*. The wave function is written as a simple product of one-electron states,

$$\psi = \psi_1(\mathbf{R}_1, s_1) \, \psi_2(\mathbf{R}_2, s_2) \cdots \psi_N(\mathbf{R}_N, s_N). \tag{9.26}$$

The energy of this wave function is minimized with respect to the Hamiltonian (9.25) using the variational method. The result is that each individual wave function obeys a one-electron Schrödinger equation of the form

$$\left[-\frac{\hbar^2}{2m}\nabla_j^2 + V_{\text{ext}}(\mathbf{R}_j) + V_{\text{H}}^{(j)}(\mathbf{R}_j) \right] \psi_j(\mathbf{R}_j) = \varepsilon_j \psi_j(\mathbf{R}_j), \qquad (9.27)$$

where the *Hartree potential energy* $V_{\text{H}}^{(j)}(\mathbf{R}_j)$ is given by

$$V_{\text{H}}^{(j)}(\mathbf{R}_j) = \frac{e^2}{4\pi\epsilon} \sum_{k,\, k \neq j} \int \frac{|\psi_k(\mathbf{R}_k)|^2}{|\mathbf{R}_j - \mathbf{R}_k|} d^3\mathbf{R}_k . \qquad (9.28)$$

The Hartree potential is the electrostatic potential generated by the total charge density from all the other electrons. No account is taken of how these vary in time, so each electron sees only the *average* potential generated by the others.

The problem is self-consistent because the Hartree potential must be calculated from the wave functions, but the wave functions are themselves solutions to Schrödinger equations containing the Hartree potential. We can start with a guess for the wave functions, calculate the Hartree potentials, solve the Schrödinger equations to find better estimates for the wave functions, and repeat this loop until the wave functions cease changing. This yields a self-consistent solution. An irritating feature is that the Hartree potential is different for each electron, because we should omit an electron's contribution to the total electrostatic potential when calculating its own behaviour. Fortunately this effect is very small if there are many electrons and we can use a common Hartree potential $V_{\text{H}}(\mathbf{R})$ for all states, omitting the restriction on the summation in equation (9.28).

A disadvantage of the wave function used in the Hartree approximation is that it does not obey the Pauli symmetry (9.24). A wave function with the correct symmetry can be constructed using a *Slater determinant* of one-electron functions,

$$\psi = \frac{1}{\sqrt{N!}} \det \begin{vmatrix} \psi_1(\mathbf{R}_1, s_1) & \psi_2(\mathbf{R}_1, s_1) & \cdots & \psi_N(\mathbf{R}_1, s_1) \\ \psi_1(\mathbf{R}_2, s_2) & \psi_2(\mathbf{R}_2, s_2) & \cdots & \psi_N(\mathbf{R}_2, s_2) \\ \vdots & \vdots & \ddots & \vdots \\ \psi_1(\mathbf{R}_N, s_N) & \psi_2(\mathbf{R}_N, s_N) & \cdots & \psi_N(\mathbf{R}_N, s_N) \end{vmatrix} . \qquad (9.29)$$

Interchanging two coordinates is the same as interchanging two rows of the determinant and this changes its sign, as required by the Pauli exclusion principle. Also, if we try to put two electrons into the same state, this would make two columns of the determinant identical and it would again vanish. This confirms the usual statement of the exclusion principle that no two electrons may occupy the same state.

The variational calculation is repeated with this wave function, and a more complicated Schrödinger equation emerges for each state. There is an extra 'non-local' potential that cannot be written simply as a function that multiplies the wave function, called the *exchange* term. It causes repulsion between electrons of the same spin, which is hardly surprising as this is the new feature that we have built into the wave function by forcing it to obey the Pauli principle. The result is the *Hartree–Fock approximation*.

It seems obvious that this should be an improvement on the Hartree approximation. In fact this is not the case for systems such as the electron gas and we need to look at what is omitted from these approximations. In both cases, the main feature is that each electron moves in the *average* potential due to the others. Thus no account is taken of how the motion of the other electrons is affected by the fact that electron j happens to be at $\mathbf{R}_j$ at a particular time other than through the exclusion principle, which affects only electrons of parallel spin. The Coulomb repulsion between electrons tends to keep them apart so an electron is surrounded by a *correlation hole* as it moves through the electron gas, where the density is below average. This is a similar effect to the way in which electrons rearrange to screen the electrostatic potential of an external charge, which will be discussed in Section 9.4. The treatment of correlation requires many-body theory, which is beyond the scope of this book. Fortunately, detailed calculations show that the Hartree approximation is adequate for many properties of extended systems such as the 2DEG, so we shall go no further. This happy situation may not apply to small systems such as quantum dots, where Hartree–Fock may be more appropriate and quantum chemical techniques developed for atoms and molecules can be applied.

Finally, this picture assumes that free electrons provide the correct starting point. The kinetic energy per electron in a 2DEG is proportional to E_F and thus to n_{2D}, whereas a typical Coulomb energy is inversely proportional to the average separation and therefore goes like $n_{2D}^{1/2}$. At low density the Coulomb repulsion dominates and it becomes energetically favourable for the electrons to form a *Wigner crystal*, discussed in Section 6.6.2, rather than propagate freely.

9.3.3 VARIATIONAL HARTREE CALCULATION OF A 2DEG

We shall now use the variational method to find the electronic states of the lowest subband of a 2DEG within the Hartree approximation, using the Fang–Howard wave function. Let the density of electrons per unit area be n_{2D} and assume, as before, that the electric field vanishes in the substrate below the 2DEG. The trial wave function for the bound state, which we used for the triangular well in Section 7.5.2, is

$$u(z) = (\tfrac{1}{2}b^3)^{1/2}z \exp(-\tfrac{1}{2}bz). \tag{9.30}$$

The main weakness of this in the GaAs–AlGaAs system is the assumption that the wave function vanishes in the barrier $z \le 0$; unfortunately the algebra is complicated if this approximation is relaxed. The three-dimensional wave functions have the form $\exp(i\mathbf{k} \cdot \mathbf{r})u(z)$, giving a total charge density that is only a function of z,

$$\rho(z) = -en_{2D}|u(z)|^2 = -\tfrac{1}{2}en_{2D}b^3z^2 \exp(-bz). \tag{9.31}$$

The Hartree potential energy $V_H(z) = -e\phi_H(z)$, where $\phi_H(z)$ satisfies Poisson's equation with the preceding charge density, $d^2\phi_H(z)/dz^2 = -\rho/\epsilon_0\epsilon_b$. Here we can simply integrate the equation twice to find $\phi_H(z)$. One boundary condition is $d\phi/dz = 0$ as $z \to \infty$ in the substrate, and for consistency with our other treatments we want $V(0_+) = 0$, which means $\phi_H(z{=}0) = 0$. The result is

$$\phi_H(z) = -\frac{en_{2D}}{2\epsilon_0\epsilon_b b}\{6 - [(bz)^2 + 4bz + 6]\exp(-bz)\}. \qquad (9.32)$$

The Hamiltonian for each electron is $\hat{H}_1 = \hat{T} + V_H$, where $\hat{T}$ is the kinetic energy operator; there is no external potential energy $V_{ext}(z)$ in this case. The obvious next step would be to calculate the expectation value $\varepsilon_1 = \langle\hat{H}_1\rangle$ of this Hamiltonian with the wave function (9.30) and minimize it with respect to b. This is wrong, however, because the variational method must be applied to the *total* energy of the system. The important feature is the factor of $\frac{1}{2}$ in front of the interaction in the many-electron Hamiltonian (9.25). This is the same factor that comes into the expression for the energy of a set of charges in classical electrostatics, $E = \frac{1}{2}\sum_j q_j\phi_j$. By minimizing $\langle H_1\rangle$ we are omitting this $\frac{1}{2}$ and double-counting the electron–electron interaction. Instead we should minimize the total energy per electron $E_T = \langle\hat{T}\rangle + \frac{1}{2}\langle V_H\rangle$.

The kinetic energy is

$$\langle\hat{T}\rangle = \frac{\hbar^2}{2m}\frac{b^3}{2}\int_0^\infty z\exp(-\tfrac{1}{2}bz)\left[-\frac{d^2}{dz^2}z\exp(-\tfrac{1}{2}bz)\right]dz = \frac{\hbar^2 b^2}{8m}. \qquad (9.33)$$

All the integrals reduce to simple factorials, an attractive feature of this wave function that we met in Section 7.5.2. The Hartree energy per electron is the integral of $\rho\phi_H$,

$$\langle V_H\rangle = \frac{e^2 n_{2D}}{2\epsilon_0\epsilon_b b}\frac{b^3}{2}\int_0^\infty \{6 - [(bz)^2 + 4bz + 6]\exp(-bz)\}z^2\exp(-bz)\,dz$$

$$= \frac{33e^2 n_{2D}}{16\epsilon_0\epsilon_b b}. \qquad (9.34)$$

Thus we have to minimize

$$E_T = \langle\hat{T}\rangle + \tfrac{1}{2}\langle V_H\rangle = \frac{\hbar^2 b^2}{8m} + \frac{33e^2 n_{2D}}{32\epsilon_0\epsilon_b b}. \qquad (9.35)$$

Straightforward calculus gives

$$b = \left(\frac{33me^2 n_{2D}}{8\hbar^2\epsilon_0\epsilon_b}\right)^{1/3}. \qquad (9.36)$$

The one-electron energy level ε_1 (*not* E_T!) is then

$$\varepsilon_1 = \left[\frac{5}{16}\left(\frac{33}{2}\right)^{2/3}\right]\left[\frac{\hbar^2}{2m}\left(\frac{e^2 n_{2D}}{\epsilon_0\epsilon_b}\right)^2\right]^{1/3}. \qquad (9.37)$$

The prefactor is 2.025, to be compared with 2.338 from the Airy function. Thus the variational calculation has given a lower energy, as we would expect from the shape of the potential well shown in Figure 9.7. The two definitions of thickness give different results:

$$\langle z \rangle = \int_0^\infty z\,|u(z)|^2\,dz = \frac{3}{b}, \qquad h = \frac{\epsilon_0\epsilon_b}{e^2}\frac{\partial\varepsilon_1}{\partial n_{2D}} = \frac{55}{32b}. \tag{9.38}$$

For the example with $n_{2D} = 3\times10^{15}\,\mathrm{m}^{-2}$, the variational method gives $\varepsilon_1 = 42\,\mathrm{meV}$ and a thickness $\langle z \rangle = 12\,\mathrm{nm}$ or $h = 7\,\mathrm{nm}$.

Nowadays a one-dimensional Hartree problem can be solved in a few seconds on a personal computer. The result with $n_{2D} = 3\times10^{15}\,\mathrm{m}^{-2}$ is shown in Figure 9.7(c). Although the general form is close to the Fang–Howard solution, there is clear penetration of the wave function into the barrier for $z < 0$. It is possible to modify the Fang–Howard wave function to include this tail, but the method then loses its attractive simplicity. The numerical solution gives $\varepsilon_1 = 34\,\mathrm{meV}$ and $\langle z \rangle = 9\,\mathrm{nm}$. Exchange, correlation, and the image forces arising from the difference in dielectric constant between GaAs and AlGaAs have all been included in fuller calculations, so one might presume that the variational method has had its day. Nevertheless, the extreme simplicity of the Fang–Howard wave function ensures that it remains widely used.

9.4 Screening by an Electron Gas

The effect of simple dielectrics in electrostatics is familiar. For example, the potential of an ionized donor in a semiconductor is reduced from its value in free space by a factor of ϵ_b, the relative permittivity or dielectric constant. This picture applies to a solid with no free carriers. The electric field of the donor polarizes each atom, inducing a dipole moment $\mathbf{p}$. This can be expressed as a polarization field $\mathbf{P} = n\mathbf{p}$, where n is the number density of dipoles. For small fields the induced dipoles are proportional to the total electric field, $\mathbf{p} = \epsilon_0\alpha\mathbf{E}$, or $\mathbf{P} = \epsilon_0\chi\mathbf{E}$, where $\chi = n\alpha$ is the susceptibility (ignoring such subtleties as local field corrections). The electric displacement $\mathbf{D} = \epsilon_0\mathbf{E} + \mathbf{P} = \epsilon_0(1+\chi)\mathbf{E} = \epsilon_0\epsilon_r\mathbf{E}$. The polarization can be replaced by a charge density $\rho_{ind} = -\mathrm{div}\,\mathbf{P}$, from which it can be shown that $\mathbf{D}$ is generated by the external charge alone (the donor in this case), whereas $\mathbf{E}$ arises from both external and induced charge.

An important feature of this picture of screening is that the charges are trapped. We can imagine each atom as a positive and negative charge connected by a spring; in fact this is the Lorentz model of optical response (Section A6.2). This means that charge cannot flow in response to an applied electric field. There is no such limitation in a metal or a semiconductor with free carriers and screening is much more effective. In fact one often says that a potential such as $e/4\pi\epsilon_0\epsilon_b R$ is 'unscreened' because it retains the long-range $1/R$ form, although it is reduced from its magnitude in free space.

In general, the fields can be Fourier transformed and each component treated independently to give $\tilde{\mathbf{D}}(\mathbf{Q}, \omega) = \epsilon_0 \epsilon_r(\mathbf{Q}, \omega) \tilde{\mathbf{E}}(\mathbf{Q}, \omega)$ with a dielectric function $\epsilon_r(\mathbf{Q}, \omega)$ that depends on $\mathbf{Q}$ and ω. Such an approach is restricted to weak fields, where the response is linear. This is often adequate in metals but is frequently overcome in semiconductors, where the attempt to screen an applied field may result in depletion.

It is more convenient to use ρ_{ind} than $\mathbf{P}$ when charge may flow. Also, perturbation theory is based on changes in potential energy so the electrostatic potential enters more directly than the fields. Let ϕ_{tot}, ϕ_{ext}, and ϕ_{ind} be the potentials generated by the total, external, and induced charge densities; they are related to $\mathbf{E}$, $\mathbf{D}$, and $\mathbf{P}$, respectively, and obey $\phi_{\text{tot}} = \phi_{\text{ext}} + \phi_{\text{ind}}$. The dielectric function is given in terms of the Fourier transforms of these potentials by $\epsilon_r(\mathbf{Q}, \omega) = \tilde{\phi}_{\text{ext}}(\mathbf{Q}, \omega)/\tilde{\phi}_{\text{tot}}(\mathbf{Q}, \omega) = 1 - \tilde{\phi}_{\text{ind}}(\mathbf{Q}, \omega)/\tilde{\phi}_{\text{tot}}(\mathbf{Q}, \omega)$.

9.4.1   SCREENING IN THREE DIMENSIONS

The theory of screening falls into two stages. First is the calculation of the induced charge density $\tilde{\rho}_{\text{ind}}$. The electrons cannot distinguish the source of the field in which they move, so $\tilde{\rho}_{\text{ind}}$ is a function of the *total* potential $\tilde{\phi}_{\text{tot}}$. Second, the potential $\tilde{\phi}_{\text{ind}}$ generated by $\tilde{\rho}_{\text{ind}}$ must be found to complete the loop. This is the easier part in three dimensions so we shall treat it first.

Consider a semiconductor, which screens the potential by a factor of ϵ_b before the free charges add their contribution. The potential and charge density are related by Poisson's equation $\nabla^2 \phi_{\text{ind}}(\mathbf{R}) = -\rho_{\text{ind}}(\mathbf{R})/\epsilon_0 \epsilon_b$. The Fourier transform of this equation is $-Q^2 \tilde{\phi}_{\text{ind}}(\mathbf{Q}) = -\tilde{\rho}_{\text{ind}}(\mathbf{Q})/\epsilon_0 \epsilon_b$ so the solution to Poisson's equation in Fourier space is $\tilde{\phi}_{\text{ind}} = \tilde{\rho}_{\text{ind}}/\epsilon_0 \epsilon_b Q^2$.

For the special case of an electron at the origin, the charge density $\rho(\mathbf{R}) = -e\delta(\mathbf{R})$. This has Fourier transform $\tilde{\rho}(Q) = -e$, so the resulting potential is $\tilde{\phi} = -e/\epsilon_0 \epsilon_b Q^2$. The corresponding potential energy for another electron is

$$\tilde{v}_{\text{3D}}(Q) = \frac{e^2}{\epsilon_0 \epsilon_b Q^2}. \tag{9.39}$$

This is the Fourier transform of the Coulomb potential energy in three dimensions.

Returning to screening, we find the induced potential is given by $\tilde{\phi}_{\text{ind}}(\mathbf{Q}, \omega) = \tilde{\rho}_{\text{ind}}(\mathbf{Q}, \omega)/\epsilon_0 \epsilon_b Q^2$ and the dielectric function becomes

$$\epsilon_r(\mathbf{Q}, \omega) = 1 - \frac{1}{\epsilon_0 \epsilon_b Q^2} \frac{\tilde{\rho}_{\text{ind}}(\mathbf{Q}, \omega)}{\tilde{\phi}_{\text{tot}}(\mathbf{Q}, \omega)}. \tag{9.40}$$

Writing $\tilde{\rho}_{\text{ind}} = -e\, d\tilde{n}$, in terms of the (infinitesimal) change in number density of the electrons, and defining $d\tilde{\varepsilon} = -e\tilde{\phi}_{\text{tot}}$ as the change in potential energy of these

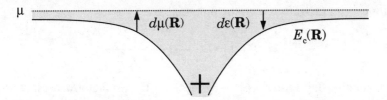

FIGURE 9.8. Thomas–Fermi screening of a donor ion.

electrons, gives

$$\epsilon_r(\mathbf{Q}, \omega) = 1 - \frac{e^2}{\epsilon_0 \epsilon_b Q^2} \frac{d\tilde{n}(\mathbf{Q}, \omega)}{d\tilde{\varepsilon}(\mathbf{Q}, \omega)} \equiv 1 - \tilde{v}_{3D}(Q) \Pi(\mathbf{Q}, \omega), \qquad (9.41)$$

where $\Pi(\mathbf{Q}, \omega)$ is called the *polarization function* of the electron gas, its change in density in response to a change in energy.

The difficult part of the theory is the calculation of $\Pi(\mathbf{Q}, \omega)$, which requires many-body techniques. We shall derive only the Thomas–Fermi approximation, which describes the static response ($\omega = 0$) to long wavelengths ($Q \ll k_F$). Now Π is the change in density in response to a change in energy, and this is easy to calculate if $d\varepsilon(\mathbf{R}, t)$ is strictly constant in both space and time. Figure 9.8 shows that a change in potential energy of $d\varepsilon(\mathbf{R})$ at a point is equivalent to an apparent change in the local Fermi level of $d\mu(\mathbf{R}) = -d\varepsilon(\mathbf{R})$ measured with respect to the conduction band; the absolute value μ of the Fermi level is of course constant throughout the sample at equilibrium. The *Thomas–Fermi approximation* is to extend this argument to potentials that are not constant but vary slowly in space. The derivative in Π with respect to potential energy becomes one with respect to Fermi level and $\Pi_{TF} = -dn/d\mu$. This derivative of the number density of electrons with respect to the Fermi level is called the *thermodynamic density of states* (yet another one!). An increase in the potential energy of the electrons leads to a decrease in their density at fixed Fermi level, hence the minus sign. The dielectric function becomes

$$\epsilon_{TF}(Q, 0) = 1 - \tilde{v}_{3D}(Q)\Pi_{TF} = 1 + \frac{e^2}{\epsilon_0 \epsilon_b Q^2} \frac{dn}{d\mu} = 1 + \frac{Q_{TF}^2}{Q^2}, \qquad (9.42)$$

where

$$Q_{TF} = \sqrt{\frac{e^2}{\epsilon_0 \epsilon_b} \frac{dn}{d\mu}} \qquad (9.43)$$

is the *Thomas–Fermi screening wave number*. Note that its definition in a semiconductor includes the factor of ϵ_b, and we should always divide the external potential by this background dielectric constant before applying $\epsilon_r(Q, \omega)$ due to the free charges.

Now consider Thomas–Fermi screening of a donor. The unscreened potential energy is that of a positive point charge, whose Fourier transform is $-\tilde{v}_{3D}(Q)$.

Dividing by $\epsilon_{\text{TF}}(Q, 0)$ to screen the potential gives

$$\tilde{V}_{\text{scr}}(Q) = -\frac{e^2}{\epsilon_0 \epsilon_b} \frac{1}{Q^2 + Q_{\text{TF}}^2}. \qquad (9.44)$$

This has an amusing limit as $Q \to 0$, because the prefactor cancels with Q_{TF}^2 to give $\tilde{V}_{\text{scr}}(0) = (dn/d\mu)^{-1}$. Screening causes the electronic charge to vanish from the potential of the donor at long wavelengths, leaving only the thermodynamic density of states.

The inverse Fourier transform of the donor's potential is

$$V_{\text{scr}}(R) = -\frac{e^2}{4\pi \epsilon_0 \epsilon_b} \frac{\exp(-Q_{\text{FT}} R)}{R}. \qquad (9.45)$$

This is called a *Yukawa potential*, which was originally used to describe the potential between nucleons mediated by mesons. The important feature is that the long-range nature of the Coulomb potential has been suppressed: it now dies off exponentially with a length scale of $1/Q_{\text{TF}}$. This was the effect described in Section 9.3.2 associated with the correlation hole of an electron, although particles in a gas are moving and a dynamic theory is needed. The screening of donors is particularly important when calculating the way in which they scatter electrons, as we shall see in Section 9.5.

We need the value of Q_{TF} to complete the analysis. The thermodynamic density of states is given by

$$\frac{dn}{d\mu} = \int n(E) \frac{\partial f(E, \mu)}{\partial \mu} dE = \int n(E) \left[-\frac{\partial f(E, \mu)}{\partial E} \right] dE, \qquad (9.46)$$

where $n(E)$ is the 'ordinary' density of states (Section 1.8), and the derivative has been changed because the occupation function depends only on the difference $E - \mu$. This is simple in the limits of low and high temperature. At low temperature, where the distribution is highly degenerate, $f(E, \mu) = \Theta(\mu - E)$ so $-\partial f/\partial E = \delta(E - \mu)$ and $dn/d\mu = n(\mu)$. Thus the thermodynamic density of states is equal to the ordinary density of states at the Fermi level. This is typical of a metal and also applies to a highly doped semiconductor. GaAs with $n_{3D} = 10^{24} \text{ m}^{-3}$ gives $Q_{\text{TF}}^{-1} \approx 5 \text{ nm}$. The range of a Coulomb potential is cut drastically, and the high density of states in a real metal reduces Q_{TF}^{-1} to atomic dimensions.

The opposite limit is at high temperatures when the distribution of electrons is non-degenerate. In this case, equation (9.46) gives $dn/d\mu = n_{3D}/k_B T$, where n_{3D} is the total density of electrons, and

$$Q_{\text{TF}} = \sqrt{\frac{e^2}{\epsilon_0 \epsilon_b} \frac{n_{3D}}{k_B T}}. \qquad (9.47)$$

This limit is known as Debye–Hückel screening and applies to lightly doped semiconductors at room temperature.

It is not too difficult to calculate the polarization function $\Pi(Q, \omega)$ with quantum mechanics using the same approach as in Section 8.5, where we found the general expression (8.76) for the optical conductivity. The main differences are that the applied field is longitudinal (which is why we have used a scalar potential rather than a vector potential), that the wave vector cannot be neglected, and that we are interested in the change in wave functions rather than the transition rate. Fortunately the two processes are intimately related and give the real and imaginary parts of a complex dielectric function known as the *Lindhard function*,

$$\tilde{\epsilon}_r(Q, \omega) = 1 - \tilde{v}_{3D}(Q)\frac{2}{\Omega}\sum_{\mathbf{K}}\frac{f(\mathbf{K}) - f(\mathbf{K} + \mathbf{Q})}{\varepsilon(\mathbf{K} + \mathbf{Q}) - \varepsilon(\mathbf{K}) - \hbar\omega + i0_+}. \quad (9.48)$$

This describes both polarization and absorption by the electron gas. It coincides with the Thomas–Fermi approximation in the limit $\omega = 0$, $Q \to 0$ but shows that screening becomes less effective as Q increases, particularly when $Q > 2k_F$. Another important limit is $Q = 0$, where it reduces to

$$\epsilon_r(0, \omega) = 1 - \frac{\omega_p^2}{\omega^2}, \qquad \omega_p^2 = \frac{e^2 n_{3D}}{\epsilon_0 \epsilon_b m}. \quad (9.49)$$

This is the dielectric function of a classical collisionless plasma (Section A6.2). It vanishes when $\omega = \omega_p$, the plasma frequency. If $\epsilon = 0$, an **E** field can exist without a **D**, so an oscillating electric field can arise without any external sources. This natural mode of oscillation is called a *plasmon* and can be excited efficiently by shooting charged objects (such as electrons) through an electron gas at high energy. Highly doped GaAs with $n_{3D} = 10^{24}\,\text{m}^{-3}$ has $\hbar\omega_p \approx 40\,\text{meV}$, which is similar to the energy of optic phonons, and both are important for cooling hot carriers.

9.4.2 SCREENING IN A 2DEG

The calculation of $\Pi(\mathbf{q}, \omega)$ in two dimensions is very similar to that in three dimensions; the main difference lies in the electrostatics. Ignore the thickness of the 2DEG for simplicity and concentrate its density into a δ-function, so the three-dimensional charge density $\rho(\mathbf{R}) = \sigma(\mathbf{r})\,\delta(z)$, where $\sigma(\mathbf{r})$ is the charge density per unit area. We need to find the relation between a Fourier component of induced charge and the corresponding potential in the plane of the 2DEG.

The reason why this is more difficult than the three-dimensional case is shown in Figure 9.9. Although we are interested only in what happens in the 2DEG, the electric field is three-dimensional and is *not* restricted to the plane $z = 0$. Moreover, the field may be distorted by a surface, metallic gate, or heterojunction close to the 2DEG, a complication that we shall ignore.

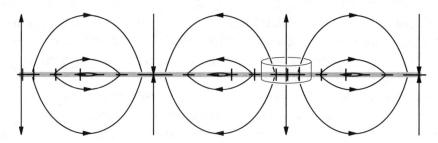

FIGURE 9.9. Electric field generated by a static charge density wave in a 2DEG, showing a Gaussian surface around the 2DEG.

Assume that we are concerned only with low frequencies and can calculate the fields as though they are static. A Fourier component of induced charge density in the plane is $\sigma_{\text{ind}}(\mathbf{r}) = \tilde{\sigma}_{\text{ind}}(\mathbf{q}) \exp(i\mathbf{q} \cdot \mathbf{r})$. It generates a potential whose value in the 2DEG is $\phi_{\text{ind}}(\mathbf{r}, z{=}0) = \tilde{\phi}_{\text{ind}}(\mathbf{q}) \exp(i\mathbf{q} \cdot \mathbf{r})$. Remember that $\mathbf{q}$ is a two-dimensional vector in the plane of the 2DEG! The induced potential spreads over all space and is given by

$$\phi_{\text{ind}}(\mathbf{r}, z) = \tilde{\phi}_{\text{ind}}(\mathbf{q}) \exp(i\mathbf{q} \cdot \mathbf{r}) \exp(-|qz|). \tag{9.50}$$

The decaying exponential satisfies Laplace's equation for $z \neq 0$ and ensures that the potential falls off for large z. The derivative along z, and therefore the electric field, is discontinuous at $z = 0$. By Gauss's theorem this discontinuity must balance the charge density there. Consideration of the Gaussian surface drawn in Figure 9.9 shows that this Fourier component obeys

$$\frac{\tilde{\sigma}_{\text{ind}}(\mathbf{q})}{\epsilon_0 \epsilon_b} = E_z(0_+) - E_z(0_-) = -\left.\frac{\partial \phi_{\text{ind}}}{\partial z}\right|_{0_+} + \left.\frac{\partial \phi_{\text{ind}}}{\partial z}\right|_{0_-} = 2q\tilde{\phi}_{\text{ind}}(\mathbf{q}). \tag{9.51}$$

Thus $\tilde{\phi}_{\text{ind}} = \tilde{\sigma}_{\text{ind}}/2\epsilon_0 \epsilon_b q$, whence the dielectric function is

$$\epsilon_r(\mathbf{q}, \omega) = 1 - \frac{e^2}{2\epsilon_0 \epsilon_b q} \Pi(\mathbf{q}, \omega) \equiv 1 - \tilde{v}_{2D}(q) \Pi(\mathbf{q}, \omega). \tag{9.52}$$

This should be multiplied by the usual factor of ϵ_b due to the surrounding host, and $v_{2D}(q)$ is the two-dimensional Fourier transform of the Coulomb energy. The Thomas–Fermi approximation again gives $\Pi_{\text{TF}} = -dn/d\mu$. At low temperature this density of states is constant at $m/\pi\hbar^2$, so

$$\epsilon_{\text{TF}}(q, 0) = 1 + \frac{q_{\text{TF}}}{q}, \qquad q_{\text{TF}} = \frac{me^2}{2\pi\epsilon_0\epsilon_b\hbar^2} = \frac{2}{a_B}, \tag{9.53}$$

where a_B is the effective Bohr radius (equation 4.67).

Now consider the screening of a point charge such as a donor. It will be useful to treat the general case of a charge situated a distance d out of the plane of the 2DEG, where $d \geq 0$. The Fourier transform of the unscreened potential energy is

$$
\begin{aligned}
\tilde{V}_{uns}(q) &= \int_0^\infty r \, dr \int_0^{2\pi} d\theta \, \frac{e^2}{4\pi\epsilon_0\epsilon_b\sqrt{r^2+d^2}} e^{iqr\cos\theta} \\
&= \frac{e^2}{4\pi\epsilon_0\epsilon_b} \int_0^\infty \frac{2\pi J_0(qr) \, r \, dr}{\sqrt{r^2+d^2}} \\
&= \frac{e^2}{2\epsilon_0\epsilon_b} \frac{\exp(-qd)}{q} \equiv \tilde{v}_{2D}(q) \exp(-qd).
\end{aligned} \tag{9.54}
$$

The integral is given as equation (6.554.1) in Gradshteyn and Ryzhik (1993). An important feature of the result is the exponential decay of the potential with respect to the separation d between the charge and the 2DEG. This is a crucial factor in the high mobility of a 2DEG, as we shall see in Section 9.5. Dividing by the Thomas–Fermi dielectric function (9.53) gives

$$
\tilde{V}_{scr}(q) = \frac{e^2}{2\epsilon_0\epsilon_b} \frac{e^{-qd}}{q + q_{TF}}. \tag{9.55}
$$

Again the electronic charge vanishes in favour of the density of states for $q = 0$. Unfortunately this Fourier transform cannot be inverted analytically to real space. The key feature is the decay of the potential at large r, which is found to be

$$
V_{scr}(r) \sim \frac{e^2}{4\pi\epsilon_0\epsilon_b} \frac{q_{TF}(1 + q_{TF}d)}{(q_{TF}r)^3}. \tag{9.56}
$$

This is a power law: there is no exponential screening in a 2DEG. The reason is that electric field lines can leave the plane of the 2DEG and avoid the screening charge, as in Figure 9.9. Screening is therefore much less effective than in three dimensions, where the induced charge completely surrounds the added charge.

An amusing feature of Thomas–Fermi screening in two dimensions is that q_{TF} is independent of the density of electrons, so it appears as though a very dilute 2DEG screens as effectively as a dense gas. This paradox is resolved by the Lindhard formula, which shows that the dielectric function for $q > 2k_F$ is reduced to

$$
\epsilon_r(q, 0) = 1 + \frac{q_{TF}}{q} \left\{ 1 - \left[1 - \left(\frac{2k_F}{q}\right)^2 \right]^{1/2} \right\}. \tag{9.57}
$$

Constant screening holds only for $q < 2k_F$, whose range of validity shrinks with the density of the 2DEG. Another limitation is that linear screening cannot hold if the 2DEG is depleted and again this restriction is more severe for a dilute system. Fortunately many phenomena require only $\epsilon_r(q, 0)$ for weak fields in the range where it is constant.

The behaviour of plasmons is also changed drastically. The dispersion relation has a square root rather than giving a constant frequency as $q \rightarrow 0$,

$$\omega_{\mathrm{p}} \sim \sqrt{\frac{e^2 n_{\mathrm{2D}} q}{2\epsilon_0 \epsilon_{\mathrm{b}} m}}. \tag{9.58}$$

An important effect of screening is to modify the interaction between electrons and scattering potentials, which we shall study next.

9.5 Scattering by Remote Impurities

The two-dimensional electron gas owes its status as the most important low-dimensional system for transport to the long mean free path of electrons at low temperature. This arises from the separation between the electrons and the donors that provided them, as discussed in Section 3.5. The mobility of two- and three-dimensional electron gases is similar at high temperature or high electric fields where phonons dominate the scattering.

The strongest scattering at low temperature in many 2DEGs arises from the ionized donors in the n-AlGaAs layer, which are separated from the electrons by a spacer. Although the specific example is the 2DEG at a heterojunction, the same principles apply to electrons or holes trapped in a quantum well; only the confined wave function is different. We shall assume that the electron gas is highly degenerate so that the scattering rate need be calculated only at the Fermi level.

Assume for simplicity that the material is δ-doped, with a plane of $n_{\mathrm{imp}}^{(2D)}$ ionized donors per unit area as shown in Figure 9.10; the extension to a doped slab is trivial. Note that the z-axis points downwards, as is the usual convention. This has the unfortunate corollary that the distance d of a plane of impurities 'above' the electrons, the usual place, is negative. We have already derived a general formula

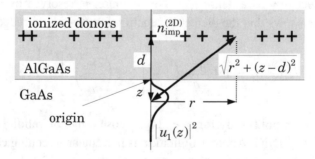

FIGURE 9.10. Geometry for scattering of electrons in a two-dimensional electron gas from remote ionized impurities. The z-axis is taken downwards, measured from the interface at which the electrons are trapped, so $d < 0$ for the impurities shown.

(8.32) for the transport lifetime,

$$\frac{1}{\tau_{tr}} = n_{imp}^{(2D)} \frac{m}{2\pi \hbar^3 k_F^3} \int_0^{2k_F} |\tilde{V}(q)|^2 \frac{q^2 \, dq}{\sqrt{1 - (q/2k_F)^2}}. \tag{9.59}$$

The single-particle rate lacks a factor of $q^2/2k_F^2$. All we need to complete this is the potential, for which we can use the Coulomb potential with Thomas–Fermi screening (equation 9.55). Thus

$$\frac{1}{\tau_{tr}} = n_{imp}^{(2D)} \frac{m}{2\pi \hbar^3 k_F^3} \left(\frac{e^2}{2\epsilon_0 \epsilon_b}\right)^2 \int_0^{2k_F} \frac{\exp(-2q|d|)}{(q + q_{TF})^2} \frac{q^2 \, dq}{\sqrt{1 - (q/2k_F)^2}}. \tag{9.60}$$

This expression contains most of the important physics but we shall make a few improvements before evaluating it.

9.5.1 MATRIX ELEMENT

Equation (9.59) was derived for a strictly two-dimensional system, neglecting the electronic wave function normal to the plane of the 2DEG. We should include the full three-dimensional wave function in the matrix element for scattering (although the wave vectors remain two-dimensional). As usual, the wave functions have the product form $\phi_{k,n}(\mathbf{R}) = A^{-1/2} u_n(z) \exp(i\mathbf{k} \cdot \mathbf{r})$, where A is the area of the 2DEG and n labels the subband. Let the initial and final states be $A^{-1/2} u_m(z) \exp(i\mathbf{k} \cdot \mathbf{r})$ and $A^{-1/2} u_n(z) \exp[i(\mathbf{k} + \mathbf{q}) \cdot \mathbf{r}]$; in general the subband may change as well as the wave vector. The matrix element takes the form

$$V_{nm}(q) = \frac{1}{A} \int u_n^*(z) e^{-i(\mathbf{k}+\mathbf{q})\cdot\mathbf{r}} V(\mathbf{r}, z) u_m(z) e^{i\mathbf{k}\cdot\mathbf{r}} \, d^3\mathbf{R}$$

$$= \frac{1}{A} \int dz \, u_n^*(z) u_m(z) \int d^2\mathbf{r} \, V(r, z) e^{i\mathbf{q}\cdot\mathbf{r}}. \tag{9.61}$$

Let $\tilde{V}(q, z)$ denote the two-dimensional Fourier transform of $V(r, z)$ with respect to r at fixed z, assuming that the potential has cylindrical symmetry. Then

$$V_{nm}(q) = \frac{1}{A} \int u_n^*(z) u_m(z) \tilde{V}(q, z) \, dz. \tag{9.62}$$

If scattering is within a subband ($n = m$) and the wave functions have zero thickness ($|u_n(z)|^2 = \delta(z)$), this reduces to $\tilde{V}(q, 0)$, which is our previous matrix element containing only the potential in the plane $z = 0$. This approximation is clearly inadequate for intersubband scattering, and in general we write

$$V_{nm}(q) = \frac{1}{A} F_{nm}(q) \tilde{V}(q, 0), \tag{9.63}$$

where $F_{nm}(q)$ is called a *form factor* and has the limit $F_{nn}(0) = 1$.

Consider intrasubband scattering with $n = m = 1$. Equation (9.54) gave the Fourier transform of the potential from an ionized donor (dropping the sign), $\tilde{V}(q, z) = (e^2/2\epsilon_0\epsilon_b q)e^{-q(z-d)}$ for $z > 0$; remember that $d < 0$ in these coordinates. With the Fang–Howard approximation to $u_1(z)$, the matrix element (9.62) becomes

$$V_{11}(q) = \frac{1}{A}\frac{e^2}{2\epsilon_0\epsilon_b q}\int_0^\infty [\tfrac{1}{2}b^3 z^2 e^{-bz}]e^{-q(z-d)}\,dz = \frac{1}{A}\frac{e^2}{2\epsilon_0\epsilon_b}\frac{e^{-q|d|}}{q}\left(\frac{b}{b+q}\right)^3.$$

(9.64)

Comparison with $\tilde{V}(q, 0)$ shows that the form factor is $F_{11}(q) = [b/(b+q)]^3$. Its main effect is to account for the shift of the mean position of the electrons away from the interface. The scattering rate contains the square of the matrix element and form factor so the rate for remote impurities becomes

$$\frac{1}{\tau_{tr}} = n_{imp}^{(2D)}\frac{m}{2\pi\hbar^3 k_F^3}\left(\frac{e^2}{2\epsilon_0\epsilon_b}\right)^2\int_0^{2k_F}\frac{e^{-2q|d|}}{(q+q_{TF})^2}\left(\frac{b}{b+q}\right)^6\frac{q^2\,dq}{\sqrt{1-(q/2k_F)^2}}.$$

(9.65)

The potential has been screened with our Thomas–Fermi dielectric function. Unfortunately this is inconsistent because we have accounted for $u_1(z)$ in the matrix element, whereas our derivation of $\epsilon_r(q, \omega)$ assumed there is a δ-function. One should use a charge density $\rho(\mathbf{R}) = \sigma(\mathbf{r})|u(z)|^2$ to calculate ϕ_{ind}, and average the resulting potential over $|u(z)|^2$ to give its effect on the electrons as we have done for the effect of an impurity. The outcome is that another form factor enters the dielectric function, giving

$$\frac{1}{\tau_{tr}} = n_{imp}^{(2D)}\frac{m}{2\pi\hbar^3 k_F^3}\left(\frac{e^2}{2\epsilon_0\epsilon_b}\right)^2\int_0^{2k_F}\frac{e^{-2q|d|}}{[q+q_{TF}G(q)]^2}\left(\frac{b}{b+q}\right)^6\frac{q^2\,dq}{\sqrt{1-(q/2k_F)^2}}$$

(9.66)

with

$$G(q) = \frac{1}{8}\left[2\left(\frac{b}{b+q}\right)^3 + 3\left(\frac{b}{b+q}\right)^2 + 3\left(\frac{b}{b+q}\right)\right].$$
(9.67)

This completes our expression for the scattering rate.

9.5.2 SCATTERING RATE

The final scattering rate, equation (9.66), looks formidable but is straightforward to evaluate numerically. Rather than do this, we shall make some approximations that lead to a transparent result. First, it is clear that screening has a major effect. The unscreened potential would contribute $1/q^2$ rather than $1/q_{TF}^2$ as $q \to 0$ giving a huge increase in scattering, and the integral for $1/\tau_i$ would diverge. The calculation of scattering from unscreened Coulomb potentials remains a difficult topic.

Now examine the quantities that set the scales of the integral over q.

(i) The separation $|d|$ between the impurity and 2DEG appears in the decaying exponential and causes the integral to die off for $q \gg 1/|d|$. Typically $d > 10\,\mathrm{nm}$, so the significant range of the integral is $q < 0.1\,\mathrm{nm}^{-1}$.

(ii) The Thomas–Fermi screening wave number $q_{TF} = 0.2\,\mathrm{nm}^{-1}$ appears as $q_{TF} + q$ in the denominator. Since the significant range of the integral is for $q < 0.1\,\mathrm{nm}^{-1}$, it is reasonable to drop the q from the denominator. This emphasizes the strong influence of screening.

(iii) The maximum wave number for scattering is set by $2k_F$, typically around $0.3\,\mathrm{nm}^{-1}$ for a 2DEG with $3 \times 10^{15}\,\mathrm{m}^{-2}$ electrons. This is again much greater than $1/|d|$. The upper limit on the integral can therefore be set to infinity, as the exponential kills the integrand off before the upper limit is reached. For the same reason, the square root in the denominator can be replaced by unity.

(iv) The Fang–Howard parameter $b \approx 0.2\,\mathrm{nm}^{-1}$, which again is (just about) larger than q throughout the important range. We can put $q = 0$ in the form factors, which then both reduce to unity and drop out.

Unfortunately these approximations, although convenient, are not of spectacular accuracy and the result can be taken only as a rough guide, particularly if the spacer is thin. The simplifications give

$$
\frac{1}{\tau_{tr}} = n_{\mathrm{imp}}^{(2D)} \frac{m}{2\pi\hbar^3 k_F^3} \left(\frac{e^2}{2\epsilon_0\epsilon_b q_{TF}} \right)^2 \int_0^\infty q^2 e^{-2q|d|} dq
$$

$$
= n_{\mathrm{imp}}^{(2D)} \frac{m}{8\pi\hbar^3 (k_F|d|)^3} \left(\frac{e^2}{2\epsilon_0\epsilon_b q_{TF}} \right)^2 = \frac{\pi\hbar n_{\mathrm{imp}}^{(2D)}}{8m(k_F|d|)^3}. \tag{9.68}
$$

The last form employs the definition (4.67) of the Bohr radius and the relation $q_{TF} = 2/a_B$. As promised, the result is remarkably simple. The mobility $\mu = e\tau_{tr}/m$ and the mean free path $l_{tr} = v_F\tau_{tr}$ are given by

$$
\mu = \frac{8e(k_F|d|)^3}{\pi\hbar n_{\mathrm{imp}}^{(2D)}}, \qquad l_{tr} = \frac{32\pi n_{2D}^2 |d|^3}{n_{\mathrm{imp}}^{(2D)}}. \tag{9.69}
$$

The relation between the density and Fermi wave number in two dimensions, $n_{2D} = k_F^2/2\pi$, has been used to simplify l_{tr}. An amusing feature is that the mass enters neither result directly, nor does the scattering rate depend on the electronic charge.

To get a rough idea of the numbers, take $n_{2D} = 3 \times 10^{15}\,\mathrm{m}^{-2}$, $k_F = 0.14\,\mathrm{nm}^{-1}$, $|d| = 30\,\mathrm{nm}$ and $n_{\mathrm{imp}}^{(2D)} = 10^{16}\,\mathrm{m}^{-2}$. The simple formulas give $l_{tr} \approx 2.4\,\mu\mathrm{m}$ and $\mu \approx 27\,\mathrm{m}^2\,\mathrm{V}^{-1}\,\mathrm{s}^{-1}$. For comparison, numerical evaluation of equation (9.66) gives $l_{tr} \approx 5.0\,\mu\mathrm{m}$ and $\mu \approx 56\,\mathrm{m}^2\,\mathrm{V}^{-1}\,\mathrm{s}^{-1}$, so the simple result is indeed approximate but useful. The mean free path for the example is not large but it depends on the cube of the distance between the 2DEG and the donors, which can easily be increased.

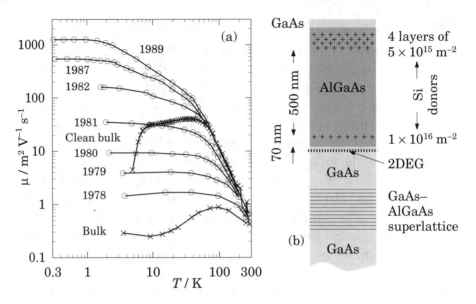

FIGURE 9.11. (a) Mobility of various 2DEGs as a function of temperature (circles), showing how the peak mobility (limited by impurity scattering) has risen over 20 years. The mobility of bulk samples is shown for comparison (crosses), for old material ('Bulk') and purer material ('Clean bulk') [Stanley et al. (1991)]. (b) Simplified structure of wafer grown in the sample of highest mobility. [Redrawn from Pfeiffer et al. (1989).]

A thicker spacer of $|d| = 100$ nm raises μ to $1000\,\mathrm{m^2\,V^{-1}\,s^{-1}}$ with $l_{\mathrm{tr}} \approx 0.1$ mm. This is a spectacular value for a semiconductor, as $\mu < 1\,\mathrm{m^2\,V^{-1}\,s^{-1}}$ at room temperature. It might appear that we can increase μ without limit, but other scattering processes to be discussed in Section 9.6 eventually dominate. The mobility is also a function of the density of the 2DEG, and equation (9.69) shows the useful rule of thumb that $\mu \propto n_{\mathrm{2D}}^{3/2}$.

Figure 9.11(a) shows how the mobility of 2DEGs has improved over the years. This is mainly due to increasing purity of the material and is also reflected in the curves for bulk GaAs. The highest mobility, exceeding $1000\,\mathrm{m^2\,V^{-1}\,s^{-1}}$ in a 2DEG of density $2.4 \times 10^{15}\,\mathrm{m^{-2}}$, was obtained with a spacer of 70 nm. Note that the doping was split into two layers (Figure 9.11(b)); only $10^{16}\,\mathrm{m^{-2}}$ donors are in the layer close to the 2DEG while the majority of the donors, needed to cancel the surface states, are much farther away from the electrons.

Similar formulas can be developed for the single-particle lifetime τ_{i} and the unweighted mean free path, which is $l_{\mathrm{i}} \approx 0.05\,\mu$m for the example. The ratio of the mean free paths is large, $l_{\mathrm{tr}}/l_{\mathrm{i}} \approx 8\pi n_{\mathrm{2D}}|d|^2 \approx 70$, which can be traced to the rapid decay of the Fourier transform $\tilde{V}(q) \propto e^{-|d|q}$. This in turn reflects a potential that varies slowly in space because of the separation between the planes of impurities and electrons. All that is left is the long $1/r$ tail of the potential, a slowly varying function in space, which screening reduces only to $1/r^3$ in two dimensions. The

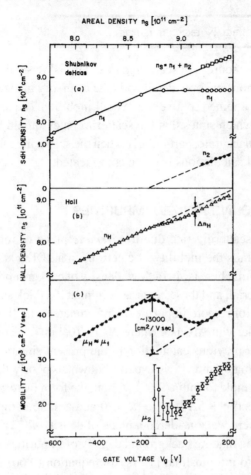

FIGURE 9.12. Effect of a second occupied subband on the mobility of a 2DEG. (a) Density of electrons in each subband, resolved from the spectrum of Shubnikov–de Haas oscillations. (b) Density of electrons deduced from Hall effect. (c) Mobility of electrons in each subband, deduced from equation (6.24). [Reprinted from Störmer, Gossard, and Wiegmann (1982). Copyright 1982, with kind permission from Elsevier Science Ltd, The Boulevard, Langford Lane, Kidlington OX5 1GB, UK.]

behaviour is quite different in a three-dimensional metal, where screening gives short-range potentials and isotropic scattering, so $\tau_{tr} \approx \tau_i$.

A second subband becomes occupied if the density of a 2DEG is raised. Electrons can then scatter between subbands and the mobility falls. This is seen in the experiment shown in Figure 9.12, where the density was changed by a substrate bias. The densities n_1 and n_2 were resolved in Figure 9.12(a) by taking the Fourier transform of the Shubnikov–de Haas oscillations. We have already found equations for the apparent Hall constant and density when two parallel conducting channels are occupied (Section 6.3), and these were used to deduce the mobility of the two subbands plotted in Figure 9.12(b). The mobility of electrons in the second subband is lower than that in the first, partly due to the lower density n_2. A curious feature of these data is that n_1 remains constant after $n_2 > 0$.

9.6   Other Scattering Mechanisms

There are many other sources of scattering in a 2DEG. We shall look at background impurities and phonons in some detail and take a cursory glance at others. Scattering rates must be combined to give the overall mobility. The simplest approach is *Mattheisen's rule*, which states that the scattering rates should be added. Unfortunately this is often inaccurate, particularly when the scattering rates depend strongly on temperature, and calculations become complicated.

9.6.1   BACKGROUND IONIZED IMPURITIES

We calculated the scattering rate due to a *remote* plane of ionized impurities in Section 9.5. This limits the mobility of electrons in a 2DEG at low temperature if the spacer layer is thin. There is also a 'low' density of background ionized impurities throughout the material, and these become dominant if a thick spacer layer is grown to reduce the effect of remote impurities. Unfortunately the 'low' density is often rather high, particularly in AlGaAs where $N_A \approx 10^{21} \text{ m}^{-3}$ is common.

We can adapt our previous calculation to the present situation with little effort. To get a rough estimate (and N_A is usually known so poorly that greater effort is pointless), use the simple formula (9.60) without the form factors; note that the form factor derived in Section 9.5.1 was for $d < 0$ and a more complicated expression holds for $d > 0$. There we considered a plane of density $n_{\text{imp}}^{(2D)}$ at a distance d from the 2DEG. Assuming that the background density of impurities is constant at $n_{\text{imp}}^{(3D)}$, we can divide it into infinitesimal layers, use equation (9.60) for each layer, and integrate over d to get the total scattering rate. This process is simple because d appears only in the exponential. Picking out the terms affected gives

$$n_{\text{imp}}^{(2D)} e^{-2q|d|} \rightarrow n_{\text{imp}}^{(3D)} \int_{-\infty}^{\infty} e^{-2q|d|} dd = \frac{n_{\text{imp}}^{(3D)}}{q}. \tag{9.70}$$

The exponential factor that suppressed scattering for high q has vanished because some impurities lie close to the 2DEG and can scatter electrons through large angles. The transport lifetime becomes

$$\frac{1}{\tau_{\text{tr}}} = n_{\text{imp}}^{(3D)} \frac{m}{2\pi \hbar^3 k_F^3} \left(\frac{e^2}{2\epsilon_0 \epsilon_b}\right)^2 \int_0^{2k_F} \frac{1}{(q + q_{\text{TF}})^2} \frac{q \, dq}{\sqrt{1 - (q/2k_F)^2}}. \tag{9.71}$$

The integral is dimensionless and depends only on the ratio $q_{\text{TF}}/2k_F$. The numbers earlier showed that this ratio is not far from unity, and this must also be true of the integral itself. Thus

$$\frac{1}{\tau_{\text{tr}}} \approx n_{\text{imp}}^{(3D)} \frac{m}{2\pi \hbar^3 k_F^3} \left(\frac{e^2}{2\epsilon_0 \epsilon_b}\right)^2. \tag{9.72}$$

Evaluating this with $n_{2D} = 3 \times 10^{15}\,\mathrm{m}^{-2}$ and $n_{\mathrm{imp}}^{(3D)} = 10^{21}\,\mathrm{m}^{-3}$ leads to $\mu \approx 70\,\mathrm{m}^2\,\mathrm{V}^{-1}\,\mathrm{s}^{-1}$. It is clear that scrupulous cleanliness is vital to achieve the mobility of $1000\,\mathrm{m}^2\,\mathrm{V}^{-1}\,\mathrm{s}^{-1}$ seen in Figure 9.11 and that it is pointless to use thick spacer layers without such efforts.

9.6.2 OTHER ELASTIC SCATTERING

Electrons in a 2DEG scatter from imperfections in the interface between GaAs and AlGaAs at which they are trapped. This interface-roughness scattering is important in MOSFETs but believed to be negligible at 'normal' interfaces grown by MBE, which have AlGaAs grown above GaAs; 'inverted' interfaces of GaAs on AlGaAs are more troublesome.

Electrons are also scattered by the random nature of the AlGaAs alloy. This is a fundamental limitation that cannot be removed (unless the alloy can be grown in an ordered form), but has a weak effect in GaAs–(Al,Ga)As because only the tail of the wave function penetrates the barrier. It is more serious in other systems, such as those using (In,Ga)As channels, where conduction takes place in the alloy.

Another possible scattering mechanism is that from the charge on the surface of a heterostructure. We saw in Section 9.1 that the surface states hold a high density of charge and these could cause significant scattering if they are random. Fortunately they are a long way from the 2DEG, so it is hoped that they can be neglected. Many steps in processing, particularly etching, damage the surface to some extent and may add to this scattering. It has also been claimed that electron-beam lithography degrades mobility. Dislocations and other defects in the crystal may pose further problems, particularly in layers grown on strain-relieved buffers (Section 3.6).

9.6.3 PHONONS

Phonons dominate scattering at high temperatures, typically above 77 K but down to lower temperatures in cleaner material. A complication in treating phonon scattering is that the electrons are quasi-two-dimensional, whereas the phonons remain three-dimensional. Even this picture is simplified because the heterojunction that traps the 2DEG also distorts the phonons and there may even be interface modes localized around the junction. The LO phonons are particularly susceptible because their frequency depends on the dielectric constant, which is discontinuous at the junction. We shall ignore these complications and assume that the phonons behave as they would in an infinite single crystal.

Consider the analogue of the rate (8.50), where an electron scatters from $(m, \mathbf{k})$ to $(n, \mathbf{k}')$ by absorbing a phonon, which sets up a potential energy $M(\mathbf{Q})\exp[i(\mathbf{Q}\cdot\mathbf{R} - \omega_Q t)]$. Remember that $\mathbf{Q} = (\mathbf{q}, q_z)$ is a three-dimensional wave vector, whereas $\mathbf{k}$,

$\mathbf{k}'$, and $\mathbf{q}$ are two-dimensional in the plane of the 2DEG. The matrix element is

$$\frac{M(\mathbf{Q})}{A} \int u_n^*(z) e^{-i\mathbf{k}'\cdot\mathbf{r}} e^{i(\mathbf{q}\cdot\mathbf{r}+q_z z)} u_m(z) e^{i\mathbf{k}\cdot\mathbf{r}} d^3\mathbf{R}. \qquad (9.73)$$

The integral factorizes into one over $\mathbf{r}$ and another over z. The former simply involves plane waves and gives A if the total wave vector is zero, and vanishes otherwise. Thus we get the usual electronic selection rule, $\mathbf{k}' = \mathbf{k} + \mathbf{q}$, but this fixes only two components of the phonon's wave vector.

The matrix element has been reduced to

$$M(\mathbf{Q})\delta_{\mathbf{k}',\mathbf{k}+\mathbf{q}} \int u_n^*(z) e^{iq_z z} u_m(z) dz = M(\mathbf{Q})\delta_{\mathbf{k}',\mathbf{k}+\mathbf{q}} F_{nm}'(q_z) \qquad (9.74)$$

with another form factor $F_{nm}'(q_z)$ that depends on the wave function of the bound states. For example, $m = n = 1$ and the Fang–Howard wave function gives

$$F_{11}'(q_z) = \tfrac{1}{2}b^3 \int_0^\infty z^2 e^{-bz} e^{iq_z z} dz = \left(\frac{b}{b - iq_z}\right)^3. \qquad (9.75)$$

This is the same as that for Coulomb scattering in equation (9.64) apart from the 'i' in front of the wave number, which arises from the propagating rather than decaying nature of the potential. The scattering rate becomes

$$W_{n\mathbf{k}',m\mathbf{k}}^+(\mathbf{Q}) = \frac{2\pi}{\hbar} N_{\mathbf{Q}} |M(\mathbf{Q})|^2 |F_{nm}'(q_z)|^2 \delta_{\mathbf{k}',\mathbf{k}+\mathbf{q}} \delta[\varepsilon(\mathbf{k}+\mathbf{q}) - \varepsilon(\mathbf{k}) - \hbar\omega_{\mathbf{Q}}]. \qquad (9.76)$$

Note the two- and three-dimensional vectors in this! The usual rule for conservation of momentum determines only two components of the phonon's wave vector; the third is not fixed but the form factor suppresses scattering when $q_z \gg b$. The δ-function for energy involves the three-dimensional wave vector $\mathbf{Q}$ so there is no unique relation between the change in two-dimensional wave vector $\mathbf{q}$ and energy $\hbar\omega_{\mathbf{Q}}$.

As an example, consider acoustic phonons coupling through the deformation potential, making the quasi-elastic approximation where the change in energy of the electron is neglected (Section 8.4.1). Again we assume that $N_{\mathbf{Q}} \approx (N_{\mathbf{Q}} + 1) \approx k_B T/\hbar\omega_{\mathbf{Q}}$, which holds at all but the lowest temperatures. Combining the rates for emission and absorption, and inserting the matrix element in equation (8.47) for the deformation potential, gives

$$W_{n\mathbf{k}',m\mathbf{k}}(\mathbf{Q}) = \frac{2\pi}{\hbar} \frac{2k_B T}{\hbar v_s Q} \frac{\hbar Q \Xi^2}{2\Omega\rho v_s} |F_{nm}'(q_z)|^2 \delta_{\mathbf{k}',\mathbf{k}+\mathbf{q}} \delta[\varepsilon(\mathbf{k}+\mathbf{q}) - \varepsilon(\mathbf{k})]. \qquad (9.77)$$

Now sum this rate over all phonons $\mathbf{Q}$ and write it as a product of sums over $\mathbf{q}$ and q_z. The former is trivial because of the δ-function. The sum over q_z includes only the form factor and can be converted into an integral in the usual way to give

$$\sum_{q_z} |F_{nm}'(q_z)|^2 \to \frac{L}{2\pi} \int_{-\infty}^\infty |F_{nm}'(q_z)|^2 dq_z = L \int_{-\infty}^\infty |u_n(z)|^2 |u_m(z)|^2 dz, \qquad (9.78)$$

where L is the length of the sample along z, the limits on q_z have been extended to infinity, and the final form follows from Parseval's theorem. For the Fang–Howard wave function with both states in the bottom subband,

$$\int_{-\infty}^{\infty} |u_1(z)|^2 |u_1(z)|^2 dz = \frac{b^6}{4} \int_0^{\infty} z^4 \exp(-2bz)dz = \frac{3b}{16}. \qquad (9.79)$$

The integral has the dimensions of (length)$^{-1}$ and is clearly related to the thickness of the wave function. The total scattering rate for an electron from $\mathbf{k}$ to $\mathbf{k'} = \mathbf{k} + \mathbf{q}$ becomes

$$W_{\mathbf{k},\mathbf{k+q}} = \frac{2\pi}{\hbar} \frac{1}{A} \frac{k_B T \Xi^2}{\rho v_s^2} \frac{3b}{16} \delta[\varepsilon(\mathbf{k} + \mathbf{q}) - \varepsilon(\mathbf{k})]. \qquad (9.80)$$

As in Section 8.4.1, this final scattering rate is identical to that for isotropic impurity scattering. The inverse lifetime is obtained by multiplying by the density of states for one spin, $m/2\pi\hbar^2$, giving

$$\frac{1}{\tau} = \frac{3mbk_B T \Xi^2}{16\rho v_s^2 \hbar^3}. \qquad (9.81)$$

The mobility $\mu = e\tau/m$, and the constant density of states gives $\mu \propto T^{-1}$ in a 2DEG. Typical values are $\rho v_s^2 = 1.4 \times 10^{11} \, \mathrm{J\,m^{-3}}$ and $\Xi = 10\,\mathrm{eV}$, although there is wide variation in the values quoted for Ξ in a 2DEG. These give $\mu = 350 \, \mathrm{m^2\,V^{-1}\,s^{-1}}$ at $10\,\mathrm{K}$. Looking at the top curve in Figure 9.11(a), we see that $\mu \propto T^{-1}$ for $2 < T < 30\,\mathrm{K}$ as we predict, and the coefficient is in remarkably good agreement too. This dependence is masked in older samples by scattering from impurities. The rapid fall in mobility above $50\,\mathrm{K}$ is due to polar optic phonons, as in three-dimensional material. Their energy is very high ($36\,\mathrm{meV}$), so their occupation number N_{LO} is small even at room temperature, but their coupling to electrons is extremely strong, as we saw in Section 8.4.2.

Unfortunately we have cheated a little. The scattering rate due to remote impurities was strongly reduced by screening in the 2DEG, which we have neglected here. The potential due to the phonon oscillates in time, unlike the static potential due to the impurities, and we ought therefore to include dynamic screening by the dielectric function $\epsilon_r(\mathbf{q}, \omega_{\mathbf{Q}})$. An important difference is that remote impurities scatter mainly through small wave vectors, whereas acoustic phonons give isotropic scattering. Since $2k_F \approx 0.3\,\mathrm{nm^{-1}}$, which is larger than $q_{TF} = 0.2\,\mathrm{nm^{-1}}$, screening has a much weaker effect for the phonons and it is a reasonable first approximation to neglect it. Its inclusion gives a weak dependence of μ on n_{2D}.

Further Reading

The classic reference on the two-dimensional electron gas is by Ando, Fowler, and Stern (1982). The emphasis is on the silicon MOSFET rather than the III–V heterojunction, so much of the material is rather dated but the basic theory remains applicable. Stern (1983) gives a useful summary of the concentration and mobility of a 2DEG at a heterojunction, using the variational approximation to the wave function.

The long mean free path of electrons in a 2DEG has encouraged the performance of a cornucopia of experiments using ballistic electrons. Many can be interpreted classically but others involve interference. A broad survey is provided by Beenakker and van Houten (1991). Vigorous research continues! Two areas that are particularly active at the time of this writing are quantum chaos in small cavities and the effect of electron–electron interaction in one-dimensional systems, which may need to be described in terms of Luttinger liquids. Correlation between electrons is discussed in books on many-body theory, and Mahan (1990) contains an introduction to the Luttinger model.

The main practical application of the two-dimensional electron gas is in MOD-FETs. These transistors and their variants are discussed by Weisbuch and Vinter (1991) and by Kelly (1995), who makes an interesting comparison of the different families of device. Sze (1990) contains a chapter on heterostructure FETs in a survey of the whole field of high-speed devices. A broad analysis of III–V devices is given by Tiwari (1992), including bipolar and field-effect transistors and tunnelling structures. There is also a good account of classical transport theory and issues in device modelling.

The program used for the numerical results shown in this chapter is described by Tan, Snider, and Hu (1990). It runs on the Apple Macintosh and is freely available on the World Wide Web at `http://www.nd.edu/~gsnider`.

EXERCISES

9.1   Extend the theory of Section 9.1 to include the difference in dielectric constant between GaAs and AlGaAs, by matching D_z at the heterojunctions rather than E_z. Is the difference likely to be significant in practice?

9.2   The composition of the structure used in Figure 9.1 was given in Section 9.1. Estimate analytically its threshold voltage and, using equation (9.14), the density of electrons at equilibrium and with $v_G = +0.2$ V, which can be compared with 3.0×10^{15} m^{-2} from the numerical simulation. What fraction of the electrons from the donors goes to the 2DEG, and what density of surface states is needed to keep the Fermi level pinned? What would happen if the doping were in error by $\pm 20\%$? (This is only too common!)

9.3   How thin a doped layer would be needed in these layers for an enhancement-mode device with $v_T = +0.2$ V?

9.4   Estimate the threshold voltage of the device in Figure 9.4 with the DX centres either occupied or empty. Can one use a 'capacitor' estimate such as equation (9.14)?

How well does equation (9.16) predict the density of electrons in Figure 9.4(a), where the DX centres are occupied? If the DX centres were removed, what would be the maximum density of electrons in the 2DEG before parallel conduction appeared? Is this consistent with Figure 9.4(d)?

9.5   Over what range of density is only one subband in a 2DEG occupied, according to the triangular-well model? Is this model appropriate?

9.6   Section 9.1 concentrated on the type of heterostructure most commonly used with a 2DEG, but others have been proposed for particular applications. Three, the inverted MODFET, SISFET, and inverted SISFET, are shown in Figure 9.13. Note that the last two are largely undoped and that the inverted SISFET combines this with a clear surface for patterned gates. Sketch the band diagram through these structures, both at equilibrium and with a bias on the gate. Why is there a thin layer of n^+-GaAs at the surface in both SISFETs (which could be included in the i-MODFET too)?

9.7   Estimate the effect of a doped substrate with $N_A = 10^{21}$ m^{-3} on the threshold voltage of the layers in Figure 9.1. Remember that the bands in the spacer are no longer flat, so there is still a potential well at threshold.

9.8   Attempt a better treatment of the finite discontinuity ΔE_c at $z = 0$ than assuming it to be infinitely high, using a triangular well for $z > 0$. A simple model is to assume that $V(z) = \Delta E_c$ for $z < 0$, a constant rather than the linear fall that prevails inside the spacer. Then the wave function is proportional to $\text{Ai}[(eFz - \varepsilon_1)/\varepsilon_0]$ for $z > 0$ and to $\exp(\kappa z)$ for $z < 0$. The value and derivative must be matched at $z = 0$ to obtain the allowed states. Unfortunately κ is a function of the eigenenergy ε_1, but it should be adequate to estimate κ using the value of ε_1 in the infinitely deep triangular

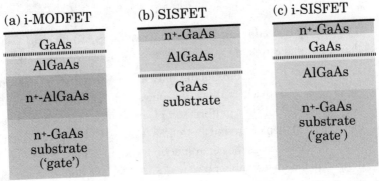

FIGURE 9.13. Layers used in (a) inverted MODFET, (b) semiconductor–insulator-semiconductor FET (SISFET), and (c) inverted SISFET. The thick dashed line marks the 2DEG.

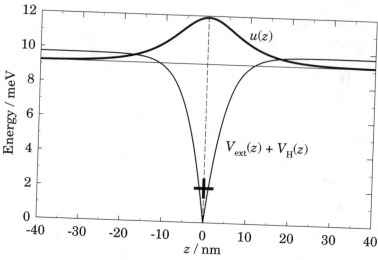

FIGURE 9.14. A δ-doped layer with $3 \times 10^{15}\,\mathrm{m}^{-2}$ donors surrounded by an equal density of electrons. The wave function and energy are variational estimates using $u(z) \propto \mathrm{sech}(bz)$.

well and to treat it as a constant, or even simply to put $\hbar^2\kappa^2/2m = \Delta E_{\mathrm{c}}$. How sensitive is ε_1 to this correction?

9.9 Extend the Fang–Howard variational calculation to include the potential of a p-type depletion layer below the 2DEG (Figure 9.6). Use the linear approximation $V_{\mathrm{dep}}(z) = eF_{\mathrm{dep}}z$ in the 2DEG, which plays the role of the external potential $V_{\mathrm{ext}}(z)$. Find the variational parameter b by minimizing $E_{\mathrm{T}} = \langle \hat{T} \rangle + \langle V_{\mathrm{ext}} \rangle + \frac{1}{2}\langle V_{\mathrm{H}} \rangle$; the energy level is then given by $\varepsilon_1 = \langle \hat{T} \rangle + \langle V_{\mathrm{ext}} \rangle + \langle V_{\mathrm{H}} \rangle$. Show that

$$b = \left[\frac{12me^2(N_{\mathrm{dep}} + \frac{11}{32}n_{\mathrm{2D}})}{\hbar^2\epsilon_0\epsilon_{\mathrm{b}}} \right]^{1/3}, \qquad (\text{E9.1})$$

$$\varepsilon_1 = \left(\frac{243}{16} \right)^{1/3} \left[\frac{\hbar^2}{2m}\left(\frac{e^2}{\epsilon_0\epsilon_{\mathrm{b}}} \right)^2 \right]^{1/3} \frac{N_{\mathrm{dep}} + \frac{55}{96}n_{\mathrm{2D}}}{(N_{\mathrm{dep}} + \frac{11}{32}n_{\mathrm{2D}})^{1/3}}. \qquad (\text{E9.2})$$

This widely quoted result reduces to equation (9.37) for $N_{\mathrm{dep}} = 0$ and to the variational estimate for a triangular well (7.76) for $n_{\mathrm{2D}} = 0$.

9.10 Although we have seen δ-doping only as an alternative to slab doping to supply electrons to a heterojunction, one can instead put a δ-layer of donors $N_{\mathrm{D}}^{(2\mathrm{D})}$ in the middle of a uniform sample of undoped semiconductor and the electrons will then lie around the donors (Figure 9.14). This can be treated in much the same way as the heterojunction, assuming that all the electrons are trapped in the lowest subband so $n_{\mathrm{2D}} = N_{\mathrm{D}}^{(2\mathrm{D})}$. The donors give a potential energy $V_{\mathrm{ext}}(z) = eF|z|$ with $F = eN_{\mathrm{D}}^{(2\mathrm{D})}/2\epsilon_0\epsilon_{\mathrm{b}}$, and the electrons generate a Hartree potential $V_{\mathrm{H}}(z)$ that cancels the electric field of the donors at large z. Calculate the energy level using a variational approximation. An obvious choice for the wave function is a Gaussian function, but this leads to error functions in the Hartree potential and $u(z) \propto \operatorname{sech}(bz)$ is simpler.

9.11 A contrasting situation is provided by electrons trapped in a GaAs quantum well. Suppose that the well is 10 nm wide with $3 \times 10^{15} \, \text{m}^{-2}$ electrons, which might arise from remote doping (undoped well) or from a uniform density of donors in the well. Use the wave function of an infinitely deep well to calculate the electrostatic potential from the electrons and show that its effect on the confining potential is small for both types of doping. Continue by finding the change in energy of the bound state with perturbation theory.

9.12 Derive the form factor $G(q)$ in equation (9.67).

9.13 The example of a 2DEG quoted in Section 9.5.2 was δ-doped with $n_{\text{imp}}^{(2D)} = 10^{16} \, \text{m}^{-2}$ at a distance $|d| = 30 \, \text{nm}$. Suppose that the same number of donors were spread along z over a slab of thickness 20 nm with the same average position. What effect would this have on the mobility?

9.14 Estimate the mobility at low temperature, due to remote ionized impurities, of the layers shown in Figure 9.11 and compare the result with experiment. What density of donors in the nearby δ-doped sheet do you expect to be ionized?

9.15 Calculate the form factor and its effect on impurity scattering within the second subband, and between the first and second subbands, for the measurements in Figure 9.12. You will need a wave function for the second subband. Show that $u_2(z) = (\tfrac{1}{6}b^3)^{1/2}z(3 - bz)\exp(-\tfrac{1}{2}bz)$ is correctly normalized and orthogonal to the Fang–Howard $u_1(z)$ and is therefore a reasonable approximation. Use these results to estimate the difference in mobility between the first and second subbands in Figure 9.12.

9.16 Calculate the form factor for the lowest state in an infinitely deep square well of width a. How do the features compare qualitatively with those of the Fang–Howard wave function?

9.17 Calculate an upper bound on the density of background ionized impurities for the best material in Figure 9.11.

9.18 Repeat the calculations of Section 9.6.3 for polar coupling to LO phonons, building on the results in Section 8.4.2. For simplicity set the form factor to unity and assume that the 2DEG is non-degenerate. Show that the rate of absorption of LO phonons by a low-energy electron, scattered to an energy only just above $\hbar\omega_{\text{LO}}$, is

$$\frac{1}{\tau_{\text{LO}}} = \frac{k_{\text{LO}}}{8\hbar} \frac{e^2}{\epsilon_0} \left(\frac{1}{\epsilon(\infty)} - \frac{1}{\epsilon(0)} \right) N_{\text{LO}}, \tag{E9.3}$$

where k_{LO} is the wave number of an electron with the energy of the optic phonon $\hbar^2 k_{\text{LO}}^2/2m = \hbar\omega_{\text{LO}}$, and N_{LO} is the thermal number of phonons in each LO mode. The coefficient of N_{LO} is 1.2×10^{13} s^{-1}, showing the strength of this scattering mechanism. How well does the mobility agree with Figure 9.11(a) and with the result for three dimensions? [Adapted from P. J. Price, *Annals of Physics (New York)* **133** (1981): 217–39.]

9.19   How well can the mobility as a function of temperature of the most recent 2DEG in Figure 9.11(a) be explained? Use Mattheisen's rule to combine the rates for ionized-impurity scattering (which you will have to estimate from the data), acoustic phonons, and longitudinal optic phonons (from the previous exercise). (A rise in temperature has several effects that we have not considered, including a broadening of the active range in energy of electrons and changes in the screening.)

10

OPTICAL PROPERTIES OF QUANTUM WELLS

The optical properties of low-dimensional systems are put to wide practical use, the semiconductor laser being an obvious example. In this chapter we shall expand the general results derived in Chapter 8 and apply them to low-dimensional structures.

First, the general theory needs to be developed further. A surprising result is that the real and imaginary parts of the complex dielectric function or conductivity are not independent functions, but can be derived from one another. This relies on the principle of causality, that a response should follow its stimulus, embodied in the Kramers–Kronig relations. Other important results follow, such as the f-sum rule that controls the total absorption of a material integrated over all frequencies.

Although transport properties often rely on one kind of carrier alone, this is not true of optical processes. We must therefore treat the valence band of semiconductors in detail, a task that we have long postponed. We shall do this with the celebrated Kane model, an extension of the $\mathbf{k} \cdot \mathbf{p}$ theory developed earlier. We also need to consider the full wave function within the effective-mass approximation. Usually we neglect the Bloch part and study only the slowly varying envelope, but both must be included in the matrix elements and each makes a contribution to the selection rules. Quite different results emerge for transitions between bands and those within the same band.

Optical absorption creates an electron and hole, and the interaction between these oppositely charged particles cannot be neglected. Their mutual attraction leads to a bound state very similar to a rescaled hydrogen atom, called an exciton. These excitons modify the strength of optical absorption strongly near the band gap, and they become of increasing importance as the number of dimensions is reduced.

10.1 General Theory

In Section 8.5.2 we derived a formula (8.76) for the real part of the optical conductivity, which is equivalent to the imaginary part of the dielectric function,

$$\sigma_1(\omega) = \frac{\pi e^2}{m_0^2 \omega} \frac{2}{\Omega} \sum_{i,j} |\langle j|\mathbf{e} \cdot \hat{\mathbf{p}}|i\rangle|^2 [f(E_i) - f(E_j)]\delta(E_j - E_i - \hbar\omega). \quad (10.1)$$

We now need to complete this with the imaginary part. Although it can be calculated directly, we shall use the Kramers–Kronig relations as these lead to several other useful results. Here we shall concentrate on the physics; more details of the mathematics are given in Appendix 6.

10.1.1 KRAMERS–KRONIG RELATIONS

The complex conductivity is defined by $\tilde{J}(\omega) = \tilde{\sigma}(\omega)\tilde{E}(\omega)$, where $\tilde{J}(\omega)$ is a Fourier component of the total current in the fourth Maxwell equation. The Kramers–Kronig relations apply to any linear-response function such as this (similar relations exist for nonlinear response). A Fourier transform turns the definition into one relating the functions in time, using the convolution theorem,

$$ J(t) = \int_{-\infty}^{\infty} \sigma(t')E(t-t')\,dt'. \tag{10.2} $$

The response function in time $\sigma(t)$ is often called the *impulse response function* because if we apply a δ-function 'force' $E(t) = E_0\,\delta(t-t_0)$ the response is

$$ J(t) = \int_0^{\infty} \sigma(t')E_0\,\delta(t-t'-t_0)\,dt = E_0\,\sigma(t-t_0). \tag{10.3} $$

Quantities such as the dielectric function also depend on the wave number, and the Kramers–Kronig equations apply to each value of the wave number separately.

The Kramers–Kronig relations arise from the principle of *causality*. This asserts that a response follows its cause in time, or that the response of a system at some time depends only on the driving force in the past. Thus $J(t)$ depends on $E(t')$ only for times $t' < t$, which in turn means that only $t' > 0$ contributes to the integral in equation (10.2). This integral is over all times, so we need $\sigma(t) = 0$ for $t < 0$ to ensure that only positive times contribute. Complex analysis, given in Appendix 6, then shows that the real and imaginary parts of $\tilde{\sigma}(\omega)$ can be written in terms of one another:

$$ \sigma_1(\omega) = \frac{1}{\pi}\mathcal{P}\int_{-\infty}^{\infty}\frac{\sigma_2(\omega')}{\omega'-\omega}\,d\omega', \tag{10.4} $$

$$ \sigma_2(\omega) = -\frac{1}{\pi}\mathcal{P}\int_{-\infty}^{\infty}\frac{\sigma_1(\omega')}{\omega'-\omega}\,d\omega'. \tag{10.5} $$

These are the *Kramers–Kronig relations*. The '$\mathcal{P}$' in these equations denotes the *Cauchy principal part* of the integral (Appendix 6), which prevents divergence when $\omega' = \omega$.

The conductivity as a function of time (equation 10.2) relates the current and electric field, both of which are physical quantities, so it must be mathematically real rather than complex. Its Fourier transform accordingly has the symmetry $\tilde{\sigma}(-\omega) = \tilde{\sigma}^*(\omega)$, so the real and imaginary parts obey $\sigma_1(-\omega) = \sigma_1(\omega)$ and $\sigma_2(-\omega) = -\sigma_2(\omega)$. This symmetry can be used to write the integrals in the Kramers–Kronig relations over positive frequencies only, giving

$$\sigma_1(\omega) = \frac{2}{\pi} P \int_0^\infty \frac{\omega' \, \sigma_2(\omega')}{\omega'^2 - \omega^2} \, d\omega', \tag{10.6}$$

$$\sigma_2(\omega) = -\frac{2}{\pi} P \int_0^\infty \frac{\omega \, \sigma_1(\omega')}{\omega'^2 - \omega^2} \, d\omega'. \tag{10.7}$$

The Kramers–Kronig relations show that the real and imaginary parts of any response function determine each other and are not independent quantities. Even though the physical significance of the two parts appears quite different, being dissipation for σ_1 and polarization for σ_2, they are inextricably related. The Kramers–Kronig relations are also called *dispersion relations* and are widely used in other branches of physics and engineering such as control theory.

A practical difficulty is that the integrals converge only slowly, which is a nuisance because it means that one component of $\tilde{\sigma}$ must be measured over a wide range of frequencies if a Kramers–Kronig relation is used to extract the other component. It is much easier to use the Kramers–Kronig relations for *differences*, which tend to be localized in frequency. For example, we saw in Section 7.2.2 that an electric field shifts the energies in a quantum well (quantum-confined Stark effect) and therefore changes σ_1 over a small range of energy. The Kramers–Kronig equations can be used to find the corresponding change in σ_2.

10.1.2 OPTICAL-RESPONSE FUNCTIONS

Now we can return to equation (10.1) for σ_1. It is convenient to move the ω from the denominator of the prefactor inside the summation. We can then use the δ-function to replace it with $(E_j - E_i)/\hbar$ and get

$$\sigma_1(\omega) = \frac{\pi e^2 \hbar}{m_0^2} \frac{2}{\Omega} \sum_{i,j} |\langle j | \mathbf{e} \cdot \hat{\mathbf{p}} | i \rangle|^2 \frac{f(E_i) - f(E_j)}{E_j - E_i} \delta(E_j - E_i - \hbar\omega). \tag{10.8}$$

Now the frequency enters only as $\delta(E_{ji} - \hbar\omega)$, where $E_{ji} = E_j - E_i$. Substituting this part alone into the Kramers–Kronig relation (10.5) gives $-(1/\pi)[1/(E_{ji} - \hbar\omega)]$, whence

$$\sigma_2(\omega) = -\frac{e^2 \hbar}{m_0^2} \frac{2}{\Omega} \sum_{i,j} |\langle j | \mathbf{e} \cdot \hat{\mathbf{p}} | i \rangle|^2 \frac{f(E_i) - f(E_j)}{E_j - E_i} P \frac{1}{E_j - E_i - \hbar\omega}. \tag{10.9}$$

The two parts of $\tilde{\sigma}$ can now be combined. For a change, the result for the complex dielectric function is

$$\tilde{\epsilon}_r(\omega) = 1 + \frac{e^2\hbar}{\epsilon_0 m_0^2\omega} \frac{2}{\Omega} \sum_{i,j} |\langle j|\mathbf{e}\cdot\hat{\mathbf{p}}|i\rangle|^2 \frac{f(E_i) - f(E_j)}{E_j - E_i} \tag{10.10}$$

$$\times \left[\mathcal{P}\frac{1}{E_j - E_i - \hbar\omega} + i\pi\delta(E_j - E_i - \hbar\omega)\right].$$

The two terms are often combined using a simple trick. This follows from taking the real and imaginary parts of $1/(x - i\varepsilon)$, where ε is small but nonzero,

$$\frac{1}{x - i\varepsilon} = \frac{x}{x^2 + \varepsilon^2} + i\pi\left(\frac{\varepsilon}{\pi}\frac{1}{x^2 + \varepsilon^2}\right). \tag{10.11}$$

The quantity in parentheses is a Lorentzian function normalized to unit area. It therefore approaches a δ-function if $\varepsilon \to 0_+$, where 0_+ is an infinitesimal positive quantity. The first term looks like $1/x$ for $|x| \gg \varepsilon$ but goes to zero rather than infinity for small x. This is like the prescription for removing the divergence of an integral by taking its principal part. Thus we can write

$$\frac{1}{x - i0_+} = \mathcal{P}\frac{1}{x} + i\pi\delta(x), \tag{10.12}$$

which can also be justified from contour integration. Then

$$\tilde{\epsilon}_r(\omega) = 1 + \frac{e^2\hbar}{\epsilon_0 m_0^2\omega} \frac{2}{\Omega} \sum_{i,j} |\langle j|\mathbf{e}\cdot\hat{\mathbf{p}}|i\rangle|^2 \frac{f(E_i) - f(E_j)}{E_j - E_i} \frac{1}{E_j - E_i - \hbar\omega - i0_+},$$

$$\tag{10.13}$$

which is the form in which the complex dielectric function is usually quoted.

Optical response is often expressed in terms of *oscillator strengths*, particularly for a discrete set of transitions, as we saw in Section 8.7. Our earlier result, equation (8.97), contained a sum over initial and final states. It can instead be written as a sum over *transitions k*, restricted to those between filled and empty states. Including the factor of 2 for spin within the summation, the result is

$$\sigma_1(\omega) = \frac{\pi e^2}{2m_0\Omega} \sum_k f_k[\delta(\omega - \omega_k) + \delta(\omega + \omega_k)]. \tag{10.14}$$

There are two terms because each transition enters equation (8.97) for both permutations of i and j. This also shows explicitly that σ_1 is an even function of ω. It can instead be written for ϵ_2 as

$$\epsilon_2(\omega) = \frac{\pi e^2}{2\epsilon_0 m_0\omega\Omega} \sum_k f_k[\delta(\omega - \omega_k) + \delta(\omega + \omega_k)]. \tag{10.15}$$

The Kramers–Kronig relation (10.4) gives the corresponding real part,

$$\epsilon_1(\omega) = 1 - \frac{e^2}{2\epsilon_0 m_0 \omega \Omega} \sum_k f_k \left[\frac{1}{\omega - \omega_k} + \frac{1}{\omega + \omega_k} \right]. \tag{10.16}$$

The structure in $\epsilon_2(\omega)$ consists of sharp δ-functions in this picture. In practice the lines are broadened by the lifetime of the states involved and a Lorentzian profile is often a good approximation. If the full width at half-maximum is γ_k, assumed to be much smaller than ω_k, the δ-functions are replaced by the substitution

$$\delta(\omega - \omega_k) \rightarrow \frac{(\frac{1}{2}\gamma_k)}{\pi} \frac{1}{(\omega - \omega_k)^2 + (\frac{1}{2}\gamma_k)^2}. \tag{10.17}$$

The complex dielectric function can then be written in the compact form

$$\tilde{\epsilon}_r(\omega) = 1 - \frac{e^2}{\epsilon_0 m_0 \Omega} \sum_k \frac{f_k}{(\omega^2 - \omega_k^2) + i\gamma_k\omega}. \tag{10.18}$$

The optical-response functions are further constrained by restrictions called *sum rules*, and we shall now derive some simple examples.

10.1.3 SUM RULES

We can now prove results such as the f-sum rule (equation 8.96). Consider expression (10.18) for $\tilde{\epsilon}_r$ in terms of oscillator strengths. At high frequencies, well above all the possible transitions ω_k, the electrons oscillate only through a tiny distance in response to an applied electric field and barely feel the forces that constrain them. Under these conditions they respond like free electrons in a plasma, whose dielectric function we saw earlier (equation 9.49). Another argument for this behaviour is given in Appendix 6. Thus

$$\tilde{\epsilon}_r(\omega) \sim 1 - \frac{\omega_p^2}{\omega^2} \quad \text{as } \omega \to \infty, \quad \text{with } \omega_p^2 = \frac{e^2 n}{\epsilon_0 m_0}. \tag{10.19}$$

Here n is the number density of electrons in the system. Equation (10.18) must obey this limit, which shows that

$$\sum_k f_k = n\Omega = N, \tag{10.20}$$

where N is the total number of electrons in the sample. This is the *f-sum rule*. It does not depend on details such as the energy levels and is therefore a powerful constraint on both experimental and theoretical results.

Other sum rules on $\tilde{\epsilon}_r$ follow from the Kramers–Kronig relations. We have to be a little careful, however, because the response of a medium to an applied electric field $\mathbf{E}$ is the polarization $\mathbf{P}$, not the electric displacement $\mathbf{D}$. Thus the susceptibility $\tilde{\chi}(\omega) = \tilde{\epsilon}_r(\omega) - 1$ is the true response function and should be used in the Kramers–Kronig relations rather than $\tilde{\epsilon}_r(\omega)$ itself.

Suppose that we evaluate the first Kramers–Kronig relation (10.6) at a very high frequency ω, much higher than any frequency for which ϵ_2 makes a significant contribution. Then

$$\epsilon_1(\omega \to \infty) - 1 \sim -\frac{2}{\pi \omega^2} \int_0^\infty \omega' \, \epsilon_2(\omega') \, d\omega'. \tag{10.21}$$

Again we use the argument that the electrons behave as a plasma at very high frequencies. Comparison with equation (10.19) shows that

$$\int_0^\infty \omega \, \epsilon_2(\omega) \, d\omega = \frac{\pi \omega_p^2}{2} = \frac{\pi n e^2}{2 \epsilon_0 m_0}. \tag{10.22}$$

This is the f-sum rule for $\epsilon_2(\omega)$.

Another application of the Kramers–Kronig relations is to find the real part at zero frequency from the imaginary part. Setting $\omega = 0$ in equation (10.6) gives

$$\epsilon_1(0) - 1 = \frac{2}{\pi} \int_0^\infty \frac{\epsilon_2(\omega)}{\omega} \, d\omega. \tag{10.23}$$

In semiconductors $\epsilon_2(\omega) = 0$ for $\hbar\omega < E_g$ and the main contributions to this integral often come from just above the band gap. This tells us that semiconductors with smaller band gaps tend to have larger static dielectric constants. A more quantitative account is given in Section A6.2.

Although we normally think of $\mathbf{D}$ or $\mathbf{P}$ as a response to $\mathbf{E}$, it is also possible to apply a $\mathbf{D}$ field. This is done by injecting free charge ρ_f, which couples through the first Maxwell equation, $\operatorname{div} \mathbf{D} = \rho_f$. A practical example is energy-loss spectroscopy in an electron microscope where one measures the energy lost by electrons as they traverse a thin specimen. It may also be important inside a semiconductor device; hot electrons crossing the base of a bipolar transistor behave in a very similar way. The response to $\mathbf{D}$ is an electric field $\mathbf{E} = \mathbf{D}/\epsilon_0 \tilde{\epsilon}_r$ in the presence of a dielectric medium and $\mathbf{E} = \mathbf{D}/\epsilon_0$ otherwise, so the response function is $[1/\tilde{\epsilon}_r(\omega)] - 1$. This must obey the Kramers–Kronig relations in the same way as $\tilde{\epsilon}_r - 1$. Just as $\operatorname{Im} \tilde{\epsilon}_r$ gives the energy lost by a wave propagating through the material, the energy lost by an injected particle is proportional to $-\operatorname{Im}(1/\tilde{\epsilon}_r)$, which is therefore called the *energy-loss function*. It obeys a sum rule, which again follows because the electrons behave as though they are free at very high frequencies, given by

$$\int_0^\infty \omega \left[-\operatorname{Im} \frac{1}{\tilde{\epsilon}_r(\omega)} \right] d\omega = \frac{\pi \omega_p^2}{2} = \frac{\pi n e^2}{2 \epsilon_0 m_0}. \tag{10.24}$$

The imaginary parts of both $\tilde{\epsilon}_r$ and $-1/\tilde{\epsilon}_r$ integrate to the same value. The energy-loss function is dominated by regions where $\tilde{\epsilon}_r$ is small, and we saw in Section 9.4.1

that it vanishes at the plasma frequency ω_p in three dimensions. Plasmons are there-fore important mechanisms by which electrons lose energy and dominate the energy-loss function.

The importance of the sum rules lies in their generality. They apply to any system, however complicated, whether it is in thermal equilibrium or not. Thus one cannot make or modify a material or device to have arbitrary strength of optical absorption: an increase at one frequency must be compensated by a decrease elsewhere. An example is the change in absorption at the edge of a band produced by an electric field, the Franz–Keldysh effect (Section 6.2.1). The strength of interband absorption mirrors the density of states. An electric field spreads out the edge of the band (Figure 6.2), which introduces absorption at frequencies below the band edge, but it is compensated by a reduction in absorption just above. It is often said that 'the absorption edge is shifted by the Franz–Keldysh effect' but this statement, taken literally, would violate the f-sum rule.

Lasers too must obey the sum rules. The population inversion in a laser gener-ates $\epsilon_2(\omega) < 0$ over some range of frequencies, representing emission rather than absorption of energy; this must be balanced by additional absorption elsewhere in the spectrum to preserve the f-sum rule.

Sum rules are also useful constraints on both experiment and theory. Suppose, for example, that $\epsilon_2(\omega)$ has been measured over 'all' frequencies. It could then be substituted into equation (10.22) to check that the integral reached the value on the right-hand side. If it fell significantly short, a band of absorption must have been missed.

10.2 Valence-Band Structure: The Kane Model

Most optical processes in semiconductors take place between the conduction and valence bands, although there is increasing interest in intraband processes for in-frared applications. We have used a simple parabolic approximation for the valence band, with light and heavy masses, but this is often inadequate to describe optical phenomena. Unfortunately the superior models are much more involved, and are further complicated by the need to treat spin–orbit coupling. The Kane model is a standard approximation that is used at various levels of sophistication. We shall start with a spinless approximation, which is rarely used in practice but makes the physics clear, before embarking on the full model with spin–orbit coupling. In the next section we shall look briefly at valence bands in a quantum well, which are even worse!

A simple chemical picture of the top of the valence band was given in Sec-tion 2.6.3, based on p orbitals. It predicted a band of light holes and a doubly degenerate band of heavy holes, which was a reasonable description of the band over the whole Brillouin zone. Unfortunately, Figure 2.18 showed that this description fails near the top of the band, which is the most important region. Also, we cheated a little. Armed with the tight-binding model (Section 7.7), we can see that the 'heavy' bands should bend upwards rather than downwards because of the sign of the tunnelling matrix element. The Kane model provides a more accurate description.

10.2.1   THE KANE MODEL WITHOUT SPIN

In Section 7.3 we derived the $\mathbf{k} \cdot \mathbf{p}$ description of band structure. The idea was to rewrite the Schrödinger equation for the periodic part $u_{n\mathbf{K}}(\mathbf{R})$ of the Bloch functions alone, factoring out the plane wave. The resulting Hamiltonian (in equation 7.42) was

$$\hat{H}_{\mathbf{k}\cdot\mathbf{p}}(\mathbf{K}) = \left[\frac{\hat{\mathbf{p}}^2}{2m_0} + V_{\mathrm{per}}(\mathbf{R}) \right] + \left[\frac{\hbar}{m_0}\mathbf{K} \cdot \hat{\mathbf{p}} + \frac{\hbar^2 K^2}{2m_0} \right]. \qquad (10.25)$$

Assuming that we knew the Bloch functions at the point of interest (usually Γ), we treated the terms in $\mathbf{K}$ as a perturbation and found $\varepsilon_n(\mathbf{K})$ to order K^2. This is satisfactory only for small values of K and certainly fails when the change in $\varepsilon_n(\mathbf{K})$ becomes similar to the separation between the bands at Γ; it is also inapplicable to the degenerate valence band.

A better approach, in the spirit of Section 7.6, is to solve the Schrödinger equation exactly within a restricted basis set. This is the foundation of the Kane model. The obvious basis set is that of all eigenfunctions at $\mathbf{K} = \mathbf{0}$, as used in the perturbation theory. These provide a complete set, so no approximation is made if all

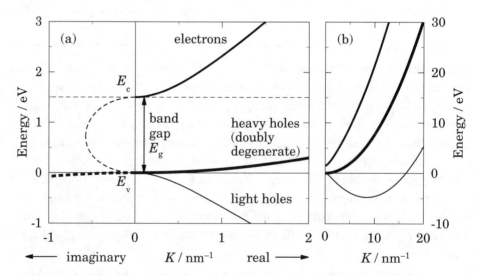

FIGURE 10.1. Energy of the conduction and valence bands to the Kane model without spin, for $E_g = 1.5\,\mathrm{eV}$ and $E_P = 22\,\mathrm{eV}$. Both real and imaginary K are shown in (a), with a wider range in (b).

the eigenfunctions are retained, as any $u_{n\mathbf{K}}(\mathbf{R})$ can be expanded in the complete set $\{u_{n0}(\mathbf{R})\}$. Only a subset of these is retained in approximate methods, perhaps 10 for numerical work but a much smaller number for analytical treatments. A minimal choice includes only those on either side of the band gap. Let $|S\rangle$ denote the wave function of the single s-like conduction band at Γ, and use $|X\rangle$, $|Y\rangle$, and $|Z\rangle$ for the three valence-band states with the symmetry of p_x, p_y, and p_z orbitals.

We must now find the matrix elements of $\hat{H}_{\mathbf{k}\cdot\mathbf{p}}(\mathbf{K})$ between these states. Most of the terms in the Hamiltonian (10.25) are diagonal; the only off-diagonal terms come from $\mathbf{K}\cdot\hat{\mathbf{p}}$ and these are also simple. For example, $\mathbf{K}\cdot\langle S|\hat{\mathbf{p}}|X\rangle = k_x\langle S|\hat{p}_x|X\rangle$; the other two components of the momentum (gradient) operator give vanishing matrix elements by symmetry. The three surviving matrix elements $\langle S|\hat{p}_x|X\rangle$, $\langle S|\hat{p}_y|Y\rangle$, and $\langle S|\hat{p}_z|Z\rangle$ are identical and denoted $(im_0/\hbar)P$ as in Section 7.3. Thus the Hamiltonian matrix, showing the basis states, becomes

$$
\begin{array}{cccc}
& |S\rangle & |X\rangle & |Y\rangle & |Z\rangle \\
\langle S| & \begin{pmatrix} E_c + \varepsilon_0(k) & iPk_x & iPk_y & iPk_z \\ \langle X| & -iPk_x & E_v + \varepsilon_0(K) & 0 & 0 \\ \langle Y| & -iPk_y & 0 & E_v + \varepsilon_0(K) & 0 \\ \langle Z| & -iPk_z & 0 & 0 & E_v + \varepsilon_0(K) \end{pmatrix}
\end{array}, \quad (10.26)
$$

where $\varepsilon_0(K) = \hbar^2 K^2/2m_0$, the kinetic energy of free electrons.

The energies are given by the usual secular equation $\det|E\mathbf{I} - \mathbf{H}| = 0$. Here the quartic equation is simple to solve and the bands have spherical symmetry:

$$
\varepsilon_e(K) = \tfrac{1}{2}(E_c + E_v) + \varepsilon_0(K) + \sqrt{\tfrac{1}{4}E_g^2 + P^2K^2},
$$

$$
\varepsilon_{lh}(K) = \tfrac{1}{2}(E_c + E_v) + \varepsilon_0(K) - \sqrt{\tfrac{1}{4}E_g^2 + P^2K^2},
$$

$$
\varepsilon_{hh}(K) = E_v + \varepsilon_0(K) \quad \text{(twice)}. \quad (10.27)
$$

These branches give electrons, light holes, and a doubly degenerate branch of heavy holes, respectively. They are plotted in Figure 10.1 for $E_g = 1.5$ eV and $E_P = 2m_0P^2/\hbar^2 = 22$ eV, rough values for GaAs. The dispersion near Γ, shown in Figure 10.1(a), is exactly as predicted by the qualitative argument. There is a single branch of light holes, one of even lighter electrons, and a double branch of heavy holes whose dispersion is unchanged from that of free electrons and therefore curves in the wrong direction. The model can also be solved for imaginary K as shown on the left of the plot, which is useful for calculating tunnelling through barriers.

Expansion near Γ gives effective masses for the electrons and light holes,

$$\frac{1}{m_{\mathrm{e}}} = \frac{E_P}{E_{\mathrm{g}}} + 1, \qquad \frac{1}{m_{\mathrm{lh}}} = \frac{E_P}{E_{\mathrm{g}}} - 1. \tag{10.28}$$

Usually $E_P/E_{\mathrm{g}} \gg 1$, so this term dominates and the effective masses are much smaller than unity. The values for the example are $m_{\mathrm{e}} = 0.064$ and $m_{\mathrm{lh}} = 0.073$, a remarkably good prediction for GaAs from so simple a theory. They can be combined to give a reduced mass of $1/m_{\mathrm{e}} + 1/m_{\mathrm{lh}} = 2E_P/E_{\mathrm{g}}$. This mass enters the optical joint density of states in interband absorption (Section 8.6), for which P is also the matrix element. Thus $\mathbf{k} \cdot \mathbf{p}$ theory is intimately related to optical phenomena.

The eigenvectors are also simple to find and depend on the direction of $\mathbf{K}$. If we choose it along z, the states $|X\rangle$ and $|Y\rangle$ decouple from the others to provide the heavy holes; $|S\rangle$ and $|Z\rangle$ are mixed by $\hat{p}_z$ as k_z increases from zero to give the repulsion between the bands, which causes the light masses. An amusing feature is that the dispersion relation (10.27) for these branches closely resembles that for particles and antiparticles in special relativity, $E^2 = p^2 c^2 + m_0^2 c^4$, if the energy is measured from the middle of the band gap and $\varepsilon_0(K)$ is ignored.

Although these results were expected, a strange problem appears at large K (Figure 10.1(b)). Equation (10.27) is eventually dominated by $\varepsilon_0(K)$ and all bands turn upwards. Although this happens at unphysically large values of K, which are outside the Brillouin zone, it may lead to troublesome 'ghost' solutions. The reason is that the Kane model preserves the rotational symmetry of the crystal, through the symmetry of the wave functions that enter the matrix elements, but the translational symmetry and therefore the periodic structure of $\varepsilon(\mathbf{K})$ are lost. Apart from this problem, we would expect the model to work well until the energy approaches bands that we have neglected.

The model is clearly oversimplified in its present form and two extensions should be made. The first is to include the influence of further bands using perturbation theory. This is particularly important for the heavy holes, whose curvature is reversed. Many extra terms appear in the Hamiltonian:

$$\begin{pmatrix} A'K^2 & Bk_yk_z & Bk_xk_z & Bk_xk_y \\ Bk_yk_z & L'k_x^2 + M(k_y^2 + k_z^2) & N'k_xk_y & N'k_xk_z \\ Bk_xk_z & N'k_xk_y & L'k_y^2 + M(k_x^2 + k_z^2) & N'k_yk_z \\ Bk_xk_y & N'k_xk_z & N'k_yk_z & L'k_z^2 + M(k_x^2 + k_y^2) \end{pmatrix}. \tag{10.29}$$

This follows the notation of Kane (1982). Although it is more complicated, all parameters are known for the best-studied semiconductors. We shall now turn to the second extension, spin–orbit coupling.

10.2.2 SPIN–ORBIT COUPLING

The second extension to the basic Kane model introduces *spin–orbit coupling*, mentioned in Section 2.6.3. We have generally ignored the spin of the electron except for the doubling of states that results. The exception was in a magnetic field (Section 6.4.3), where the magnetic moment associated with the spin can align to be either parallel or antiparallel to the field, giving different energies. Spin–orbit coupling can be viewed crudely as a similar interaction but with a magnetic field produced by the electron itself, although a rigorous treatment must be relativistic. Suppose that an electron has angular momentum $\mathbf{l}$, which means that the electron is orbiting about the axis defined by this vector. The electron is of course charged so this orbital motion generates a circulating current that in turn produces a magnetic field. As in the case of the external magnetic field, the energy of the electron depends on the orientation of its spin $\mathbf{s}$ with respect to this field, giving a coupling between the spin and orbital motion. The usual expression for the energy of a magnetic dipole shows that this coupling is proportional to $\mathbf{l} \cdot \mathbf{s}$.

Angular momentum obeys a set of rules in quantum mechanics that can be described only very briefly here. First, one can specify both the magnitude of a particle's angular momentum and its component along one axis, which we shall take to be z. It is not possible to know all three components simultaneously because the corresponding operators $\hat{l}_x$, $\hat{l}_y$, and $\hat{l}_z$ do not commute. Both the magnitude and component are quantized in units of $\hbar$. We shall be concerned with spin, which has magnitude $s = \frac{1}{2}$ and component $s_z = \pm\frac{1}{2}$; s orbitals, which have $l = 0$; and p orbitals, which have magnitude $l = 1$ and component $m \equiv l_z = -1, 0, 1$.

These p states with particular m are not the same as the p orbitals that we have used previously, which were oriented along the Cartesian axes. The new states are labelled $|m\rangle$ and are related to the previous states by

$$|0\rangle = |Z\rangle, \qquad |\pm 1\rangle = \sqrt{\tfrac{1}{2}}\big(|X\rangle \pm i|Y\rangle\big). \qquad (10.30)$$

To each of these can be added a spin, so, for example, $|+1\uparrow\rangle$ is the state with $m = +1$ and $s_z = +\frac{1}{2}$. There are six p orbitals in both descriptions. Similarly the s orbital in the conduction band, with two directions of spin, is denoted $|S\uparrow\downarrow\rangle$.

We now have to find the total angular momentum $\mathbf{j} = \mathbf{l} + \mathbf{s}$ for the p orbitals according to the rules of angular momentum in quantum mechanics, remembering that $\mathbf{j}$ must itself be quantized. In the case of $l = 1$ and $s = \frac{1}{2}$ the rules state that j may take the value $\frac{3}{2}$ or $\frac{1}{2}$, with four values of $j_z = \pm\frac{3}{2}, \pm\frac{1}{2}$ in the first case and two of $j_z = \pm\frac{1}{2}$ in the second. These states are labelled by $|j, j_z\rangle$. There are again six of them, and the two sets of states must be related. The connection is

$$|\tfrac{3}{2}, +\tfrac{3}{2}\rangle = |+1\uparrow\rangle,$$

$$|\tfrac{3}{2}, +\tfrac{1}{2}\rangle = \sqrt{\tfrac{1}{3}}|+1\downarrow\rangle - \sqrt{\tfrac{2}{3}}|0\uparrow\rangle,$$

$$|\tfrac{3}{2}, -\tfrac{1}{2}\rangle = -\sqrt{\tfrac{1}{3}}|-1\uparrow\rangle - \sqrt{\tfrac{2}{3}}|0\downarrow\rangle,$$

$$|\tfrac{3}{2}, -\tfrac{3}{2}\rangle = |-1\downarrow\rangle,$$

$$|\tfrac{1}{2}, +\tfrac{1}{2}\rangle = \sqrt{\tfrac{2}{3}}|+1\downarrow\rangle + \sqrt{\tfrac{1}{3}}|0\uparrow\rangle,$$

$$|\tfrac{1}{2}, -\tfrac{1}{2}\rangle = -\sqrt{\tfrac{2}{3}}|-1\uparrow\rangle + \sqrt{\tfrac{1}{3}}|0\downarrow\rangle. \tag{10.31}$$

The z-component of angular momentum must be the same in both descriptions, $j_z = m + s_z$. This shows that $|\tfrac{3}{2}, +\tfrac{3}{2}\rangle$ can arise only from $|+1\uparrow\rangle$. In contrast, both $|+1\downarrow\rangle$ and $|0\uparrow\rangle$ can contribute to $|\tfrac{3}{2}, +\tfrac{1}{2}\rangle$ since all have $j_z = +\tfrac{1}{2}$. It is too complicated to derive here the Clebsch–Gordon coefficients that give the amplitudes of the two states. Note that most of the $|j, j_z\rangle$ states do not have a definite spin.

Having found the states, we can set up the new Hamiltonian matrix. The eigenvalues of the original matrix (10.26) were functions only of $|\mathbf{K}|$, not the direction, and this continues to be true when spin–orbit coupling is included. The matrix elements are greatly simplified if we choose $\mathbf{K}$ along z, the same direction used to quantize the angular momentum, and the 8×8 matrix decouples into two 4×4 matrices. Most of the entries in the matrix follow from the composition (10.31) of the wave functions and the original matrix elements. The exceptions are the diagonal matrix elements in the valence band, which now include the spin–orbit coupling. This is proportional to $\mathbf{l} \cdot \mathbf{s}$, and expanding $\mathbf{j}^2 = (\mathbf{l} + \mathbf{s})^2$ shows that $\mathbf{l} \cdot \mathbf{s} = \tfrac{1}{2}(\mathbf{j}^2 - \mathbf{l}^2 - \mathbf{s}^2)$. The spin–orbit coupling is therefore different for $j = \tfrac{3}{2}$ and $j = \tfrac{1}{2}$. In the usual semiconductors the four states with $j = \tfrac{3}{2}$ provide the top of the valence band with energy E_v at Γ, while the spin–orbit coupling pulls the pair of $j = \tfrac{1}{2}$ states down to $E_\mathrm{v} - \Delta$. The magnitude of Δ goes roughly as Z^4, where Z is the atomic number; thus $\Delta = 0.044\,\mathrm{eV}$ in silicon and is often neglected, whereas $\Delta = 0.34\,\mathrm{eV}$ in GaAs, which must be retained. An extreme case is InSb with $\Delta = 0.98\,\mathrm{eV}$, much larger than the band gap $E_\mathrm{g} = 0.18\,\mathrm{eV}$ at room temperature. We finally get the matrix

$$
\begin{array}{c}
 \quad |S\uparrow\rangle \qquad\qquad |\tfrac{3}{2},\tfrac{3}{2}\rangle \qquad\qquad |\tfrac{3}{2},\tfrac{1}{2}\rangle \qquad\qquad |\tfrac{1}{2},\tfrac{1}{2}\rangle \\[4pt]
\begin{array}{c}
\langle S\uparrow| \\[6pt]
\langle \tfrac{3}{2},\tfrac{3}{2}| \\[6pt]
\langle \tfrac{3}{2},\tfrac{1}{2}| \\[6pt]
\langle \tfrac{1}{2},\tfrac{1}{2}|
\end{array}
\left(
\begin{array}{cccc}
E_\mathrm{c} + \varepsilon_0(K) & 0 & -i\sqrt{\tfrac{2}{3}}PK & i\sqrt{\tfrac{1}{3}}PK \\[6pt]
0 & E_\mathrm{v} + \varepsilon_0(K) & 0 & 0 \\[6pt]
i\sqrt{\tfrac{2}{3}}PK & 0 & E_\mathrm{v} + \varepsilon_0(K) & 0 \\[6pt]
-i\sqrt{\tfrac{1}{3}}PK & 0 & 0 & E_\mathrm{v} - \Delta + \varepsilon_0(K)
\end{array}
\right).
\end{array}
$$

$$\tag{10.32}$$

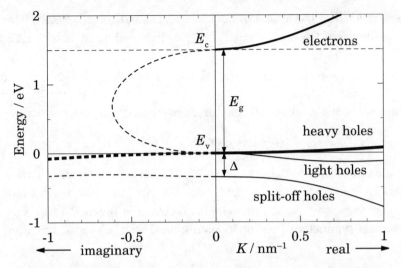

FIGURE 10.2. Band structure of the Kane model including spin–orbit coupling of $\Delta = 0.34\,\mathrm{eV}$, for real and imaginary K.

The $|\frac{3}{2}, \frac{3}{2}\rangle$ state has no coupling to the other states and therefore gives a heavy-hole band with $\varepsilon_{\mathrm{hh}}(K) = E_{\mathrm{v}} + \varepsilon_0(K)$ as in the simple model. The secular equation for the remaining three states is

$$(E' - E_{\mathrm{c}})(E' - E_{\mathrm{v}})(E' - E_{\mathrm{v}} - \Delta) - P^2 k^2 (E' - E_{\mathrm{v}} + \tfrac{2}{3}\Delta) = 0, \qquad (10.33)$$

where $E' = E - \varepsilon_0(K)$. The solutions are shown in Figure 10.2 and give bands for electrons, light holes, and split-off holes. The light holes live up to their name for small K but the band flattens for large K and curves up again close to the heavy holes. The split-off band is relatively heavy for small K but curves downwards more rapidly when the light holes flatten out. The effective masses for small K are

$$\frac{1}{m_{\mathrm{e}}} = 1 + \frac{2E_P}{3E_{\mathrm{g}}} + \frac{E_P}{3(E_{\mathrm{g}} + \Delta)}, \quad \frac{1}{m_{\mathrm{lh}}} = \frac{2E_P}{3E_{\mathrm{g}}} - 1, \quad \frac{1}{m_{\mathrm{so}}} = \frac{E_P}{3(E_{\mathrm{g}} + \Delta)} - 1.$$
$$(10.34)$$

The contributions to the valence and conduction bands are again mirrored.

Spin–orbit coupling modifies the bands strongly for $K < 0.5\,\mathrm{nm}^{-1}$ but for larger K the bands approach the previous model of Figure 10.1. The description of the heavy holes remains particularly unsatisfactory, as does the light-hole band at large K. Both defects can be rectified by including the coupling to remote bands through perturbation theory, giving the additional terms in the matrix (10.29). Numerical solution is then required, and a number of less unwieldy approximations have been developed.

The simplest approach is to note that the heavy holes are decoupled, so we can merely adjust their mass without affecting the others. Beyond this, the Hamiltonian can be restricted to the troublesome $j = \frac{3}{2}$ states and solved exactly. This is the Luttinger model of the valence band and gives light and heavy holes with energies

$$\varepsilon(K) = E_{\mathrm{v}} - \frac{\hbar^2}{2m_0} \left[AK^2 \pm \sqrt{(BK^2)^2 + C^2(k_x^2 k_y^2 + k_y^2 k_z^2 + k_z^2 k_x^2)} \right]. \quad (10.35)$$

The constants are conventionally expressed in terms of Luttinger parameters:

$$A = \gamma_1, \quad B = 2\gamma_2, \quad C^2 = 12(\gamma_3^2 - \gamma_2^2). \quad (10.36)$$

These are also related to the parameters L', M, and N' used in the matrix (10.29). For GaAs, $\gamma_1 = 6.85$, $\gamma_2 = 2.1$, and $\gamma_3 = 2.9$. An important feature is that the bands now show only cubic symmetry, as sketched in Figure 2.18(b), rather than the spherical symmetry of our previous results. The effective masses along [100] are

$$m_{\mathrm{hh}} = \frac{1}{\gamma_1 - 2\gamma_2}, \quad m_{\mathrm{lh}} = \frac{1}{\gamma_1 + 2\gamma_2}, \quad (10.37)$$

but γ_3 replaces γ_2 along [111]. The constant-energy surfaces become warped spheres, but the approximation $\gamma_3 = \gamma_2$ is sometimes made to restore spherical symmetry.

There are many more models of varying sophistication used to describe valence bands, but it is now time to leave the bulk materials behind and explore the energy levels in quantum wells.

10.3　Bands in a Quantum Well

We have already dealt with energy levels for electrons in a quantum well. The previous picture does not change much with the more sophisticated models described in this chapter, although non-parabolicity of the bands becomes important at high energies. In contrast, the degenerate nature of valence bands leads to more complicated behaviour. Unfortunately there are few analytic results available, but a simple picture provides a reasonable foundation.

Consider a quantum well of GaAs surrounded by AlGaAs, with growth along the z-axis, which is also the axis used for the component j_z. Assume that the heterojunction can be treated simply as a potential step, ignoring other differences in the band structure. The wave functions in the bulk are then given in terms of p orbitals by equation (10.31). The spin–orbit coupling in GaAs is large enough that the split-off band $j = \frac{1}{2}$ can be neglected to leave the light and heavy holes.

The heavy holes of the bulk are the $|\frac{3}{2}, \frac{3}{2}\rangle$ states. Equation (10.31) shows that they are the same as the $|+1 \uparrow\rangle p$ orbitals, which are in turn linear combinations of $|X \uparrow\rangle$ and $|Y \uparrow\rangle$ (equation 10.30). Thus the p orbitals in this state are all oriented normal to z. The picture of the bands in the bulk based on Figure 2.17 shows that

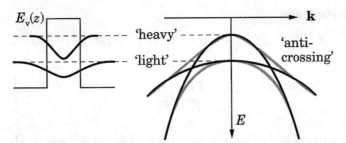

FIGURE 10.3. Simple model of holes in a quantum well. The 'heavy' holes have the lower energy at $k = 0$ but their energy rises more rapidly with k so the two bands cross. Coupling between the bands leads to 'anticrossing' behaviour instead, shown by the grey line.

the resulting band should be heavier along z and lighter in the xy-plane. If we can decouple the motion along z and normal to it, as we did for electrons, the heavy mass along z means that the bound state is deep in the quantum well, while the light mass in the xy-plane means that the kinetic energy rises rapidly as a function of the wave vector k in the plane of the well.

The largest contribution to the wave function of light holes such as $|\frac{3}{2}, \frac{1}{2}\rangle$ is from $|0\uparrow\rangle$. This is the same as $|Z\uparrow\rangle$, whose p orbitals are oriented along z. The masses of the light holes are therefore reversed, with the lighter along z and the heavier normal to it. There is therefore a high kinetic energy in the well (lower binding energy), but the energy rises more slowly with k. The two states are illustrated in Figure 10.3. The masses for motion in the xy-plane within the Luttinger model are

$$m_{\perp hh} = \frac{1}{\gamma_1 + \gamma_2}, \qquad m_{\perp lh} = \frac{1}{\gamma_1 - \gamma_2}. \qquad (10.38)$$

Their order is reversed in comparison with the bulk masses (equation 10.37).

An important feature is that the heavy and light holes are no longer degenerate at $k = 0$; the quantum well has lifted the remaining degeneracy of the band structure in the bulk. The well also reveals the anisotropy associated with each p orbital seen in Figure 2.17, where our picture of valence bands started. This anisotropy means that the 'heavy' holes are actually lighter for transverse motion and the light and heavy bands therefore cross. A less simplified picture would include coupling between these bands and 'anticrossings' instead of crossings. Clearly the 'heavy' and 'light' character is strongly mixed.

More realistic calculations may be based on the Kane or Luttinger models. These are extended from the perfect crystal to inhomogeneous structures by using the effective-mass approximation, with $K = -i\nabla$. Taking our simplest Kane model without spin–orbit coupling (10.26) as an example, we get four coupled Schrödinger differential equations to solve. Assume that the composition changes only along z so that we can factor the wave function into a transverse plane wave $\exp(i k \cdot r)$ and an unknown set of functions along z. Then k_x and k_y can be left as numbers and the

Hamiltonian becomes

$$\left[\varepsilon_0(k) - \frac{\hbar^2}{2m_0}\frac{d^2}{dz^2}\right]1 + \begin{pmatrix} E_c(z) & iPk_x & iPk_y & Pd/dz \\ -iPk_x & E_v(z) & 0 & 0 \\ -iPk_y & 0 & E_v(z) & 0 \\ -Pd/dz & 0 & 0 & E_v(z) \end{pmatrix}. \quad (10.39)$$

The identical terms in the diagonal have been taken out separately. The wave function along z is a four-component vector containing contributions from all four basis states. The edges of the conduction and valence bands, E_c and E_v, are functions of z, although P can often be assumed constant throughout. To this Hamiltonian must be added the boundary conditions at a heterojunction. We saw earlier (equation 3.19) that the straightforward 'value and slope' matching must be modified at a heterojunction where the effective mass changes. Here we require a matrix of boundary conditions, 4×4 for the simple model, because the different components of the wave function may mix at a heterojunction.

Valence bands in a quantum well are complicated and it is not surprising that most results are numerical. An example is shown in Figure 10.4. Although some of the features, such as anticrossings and the transverse effective masses of the light and heavy holes, were expected from the simple arguments given earlier, the numerical solution has even richer structure. Some bands are extremely heavy, whereas others have a negative effective mass (from the point of view of holes). A slight anisotropy is visible in the figure, as is a small splitting in all the states due to terms such as Bk_xk_y in equation (10.29).

Uniaxial strain has an effect on the valence bands similar to that of confinement in a quantum well. This is of great practical importance because large strains can be

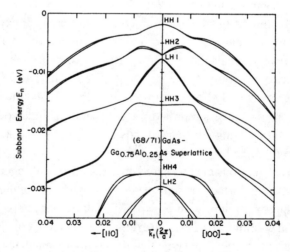

FIGURE 10.4. Valence-band structure in a multiquantum well as a function of **k** along two directions. The wells comprise 68 atomic layers of GaAs with barriers of 71 atomic layers of $Al_{0.25}Ga_{0.75}As$. [From Chang and Schulman (1985).]

induced by growing layers with deliberately mismatched lattice constants to produce a pseudomorphic structure (Section 3.6). Confinement and strain then work together, and the splitting of the valence bands has large beneficial effects on the performance of quantum-well lasers. Sadly, details of these applications are beyond the scope of this book.

Having derived some detailed models of the valence bands in a quantum well, we shall now abandon them! To calculate the optical properties we shall return to a model of parabolic bands with the sole complication that the effective masses for motion in the plane of the well and normal to it may be different. Although more sophisticated models are undoubtedly needed for accurate numbers, it is possible to derive only simple analytic results within this parabolic approximation.

10.4 Interband Transitions in a Quantum Well

We can now put these results together to calculate the optical properties of quantum wells. Here we shall look at transitions from the valence band to the conduction band. This was treated in Section 8.6 for the bulk, but now we allow the initial and final states to be confined within a quantum well. This is shown in Figure 10.5 and was discussed at a simplified level in Section 1.3.1.

Once we have solved the Kane model or something similar for the energy levels, the next step is to evaluate the matrix element. This requires the full wave function,

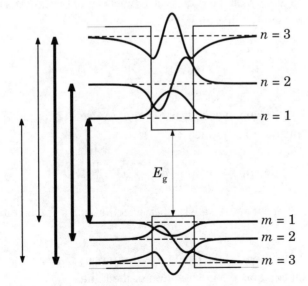

FIGURE 10.5. Transitions between bound states in the valence and conduction bands of a quantum well. Only the envelope functions are shown, but the Bloch functions are also vital in determining selection rules. Thick lines indicate strong transitions (the '$\Delta n = 0$' rule); thin lines mark transitions that are weak but allowed. Only one set of holes is shown.

which is a product of the envelope function and the appropriate Bloch function at Γ, $\psi(\mathbf{R}) \propto \chi(\mathbf{R})u_{n0}(\mathbf{R})$. This must first be normalized. The two components of the wave function are usually treated independently and normalized separately, so

$$\int |\chi(\mathbf{R})|^2 d^3\mathbf{R} = 1, \qquad \int |u_{n0}(\mathbf{R})|^2 d^3\mathbf{R} = 1, \qquad (10.40)$$

where both integrals run over the volume Ω of the sample. We should set $\psi(\mathbf{R}) = \Omega^{1/2}\chi(\mathbf{R})u_{n0}(\mathbf{R})$ so that the product obeys the same normalization.

Now construct the matrix element, which becomes

$$\langle j|\mathbf{e} \cdot \hat{\mathbf{p}}|i\rangle = \Omega \int \chi_c^*(\mathbf{R})u_c^*(\mathbf{R})(\mathbf{e} \cdot \hat{\mathbf{p}})\chi_v(\mathbf{R})u_v(\mathbf{R})d^3\mathbf{R}. \qquad (10.41)$$

The integral runs over the whole sample but can be broken up because it contains functions that vary on two length scales. The Bloch functions vary within each unit cell (and are of course the same within each unit cell), whereas the envelope functions vary only on a much longer scale and are almost constant within each unit cell. Thus we can divide the integral into unit cells and pull the envelope functions out as being constant within each cell, giving

$$\langle j|\mathbf{e} \cdot \hat{\mathbf{p}}|i\rangle \approx \Omega \sum_j^{\text{cells}} \chi_c^*(\mathbf{R}_j)\chi_v(\mathbf{R}_j) \int_{\text{cell } j} u_c^*(\mathbf{R})(\mathbf{e} \cdot \hat{\mathbf{p}})u_v(\mathbf{R})d^3\mathbf{R}. \qquad (10.42)$$

The integral within each unit cell can be denoted $(\Omega_{\text{cell}}/\Omega)\mathbf{e} \cdot \mathbf{p}_{cv}(\mathbf{0})$, where $\mathbf{p}_{cv}(\mathbf{0})$ is the matrix element between the Bloch wave functions at the extrema of the bands, which has appeared many times already. Then

$$\langle j|\mathbf{e} \cdot \hat{\mathbf{p}}|i\rangle \approx \mathbf{e} \cdot \mathbf{p}_{cv}(\mathbf{0})\Omega_{\text{cell}} \sum_j^{\text{cells}} \chi_c^*(\mathbf{R}_j)\chi_v(\mathbf{R}_j). \qquad (10.43)$$

Finally, we can turn the sum over cells back into an integral over the whole sample. This absorbs the factor of Ω_{cell} to leave

$$\langle j|\mathbf{e} \cdot \hat{\mathbf{p}}|i\rangle \approx \mathbf{e} \cdot \mathbf{p}_{cv}(\mathbf{0}) \int \chi_c^*(\mathbf{R})\chi_v(\mathbf{R})d^3\mathbf{R}. \qquad (10.44)$$

Thus the matrix element must satisfy *two* selection rules for a transition to be allowed:

   (i) The Bloch functions of the band edges must obey the usual dipole selection rule, which contains the polarization of the light.
   (ii) The integral of the product of the envelope functions, with no operator between, must not vanish. This gives quite different selection rules for the envelope functions.

Within a simple picture, the envelope functions in a quantum well have the separable form of a plane wave in x and y and a bound state along z, such as $\chi_c(\mathbf{r}) = A^{-1/2} \exp(i\mathbf{k} \cdot \mathbf{r})\phi_{cn}(z)$, which we shall denote $|cn\mathbf{k}\rangle$. The selection rule imposes vertical transitions, meaning that the transverse wave vector $\mathbf{k}$ must not change, as for absorption within a quantum well (Section 8.7). Thus the matrix element reduces to

$$\langle cn\mathbf{k}'|\mathbf{e} \cdot \hat{\mathbf{p}}|vm\mathbf{k}\rangle \approx \mathbf{e} \cdot \mathbf{p}_{cn,vm}\delta_{\mathbf{k},\mathbf{k}'} \int \phi_{cn}^*(z)\phi_{vm}(z)dz$$

$$\equiv \mathbf{e} \cdot \mathbf{p}_{cn,vm}\delta_{\mathbf{k},\mathbf{k}'}\langle cn|vm\rangle. \tag{10.45}$$

Here 'c' and 'v' indicate the conduction and valence bands, n and m label the bound states, and $\mathbf{k}'$ and $\mathbf{k}$ are the transverse wave vectors. We shall now consider the influence of the two selection rules.

10.4.1 INTERBAND MATRIX ELEMENT

This matrix element $\mathbf{e} \cdot \mathbf{p}_{cn,vm}$ depends on the nature of the Bloch functions and on the polarization $\mathbf{e}$. We know that the only non-zero matrix elements are those such as $\langle S{\uparrow}|p_x|X{\uparrow}\rangle$, which were previously denoted $(im_0/\hbar)P$. An arrow has been added as a reminder that spin is not changed in an optical transition.

Consider transitions from the lowest state in the valence band. These are the heavy holes or $|\frac{3}{2}, \frac{3}{2}\rangle$ states, which are heavy along z but light in the xy-plane. Equation (10.31) shows that these are the same as $|+1{\uparrow}\rangle$, which in turn can be written as $(|X{\uparrow}\rangle + i|Y{\uparrow}\rangle)/\sqrt{2}$. Now let $\mathbf{e}$ be directed along the x-axis, so $\mathbf{e} \cdot \hat{\mathbf{p}} = p_x$. The matrix element between the heavy holes and the conduction band is

$$\langle S{\uparrow}|p_x|\tfrac{3}{2}, \tfrac{3}{2}\rangle = \sqrt{\tfrac{1}{2}}\big(\langle S{\uparrow}|p_x|X{\uparrow}\rangle + i\langle S{\uparrow}|p_x|Y{\uparrow}\rangle\big) = \frac{im_0P}{\hbar\sqrt{2}}. \tag{10.46}$$

Only $|X{\uparrow}\rangle$ contributes to the integral. The same result (apart from a phase) is found if the light is polarized along y. For polarization along z, however, both contributions vanish and the matrix element is zero; no absorption from heavy holes is possible with this polarization.

These results can also be expressed in terms of the direction of propagation. There are three cases:

(i) Light propagating along z, normal to the layers, can be polarized along x or y with equal absorption.

(ii) Light propagating in the plane of the well, along x, say, can have two possible polarizations. A transverse electric (TE) mode also has its electric field parallel to the layer (along y here), and absorption is again possible.

(iii) No absorption by heavy holes is seen for the other polarization, a transverse magnetic (TM) mode with its electric field normal to the quantum well (along z). (This is not precisely true, because the TM mode has a component of electric field along the direction of propagation x, but it is small in a weakly guiding structure.)

Selection rules clearly place important constraints on the design of low-dimensional optical devices.

Different results are obtained for the light holes. The wave functions are

$$|\tfrac{3}{2}, \tfrac{1}{2}\rangle = \sqrt{\tfrac{1}{3}}|+1\downarrow\rangle - \sqrt{\tfrac{2}{3}}|0\uparrow\rangle = \sqrt{\tfrac{1}{6}}|X\downarrow\rangle + i\sqrt{\tfrac{1}{6}}|Y\downarrow\rangle - 2\sqrt{\tfrac{1}{6}}|Z\uparrow\rangle. \quad (10.47)$$

All polarizations can be absorbed in this case, although the matrix element is twice as large for $\mathbf{e}$ along z. Thus the TM mode should show stronger absorption by light holes than the TE mode. The split-off band can be treated in a similar way but is rarely measured.

These two sets of results can be used to predict the relative contributions of light and heavy holes to absorption (if it is assumed that the envelope functions are much the same). For light polarized along x, the matrix element is $im_0 P/\hbar\sqrt{2}$ from the heavy holes and $im_0 P/\hbar\sqrt{6}$ from the light holes. The absorption, proportional to the square of the matrix element, should therefore be in the ratio $3 : 1$. There is no contribution from heavy holes if the light is polarized along z, and the ratio is $0 : 4$.

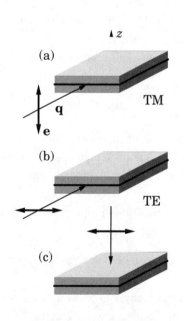

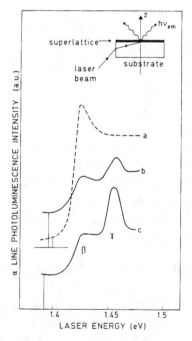

FIGURE 10.6. Photoluminescence excitation spectra (roughly equivalent to absorption) from a quantum well for the three polarizations illustrated on the left. [From Marzin, Charasse, and Sermage (1985).]

These results are demonstrated in Figure 10.6, where absorption is shown for the three possible polarizations in a quantum well. The transition β involves the lowest level for light holes, whereas γ arises from the *second* level for heavy holes, so the transition for heavy holes is at higher energy. Curve (a) is a TM mode with **e** along z and light propagating in the plane of the well; only the light holes are seen. The direction of propagation is the same in (b) but the polarization is now TE, in the plane of the well; both light and heavy holes are seen, with the heavy holes being stronger. The light propagates normal to the well in (c), with similar results to (b).

This analysis of the selection rules can be carried further, notably by using circularly polarized light, which couples to the $|\pm 1\rangle$ states directly rather than to $|X\rangle$ and $|Y\rangle$. The accuracy of the rules is limited by mixing between the branches of the valence band when $k > 0$, which led to the structure shown in Figure 10.4. The usual nomenclature of 'light' and 'heavy' holes is only approximate in a quantum well. We shall now look at the restrictions on the envelope functions.

10.4.2 MATRIX ELEMENT OF ENVELOPE FUNCTIONS

The matrix element for the envelope functions is an integral over their product with no dependence on polarization. If the quantum well is symmetric, the envelope functions will as usual be symmetric or antisymmetric and their matrix element will vanish unless both have the *same* parity. This is the opposite of the selection rule for transitions between levels in the same band, derived in Section 8.7! An even more restrictive condition is found if we make the common approximation that the potential in both conduction and valence bands can be modelled as an infinitely deep square well. Then the two sets of envelope functions are identical and the integral becomes

$$\int \phi_{cn}^*(z)\phi_{vm}(z)\,dz = \int \phi_n^*(z)\phi_m(z)\,dz = \delta_{nm}. \qquad (10.48)$$

Only transitions between levels with the same index are allowed, the '$\Delta n = 0$' rule. This rule does not survive a more accurate calculation with finite wells, where $\phi_{cn}(z) \neq \phi_{vm}(z)$ because the effective depth of the well is different for electrons and holes. Nevertheless, $\Delta n = 0$ gives the strongest transitions.

Parabolic wells provide a step further away from $\Delta n = 0$ because the curvature in the conduction and valence bands is different. The selection rule due to parity remains but all other transitions can be seen. An example was shown in Figure 4.5. Experiments on these systems were important in determining the ratio $\Delta E_c / \Delta E_g$ in $Al_x Ga_{1-x} As$. There is no restriction on the allowed values of m and n for a well that is asymmetric, either through growth or because it has been distorted by an electric field.

10.4.3   ABSORPTION SPECTRUM

We can now put the matrix element into the general formula (8.76) for optical absorption. The sum over the initial state i becomes one over m, $\mathbf{k}$ here, and similarly for f. The restriction to vertical transitions removes one sum over $\mathbf{k}$. Thus

$$\sigma_1(\omega) = \frac{\pi e^2}{m_0^2 L \omega} \sum_{n,m} |\mathbf{e} \cdot \mathbf{p}_{cn,vm}|^2 |\langle cn | vm \rangle|^2 \frac{2}{A} \sum_{\mathbf{k}} \delta[E_{cn}(\mathbf{k}) - E_{vm}(\mathbf{k}) - \hbar\omega].$$

(10.49)

The volume $\Omega = AL$, where A is the area of the sample and L is its thickness. The sum over $\mathbf{k}$ includes only the δ-function as in the three-dimensional case (equation 8.84) and will again give a density of states. Although we know that the valence bands are ghastly, we shall use the parabolic approximation as mentioned earlier. Thus $E_{vm}(\mathbf{k}) = E_v - \varepsilon_{vm} - \hbar^2 k^2 / 2m_0 m_{vm}$, where the mass m_{vm} is for *transverse* motion in subband m. Then the sum over $\mathbf{k}$ becomes

$$\frac{2}{A} \sum_{\mathbf{k}} \delta[E_{cn}(\mathbf{k}) - E_{vm}(\mathbf{k}) - \hbar\omega]$$

$$= \frac{2}{A} \sum_{\mathbf{k}} \delta\left(E_c + \varepsilon_{cn} + \frac{\hbar^2 k^2}{2m_0 m_{cn}} - E_v + \varepsilon_{vm} + \frac{\hbar^2 k^2}{2m_0 m_{vm}} - \hbar\omega \right)$$

$$= \frac{2}{A} \sum_{\mathbf{k}} \delta\left(E_g + \varepsilon_{cn} + \varepsilon_{vm} + \frac{\hbar^2 k^2}{2m_0 m_{nm}} - \hbar\omega \right)$$

$$= \frac{m_0 s m_{nm}}{\pi \hbar^2} \Theta[\hbar\omega - (E_g + \varepsilon_{cn} + \varepsilon_{vm})].$$

(10.50)

The sum reduces to an optical joint density of states as in equation (8.85). Here there is a two-dimensional band that starts at $E_g + \varepsilon_{cn} + \varepsilon_{vm}$ with the reduced mass of the two subbands involved, $1/m_{nm} = 1/m_{cn} + 1/m_{vm}$. The optical conductivity becomes

$$\sigma_1(\omega) = \frac{\pi e^2}{m_0^2 L \omega} \sum_{n,m} |\mathbf{e} \cdot \mathbf{p}_{cn,vm}|^2 |\langle cn | vm \rangle|^2 \frac{m_0 m_{nm}}{\pi \hbar^2} \Theta[\hbar\omega - (E_g + \varepsilon_{cn} + \varepsilon_{vm})].$$

(10.51)

A factor of $1/L$ remains because there is only one quantum well in the thickness of the sample, as in Section 8.7.

The absorption shows a series of steps at the difference in energy of each pair of subbands, reflecting the two-dimensional density of states in which the carriers are free. There is a 'blue shift' in the onset of absorption compared with the bulk because the energy of the lowest step, $E_g + \epsilon_{c1} + \epsilon_{v1}$, contains the energies of the two bound states.

This theory explains the overall structure of the data for a multiquantum well in Figure 8.4, but the features near the edge of each subband are as usual dominated by excitons. The result for a quantum well is very close to that for bulk material, equation (8.85). The same matrix elements of Bloch functions appear, and the only difference lies in the density of states if we take infinitely deep wells with $\langle cn|vm\rangle = \delta_{nm}$. Figure 4.7(c) showed that the density of states of the two-dimensional system is lower than that of a three-dimensional one of the same volume, except at the steps where they coincide. Thus we expect the absorption of a low-dimensional sample to be generally lower than that of a three-dimensional one. The data in Figure 8.4 do not support this, because the multiquantum well has only half the active volume of the bulk sample. Again we must invoke excitons.

Finally, let's estimate the magnitude of the steps in σ_1. Put $\mathbf{e} \cdot \mathbf{p}_{cn,vm} \approx im_0 P/\hbar$, $\langle cn|vm\rangle \approx 1$, and $\hbar\omega \approx E_g$. Then

$$\sigma_1 \approx \frac{\pi e^2 \hbar}{m_0^2 L E_g} \left(\frac{m_0 P}{\hbar}\right)^2 \frac{m_0 m_{nm}}{\pi \hbar^2} = \frac{e^2 m_{nm} E_P}{2\hbar E_g L} \approx \frac{\pi e^2}{2hL}, \tag{10.52}$$

or $\epsilon_2 \approx e^2/4\epsilon_0 E_g L$. The final step uses the relation $1/m_{nm} = 2E_P/E_g$ between the reduced mass and the interband matrix element from equation (10.28). It is amusing that this estimate of σ_1, admittedly rough, contains the quantum of conductance but no parameters of the material!

10.5 Intersubband Transitions in a Quantum Well

A second class of optical transitions takes place between different levels of a quantum well in the *same* band of the crystal, which we shall take to be the conduction band. Most of the theory and some measurements have already been given in Section 8.7 but we need to verify the matrix element.

The approximation (10.44) for the matrix element, on which our treatment of interband transitions was based, is useless for intraband transitions. It includes the matrix element of $\hat{\mathbf{p}}$ between the two Bloch states involved; here they are the same so the matrix element vanishes. We must go back to equation (10.41). The operator $\hat{\mathbf{p}}$ is just a first derivative so the product rule gives

$$\langle j|\mathbf{e} \cdot \hat{\mathbf{p}}|i\rangle = \Omega \int [\chi_c^* \chi_v][u_c^*(\mathbf{e} \cdot \hat{\mathbf{p}})u_v]d^3\mathbf{R} + \Omega \int [u_c^* u_v][\chi_c^*(\mathbf{e} \cdot \hat{\mathbf{p}})\chi_v]d^3\mathbf{R}. \tag{10.53}$$

Each of the integrals may be summed over contributions from each unit cell, the envelope functions may be pulled out, and the summation may be converted back to an integral, as before. Thus

$$\langle j|\mathbf{e} \cdot \hat{\mathbf{p}}|i\rangle \approx \int \chi_c^* \chi_v \, d^3\mathbf{R} \int u_c^*(\mathbf{e} \cdot \hat{\mathbf{p}})u_v \, d^3\mathbf{R} + \int u_c^* u_v \, d^3\mathbf{R} \int \chi_c^*(\mathbf{e} \cdot \hat{\mathbf{p}})\chi_v \, d^3\mathbf{R}. \tag{10.54}$$

For intersubband transitions we set $u_c = u_v$ and the first term vanishes. The first integral of the surviving term gives unity as it is simply the normalization of the Bloch states, and we are left with the usual matrix element of $\mathbf{e} \cdot \hat{\mathbf{p}}$ between the initial and final envelope functions. Thus the theory of Section 8.7 appears to need no modification.

Actually there is a subtle defect in this argument that has arisen from careless use of the effective-mass approximation. It was noted in Section 10.2.1 on the Kane model that the electrons and light holes repel each other to give their low effective masses, and that this arose from mixing between the states induced by the $\mathbf{k} \cdot \hat{\mathbf{p}}$ coupling. We have included the effect of this mixing on the energy levels, by using the effective mass in the Schrödinger equation that defines the envelope functions. However, we have omitted the corresponding change in the Bloch functions and must include both to be consistent. Fortunately there is a simple way around this, which is to start the calculation of optical absorption from the effective-mass Hamiltonian for the conduction band, rather than the full Hamiltonian of the crystal. All this means is that the mass of the electron should be the effective mass *throughout*, and equation (8.94) is modified to

$$\sigma_1(\omega) = \frac{\pi e^2}{2 m_0 m_e L} \sum_{\substack{i,j \\ j \neq i}} f_{ji}(n_i - n_j)\delta(\omega - \omega_{ji}). \tag{10.55}$$

Similarly, the effective mass appears in the definitions of oscillator strengths and sum rules.

It is interesting to compare the strength of intraband and interband transitions. Suppose that only the lowest subband is occupied, so $i = 1$. We found that most of the oscillator strength goes into $j = 2$ so put $f_{21} = 1$ and ignore the rest. To make the absorption as strong as possible, put as many electrons as possible into the lowest subband while keeping the next empty. This gives a range in energy of $\varepsilon_2 - \varepsilon_1 = \hbar\omega_{21}$ and the density of states is $m_0 m_e/\pi\hbar^2$. Thus

$$\sigma_1(\omega) \approx \frac{\pi e^2}{2 m_0 m_e L} \frac{m_0 m_e \hbar\omega_{21}}{\pi\hbar^2}\delta(\omega - \omega_{21}) = \frac{\pi e^2}{hL} \omega_{21}\delta(\omega - \omega_{21}). \tag{10.56}$$

Alternatively, $\epsilon_2 \approx (e^2/2\epsilon_0 L)\delta(E - \hbar\omega_{21})$. Again most parameters of the material vanish. The prefactor is almost identical to the interband result (10.52), but the absorption is now concentrated into a single frequency instead of being spread over a range ω_{21}. This result can also be derived from the f-sum rule.

A limitation of this calculation, as in the case of interband absorption, is that we have ignored electron–electron interaction. This becomes more significant as the density of electrons increases, and it shifts the energy of the absorption lines. There is a simple result known as Kohn's theorem for parabolic wells, which is mentioned briefly at the end of the exercises.

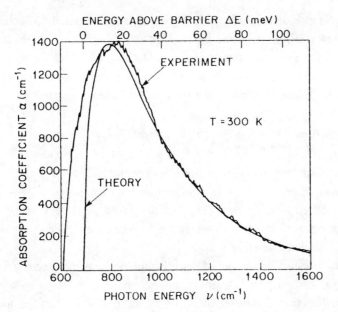

FIGURE 10.7. Absorption between bound and extended states in a multiquantum well with 4.5 nm GaAs wells separated by 14 nm $Al_{0.2}Ga_{0.8}As$ barriers. [Reprinted with permission from Levine et al. (1988). Copyright 1988 American Institute of Physics.]

Finally, it has been assumed that all the states of interest are bound in the well. We saw that most of the oscillator strength from the lowest state went into the $1 \to 2$ transition in an infinitely deep well. Real quantum wells are of course finite, and the number of bound states falls if the well is made narrower (Section 4.2) until only the lowest state is left. All the oscillator strength from this state must now go into the continuum of transitions from the bound state to those above the plateau. An example is shown in Figure 10.7. The shape of the absorption spectrum does not reflect the density of states for *free* electrons, particularly at low energy, because the states are distorted by the presence of the well even though they are not bound. The same effect was seen in the local density of states above a square well, plotted in Figure 4.3.

10.6 Optical Gain and Lasers

We noted at the end of Section 8.5.2 that the optical conductivity is positive, giving dissipation, for a system in equilibrium because the occupation function $f(E)$ decreases monotonically with energy. A laser requires optical amplification so the population of the active states must be inverted.

The amplitude of a light wave propagating in the z-direction varies like $\exp(ikz)$. Optical properties enter through the complex refractive index $\tilde{n}_r = n_r + i\kappa_r$, which gives $k = \tilde{n}_r\omega/c$. Thus the amplitude decays as $\exp(-\omega\kappa_r z/c)$, and the intensity goes as the square of this. An optical gain of g per unit length would cause the intensity to grow like $\exp(gz)$ so $g = -2\omega\kappa_r/c = -\omega\epsilon_2/cn_r = -\sigma_1/\epsilon_0 cn_r$.

A simple expression for the optical conductivity of a bulk semiconductor was given in equation (8.85). We assumed a full valence band and an empty conduction band in the derivation. Restoring the occupation functions gives

$$g(\omega) = \frac{\pi e^2}{\epsilon_0 cn_r m_0^2 \omega}|p_{cv}(\mathbf{0})|^2 n_{\mathrm{opt}}(\hbar\omega)[f(\varepsilon_e) - f(\varepsilon_h)]. \tag{10.57}$$

Here ε_e and ε_h are the energies of the electron and hole taking part in the optical transition, with $\varepsilon_e = \varepsilon_h + \hbar\omega$. In equilibrium both ε_e and ε_h are governed by the same Fermi–Dirac function, so $f(\varepsilon_e) < f(\varepsilon_h)$ and $g(\omega) < 0$ always. Note that the occupation function is that for electrons in *both* cases, even in the valence band where that for holes is usually used.

When the system is driven out of equilibrium, perhaps by injecting a current into a p–n junction or by optical pumping, the occupations are no longer given by the Fermi–Dirac function. However, an approximation described in Section 1.8.3 is to use a Fermi function with a different quasi-Fermi level or imref for each band. Thus the probability of a state in the conduction band being occupied by an electron is given by $f_n(\varepsilon_e) = f(\varepsilon_e - E_F^{(n)})$, where $E_F^{(n)}$ is the imref for electrons. Similarly, the probability of a state in the valence band being occupied by an electron is $f_p(\varepsilon_h) = f(\varepsilon_h - E_F^{(p)})$. The difference in occupation functions becomes

$$f_n(\varepsilon_e) - f_p(\varepsilon_h) = f(\varepsilon_e - E_F^{(n)}) - f(\varepsilon_h - E_F^{(p)}). \tag{10.58}$$

Now $f(E)$ is a decreasing function, so the difference will be positive if $(\varepsilon_e - E_F^{(n)}) < (\varepsilon_h - E_F^{(p)})$ or

$$E_F^{(n)} - E_F^{(p)} > \varepsilon_e - \varepsilon_h = \hbar\omega. \tag{10.59}$$

The separation of the quasi-Fermi levels must exceed the energy of the optical transition, which must itself be greater than the band gap. This is the *Bernard–Durrafourg condition* for population inversion. It implies a large density of electrons in the conduction band at the same time as a large density of holes in the valence band, which is obviously far from equilibrium.

The optical gain as a function of frequency is given by competition between the occupation functions and the optical joint density of states in equation (10.57), as the other quantities vary only slowly. This is illustrated in Figure 10.8(a). There is no gain at all unless $E_F^{(n)} - E_F^{(p)} > E_g$. A larger separation of the quasi-Fermi levels makes $g(\omega)$ positive for energies below $E_F^{(n)} - E_F^{(p)}$. In three dimensions the density of states forces $g(\omega)$ to be zero at the bottom of the band so the gain reaches a peak g_{pk} between E_g and $E_F^{(n)} - E_F^{(p)}$. Increasing the population inversion, perhaps by increasing the current through an injection laser, increases both g_{pk} and the frequency at which it occurs. Eventually g_{pk} may reach a level at which it overcomes optical losses in the structure and laser action commences. An aim

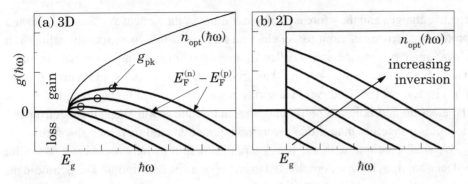

FIGURE 10.8. Optical gain $g(\omega)$ as a function of frequency in three and two dimensions. This is dominated by the product of the density of states $n_{opt}(\hbar\omega)$ and difference of occupation functions.

of optoelectronic engineering is to design devices where this occurs with a low threshold-current density. This involves many issues such as optical confinement that cannot be covered here.

The picture in two dimensions is very similar except for the form of the density of states, with a step at E_g. This is shown in Figure 10.8(b). The peak gain always occurs at E_g, and it is possible to induce laser action with a smaller threshold current than in devices with three-dimensional active regions. A further reduction in dimensionality gives devices based on quantum wires. Their density of states diverges at the bottom of the band, and this is expected to contribute to another fall in the threshold current of lasers. This trend continues with quantum dots, and vigorous research is in progress. There has also been strong interest in developing intersubband transitions to give lasers that operate in the far infrared. This has recently been achieved as a *quantum-cascade laser*, which also employs resonant tunnelling between the wells of a superlattice in a large electric field, rather like Figure 6.4.

10.7 Excitons

We have already seen that a one-electron picture fails to give the correct behaviour of $\sigma_1(\omega)$ near direct band gaps, and we have blamed this on excitons. The problem is that we have been selective in treating the electric charge of electrons. An electron interacts with a light wave because it carries a charge, but the same charge gives rise to electron–electron interaction, which we have generally ignored. This selective vision fails for excitons.

In absorption across the band gap, an electron is excited from a state in the valence band to one in the conduction band. Another way of describing this is to say that an electron in the conduction band and a hole in the conduction band have been superposed on the ground state of a filled valence band and an empty conduction band: an *electron–hole pair* has been created. The electron and hole carry opposite electric charges and therefore attract one another in the same way as an electron and proton in a hydrogen atom (or an electron and positron in positronium, which is a closer analogy). Bound states called *excitons* can form in the attractive potential. These have a lower total energy than free electron–hole pairs and therefore give lines just below the band gap in absorption spectra. Even at higher energies, where the electron and hole are not bound, the mutual attraction keeps the electron and hole closer together than if there were no interaction, and increases absorption.

We shall continue with the simple approximation of light and heavy parabolic valence bands as it is too complicated to go beyond this description. The Schrödinger equation for excitons therefore takes the form

$$
\left[\left(E_c - \frac{\hbar^2}{2m_0 m_e} \nabla_e^2 \right) - \left(E_v + \frac{\hbar^2}{2m_0 m_h} \nabla_h^2 \right) - \frac{e^2}{4\pi\epsilon_0\epsilon_b |\mathbf{R}_e - \mathbf{R}_h|} \right.
$$

$$
\left. + V_e(\mathbf{R}_e) + V_h(\mathbf{R}_h) \right] \psi(\mathbf{R}_e, \mathbf{R}_h) = E\psi(\mathbf{R}_e, \mathbf{R}_h). \qquad (10.60)
$$

The wave function depends on the coordinates of both electron and hole, $\mathbf{R}_e$ and $\mathbf{R}_h$. This is an example of the general many-electron Schrödinger equation (9.25); there are only two particles, with a single interaction, but the kinetic energy is complicated because two bands are involved. The potentials V_e and V_h confine the electron and hole in a quantum well (and are functions only of z in layered structures). We shall solve this problem for free excitons in three dimensions before returning to the more complicated problem of excitons in a quantum well.

10.7.1 EXCITONS IN THREE DIMENSIONS

The confinement potentials V_e and V_h are absent in this case, and the electron and hole are free to move throughout the crystal except for their interaction. This depends only on the *difference* in their coordinates; moving the exciton bodily has no effect. Under these conditions we can use a standard method for separating the motion into that of the centre of mass $\mathbf{R}_{CM}$ and the relative coordinate $\mathbf{R}$ of the electron and hole. The coordinate of the centre of mass is defined by

$$
\mathbf{R}_{CM} = \frac{m_e \mathbf{R}_e + m_h \mathbf{R}_h}{M}, \qquad M = m_e + m_h, \qquad (10.61)
$$

where M is the total mass, while the relative coordinate and reduced mass are

$$
\mathbf{R} = \mathbf{R}_e - \mathbf{R}_h, \qquad \frac{1}{m_{eh}} = \frac{1}{m_e} + \frac{1}{m_h}. \qquad (10.62)
$$

The Schrödinger equation can be transformed from the coordinates $(\mathbf{R}_e, \mathbf{R}_h)$ to $(\mathbf{R}_{CM}, \mathbf{R})$, giving

$$
\left[\left(-\frac{\hbar^2}{2m_0 M} \nabla_{CM}^2 \right) + \left(E_g - \frac{\hbar^2}{2m_0 m_{eh}} \nabla^2 - \frac{e^2}{4\pi\epsilon_0\epsilon_b R} \right) \right] \psi = E\psi. \qquad (10.63)
$$

The operators with $\mathbf{R}$ and $\mathbf{R}_{CM}$ appear additively and the wave function can therefore be factored into a product of functions of $\mathbf{R}_{CM}$ and $\mathbf{R}$. The part for the centre of mass is simply a free-particle Schrödinger equation with mass M, so the solutions are plane waves $\exp(i\mathbf{K}_{CM} \cdot \mathbf{R}_{CM})$ with energy $\hbar^2 K_{CM}^2/2m_0 M$. The plane wave means that the exciton can be found anywhere in the crystal, or we can make a wave packet by mixing a range of $\mathbf{K}_{CM}$. However, the tiny wave vector of light means that $K_{CM} \approx 0$ for optically generated excitons and the motion of the centre of mass can be ignored.

The equation for relative motion is

$$\left(-\frac{\hbar^2}{2m_0 m_{eh}}\nabla^2 - \frac{e^2}{4\pi\epsilon_0\epsilon_b R}\right)\phi(\mathbf{R}) = (E - E_g)\phi(\mathbf{R}). \qquad (10.64)$$

This is just like that for a donor, equation (3.15), except for the parameters. The energies of the bound states are therefore given by

$$\varepsilon_n^{(3D)} = E_g - \frac{\mathcal{R}_{eh}}{n^2}, \qquad \mathcal{R}_{eh} = \frac{m_0 m_{eh}}{2}\left(\frac{e^2}{4\pi\epsilon_0\epsilon_b\hbar}\right)^2, \qquad (10.65)$$

where $\mathcal{R}_{eh}$ is the Rydberg energy scaled by the dielectric constant of the host and the reduced mass of the electron–hole pair (which is different for light and heavy holes). Similarly, the scale of the wave function (equation 4.69) is set by a scaled Bohr radius $a_{eh} = 4\pi\epsilon_0\epsilon_b\hbar^2/e^2m_0 m_{eh}$. The binding is even weaker than that of a donor since $m_{eh} < m_e$ or m_h.

Equation (10.65) shows that there should be a series of discrete absorption lines due to excitons below the band gap. This is shown schematically in Figure 10.9(a). These lines are seen at low temperature, but the weak binding means that excitons are rapidly ionized at room temperature by optic phonons whose energy is much greater than $\mathcal{R}_{eh}$. Usually only $n = 1$ is seen in III–V materials.

It seems reasonable that the electron and hole should be at the same point in space for an optical transition, so the strength should depend on the wave function when $\mathbf{R}_e = \mathbf{R}_h$, or $\mathbf{R} = \mathbf{0}$. It can indeed be shown that the oscillator strength is proportional to $\Omega|\phi(\mathbf{R} = \mathbf{0})|^2$, where Ω is the volume of the crystal. This is $\Omega/\pi a_{eh}^3$ for the ground states of the exciton. The appearance of the volume here is unusual, but we expect the total absorption to be proportional to the volume of the crystal. This factor normally appears in the sum over states since their density is proportional to the volume for extended states. Here the states are discrete and their number does not depend on the volume, which therefore appears in the strength of each transition instead.

Even at energies above the band gap, where excitons are no longer bound, we still need to use wave functions that obey equation (10.64) rather than simple plane waves because the Coulomb attraction distorts the wave functions and enhances the density $|\phi(\mathbf{R} = \mathbf{0})|^2$. Thus our previous expression for interband absorption should

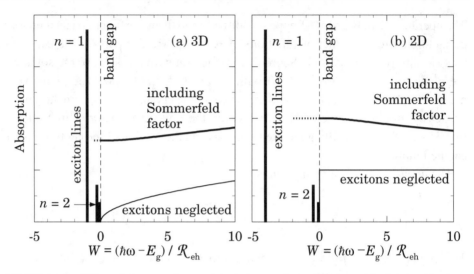

FIGURE 10.9. Effect of electron–electron interaction on absorption around the band gap in three and two dimensions. The horizontal axis W is energy as measured from the band gap E_g in units of the Rydberg $\mathcal{R}_{eh}$, so the scale is highly expanded in comparison with E_g.

be multiplied by a *Sommerfeld factor*. In three dimensions this is

$$\Omega|\phi^{(3D)}(\mathbf{R}=\mathbf{0})|^2 = \frac{2\pi/\sqrt{W}}{1-\exp(-2\pi/\sqrt{W})}, \qquad W = \frac{\hbar\omega - E_g}{\mathcal{R}_{eh}}. \qquad (10.66)$$

The denominator may be neglected for energies just above the band gap and $|\phi_k(\mathbf{R}=\mathbf{0})|^2 \approx 2\pi/\sqrt{W} = 2\pi[\mathcal{R}_{eh}/(\hbar\omega - E_g)]^{1/2}$. This square root cancels that in the optical joint density of states that enters the result for independent electrons and holes, equation (8.85). Thus absorption does not rise as a square root, reflecting the joint density of states, but is roughly constant just above the band gap as shown in Figure 10.9(a). This is much closer to experiment (Figure 8.4). There is no step at E_g because the absorption blends smoothly into that due to the weakly bound higher states of the exciton.

10.7.2 EXCITONS IN TWO DIMENSIONS

The theory of excitons is not greatly changed in the strictly two-dimensional limit, where the wave functions are assumed to have zero thickness along z. The separation into relative and centre-of-mass coordinates proceeds as before, and the 'two-dimensional hydrogen atom' (Section 4.7.3) gives energy levels $\varepsilon_n^{(2D)} = E_g - \mathcal{R}_{eh}/(n-\frac{1}{2})^2$. Thus the lowest state has four times the binding energy of the three-dimensional case. The radius of the wave function is halved, giving $\phi_1^{(2D)} \propto \exp(-2r/a_{eh})$. The Sommerfeld factor in two dimensions is

$$|\phi_k^{(2D)}(\mathbf{r}=\mathbf{0})|^2 = \frac{2}{1+\exp(-2\pi/\sqrt{W})}. \qquad (10.67)$$

The definitions of $\mathcal{R}_{eh}$ and a_{eh} are as before but W should be measured from the energy at which absorption starts in the independent-electron picture; this includes the energies of the bound states in the quantum well as well as the band gap. The effect on the edge of the band is less dramatic than in three dimensions, but absorption is increased by a factor of 2 and this enhancement persists for many Rydbergs into the band.

These results show that excitons should be more significant in low dimensions because of the increased binding energy. Ideally the binding energy of a heavy-hole exciton rises from about 5 to 20 meV in GaAs. Thus excitons should remain visible at room temperature in a two-dimensional sample, and this is confirmed by Figure 8.4; an exciton is barely visible as a bump at the onset of absorption in the bulk, but the lowest subband of the quantum well shows clear heavy-hole and light-hole excitons.

Unfortunately the strictly two-dimensional limit is not realistic because typical quantum wells are some 10 nm wide, which is close to the Bohr radius that defines the size of an exciton. A more complete model is needed for practical calculations. The problem is worse in quantum wires, where the lowest state of a 'one-dimensional hydrogen atom' has infinite binding energy.

10.7.3 EXCITONS IN A QUANTUM WELL

We must return to the full three-dimensional Schrödinger equation (10.60) to treat excitons in a quantum well, with potentials $V_e(z_e)$ and $V_h(z_h)$ to confine the electrons and holes. The separation into centre-of-mass and relative coordinates is now possible only for motion in the plane of the well. Again we shall ignore the kinetic energy of the centre of mass. The remaining wave function depends on the relative coordinate $\mathbf{r}$ in the plane of the well and both original z-coordinates. It obeys the following Schrödinger equation:

$$\left\{ \left[-\frac{\hbar^2}{2m_0 m_e} \frac{\partial^2}{\partial z_e^2} + V_e(z_e) \right] + \left[-\frac{\hbar^2}{2m_0 m_h} \frac{\partial^2}{\partial z_h^2} + V_h(z_h) \right] - \frac{\hbar^2}{2m_0 m_{eh}} \nabla^2 \right.$$
$$\left. - \frac{e^2}{4\pi\epsilon_0\epsilon_b\sqrt{|\mathbf{r}|^2 + (z_e - z_h)^2}} \right\} \psi(\mathbf{r}, z_e, z_h) = E\psi(\mathbf{r}, z_e, z_h). \quad (10.68)$$

This includes the individual masses m_e and m_h for motion along z, normal to the well, as well as the reduced mass m_{eh} for motion in the plane of the well. Recall from Section 10.3 that the effective mass of holes is anisotropic in a quantum well. For example, the 'heavy' holes are heavy along z but light for motion in the plane of the well. Thus the mass m_h should be heavy, whereas the reduced mass m_{eh} should contain a lighter mass for the hole.

The similar length scales of the confinement due to the quantum well and the Coulomb interaction mean that a solution of the foregoing Schrödinger equation must treat both effects on an equal footing. This requires a variational or numerical treatment. Two wave functions with a single variable parameter λ that have been used are

$$\psi_1(\mathbf{r}, z_e, z_h) \propto \phi_{c1}(z_e)\phi_{v1}(z_h) \exp\left(-\frac{|\mathbf{r}|}{\lambda}\right), \qquad (10.69)$$

$$\psi_2(\mathbf{r}, z_e, z_h) \propto \phi_{c1}(z_e)\phi_{v1}(z_h) \exp\left(-\frac{|\mathbf{R}|}{\lambda}\right). \qquad (10.70)$$

Here $\phi_{c1}(z_e)$ and $\phi_{v1}(z_h)$ are the lowest bound states for electrons and holes, which are identical for infinitely deep wells. The difference is that the envelope of the exciton depends only on the transverse coordinate $\mathbf{r}$ in the wave function given in equation (10.69). It is easier to handle but is accurate only in very narrow wells because its behaviour along z is unaffected by the Coulomb attraction. This approximation fails in a wide well, where the relative motion in all directions should be dominated by the binding of the exciton. The second wave function, from equation (10.70), contains the full difference $R = \sqrt{r^2 + (z_e - z_h)^2}$. It gives very similar behaviour in a narrow well but all three components of the relative motion are constrained by the exponential function in a well whose width exceeds the radius of the exciton. It therefore approaches the three-dimensional result in very wide wells.

Results obtained using these approximations to the wave function are plotted in Figure 10.10. The inseparable wave function is clearly superior as expected, particularly for wide wells. The binding energy interpolates smoothly between its

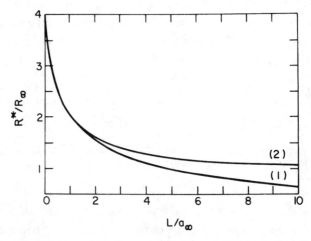

FIGURE 10.10. Binding energy R^* of an exciton in an infinitely deep well as a function of the width L of the well. Both quantities are measured in terms of the effective Rydberg energy R_∞ and Bohr radius a_∞ for a three-dimensional exciton in the same material. Curve (1) refers to the separable wave function from equation (10.69) and (2) is the inseparable approximation, equation (10.70). [From Bastard et al. (1982).]

value $\mathcal{R}_{eh}$ for very wide wells and $4\mathcal{R}_{eh}$ as the width L goes to zero, doubling when $L \approx a_{eh}$, the Bohr radius. An amusing feature is that the wave function shrinks in **r** as the well gets narrower and the wave function is squashed along z – it does not behave like rubber!

Unfortunately we know that the infinitely deep well is an unreliable model for narrow wells, where it predicts energy levels that lie above the plateau outside the well. At the same time, the tail of the wave function that tunnels into the plateau becomes larger and the particles are less well confined. We can guess, on dimensional grounds, that the best confinement occurs when the width a and depth V_0 obey $\hbar^2 \pi^2 / 2ma^2 \approx V_0$; the confinement is set by a in wider wells and by tunnelling into the barrier in narrower wells. The depth $V_0 \approx 0.3\,\text{eV}$ in GaAs–AlGaAs systems, giving $a \approx 5\,\text{nm}$ for optimum confinement. Thus we cannot approach the limit of $4\mathcal{R}_{eh}$ in a real structure.

Numerical results for finite wells are shown in Figure 10.11. The binding energy of the exciton now has a peak when $a \approx 5\,\text{nm}$ and falls for narrower wells; it returns to $\mathcal{R}_{eh}$ as $a \to 0$ when all the wave function is in the barriers rather than the well. The peak of around 10 meV is about double that of the bulk, rather than four times as high as for an ideal two-dimensional exciton.

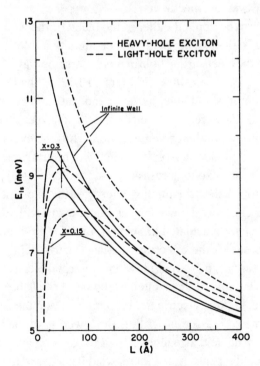

FIGURE 10.11. Binding energy of heavy-hole and light-hole excitons in GaAs–Al$_x$Ga$_{1-x}$As quantum wells as a function of width L. Both light- and heavy-hole excitons are shown, with $x = 0.15$ and $x = 0.30$. The ratio $\Delta E_c / \Delta E_g = 0.85$ was used, higher than the currently accepted range of 0.60–0.65. [From Greene, Bajaj, and Phelps (1984).]

There is an interesting competition between the in-plane and normal effective masses for the light and heavy holes. The light holes are heavier for motion in the plane of the well, and this is reflected in the higher binding energy in an infinite well. In the finite well, by contrast, the light holes cannot be confined so well along z (where they are 'light'!) and the maximum binding energy of the light-hole exciton is therefore smaller because the excitons are less two-dimensional. There is a clear crossover in Figure 10.11. In wide wells, confinement is set by the width of the well and the transverse mass dominates so light-hole excitons are better bound; confinement is controlled by tunnelling in narrow wells, which means that the heavy holes are more compact and their excitons are better bound. This competition is important in materials such as InGaAs, where there is a lower effective mass in the well but higher barriers confining the carriers.

10.7.4 QUANTUM-CONFINED STARK EFFECT

We saw in Section 7.2.2 that an electric field along z lowered the energy of both electrons and holes in a quantum well, and we argued that this should be seen in optical absorption. As we now know that absorption near thresholds is dominated by excitons, we must review this theory.

An electric field can be applied either in the plane of the well or normal to it. A field in the plane adds $e\mathbf{F} \cdot \mathbf{r}$ to the Schrödinger equation (10.68). This modifies the interaction between the electron and hole in much the same way as it did the quantum well itself in Section 7.2.2. As shown in Figure 10.12(e), the energy of an exciton is slightly lowered but it is no longer a true bound state and can tunnel through the barrier, which has been lowered by the electric field. The electron and hole are then no longer bound, a process called *field ionization*. Thus the exciton acquires a finite lifetime, reflected in a broader absorption line. The lifetime decreases as the field becomes stronger, and the exciton ceases to exist for all practical purposes when the top of the barrier drops below the binding energy of the exciton, $eF(2a_{\mathrm{eh}}) > \mathcal{R}_{\mathrm{eh}}$. For larger fields we can treat the electron and hole as individual free particles and use the densities of states calculated in Section 6.2.1 for the Franz–Keldysh effect. Figure 10.12(a)–(c) shows the effect of such an electric field on a quantum well with the same structure as in Figure 7.3, where a normal electric field was applied. The main feature is the loss of the exciton as the onset of absorption is blurred out.

Two effects compete if the electric field is applied along the direction of growth. The first, already treated in Section 7.2.2, is the lowering of the energy of the individual bound states. The field also tends to pull the exciton apart, as just discussed, which reduces the binding energy and increases the energy of the optical transition. In this case, however, the exciton cannot ionize because the electron and hole are trapped in the quantum well. Thus excitons remain visible in Figure 7.3 at much higher electric fields than those in Figure 10.12, even though the quantum wells are identical. The change in the energy levels of the electron and hole usually dominates

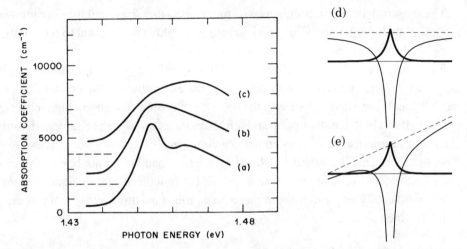

FIGURE 10.12. Absorption of a 9.5 nm GaAs quantum well for electric fields in the plane of the well of (a) $0\,\mathrm{V\,m^{-1}}$, (b) $1.6 \times 10^6\,\mathrm{V\,m^{-1}}$, and (c) $4.8 \times 10^6\,\mathrm{V\,m^{-1}}$. The zeros have been displaced, as shown by the dashed lines. [From Miller et al. (1985).] The potential energy and wave function of the exciton are sketched for (d) zero field and (e) a large field in the plane of the well.

that of the exciton, as the change in energy due to the latter cannot exceed its original binding energy. Thus the calculation in Section 7.2.2 remains qualitatively correct but the shifts should be corrected for the effect of excitons at low electric fields. At the same time, the separation of the electron and hole due to the field reduces the overlap between them and hence weakens the absorption, proportionally to the matrix element of z between the two states. The field also destroys the symmetry of the quantum well and weakens the selection rules.

10.7.5 EXCITONS, FLUCTUATIONS, AND PHONONS

We saw in Chapter 9 that the behaviour of electrons in a two-dimensional electron gas is strongly influenced by charged impurities and defects such as interface roughness. Excitons are subject to similar perturbations. Of particular importance, because excitons are trapped in a well rather than against a single interface, are fluctuations in the width of the well. This affects the energy of both the electron and hole, and the binding energy of the exciton. The effect of such fluctuations depends on their length scale relative to the radius of the exciton a_{eh}. Fluctuations that are rapid on the scale of a_{eh} tend to average out, whereas those that are slow on this scale give a well-defined energy that varies from place to place and broadens optical lines for the sample as a whole. Measurements of the linewidth of excitons have therefore been widely used to optimize the conditions of growth; temperature has a strong influence, and many samples are now grown with interrupts to give the surface an opportunity to smooth at interfaces.

The linewidth of excitons increases as the temperature rises, and they are ionized more rapidly by optic phonons. The increase is roughly proportional to the number of optic phonons $[\exp(\hbar\omega_{LO}/k_B T) - 1]^{-1}$.

Excitons give rise to many effects in nonlinear optics. For example, intense light may create so many excitons that states near the extremities of the conduction and valence bands are filled. Effectively the band gap has been pushed to a higher energy and this affects both the absorption and refractive index through the Kramers–Kronig relations. The optical properties return to their normal values only after the excitons have decayed. This is a relatively slow process and many other nonlinearities have been investigated to achieve a fast response for optoelectronic devices. Another book could easily be written about these phenomena, so this seems an appropriate point at which to stop.

Further Reading

The optical properties of low-dimensional systems are described well in several books and review articles. The textbook by Chuang (1995) covers the physics of optoelectronic devices from the fundamental theory to lasers and modulators. A rather different perspective, oriented towards the practical aspects of devices and overall systems for communication, is given by Gowar (1993).

Bastard (1988) presents a characteristically elegant description of the optical properties of semiconductors, concentrating on the physics. Optical applications of heterostructures are well described by Weisbuch and Vinter (1991) with a particularly good account of quantum-well lasers. Kelly (1995) also covers a broad range of applications.

There are also many accounts of the theoretical aspects of the optical properties of quantum wells. These include the textbook by Haug and Koch (1993) and the review by Schmitt-Rink, Chemla, and Miller (1989). The treatment of envelope functions in this chapter is far from rigorous, and Burt (1995) shows the problems that can arise.

A good account of the Kane model is given by its inventor (Kane 1982). I have followed the notation of this review.

Finally, there are many relevant articles in Willardson and Beer (1966–).

EXERCISES

10.1　　The Lorentz model of optical response is described in Section A6.2.2 and gives a peak in ϵ_2 around ω_0. Plot out its energy-loss function $-\mathrm{Im}\,(1/\tilde{\epsilon}_r)$. Is the dominant feature still at ω_0?

10.2   Verify that the Lorentz model behaves like a plasma at high frequencies, and confirm the sum rules for $\epsilon_2(\omega)$ and $-\mathrm{Im}\,(1/\bar{\epsilon}_r)$.

10.3   Examine the effect of strain on band structure within the Kane model, using the spinless version (10.26) for simplicity. Suppose that uniaxial strain along the z-axis adds a term $\frac{2}{3}S$ to E_v between the $|Z\rangle$ orbitals, and $-\frac{1}{3}S$ between both the $|X\rangle$ and $|Y\rangle$ orbitals. What happens to the valence bands?

10.4   The results of the Kane model are somewhat different in InSb because this material has a spin-orbit splitting larger than its band gap. Use the data in Appendix 2 to examine the accuracy of the Kane model in this case.

10.5   Sketch $\varepsilon(\mathbf{K})$ for light and heavy holes in GaAs according to the Luttinger model, equation (10.35). A surface plot is most illuminating; alternatively take $\mathbf{K}$ along [100], [110], and [111]. Does this leave you happy with the spherical approximation?

10.6   Consider holes in a 5 nm quantum well of GaAs surrounded by AlGaAs. Estimate the lowest energy levels for light and heavy holes and their splitting using a simple model. How many holes can be put in the lowest 'heavy' band before the 'light' band becomes occupied?

10.7   Suppose that a narrow absorption line is shifted from ω_1 to ω_2. An example might be the shift of an interband transition in a quantum well due to an applied electric field (the quantum-confined Stark effect, calculated in Section 7.2.2). Write down the change $\Delta\epsilon_2(\omega)$ and calculate the corresponding change $\Delta\epsilon_1(\omega)$. Often one aims to use the change in $\epsilon_1(\omega)$ at a frequency where $\epsilon_2(\omega)$ is small. What demands does this make on the shift in the line?

10.8   Estimate as follows the accuracy of the '$\Delta n = 0$' rule (Section 10.4) for interband transitions in a GaAs quantum well of width 10 nm. For electrons the depth is about 0.30 eV and the lowest state has energy 34 meV. The same energy would be found in an infinitely deep well if its width were 12.8 nm. Similarly the holes sit in a well of depth 0.18 eV, which gives an energy of 5.9 meV, equivalent to an infinitely deep well of width 11.3 nm. Calculate the matrix element between the envelope functions (just their product!) using the wave functions in the equivalent infinitely deep well.

For comparison, repeat this for the transition between the lowest electron and third hole states, using the same effective widths.

If you can calculate the wave functions in a finite well numerically, use these to evaluate the matrix element. How good is the rough approximation of adjusting the width of the well?

10.9   Estimate the form of the absorption spectrum shown in Figure 10.7 for transitions between a bound state in a quantum well and the continuum above it. Assume for simplicity that the initial state is half a sine wave, as in an infinitely deep well, and that the final states are unperturbed plane waves. The main discrepancy between this estimate and the experimental result is around the onset of absorption; why does this arise? 'Prove' that your result is an approximation by demonstrating that the integrated absorption over all frequencies falls short of the f-sum rule.

10.10 The energies of bound three-dimensional excitons are $\varepsilon_n = -\mathcal{R}_{eh}/n^2$, while their densities at the origin are $1/\pi a_{eh}^3 n^3$. Show that the absorption lines blur together for large n, if they acquire a nonzero width, to the same value that the Sommerfeld factor gives just above the onset of the continuum. Thus there is no discontinuity in the optical properties if $\hbar\omega = E_g$ if excitons are taken into account.

10.11 An exercise in Chapter 8 asked for the decay length α^{-1} for light of energy $\hbar\omega = E_g + 0.05\,\text{eV}$ incident on GaAs. Now include the Sommerfeld factor. Does this explain the scale of Figure 8.4?

10.12 Estimate α, the rate at which light decays with distance, for light incident on a multiquantum well. This comprises alternate wells of GaAs and barriers of $Al_{0.3}Ga_{0.7}As$, each 10 nm thick, as in Figure 8.4. The energy of the incident light is 0.05 eV above the threshold for absorption. Assume that the holes are heavy, and use data from Appendix 2 with $n_r \approx 3.5$.

How large an effect does the Sommerfeld factor have?

10.13 A thick layer of GaAs, which can be treated as three-dimensional, is pumped in such a way that $f_n(\varepsilon_e) - f_p(\varepsilon_h) = 0.5$ at the band gap. Sketch the optical gain as a function of frequency $g(\omega)$ and estimate its maximum.

The high values of optical gain that can be achieved with semiconductor lasers are another ingredient in their success.

10.14 Although the relative motion and that of the centre of mass cannot in general be separated for an exciton in a quantum well, there is an amusing special case where this is possible. Let V_e and V_h both be parabolic with the same frequency, so $V_e(z_e) = \frac{1}{2}m_e\omega_0^2 z_e$ and $V_h(z_h) = \frac{1}{2}m_h\omega_0^2 z_h$. Show that separation into $\mathbf{R}_{CM}$ and $\mathbf{R}$ is now possible. Unfortunately one still has to solve the equation involving the relative coordinate with both the parabolic potential and the interaction. Sometimes the further approximation is made that the interaction is parabolic in R, rather than $1/R$; show that this makes the problem simple.

The results are more useful when applied to an electron (or hole) gas in a parabolic potential, because they show that the motion of the centre of mass is independent of interactions between the electrons. This is known as *Kohn's theorem* and is important in determining the infrared response of doped quantum wells, dots, or superlattices.

TABLE OF PHYSICAL CONSTANTS

Charge of proton	e	1.602×10^{-19}	C
Mass of electron	m_0	9.109×10^{-31}	kg
Mass of proton	m_p	1.673×10^{-27}	kg
Velocity of light in free space	c	2.998×10^8	$\mathrm{m\,s^{-1}}$
Boltzmann's constant	k_B	$\begin{cases} 1.381 \times 10^{-23} \\ 0.0862 \end{cases}$	$\mathrm{J\,K^{-1}}$ $\mathrm{meV\,K^{-1}}$
Permittivity of free space	ϵ_0	8.854×10^{-12}	$\mathrm{F\,m^{-1}}$
	$e/4\pi\epsilon_0$	1.440	V nm
Permeability of free space	μ_0	$4\pi \times 10^{-7}$	$\mathrm{H\,m^{-1}}$
Planck's constant	h	6.626×10^{-34}	J s
	$\hbar = h/2\pi$	1.055×10^{-34}	J s
Avogadro's constant	N_A	6.022×10^{26}	$\mathrm{kmol^{-1}}$
Bohr radius	a_0	5.292×10^{-11}	m
Rydberg energy	$\mathcal{R}$	$\begin{cases} 2.180 \times 10^{-18} \\ 13.61 \end{cases}$	J eV
Bohr magneton	$\mu_B = e\hbar/2m_0$	9.274×10^{-24}	$\mathrm{J\,T^{-1}}$
Quantum of magnetic flux	$\Phi_0 = h/e$	4.136×10^{-15}	Wb
Fine structure constant	$\alpha = e^2/4\pi\epsilon_0\hbar c$	$(137.04)^{-1}$	
Electron volt	eV	1.602×10^{-19}	J
	$1\,\mathrm{meV}/h$	241.8	GHz
	$1\,\mathrm{meV}/k_B$	11.60	K

2

PROPERTIES OF IMPORTANT SEMICONDUCTORS

Notation for this and the following table:

a	lattice constant
ρ	mass density
$\hbar\omega_{LO}$	energy of longitudinal optic phonon
$\hbar\omega_{TO}$	energy of transverse optic phonon
E_g	minimum band gap
$E_c^{(min)}$	nature of lowest minimum in conduction band
E_g^{Γ}	direct band gap at Γ (if not E_g)
Δ	spin–orbit splitting of valence band
χ	electron affinity
m_{hh}	(relative) mass of heavy holes
m_{lh}	mass of light holes
m_{so}	mass of holes in split-off band
m_{Γ}	mass of electrons in Γ-valley
m_L	longitudinal mass of electrons in lowest X- or L-valley
m_T	corresponding transverse mass
ϵ_b	static dielectric constant (relative permittivity)
μ_n	mobility of electrons in lightly doped material
μ_p	mobility of holes in lightly doped material

Values refer to room temperature unless indicated otherwise. Most data are taken from Madelung (1996) and Adachi (1985), supplemented by Bastard (1988), Sze (1981), and Milnes and Feucht (1972).

There is a surprisingly large range quoted for some quantities, and others are ill defined. For example, Bastard quotes a range of values for the mass m_{hh} of heavy holes in GaAs from 0.45 to 0.57. This arises from warping of the bands so a simple parabolic approximation with a unique effective mass is poor.

Property	Si	Ge	GaAs	AlAs	InAs	GaP	InP	GaSb	AlSb	InSb	Units
a	0.5431	0.5658	0.5653	0.5660	0.6058	0.5451	0.5869	0.6096	0.6136	0.6479	nm
ρ	2.329	5.323	5.318	3.760	5.67	4.138	4.81	5.61	4.26	5.77	$\mathrm{Mg\,m^{-3}}$
$\hbar\omega_{LO}$	64	37	36	50	30	51	43	29	42	24	meV
$E_g^{(300\,\mathrm{K})}$	1.12	0.66	1.42	2.15	0.35	2.27	1.34	0.75	1.62	0.18	eV
$E_g^{(0\,\mathrm{K})}$	1.17	0.74	1.52	2.23	0.42	2.35	1.42	0.81	1.69	0.23	eV
$E_c^{(\min)}$	X	L	Γ	X	Γ	X	Γ	Γ	X	Γ	
E_g^Γ	3.5	0.80	—	3.02	—	2.78	—	—	2.30	—	eV
Δ	0.044	0.29	0.34	0.28	0.38	0.08	0.11	0.75	0.67	0.98	eV
χ	4.01	4.13	4.07	3.51	4.92	4.3	4.38	4.06	3.65	4.59	eV
E_P		26.3	25.7	21.1	22.2	22.2	20.4	22.4	18.7	23.1	eV
m_{hh}	0.54	0.28	0.5	0.5	0.41	0.67	0.6	0.28	0.4	0.4	
m_{lh}	0.15	0.044	0.082	0.15	0.026	0.17	0.12	0.050	0.11	0.02	
m_Γ			0.067	0.150	0.022		0.077	0.04		0.014	
m_L	0.92	1.64	1.3	1.1		4.8			1.0		
m_T	0.19	0.082	0.23	0.19		0.25			0.26		
ϵ_b	11.9	16.2	13.2	10.1	15.1	11.1	12.6	15.7	12.0	16.8	
μ_n	0.15	0.39	0.92		3.3	0.02	0.5	0.77	0.02	8	$\mathrm{m^2\,V^{-1}\,s^{-1}}$
μ_p	0.045	0.19	0.04		0.05	0.012	0.01	0.1	0.04	0.13	$\mathrm{m^2\,V^{-1}\,s^{-1}}$

PROPERTIES OF GaAs–AlAs ALLOYS AT ROOM TEMPERATURE

Property	GaAs	AlAs	Interpolation	Range	Units
a	0.56533	0.56611	$0.56533 + 0.00078x$		nm
$\hbar\omega_{LO}$	36.2	50.1	(does not interpolate)		meV
$\hbar\omega_{TO}$	33.3	44.9	(does not interpolate)		meV
E_g^Γ	1.424	3.018	$1.424 + 1.247x$	$(x < 0.45)$	eV
			$1.656 + 0.215x + 1.147x^2$	$(x > 0.45)$	eV
E_g^X	1.900	2.168	$1.900 + 0.125x + 0.143x^2$		eV
E_g^L	1.708	2.350	$1.708 + 0.642x$		eV
ΔE_c^Γ	0	1.120	$0.773x$	$(x < 0.45)$	eV
			$0.232 - 0.259x + 1.147x^2$	$(x > 0.45)$	eV
ΔE_c^X	0.476	0.270	$0.476 - 0.349x + 0.143x^2$		eV
ΔE_c^L	0.284	0.452	$0.284 + 0.168x$		eV
ΔE_v	0	0.474	$0.474x$		eV
χ	4.07	3.5	$4.07 - 1.1x$	$(x < 0.45)$	eV
			$3.65 - 0.14x$	$(x > 0.45)$	eV
m_Γ	0.067	0.150	$0.067 + 0.083x$		
m_T^X	0.23	0.19			
m_L^X	1.3	1.1			
m_{lh}	0.082	0.153	$0.082 + 0.071x$		
m_{hh}	0.5	0.5			
m_{so}	0.15	0.24	$0.15 + 0.09x$		
$\epsilon_b(\omega = 0)$	13.18	10.06	$13.18 - 3.12x$		

The crossover from Γ to X in the conduction band is assumed to take place at $x = 0.45$. The discontinuities ΔE_c^Γ, ΔE_c^X, and ΔE_c^L in the conduction band are measured from the bottom (Γ_6) conduction band in GaAs.

HERMITE'S EQUATION: HARMONIC OSCILLATOR

The solution of the Schrödinger equation with a parabolic potential (Section 4.3) reduces to Hermite's equation (4.32) for $u(z)$,

$$u''(z) - 2zu'(z) + (2E - 1)u(z) = 0, \tag{A4.1}$$

where the overbars on z and E have been dropped for clarity. Proceed by expanding the solution as a power series,

$$u(z) = \sum_{m=0}^{\infty} a_m z^m. \tag{A4.2}$$

Substituting into (A4.1) and carrying out the derivatives gives

$$\sum_{m=2}^{\infty} a_m m (m-1) z^{m-2} - 2z \sum_{m=1}^{\infty} a_m m z^{m-1} + (2E - 1) \sum_{m=0}^{\infty} a_m z^m = 0. \tag{A4.3}$$

Rewrite this by shifting the dummy index of m in the first summation so that the power of z appears the same in each term:

$$\sum_{m=0}^{\infty} a_{m+2}(m+2)(m+1) z^m - 2 \sum_{m=1}^{\infty} a_m m z^m + (2E - 1) \sum_{m=0}^{\infty} a_m z^m = 0. \tag{A4.4}$$

This must hold for all z, so the coefficient of each power of z must be zero. The general result is

$$(m+2)(m+1)a_{m+2} - 2m a_m + (2E - 1)a_m = 0. \tag{A4.5}$$

Thus

$$a_{m+2} = \frac{2m - 2E + 1}{(m+2)(m+1)} a_m. \tag{A4.6}$$

Note that this links coefficients only where m differs by 2, so there is one series for even m and another for odd m.

What is the behaviour of the series whose coefficients are given by equation (A4.6)? For comparison, consider the expansion of $\exp(z^2)$:

$$\exp(z^2) = 1 + \frac{z^2}{1!} + \frac{(z^2)^2}{2!} + \cdots + \frac{(z^2)^m}{m!} + \frac{(z^2)^{m+1}}{(m+1)!} + \cdots. \tag{A4.7}$$

This is an expansion in even powers of z, just like the even terms in our series for Hermite's equation. The recurrence relation for $\exp(z^2)$, the analogue of equation (A4.6), is

$$c_{2m+2} = \frac{m!}{(m+1)!}c_{2m} = \frac{1}{m+1}c_{2m}, \tag{A4.8}$$

where c_{2m} is the coefficient of z^{2m}. To make comparison easier, write (A4.6), assuming that m is even, as

$$a_{2m+2} = \frac{4m - 2E + 1}{(2m+2)(2m+1)}a_{2m}. \tag{A4.9}$$

Both (A4.8) and (A4.9) go like $1/m$ for large m. This means in turn that our solution to Hermite's equation must diverge in the same way, like $\exp(z^2)$. Unfortunately this means that the resulting wave function will diverge, because the series will overcome the factor of $\exp(-\frac{1}{2}z^2)$ that we removed from equation (4.30) before reaching Hermite's equation. Thus these solutions are unacceptable. The escape is to note that the numerator of the recurrence relation (A4.6) vanishes if the energy takes an appropriate value. This requires

$$2m - 2E + 1 = 0, \qquad E = m + \tfrac{1}{2} \quad (m = 0, 1, 2, \ldots). \tag{A4.10}$$

In this case all the higher terms of the series will be identically zero, the series terminates to leave a polynomial, and the divergence is suppressed. Thus the dimensionless energy can take on only the discrete set of values permitted by equation (A4.10) for normalizable wave functions, and this gives the quantized energies of the harmonic oscillator. Equation (A4.10) leads to the result in the main text except for the trivial difference that n starts from 1 there, whereas m starts from 0 here.

AIRY FUNCTIONS: TRIANGULAR WELL

Airy functions arise in the solution of Schrödinger's equation with a linear potential, notably for the two-dimensional electron gas trapped at a heterojunction (Chapter 9) and for an electron in an electric field (Section 6.2). They are solutions to the Stokes or Airy equation,

$$\frac{d^2y}{dx^2} = xy. \tag{A5.1}$$

This second-order equation has two linearly independent solutions, conventionally taken as the *Airy integral functions* $\mathrm{Ai}(x)$ and $\mathrm{Bi}(x)$ plotted in Figure A5.1. They have no simple expressions in terms of elementary functions. Airy functions can be written as modified Bessel functions of order $\frac{1}{3}$, which is useful for finding them in tables such as those of Abramowitz and Stegun (1972). Slightly different functions may be used in the Russian literature.

The behaviour of $\mathrm{Ai}(x)$ and $\mathrm{Bi}(x)$ can be inferred by comparing equation (A5.1) with the wave equation in a constant potential, $d^2y/dx^2 = -k^2y$. For $x < 0$, the Stokes equation is like the wave equation with $k^2 > 0$, and the solutions are oscillating trigonometric functions like $\cos kx$. The effective wave number grows like $k = \sqrt{-x}$ so the function oscillates more rapidly as x increases along the negative axis. Both $\mathrm{Ai}(x)$ and $\mathrm{Bi}(x)$ show this behaviour, oscillating in quadrature like sine and cosine. For $x > 0$ the effective value of k^2 becomes negative, which implies exponentially growing or decreasing solutions. The solutions are chosen such that $\mathrm{Ai}(x) \to 0$ as $x \to +\infty$, while $\mathrm{Bi}(x)$ diverges. The behaviour for large $|x|$ is given by asymptotic expansions for the functions. Those for $\mathrm{Ai}(x)$ are

$$\mathrm{Ai}(x) \sim \pi^{-1/2} x^{-1/4} \exp(-\tfrac{2}{3}x^{3/2}), \tag{A5.2}$$

$$\mathrm{Ai}(-x) \sim \tfrac{1}{2}\pi^{-1/2} x^{-1/4} \cos(\tfrac{2}{3}x^{3/2} - \tfrac{1}{4}\pi). \tag{A5.3}$$

The dependence on $x^{3/2}$ follows from the analogy with the simple wave equation.

The Airy functions oscillate for negative x and the points where they pass through zero are needed to find the energies of bound states in a triangular well (Section 4.4). They must be computed numerically, and the first few zeros (Abramowitz and Stegun 1972, table 10.13) are

$$a_1 = -2.338, \quad a_2 = -4.088, \quad a_3 = -5.521, \quad a_4 = -6.787. \tag{A5.4}$$

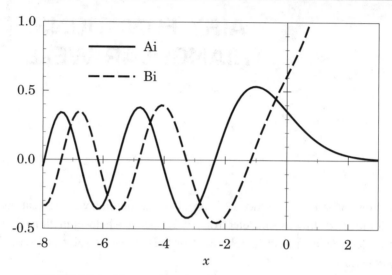

FIGURE A5.1. The two Airy (integral) functions $Ai(x)$ and $Bi(x)$.

Occasionally the zeros of the derivative $Ai'(x)$ are needed. The first few are

$$a_1' = -1.019, \quad a_2' = -3.248, \quad a_3' = -4.820, \quad a_4' = -6.163. \quad (A5.5)$$

The simple form of the differential equation satisfied by Airy functions allows many integrals over them to be simplified. An example that arises in both the normalization of bound states and the density of states in an electric field is the integral of $Ai^2(x)$. Inserting a factor of 1 and integrating by parts gives

$$\int Ai^2(x)\, dx = x Ai^2(x) - 2 \int x Ai(x) Ai'(x)\, dx$$
$$= x Ai^2(x) - [Ai'(x)]^2. \quad (A5.6)$$

The final step follows from the replacement $x Ai(x) = Ai''(x)$, which makes the integral trivial.

KRAMERS–KRONIG RELATIONS AND RESPONSE FUNCTIONS

In this appendix we shall derive the Kramers–Kronig relations. This requires an excursion into the complex frequency plane and Cauchy's integral theorem; readers unfamiliar with complex analysis can skip Section A6.1 without loss of continuity. We shall then derive two common models for the optical response of semiconductors.

A6.1 Derivation of the Kramers–Kronig Relations

The complex conductivity $\tilde{\sigma}(\omega)$, defined in Section 10.1.1, is a typical response function. We shall use this as an example but the same theory applies to any linear-response function. It can also be written in the time domain as $\sigma(t)$ and the two are related by

$$\tilde{\sigma}(\omega) = \int_{-\infty}^{\infty} \sigma(t)\, e^{i\omega t} dt, \qquad (A6.1)$$

using the quantum mechanical convention for the sign of the frequency. This integral runs from $-\infty$ to ∞ as in the usual Fourier transform. However, we saw in Section 10.1.1 that the physical principle of *causality* requires $\sigma(t) = 0$ for $t < 0$, and we may therefore set the lower limit to $t = 0$. The Kramers–Kronig relations follow from this step.

Usually we think of ω as real but mathematically it can be extended into the complex plane, and sometimes ω is given a small positive imaginary part to simulate the gentle switching on of a stimulus in the distant past. After we split ω into real and imaginary parts, $\omega = \omega_1 + i\omega_2$, the Fourier transform becomes

$$\tilde{\sigma}(\omega_1 + i\omega_2) = \int_0^{\infty} \sigma(t)\, e^{i(\omega_1 + i\omega_2)t} dt = \int_0^{\infty} \sigma(t)\, e^{i\omega_1 t}\, e^{-\omega_2 t} dt. \qquad (A6.2)$$

This integral converges when $\omega_2 > 0$ unless $\sigma(t)$ diverges exponentially as $t \to \infty$. This would imply that the system was unstable, which we assume not to be the case. Thus $\tilde{\sigma}(\omega)$ is analytic throughout the upper half-plane (the argument is the same as that for the convergence of Laplace transforms). It does not follow that $\sigma(\omega)$ is necessarily analytic *on* the real axis.

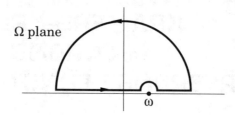

FIGURE A6.1. Contour in the complex Ω plane used to prove the Kramers–Kronig relations.

We shall next use this analytic property to apply Cauchy's integral theorem to a contour in the upper half-plane. This theorem states that the integral of a function taken around a closed loop is zero provided that there are no singularities on or within the contour. Consider the integral shown in Figure A6.1,

$$\oint \frac{\tilde{\sigma}(\Omega)}{\Omega - \omega}\, d\Omega, \tag{A6.3}$$

where ω is real. The contour is a large semicircle in the upper half-plane, closed by a line just above the real axis except for a small semicircle of radius ε where it skirts the singularity at $\Omega = \omega$ that comes from the denominator. The contribution from the large semicircle vanishes provided that $\tilde{\sigma}(\Omega)$ goes to zero faster than $1/\Omega$ as $\Omega \to \infty$. We shall assume that there are no singularities of σ on the real axis. This is not always true; $\epsilon_2(\omega)$ diverges at the origin as $\sigma_1(0)/\epsilon_0\omega$ in a metal, for example, and modified dispersion relations with subtractions are required in such cases. If we ignore this problem, the sum of the contributions from the remaining two parts of the integral must be zero. The contribution from the straight line is

$$\left[\int_{-\infty}^{\omega-\varepsilon} + \int_{\omega+\varepsilon}^{\infty} \right] \frac{\tilde{\sigma}(\Omega)}{\Omega - \omega}\, d\Omega. \tag{A6.4}$$

As $\varepsilon \to 0$, this becomes by definition the *Cauchy principal part* of the integral, denoted by $\mathcal{P}$:

$$\lim_{\varepsilon \to 0} \left[\int_{-\infty}^{\omega-\varepsilon} + \int_{\omega+\varepsilon}^{\infty} \right] \frac{\tilde{\sigma}(\Omega)}{\Omega - \omega}\, d\Omega = \mathcal{P} \int_{-\infty}^{\infty} \frac{\tilde{\sigma}(\Omega)}{\Omega - \omega}\, d\Omega. \tag{A6.5}$$

Sometimes a bar is put through the integral sign instead. The principal part controls the singularity that would otherwise occur at $\Omega = \omega$.

Now consider the small semicircle. Since $\tilde{\sigma}(\Omega)$ is analytic at $\Omega = \omega$, we can take it outside the integral as $\varepsilon \to 0$. The remaining integral is just of $d\Omega/(\Omega - \omega)$ over the semicircle. If we put $\Omega = \varepsilon e^{i\theta}$, this becomes

$$\tilde{\sigma}(\omega) \int \frac{d\Omega}{\Omega - \omega} = \tilde{\sigma}(\omega) \int_0^{\pi} \frac{i\varepsilon e^{i\theta}\, d\theta}{\varepsilon e^{i\theta}} = -i\pi\tilde{\sigma}(\omega). \tag{A6.6}$$

Thus we obtain

$$0 = \oint \frac{\tilde{\sigma}(\Omega)}{\Omega - \omega}\, d\Omega = \mathcal{P} \int_{-\infty}^{\infty} \frac{\tilde{\sigma}(\Omega)}{\Omega - \omega}\, d\Omega - i\pi\tilde{\sigma}(\omega), \tag{A6.7}$$

or

$$\tilde{\sigma}(\omega) = \frac{1}{i\pi} \mathcal{P} \int_{-\infty}^{\infty} \frac{\tilde{\sigma}(\Omega)}{\Omega - \omega} d\Omega. \qquad (A6.8)$$

This is an integral equation connecting the real and imaginary parts of $\tilde{\sigma}(\omega)$. Separating it into real and imaginary parts gives the pair of Kramers–Kronig relations:

$$\sigma_1(\omega) = \frac{1}{\pi} \mathcal{P} \int_{-\infty}^{\infty} \frac{\sigma_2(\Omega)}{\Omega - \omega} d\Omega, \qquad (A6.9)$$

$$\sigma_2(\omega) = -\frac{1}{\pi} \mathcal{P} \int_{-\infty}^{\infty} \frac{\sigma_1(\Omega)}{\Omega - \omega} d\Omega. \qquad (A6.10)$$

Physical applications of these are discussed in Chapter 10.

A6.2 Model Response Functions

Several simple models are used for $\tilde{\sigma}(\omega)$ and $\tilde{\epsilon}_r(\omega)$, of which the Drude model of free electrons and the Lorentz model of insulators are the most common. We shall now derive the response functions of these models.

The conductivity $\tilde{\sigma}$ was defined in Section 10.1.1 through $\tilde{J} = \tilde{\sigma}\tilde{E}$. The traditional way of deriving $\tilde{\sigma}(\omega)$ is to solve for the motion of a simple damped particle or oscillator in an oscillating electric field, but we shall look at the behaviour in time instead. This means that we want $\sigma(t)$, the response to an impulse of electric field $E_0\delta(t)$ applied at $t = 0$. This was defined in equation (10.3):

$$J(t) = \int_0^{\infty} \sigma(t')E_0\,\delta(t - t')dt' = E_0\,\sigma(t). \qquad (A6.11)$$

The field gives an impulse of force $-eE_0$ to each electron, which accelerates it instantly to a velocity $v = -eE_0/m$. The current density $J = -nev = ne^2E_0/m$ immediately afterwards, where n is the number density of electrons. This behaviour is true in any system because it does not depend on the environment, but the subsequent relaxation to equilibrium depends on the scattering of the electrons and the forces that restore them to equilibrium.

A6.2.1 FREE ELECTRONS: THE DRUDE MODEL

For free electrons in a metal, the Drude model makes the simplest assumption of exponential decay with a time constant τ (really the transport lifetime, discussed in Section 8.2). Thus

$$J(t) = E_0 \frac{ne^2}{m} e^{-t/\tau} \Theta(t), \qquad (A6.12)$$

where the Heaviside unit step function $\Theta(t)$ ensures that the response follows the impulse. Comparison with equation (A6.11) shows that the response function is $\sigma(t) = (ne^2/m)e^{-t/\tau}\Theta(t)$. Its Fourier transform is

$$\tilde{\sigma}(\omega) = \int_{-\infty}^{\infty} \sigma(t)e^{i\omega t}\,dt = \frac{ne^2}{m}\int_{0}^{\infty} e^{-t/\tau}e^{i\omega t}\,dt = \frac{ne^2\tau}{m}\frac{1}{1-i\omega\tau}. \qquad (A6.13)$$

This is analytic in the upper half-plane as expected; the only singularity is the pole at $\omega = -i/\tau$. Splitting this into real and imaginary parts gives

$$\sigma_1(\omega) = \frac{ne^2\tau}{m}\frac{1}{1+(\omega\tau)^2}, \qquad \sigma_2(\omega) = \frac{ne^2\tau}{m}\frac{\omega\tau}{1+(\omega\tau)^2}. \qquad (A6.14)$$

Note that $\tilde{\sigma}(0) = ne^2\tau/m$, a familiar result.

At very high frequencies, $\tilde{\sigma} \sim ine^2/m\omega$, giving $\tilde{\epsilon}_r \sim 1 - ne^2/\epsilon_0 m\omega^2 \equiv 1 - \omega_p^2/\omega^2$, where ω_p is the plasma frequency. This behaviour of electrons, like that of a plasma at high frequencies, was asserted in Section 10.1.3 and used to derive sum rules. It follows from $\sigma(t)$ at small times, just after the impulse of electric field has been applied and before any relaxation has occurred, and is therefore universal.

The Drude formula is plotted in Figure A6.2 with rough values for highly doped n-GaAs. The real part of the conductivity falls monotonically as a function of frequency, while the imaginary part rises to a broad peak $\omega = 1/\tau$ before falling slowly. It is a reasonable model for the optical response of systems with free electrons, such as simple metals or highly doped semiconductors.

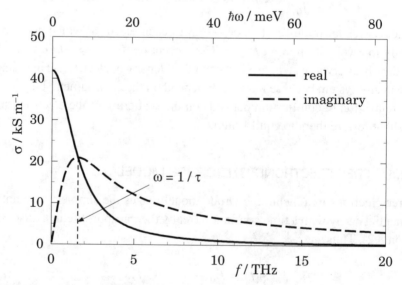

FIGURE A6.2. Real and imaginary parts of the conductivity within the Drude model, for GaAs with $10^{24}\,\mathrm{m}^{-3}$ electrons and $\tau = 0.1\,\mathrm{ps}$.

The limit $\tau \to \infty$ gives $\tilde{\sigma}(\omega) = ine^2/m\omega$ and $\tilde{\epsilon}_r(\omega) = 1 - \omega_p^2/\omega^2$ again. There is now no frictional force to dissipate the energy that the electrons acquire from the applied field so this describes completely free electrons, or a collisionless plasma (Section 9.4.1).

A6.2.2 BOUND ELECTRONS: THE LORENTZ MODEL

Electrons in an insulator are tightly bound to atoms so they 'ring' at their resonant frequency ω_0 in response to an electric field, rather than flow as in a metal. This defines the Lorentz model. The conductivity as a function of time becomes

$$\sigma(t) = \frac{ne^2}{m} \cos \omega_0 t \, e^{-t/\tau} \, \Theta(t), \qquad (A6.15)$$

whose Fourier transform is

$$\tilde{\sigma}(\omega) = \frac{ne^2\tau}{m} \frac{1}{2} \left[\frac{1}{1 - i(\omega - \omega_0)\tau} + \frac{1}{1 - i(\omega + \omega_0)\tau} \right]. \qquad (A6.16)$$

The model is used more often for the dielectric constant. Put $\omega_p^2 = ne^2/\epsilon_0 m$ for brevity and $\gamma = 2/\tau$ as the full width at half-maximum of the Lorentzian peak. Then

$$\tilde{\epsilon}_r(\omega) = 1 - \frac{\omega_p^2}{2\omega} \left[\frac{1}{(\omega - \omega_0) + \frac{1}{2}i\gamma} + \frac{1}{(\omega + \omega_0) + \frac{1}{2}i\gamma} \right] \qquad (A6.17)$$

$$\approx 1 - \frac{\omega_p^2}{(\omega^2 - \omega_0^2) + i\gamma\omega}. \qquad (A6.18)$$

The compact second form holds if the damping is small, which is almost always the case. These results can also be derived from the response of a damped harmonic oscillator.

The real and imaginary parts of $\tilde{\epsilon}_r$ are plotted in Figure A6.3 for $\hbar\omega_0 = 4\,\text{eV}$, $\hbar\gamma = 1\,\text{eV}$, and $\hbar\omega_p = 19\,\text{eV}$, values that provide a crude model of the optical response of GaAs. The imaginary part has a Lorentzian peak at $\omega = \omega_0$, typical of the response of a forced, damped oscillator. There is a region of 'anomalous dispersion' around this peak where ϵ_1 decreases as a function of frequency, including a band where it is negative.

A single oscillator is usually too simple to describe the optical properties of real materials. A number of different resonant frequencies are used with different weights f_j, giving

$$\tilde{\epsilon}_r(\omega) = 1 - \omega_p^2 \sum_j \frac{f_j}{(\omega^2 - \omega_j^2) + i\gamma_j\omega}. \qquad (A6.19)$$

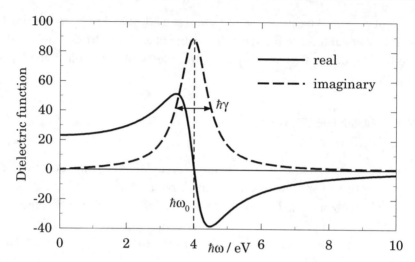

FIGURE A6.3.   Real and imaginary parts of the dielectric function of the Lorentz model for a single oscillator at $\hbar\omega_0 = 4\,\text{eV}$ damped by $\hbar\gamma = 1\,\text{eV}$, with plasma frequency $\hbar\omega_\text{p} = 19\,\text{eV}$.

This is practically identical to equation (10.18). To confirm this, note that we need $\sum_j f_j = 1$ to preserve the limit at high frequencies of $\tilde{\epsilon}_\text{r} \sim 1 - \omega_\text{p}^2/\omega^2$. The weights f_j are just the oscillator strengths introduced in Section 8.7, and this condition is the f-sum rule of Section 10.1.3 again.

As a use of the Lorentz model, we can estimate the static dielectric constant of a semiconductor $\epsilon_1(\omega = 0)$. The Lorentz model with a single oscillator at ω_0, equation (A6.18), gives

$$\epsilon_1(0) = 1 + \frac{\omega_\text{p}^2}{\omega_0^2}. \tag{A6.20}$$

Since strong absorption occurs just above the band gap, we can put $\hbar\omega_0 \approx \bar{E}_\text{g}$. Here $\bar{E}_\text{g}$ is the *average* of the optical band gap $E_\text{opt}(\mathbf{K}) = \varepsilon_\text{c}(\mathbf{K}) - \varepsilon_\text{v}(\mathbf{K})$ over the Brillouin zone, not its minimum value, which occurs only at a single point. Now $\omega_\text{p}^2 = ne^2/\epsilon_0 m$, so we need a value for n. Only the four valence electrons can respond in this range of energies, so $n = 4n_\text{atoms}$; the other electrons are much more tightly bound and higher energies (X-rays) are needed to dislodge them. The lattice constant of the common semiconductors is near 0.5 nm with 8 atoms per cubic unit cell, so $n_\text{atoms} \approx 64 \times 10^{27}\,\text{m}^{-3}$. We should use the mass of free electrons rather than any effective mass because we are considering the whole Brillouin zone rather than the lowest minimum. These values were adopted for Figure A6.3. Substitution gives

$$\epsilon_1(0) \approx 1 + \left(\frac{19\,\text{eV}}{\bar{E}_\text{g}}\right)^2. \tag{A6.21}$$

A rough value is $\bar{E}_\text{g} \approx 4\,\text{eV}$, which gives $\epsilon_1(0) \approx 23$. This is rather high, but not bad for such a crude calculation. It is probably more valuable for the trend that $\epsilon_1 \propto \bar{E}_\text{g}^{-2}$: the dielectric constant goes up when the band gap goes down.

BIBLIOGRAPHY

General References on Low-dimensional Systems

Bastard, G. (1988). *Wave mechanics applied to semiconductor heterostructures*. New York: Halsted; Les Ulis: Les Editions de Physique.

Kelly, M. J. (1995). *Low-dimensional semiconductors: materials, physics, technology, devices*. Oxford: Oxford University Press.

Weisbuch, C., and B. Vinter (1991). *Quantum semiconductor structures*. Boston: Academic Press.

Willardson, R. K., and A. C. Beer, eds. (1966–). Semiconductors and semimetals. New York: Academic Press. Vol. 24 (1987) is particularly relevant.

The journal *Surface Science* publishes the proceedings of several conferences in this field, notably *Electronic properties of two-dimensional systems*, which was held most recently in Nottingham, UK, in 1995.

Mathematical and Computational Methods

Abramowitz, M., and I. A. Stegun, eds. (1972). *Handbook of mathematical functions*. New York: Dover.

Gradshteyn, I. S., and I. M. Ryzhik (1993). *Table of integrals, series, and products*. 5th ed. New York: Academic Press.

Mathews, J., and R. L. Walker (1970). *Mathematical methods of physics*. 2d ed. Menlo Park, CA: Benjamin.

Tan, I.-H., G. L. Snider, and E. L. Hu (1990). A self-consistent solution of the Schrödinger–Poisson equations using a nonuniform mesh. *Journal of Applied Physics* **68**: 4071. See also http://www.nd.edu/~gsnider on the World Wide Web.

Data on Semiconductors

Adachi, S. (1985). GaAs, AlAs, and $Al_x Ga_{1-x}$ As: material parameters for use in research and device applications. *Journal of Applied Physics* **58**: R1–29.

Blakemore, J. S., ed. (1987). *Gallium arsenide: key papers in physics*. New York: American Institute of Physics.

Madelung, O., ed. (1996). *Semiconductors – basic data*. 2d ed. Berlin: Springer-Verlag.

Moss, T. S., ed. (1993–4). *Handbook on semiconductors*. 2d ed. 4 vols. Amsterdam: North-Holland.

References

Anderson, P. W. (1963). *Concepts in solids*. Reading, MA: Benjamin-Cummings.

Ando, T., A. B. Fowler, and F. Stern (1982). Electronic properties of two-dimensional systems. *Reviews of Modern Physics* **54**: 437–672.

Ashcroft, N. W., and N. D. Mermin (1976). *Solid state physics*. New York: Holt, Rinehart & Winston.

Bastard, G., J. A. Brum, and R. Ferreira (1991). Electronic states in semiconductor nanostructures. In *Solid state physics*, ed. H. Ehrenreich and D. Turnbull, vol. 44, 229–415. San Diego: Academic Press.

Bastard, G., E. E. Mendez, L. L. Chang, and L. Esaki (1982). Exciton binding energy in quantum wells. *Physical Review B* **26**: 1974–9.

Beenakker, C. W. J., and H. van Houten (1991). Quantum transport in semiconductor nanostructures. In *Solid state physics*, ed. H. Ehrenreich and D. Turnbull, vol. 44, 1–228. San Diego: Academic Press.

Berggren, K.-F., T. J. Thornton, D. J. Newson, and M. Pepper (1986). Magnetic depopulation of 1D subbands in a narrow 2D electron gas in a GaAs:AlGaAs heterojunction. *Physical Review Letters* **57**: 1769–72.

Bransden, B. H., and C. J. Joachain (1989). *Introduction to quantum mechanics*. Harlow: Longman; New York: Wiley.

Brown, E. R. (1994). High-speed resonant tunnelling diodes. In *Heterostructures and quantum devices*, ed. N. G. Einspruch and W. R. Frensley, VLSI electronics: microstructure science, vol. 24, 305–50. San Diego: Academic Press.

Bube, R. H. (1992). *Electrons in solids: an introductory survey*. 3d ed. Boston: Academic Press.

Burt, M. G. (1994). On the validity and range of applicability of the particle in a box model. *Applied Physics Letters* **65**: 717–19.

Burt, M. G. (1995). Breakdown of the atomic dipole approximation for the quantum well interband dipole matrix element. *Semiconductor Science and Technology* **10**: 412–15. See also the earlier work by this author cited therein.

Büttiker, M. (1988). Symmetry of electrical conduction. *IBM Journal of Research and Development* **32**: 317–34.

Chang, Y.-C., and J. N. Schulman (1985). Interband optical transitions in GaAs–$Ga_{1-x}Al_xAs$ and InSb–GaSb superlattices. *Physical Review B* **31**: 2069–79.

Chuang, S. L. (1995). *Physics of optoelectronic devices*. New York: Wiley.

Crommie, M. F., C. P. Lutz, and D. M. Eigler (1993). Confinement of electrons to quantum corrals on a metal surface. *Science* **262**: 218–20.

Datta, S. (1989). *Quantum phenomena*. Modular series on solid state devices, ed. R. F. Pierret and F. W. Neudeck, vol. 8. Reading, MA: Addison-Wesley.

Datta, S. (1995). *Electronic transport in mesoscopic systems*. Cambridge: Cambridge University Press.

Davies, J. H., and A. R. Long, eds. (1992). *Physics of nanostructures: proceedings of the thirty-eighth Scottish universities summer school in physics*. Bristol: IoP Publishing.

Einspruch, N. G., and W. R. Frensley, eds. (1994). *Heterostructures and quantum devices. VLSI electronics: microstructure science*, vol. 24. San Diego: Academic Press.

Ford, C. J. B., P. J. Simpson, I. Zailer, J. D. F. Franklin, C. H. W. Barnes, J. E. F. Frost, D. A. Ritchie, and M. Pepper (1994). The Aharonov–Bohm effect in the fractional quantum Hall regime. *Journal of Physics: Condensed Matter* **6**: L725–30.

Ford, C. J. B., T. J. Thornton, R. Newbury, M. Pepper, H. Ahmed, C. T. Foxon, J. J. Harris, and C. Roberts (1988). The Aharonov–Bohm effect in electrostatically defined heterojunction rings. *Journal of Physics C* **21**: L325.

Foxman, E. B., P. L. McEuen, U. Meirav, N. S. Wingreen, Y. Meir, P. A. Belk, N. R. Belk, M. A. Kastner, and S. J. Wind (1993). Effects of quantum levels on transport through a Coulomb island. *Physical Review B* **47**: 10020–3.

Frensley, W. R. (1994). Heterostructure and quantum well physics. In *Heterostructures and quantum devices*, ed. N. G. Einspruch and W. R. Frensley, VLSI electronics: microstructure science, vol. 24, 1–24. San Diego: Academic Press.

Gasiorowicz, S. (1974). *Quantum physics*. New York: Wiley.

Geerligs, L. J. (1992). Coulomb blockade. In *Physics of nanostructures: proceedings of the thirty-eighth Scottish universities summer school in physics*, ed. J. H. Davies and A. R. Long, 171–204. Bristol: IoP Publishing.

Gowar, J. (1993). *Optical communication systems*. 2d ed. Englewood Cliffs, N.J.: Prentice Hall.

Greene, R. L., K. K. Bajaj, and D. E. Phelps (1984). Energy levels of Wannier excitons in GaAs–Ga$_{1-x}$Al$_x$As quantum-well structures. *Physical Review B* **29**: 1807–12.

Haug, H., and S. W. Koch (1993). *Quantum theory of the optical and electronic properties of semiconductors*. Singapore: World Scientific.

Hofstadter, D. R. (1976). Energy levels and wave functions of Bloch electrons in rational and irrational magnetic fields. *Physical Review B* **14**: 2239–49.

Jaros, M. (1989). *Physics and applications of semiconductor microstructures*. Oxford: Oxford University Press (Clarendon Press).

Kane, E. O. (1982). Energy band theory. In vol. 1 of *Handbook on semiconductors*, ed. T. S. Moss, 193–217. Amsterdam: North-Holland.

Kelly, M. J., and C. Weisbuch, eds. (1986). *The physics and fabrication of microstructures and microdevices: proceedings of the Winter School, Les Houches*. Berlin: Springer-Verlag.

Kittel, C. (1995). *Introduction to solid state physics*. 7th ed. New York: Wiley.

Landau, L. D., E. M. Lifshitz, and L. P. Pitaevskii (1977). *Quantum mechanics*. 3d ed. Vol. 3 of *Course of theoretical physics*. Oxford: Pergamon Press.

Levine, B. F., C. G. Bethea, K. K. Choi, J. Walker, and R. J. Malik (1988). Bound-to-extended state absorption GaAs superlattice transport infrared detectors. *Journal of Applied Physics* **64**: 1591–3.

Mahan, G. D. (1990). *Many-particle physics*. 2d ed. New York: Plenum.

Marzin, J.-Y., M. N. Charasse, and B. Sermage (1985). Optical investigation of a new type of valence-band configuration in In$_x$Ga$_{1-x}$As–GaAs strained superlattices. *Physical Review B* **31**: 8298–301.

Merzbacher, E. (1970). *Quantum mechanics*. 2d ed. New York: Wiley.

Miller, D. A. B., D. S. Chemla, T. C. Damen, A. C. Gossard, W. Wiegmann, T. H. Wood, and C. A. Burrus (1985). Electric field dependence of optical absorption near the band gap of quantum-well structures. *Physical Review B* **32**: 1043–60.

Miller, R. C., A. C. Gossard, D. A. Kleinman, and O. Munteanu (1984). Parabolic quantum wells with the GaAs–Al$_x$Ga$_{1-x}$As system. *Physical Review B* **29**: 3740–3.

Milnes, A. G., and D. L. Feucht (1972). *Heterojunctions and metal-semiconductor junctions*. New York: Academic Press.

Myers, H. P. (1990). *Introductory solid state physics*. London: Taylor and Francis.

Nixon, J. A., J. H. Davies, and H. U. Baranger (1991). Breakdown of quantized conductance in point contacts calculated using realistic potentials. *Physical Review B* **43**: 12638–41.

O'Reilly, E. P. (1989). Valence band engineering in strained-layer structures. *Semiconductor Science and Technology* **4**: 121–37.

Pfeiffer, L., K. W. West, H. L. Störmer, and K. W. Baldwin (1989). Electron mobilities exceeding 10^7 cm^2/V s in modulation-doped GaAs. *Applied Physics Letters* **55**: 1888–90.

Prange, R. E., and S. M. Girvin, eds. (1990). *The quantum Hall effect*. 2d ed. New York: Springer-Verlag.

Reif, F. (1965). *Fundamentals of statistical and thermal physics*. New York: McGraw-Hill.

Rickayzen, G. (1980). *Green's functions and condensed matter*. London: Academic Press.

Ridley, B. K. (1993). *Quantum processes in semiconductors*. 3d ed. Oxford: Oxford University Press.

Rieger, M. M., and P. Vogl (1993). Electronic-band parameters in strained $Si_{1-x}Ge_x$ alloys on $Si_{1-y}Ge_y$ substrates. *Physical Review B* **48**: 14276–87.

Schmitt-Rink, S., D. S. Chemla, and D. A. B. Miller (1989). Linear and nonlinear optical properties of semiconductor quantum wells. *Advances in Physics* **38**: 89–188.

Seeger, K. (1991). *Semiconductor physics: an introduction*. 5th ed. Berlin: Springer-Verlag.

Stanley, C. R., M. C. Holland, A. H. Kean, M. B. Stanaway, R. T. Grimes, and J. N. Chamberlain (1991). Electrical characterization of molecular beam epitaxial GaAs with peak electron mobilities up to $\approx 4 \times 10^5$ cm^2 V^{-1} s^{-1}. *Applied Physics Letters* **58**: 478–80.

Stern, F. (1983). Doping considerations for heterojunctions. *Applied Physics Letters* **43**: 974–6.

Störmer, H. L., A. C. Gossard, and W. Wiegmann (1982). Observation of intersubband scattering in a 2-dimensional electron system. *Solid State Communications* **41**: 707–9.

Stradling, R. A., and P. C. Klipstein, eds. (1990). *Growth and characterization of semiconductors*. Bristol: Adam Hilger.

Sze, S. M. (1981). *Physics of semiconductor devices*. New York: Wiley.

Sze, S. M., ed. (1990). *High-speed semiconductor devices*. New York: Wiley.

Tiwari, S. (1992). *Compound semiconductor device physics*. Boston: Academic Press.

West, L. C., and S. J. Eglash (1985). First observation of an extremely large-dipole infrared transition within the conduction band of a GaAs quantum well. *Applied Physics Letters* **46**: 1156–8.

Willett, R., J. P. Eisenstein, H. L. Störmer, D. C. Tsui, A. C. Gossard, and J. H. English (1987). Observation of an even-denominator quantum number in the fractional quantum Hall effect. *Physical Review Letters* **59**: 1776–9.

Wolfe, C. M., N. Holonyak, and G. Stillman (1989). *Physical properties of semiconductors*. Englewood Cliffs, N.J.: Prentice Hall.

Yu, E. T., J. O. McCaldin, and T. C. McGill (1992). *Band offsets in semiconductor heterojunctions*. In *Solid state physics*, ed. H. Ehrenreich and D. Turnbull, vol. 46, 1–146. San Diego: Academic Press.

Yu, P. Y., and M. Cardona (1996). *Fundamentals of semiconductors*. Berlin: Springer-Verlag.

INDEX